Low-Voltage SOI CMOS VLSI Devices and Circuits

James B. Kuo
Shih-Chia Lin
University of Waterloo
NTUEE

A Wiley-Interscience Publication
JOHN WILEY & SONS, INC.

Preface

Silicon-on-Insulator (SOI) CMOS technology is becoming another mainstream technology for VLSI. Owing to its inherited characteristics, SOI CMOS technology is especially capable of providing deep-submicron VLSI devices for next-generation high-speed, low-power, system applications using a low-power supply voltage. Thanks to progresses in processing technology, SOI CMOS technology has been used to implement multi-giga-bit DRAM, 1 GHz microprocessors, and other high-speed low-power computer-related VLSI circuits. Owing to much smaller parasitic capacitances, SOI CMOS devices have also been used to integrate high-speed low-power network-related telecommunication VLSI circuits. Recently, the demands on low-voltage VLSI circuit designs using deep-submicron SOI CMOS technology have grown dramatically. However, nowadays the development of the supporting environment for meeting the demands on the growth of the SOI CMOS IC designs for VLSI system applications is not paced accordingly. The microelectronics industry is interested in becoming familiar with the SOI CMOS device behaviors and SOI CMOS circuit design skills and needs concise device models for SOI SPICE CAD and breakthroughs in SOI CMOS circuits. In order to help overcome the bottlenecks for the escalated development of the SOI CMOS technology for VLSI system applications, this book is the first on SOI, that provides a comprehensive description of low-voltage SOI CMOS VLSI devices and circuits. This book includes up-to-date structures and behaviors of the state-of-the-art SOI CMOS devices. In addition, a wide spectrum of SOI CMOS digital and analog circuits targeted for low-voltage, low-power, high-speed VLSI system applications such as logic, memory, CPU, telecommunications, etc, are described. In the final portion, CMOS SOI Technology SPICE models for circuit simulations are presented. This book is written for high-tech professionals interested in microelectronics and/or CMOS VLSI and is also suitable for first-year graduate students interested in VLSI. The arrangement of the book is designed for a three semester unit course.

JAMES B. KUO

SHIH-CHIA LIN

Waterloo, Canada
August 2001

Acknowledgments

This book is dedicated to the memory of Professor Chih-Chin Ma, who passed away January 2001. We would like to thank all of their teachers from kindergarten to graduate school and would like to express our gratitude to Professor Chih-Chin Ma of NTUEE for his encouragement during the past 28 years. The authors would like to thank Professor Robert W. Dutton at Stanford for his encouragement during the past 20 years, and Professors Hwei-Chung Yu, Powen Hsu, Way-Seen Wang, and Hsu-Chun Yen for their consistent encouragement during the past 14 years. We would like to express our appreciation to our colleagues in Taiwan for their support in the past and would also like to express our gratitude to colleagues at University of Waterloo, Canada for providing us with the opportunity to finish this book. Special thanks to Professors Anthony Vannelli, Arokia Nathan, and Suject K. Chaudhuri for their support and encouragement. The author would like to thank his students–Mr. C. H. Chang, Mr. I. J. Chen, Mr. T. Y. Chiang, Mr. F. A. Wu, Mr. C. L. Young, Mr. Y. T. Chang, Mr. H. H. Chen, Mr. P. C. Chen, and Mr. H. K. Sung for their intensive help drawings. We would like to thank Dr. K. W. Su for his help during the initial phase of this work. We would like to thank Mr. George Telecki, Mr. Andrew Prince, and Ms. Angioline Loredo of Wiley-Interscience for their help in making this book a reality.

About the Authors

Professor James B. Kuo received a BSEE degree from National Taiwan University in 1977, an MSEE degree from the Ohio State University in 1978, and a PhDEE degree from Stanford University in 1985. Since 1987, he has been with the National Taiwan University. In 2000, he joined the University of Waterloo as the NSERC Canada Research Chair Professor. He has published numerous international journal articles. In 1999, he was elected to be an IEEE fellow for his contributions to modeling of CMOS VLSI devices. He serves as an associate editor for the *IEEE Circuits and Devices Magazine* and a membership chair for the IEEE Electron Devices Society.

Dr. Shih-Chia Lin received a BSEE degree in 1992 and a PhDEE degree in 2001 from the National Taiwan University. His research specialty is compact modeling of SOI CMOS devices. Currently, he is a postdoctoral researcher at the University of Waterloo.

Low-Voltage
SOI CMOS VLSI
Devices and Circuits

1

Introduction

SOI CMOS VLSI technology has become a major technology for integrating VLSI systems. This technology is especially suitable for integrating low-power VLSI systems using a low-supply voltage. In this chapter, the evolution of CMOS VLSI is described, followed by comparison of SOI and bulk CMOS VLSI. Then, the trends on low-power VLSI using a low-voltage power supply are described. Finally, the objectives of this book are highlighted.

1.1 EVOLUTION OF CMOS VLSI

Along with progress in the CMOS processing technology, CMOS devices have been scaled down continuously, which triggered advances in the VLSI circuit design techniques. By using advanced CMOS VLSI technology, high-performance computer system chips using advanced architectures such as reduced instruction set computing (RISC) and sever-client system connections have been integrated as a result of the computer evolution, as shown in Fig. 1.1 [1]. Based on the VLSI CPU chips using the advanced architectures, high-performance, low-price, low-power desktop, laptop, and hand-held computers have been derived for use in the information-related systems. Along with the shrinkage of the CMOS devices, the corresponding power consumption has also been decreased– low-power has been a requirement for today's VLSI systems. In addition, the price of computer and communication (C&C) related VLSI systems has been dropping dramatically. In the past, a 10-MIPS mainframe computer consuming 10 kW of power cost 10 million. Nowadays, a 1000-MIPS computer chip consuming 1 W of power costs only 10. The power consumption per MIPS and the price per MIPS have been evolving amazingly.

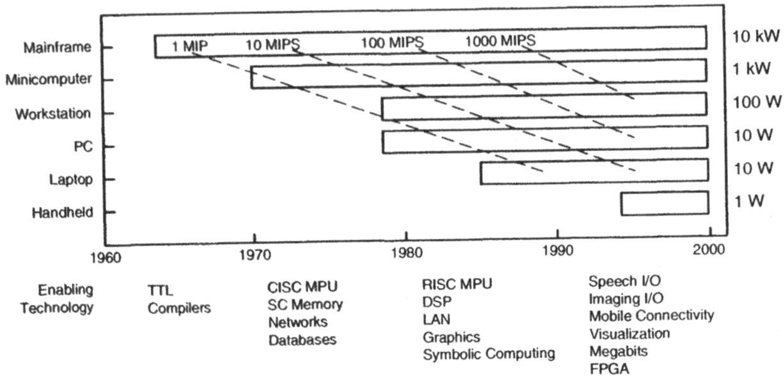

Fig. 1.1 Trends on power consumption and price per MIPS of VLSI systems. (Adapted from Stork [1].)

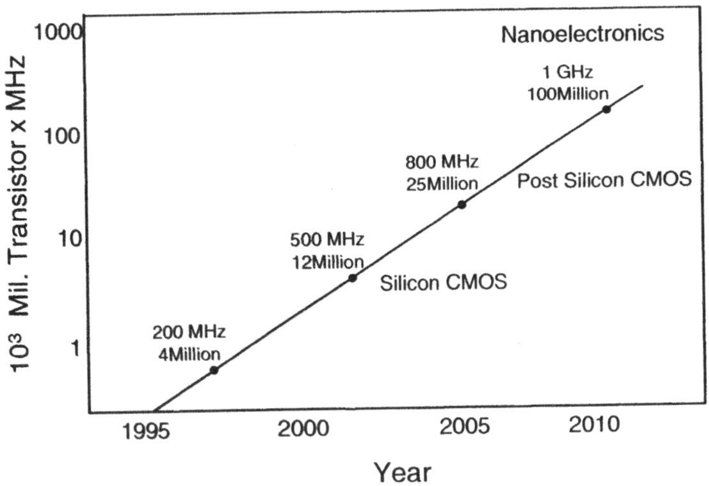

Fig. 1.2 Roadmap on the performance of CPU chips in terms of the product of the transistor number and the clock frequency predicted by SIA. (Adapted from Goser et al. [2].)

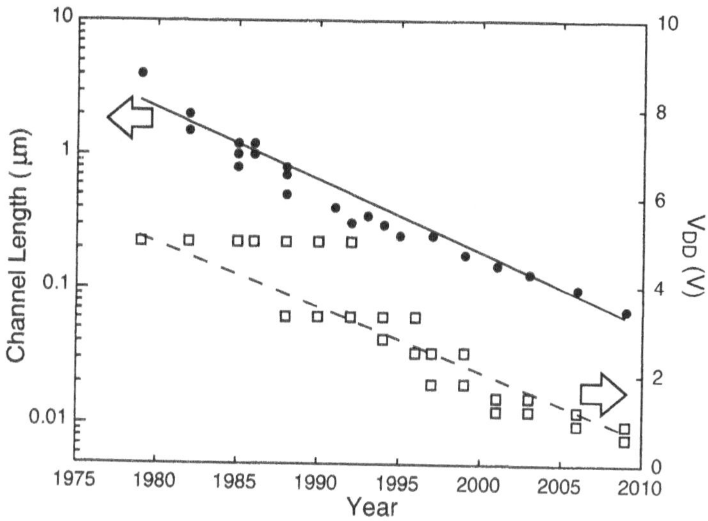

Fig. 1.3 Evolution of CMOS technology. (Adapted from Kuo and Lou [3].)

Figure 1.2 shows the roadmap of the performance of CPU chips in terms of the product of the transistor number and the clock frequency predicted by Semiconductor Industries Association (SIA) [2]. As shown in this figure for a computer-related VLSI chip, which is based on past trends, the number of transistors in a chip will grow from 12 million in 2001 to 40 billion in 2010. The clock frequency for the computer chip in 2010 will be over 1 GHz using 0.05 μm CMOS technology. Based on the prediction by the semiconductor industry, when a VLSI technology has provided an improvement in the device performance over two orders of magnitude, a revolution in the technology occurs. In the future, for a nanoelectronics system chip, there will be over 1 billion transistors. When the semiconductor industry goes toward nanoelectronics, the progress of VLSI systems and circuits also moves along with technology development as well.

Figure 1.3 shows the evolution of CMOS technology [3]. As shown in this figure, the channel length of CMOS technology will be scaled down from 0.18 μm in 2000 to 0.05 μm in 2010. From the consideration of the electric field limit, the power supply voltage will also need to be scaled down from 1.8 to 0.7 V. Hence, low voltage is a necessity for the next-generation low-power VLSI systems.

The most straightforward way to meet the low-power requirement for a VLSI system is to lower the power supply voltage, which is also pertinent for a down-scaled CMOS technology considering the internal electric field distribution. Figure 1.4 shows the trends on the power supply voltage of CMOS technology [4]. In the low-voltage regime targeted for the deep-submicron CMOS technology using a low supply voltage, SOI CMOS emerges as a dominant technology. Along with the progress in CMOS technology, its power supply voltage is also scaled down. For bulk CMOS devices, when the power supply voltage is shrunk, its threshold voltage is reduced accordingly. Therefore, the leakage current of the down-scaled CMOS devices may

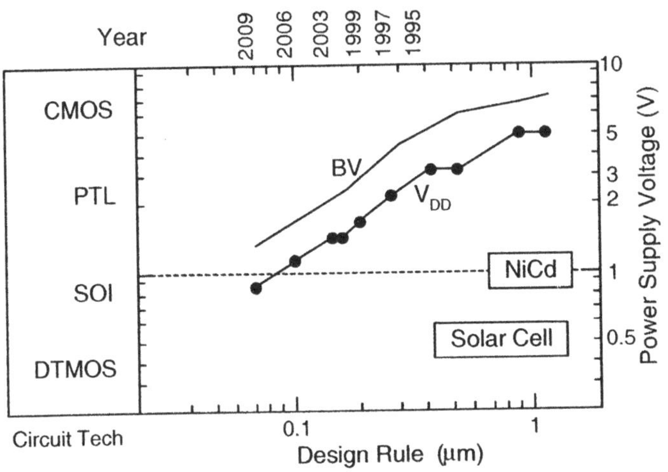

Fig. 1.4 Trends on the power supply voltage of CMOS technology. (Adapted from Adan et al. [4].)

increase, which is not good for VLSI systems from the power consumption point view. For SOI CMOS devices, owing to the buried oxide layer used for isolation, body controlled circuit techniques such as dynamic threshold (DTMOS) or a multithreshold scheme can be used to lower their threshold voltage as needed. When SOI DMOS devices are turned off, their leakage current still maintains at a low level such that the low-power requirement can be met. The body-controlled scheme is not complicated from a technology point of view. As shown in this figure, with a sub-0.1 μm technology, the power supply voltage is < 1V, which is suitable for use in portable systems using solar cells as a power supply. Note that from the speed-power performance–low-power and high-speed SOI CMOS technology is ahead of bulk CMOS by a generation.

1.2 SOI VERSUS BULK

Figure 1.5 shows the cross-section of the bulk and SOI MOS devices. As shown in this figure, owing to the oxide isolation structure, SOI devices have superior capability in good radiation hardness, no latchup, and high device density. Without the reverse biased junctions used for isolation as in bulk CMOS, the leakage current is small in the SOI CMOS devices. Owing to the oxide isolation, the source/drain parasitic capacitances are smaller, which results in a higher speed performance of the SOI CMOS devices as compared to bulk counterparts at the down-scaled power supply voltage. The SOI CMOS devices have a unique buried oxide layer, which is used to isolate the body from the substrate. During operation, the body of the SOI devices may be floating if no extra body contact is added. Floating body effects may bring in

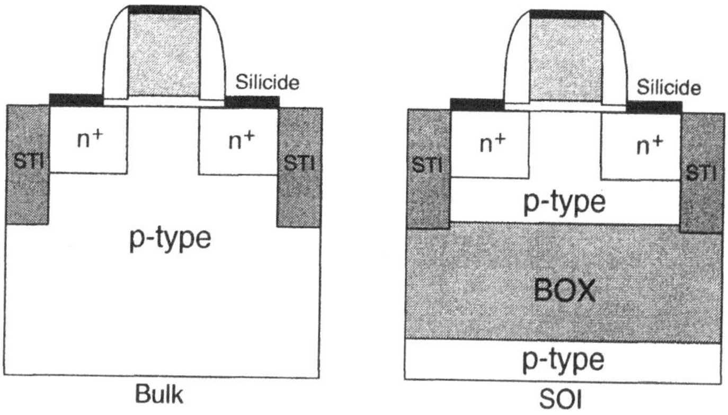

Fig. 1.5 Cross-section of bulk and SOI MOS devices.

serious problems in the SOI CMOS devices, which may be suppressed using body-control techniques described later.

Figure 1.6 shows the propagation delay versus the normalized power consumption of a CMOS inverter using 0.35 μm bulk and SOI CMOS technologies at various power supply voltages [4]. As shown in this figure, when the power supply voltage is scaled down, the speed of the CMOS inverter is slowed down. In addition, the power dissipation is shrunk accordingly. Compared to bulk, the SOI CMOS inverter is less affected by the scale-down of the power supply voltage in terms of the speed performance, which implies that SOI CMOS technology is more suitable for circuits using a low-power supply voltage.

Figure 1.7 shows the design of a 64b CPU chip using 0.2 μm bulk and SOI CMOS technologies with copper interconnects [5]. In this 64b CPU chip, multithreshold circuit techniques have been adopted in addition to SRAM, latch-based array, control logic, full-custom static logic, and domino dynamic logic. As shown in this figure, due to the lower threshold voltage and diffusion capacitances of the SOI devices, the clock frequency of the SOI RISC microprocessor is faster than the bulk counterpart by 100 MHz. Due to the increase in the operating frequency, the power consumption of the SOI CPU is 10% higher.

Quality and yield have become the key factors in determining the superiority of CMOS technologies. Quality of the thin film determines the breakdown, the leakage, and the ESD performance of an SOI device. Figure 1.8(a) shows the SOI-to-bulk wafer cost ratio to achieve the same die cost versus the chip size for various SOI-to-bulk defect density ratios k [4]. As shown in this figure, when the die area of the circuit is small, the SOI wafer cost is close to the bulk one, which implies that despite the high SOI wafer cost, SOI circuits still have the same die cost as compared to the bulk counterparts. However, when the chip size becomes as large as 1 cm^2, the SOI wafer cost needs to be much cheaper than the bulk wafer in order to be competitive if the SOI defect density is high. As shown in Fig. 1.8(b), for a 1 cm^2 chip with a large

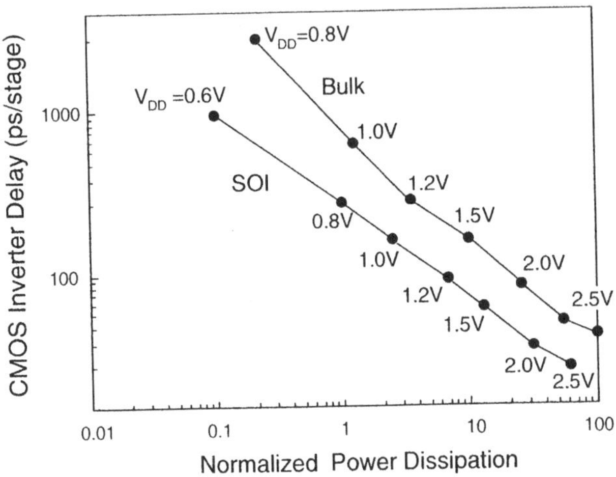

Fig. 1.6 Propagation delay versus normalized power consumption of a CMOS inverter using 0.35 μm bulk and SOI CMOS technologies at various power supply voltages. (Adapted from Adan et al. [4].)

Item	Bulk	SOI
Clock Frequency	450 MHz	550 MHz
Supply Voltage	1.8 V	1.8 V
Transistors	34 M	34 M
Die Size	139 mm^2	139 mm^2
Power	22 W	24 W
L_{eff}(NMOS)	0.12 μm	0.12 μm
Metallization	6 layers Cu	6 layers Cu

Fig. 1.7 Design of a 64b CPU chip using 0.2 μm bulk and SOI CMOS technologies with copper interconnects. (Adapted from Allen et al. [5].)

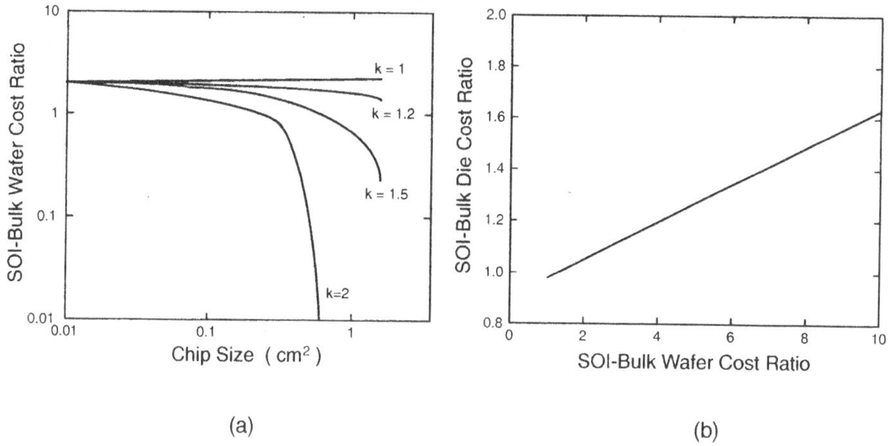

Fig. 1.8 (a) SOI-to-bulk wafer cost ratio to achieve the same die cost versus chip size for various SOI-to-bulk defect density ratios. (b) SOI-to-bulk die cost ratio versus the SOI to bulk wafer cost ratio for a 1 cm^2 chip using a $0.25\ \mu m$ technology. (Adapted from Adan et al. [4].)

amount of devices using $0.25\ \mu m$ technology, in order for the SOI technology to be competitive, the SOI wafer cost needs to be $< 1.5\times$ the bulk wafer cost.

1.3 LOW-VOLTAGE SOI VLSI

In general, when a bulk CMOS digital circuit is replaced by a compatible SOI circuit, its speed performance can be improved $\sim 25\%$ at a reduced power dissipation. Along with the down-scale of the power supply voltage, both bulk and SOI CMOS technologies are targeted for low-power consumption. Figure 1.9 shows the power consumption versus the power supply voltage of a CMOS logic gate using bulk and SOI CMOS devices [6]. The power consumption of a CMOS logic gate is composed of static and dynamic portions. The static power consumption is referred to the power consumption during standby, which is caused by leakage current ($I_{off} V_{DD}$). The dynamic power consumption is related to the power consumed during switching, which is a function of the clock frequency, the load capacitance, and the supply voltage ($fC_L V_{DD}^2$). Owing to the buried oxide isolation, the source/drain parasitic capacitances of the SOI devices are $\sim 20\%$ smaller as compared to bulk. Therefore, the dynamic power consumption of the SOI circuit is smaller. Along with the shrinkage of the power supply voltage, the threshold voltage needs to be scaled down, which may result in an increase in the leakage current. Therefore, the static power consumption is increased when the power supply voltage is scaled down. Since the SOI devices have better subthreshold characteristics, their leakage currents are also smaller as compared to bulk. Thus, the static power consumption of the SOI circuits is smaller than the bulk counterpart. From the above reasoning, the total power consumption

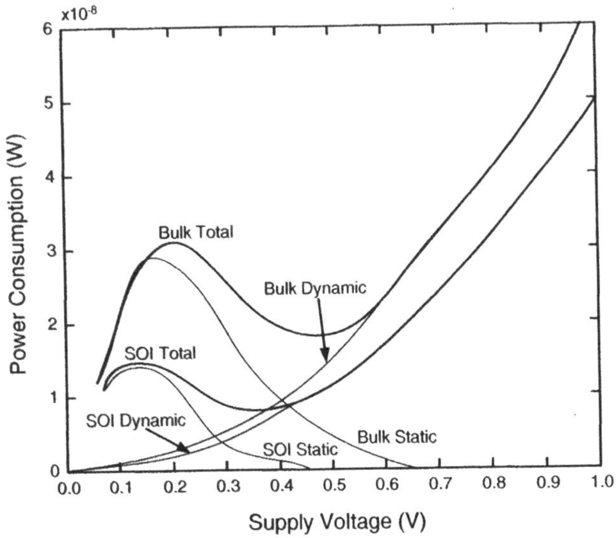

Fig. 1.9 Power consumption versus power supply voltage of a CMOS logic gate using bulk and SOI CMOS devices. (Adapted from Colinge [6].)

of the SOI circuits is smaller than that of the bulk circuits when the power supply voltage is shrunk.

Figure 1.10 shows a list of the low-voltage, low-power, high-speed CMOS VLSI circuits recently made by SOI technology. As shown in this figure, SOI technology has been used to integrate digital circuits such as gate array, DRAM, SRAM, cache memory, multiplexer, CPU, etc. In addition, SOI technology has also been used to realize analog circuits such as the phase-locked loop (PLL), the radio frequency (RF) receiver, and the tuned amplifier. As listed in this figure, the most popular SOI CMOS technology nowadays is 0.18 μm using a power supply voltage of 1~1.5 V. Owing to the small leakage current, small soft-error rate, and small parasitic capacitances, SOI CMOS technology is suitable to integrate low-voltage DRAM, SRAM, and CPU, etc. Along with improvement in the gate sheet resistance, SOI CMOS technology already could provide RF devices with f_T and f_{max} in the order of GHz for implementing RF analog circuits. In addition, due to the high-resistivity substrate in the SOI technology, the loss of the passive elements, which are required in making RF circuits, can be reduced. Owing to the buried oxide layer, the cross talk among devices can be lowered, which is important when making RF circuits.

1.4 OBJECTIVES

This book provides a comprehensive description of low-voltage SOI CMOS VLSI devices and circuits. Up-to-date structures and behaviors of the state-of-the-art SOI CMOS devices are also included. A wide spectrum of SOI CMOS digital and analog

Circuit	L (μm)	V_{DD} (V)	Frequency (Hz)	Ref.
560 kG gate array	0.35	1	70M	'97 [7]
16Mb DRAM	0.5	1	46 ns	'97 [8]
32x64 register file	0.25	1.3	660M	'98 [9]
8b MUX/DEMUX	0.32	0.5	320M	'98 [10]
128k SRAM	0.35	1.5	8.3 ns	'98 [11]
PLL	0.7	3	1.1 G	'98 [12]
tuned amp	0.1	1.5	4G/13G	'98 [13]
4x2 ATM switch	0.25	2	10 Gb/s	'99 [14]
54x54 multiplier	0.25	0.5	32M	'99 [15]
8b microprocessor	0.25	2	33M	'99 [16]
8b ASK RF receiver	0.25	2	400M	'99 [16]
16:1 MUX	0.18	2	3.6G	'99 [17] '00[18]
4M SRAM	0.2	1.8		'00 [19]
RF power amp	1.5	5	900M	'00 [20]
embedded DRAM	0.35	1.2		'00 [21]
128k embedded SRAM	0.35	1	100M	'00 [22]
2.25 Mb Cache	0.18		900M	'01 [23]
receiver		1	2G	'01 [24]
64b ALU	0.18	1.5	380~400 ns	'01 [25]

Fig. 1.10 List of low-voltage, low-power, high-speed SOI CMOS VLSI circuits recently made by SOI technology.

circuits targeted for low-voltage, low-power, high-speed VLSI system applications such as logic, memory, CPU, telecommunications, etc, are described. In addition, partially depleted SOI CMOS device models for SPICE circuit simulations are also included.

In Chapter 2, behaviors of SOI CMOS devices are described. Starting from the fundamental SOI CMOS technology, the back gate bias effects of SOI CMOS devices are introduced. Then, short and narrow channel effects are described, followed by mobility including the velocity overshoot phenomenon. Unique phenomena of SOI CMOS devices caused by their floating body structure may generate peculiar drain current characteristics–kink effects. In Chapter 2, floating body effects and subthreshold characteristics of SOI CMOS devices are analyzed. Also, impact ionization, snapback, bipolar leakage, and bipolar history effects derived from the floating body structure are depicted. Because of the insulating buried oxide below the active thin-film region, SOI CMOS devices are susceptible to thermal effects—self-heating. In the final portion of Chapter 2, self-heating of SOI CMOS devices is presented, followed by the transient analysis of SOI CMOS devices. Chapter 3 has more topics on SOI CMOS devices. Starting from the hot carriers of the SOI CMOS devices, accumulation-mode and double-gate SOI devices are analyzed. Next, DTMOS devices are introduced, followed by the scaling trends of SOI CMOS devices and the

SOI single electron transistors (SET). Then, the temperature dependence of the SOI devices is analyzed. Finally, sensitivities and radiation effects of SOI CMOS devices are described.

In Chapter 4, basic knowledge of SOI CMOS circuits is described. Starting from the basic circuit issues, the floating body effects on the performance of the SOI CMOS circuits are explained, followed by the low-voltage SOI CMOS circuits, SOI dynamic-threshold MOS (DTMOS) circuits, and SOI multithreshold MOS (MTMOS) circuits. Then, noise and self-heating problems of SOI CMOS circuits are analyzed. Finally, the SOI ESD circuits and the SOI system-on-a chip (SOC) technology are presented. In Chapter 5, starting from fundamental SOI CMOS static and dynamic logic circuits, DRAM and SRAM circuits using SOI CMOS technology are described. Then, SOI cache memory and content addressable memory (CAM) are depicted, followed by SOI gate arrays. SOI CPU and embedded memory are introduced, and finally SOI multipliers/digital signal processing (DSP) circuits and SOI frequency dividers are described. In Chapter 6, op amps, filters, analog-to-digital converter (ADC) and digital-to-analog converter (DAC), sigma-delta ADC, RF circuits, low noise amplifiers (LNA), mixers, voltage-control oscillator (VCO), and high-temperature analog circuits using SOI CMOS technology are described. In Chapter 7, partially depleted SOI CMOS device models for SPICE circuit simulation purposes using a BiCMOS device approach are described. In addition, the floating body effects of various SOI circuits using the PD-SOI Technology SPICE models are analyzed.

REFERENCES

1. J. M. C. Stork, "Technology Leverage for Ultra-Low Power Information Systems," *Proc. IEEE*, **83**(4), 607–618 (1995).

2. K. F. Goser, C. Pacha, A. Kanstein, and M. L. Rossmann, "Aspects of Systems and Circuits for Nanoelectronics," *Proc. IEEE*, **85**(4), 558–573 (1997).

3. J. B. Kuo and J. H. Lou, "Low-Voltage CMOS VLSI Circuits," *New York: Wiley*, 1999.

4. A. O. Adan, T. Naka, A. Kagisawa, and H. Shimizu, "SOI As a Mainstream IC Technology," *SOI Conf. Dig.*, 9–12 (1998).

5. D. Allen, D. Behrends, and B. Stanisic, "Converting a 64b PowerPC Processor from CMOS Bulk to SOI Technology," *Design Automation Conf. Dig.*, 892–897 (1999).

6. J.-P. Colinge, "Performances of Low-Voltage, Low-Power SOI CMOS Technology," *MIEL Dig.*, 229–236 (1997).

7. K. Mashiko, K. Ueda, K. Nii, Y. Wada, T. Hirota, S. Maeda, T. Iwamatsu, Y. Yamaguchi, T. Ipposhi, S. Maegawa, and H. Hamano, "A 0.35μm 560KG SOI/CMOS

Gate Array using Field-Shield Isolation Technique," *SOI Conf. Dig.*, 166–167 (1997).

8. K. Shimomura, H. Shimano, N. Sakashita, F. Okuda, T. Oashi, Y. Yamaguchi, T. Eimori, M. Inuishi, K. Arimoto, S. Maegawa, Y. Inoue, S. Komori, and K. Kyuma, "A 1-V 46-ns 16-Mb SOI-DRAM with Body Control Technique," *IEEE J. Sol. St. Ckts.*, **32**(11), 1712–1720 (1997).

9. R. V. Joshi, W. Hwang, W. H. Henkels, S. Wilson, W. Rausch, and G. Shahidi, "A 660MHz Self-Resetting 8 Port, 32x64 Bits Register File and Latch in $0.25\mu m$ SOI Technology," *SOI Conf. Dig.*, 131–132 (1998).

10. T. Hirota, K. Ueda, Y. Wada, K. Mashiko, and H. Hamano, "0.5V 320MHz 8b Multiplexer/Demultiplexer Chips Based on a Gate Array with Regular-Structured DTMOS/SOI," *ISSCC Dig.*, 188–189 (1998).

11. Y. Wada, K. Nii, H. Kuriyama, S. Maeda, K. Ueda, and Y. Matsuda, "A 128Kb SRAM with Soft Error Immunity for $0.35\mu m$ SOI-CMOS Embedded Cell Arrays," *SOI Conf. Dig.*, 127–128 (1998).

12. G. Lyons, "1.1GHz Integer N Phase Lock Loop with Superior Single Event Upset and Total Dose Properties Suitable for Commercial Space Applications," *SOI Conf. Dig.*, 101–102 (1998).

13. K.-H. Kim, Y.-C. Ho, B. Floyd, C. Wann, Y. Taur, I. Lagnado, and K. O, "4GHz and 13GHz Tuned Amplifiers Implemented in a $0.1\mu m$ CMOS Technology on SOI And SOS Substrates," *ISSCC Dig.*, 134–135, 1998.

14. E. Oki, N. Yamanaka, Y. Ohtomo, K. Okazaki, and R. Kawano, "A 10-Gb/s $(1.25Gb/s \times 8) 4 \times 2\, 0.25-\mu m$ CMOS/SIMOX ATM Switch Based on Scalable Distributed Arbitration," *IEEE J. Sol. St. Ckts.*, **34**(12), 1921–1934 (1999).

15. K. Fujii, and T. Douseki, "A 0.5-V, 3-mW,$54 \times 54b$ Multiplier with a Triple-V_{th} CMOS/SIMOX Circuit Scheme," *SOI Conf. Dig.*, 73–74 (1999).

16. E. McShane, K. Shenai, L. Alkalai, E. Kolawa, V. Boyadzhyan, B. Blaes, and W. C. Fang, "Monolithic Microprocessor and RF Transceiver in 0.25-micron FDSOI CMOS," *Symp. VLSI*, 332–333 (1999).

17. T. Nakura, K. Ueda, K. Kubo, W. Fernandez, Y. Matsuda, and K. Mashiko, "A 3.6Gb/s 340mW 16:1 Pipe-Lined Multiplexer using SOI-CMOS Technology," *Symp. VLSI Ckts. Dig.*, 27–30 (1999).

18. , T. Nakura, K. Ueda, K. Kubo, Y. Matsuda, K. Mashiko, and T. Yoshihara, "A 3.6-Gb/s 340-mW 16:1 Pipe-Lined Multiplexer using $0.18\mu m$ SOI-CMOS Technology," *IEEE J. Sol. St. Ckts.*, **35**(5), 751–756, (2000).

19. K. Cox, J. Scott, S. Bishop, M. Bhat, B. Nettleton, D. Pan, M. Hamilton, D. Chang, L. Day, and P. Schani,"A Partially Depleted 1.8V SOI CMOS SRAM Technology Featuring a $3.77\mu m^2$ Cell," *Symp. VLSI Tech. Dig.*, 170–171 (2000).

20. Y. Tan, M. Kumar, J. K. O. Sin, L. Shi, and J. Lau, "A 900-MHz Fully Integrated SOI Power Amplifier for Single-Chip Wireless Transceiver Applications," *IEEE J. Sol. St. Ckts.*, **35**(10), 1481–1486 (2000).

21. T. Yamauchi, F. Morisita, S. Maeda, K. Arimoto, K. Fujishima, H. Ozaki, and T. Yoshihara, "High-Performance Embedded SOI DRAM Architecture for the Low-Power Supply," *IEEE J. Sol. St. Ckts.*, **35**(8), 1169–1178 (2000).

22. N. Shibata, H. Morimura, and M. Harada, "1-V 100-MHz Embedded SRAM Techniques for Battery-Operated MTCMOS/SIMOX ASICs," *IEEE J. Sol. St. Ckts.*, **35**(10), 1396–1407 (2000).

23. J. M. Hill and J. Lachman, "A 900MHz 2.25MB Cache with On-Chip CPU – Now in Cu SOI," *ISSCC Dig.*, 176–177 (2001).

24. M. Ugajin, J. Kodate, and T. Tsukahara, "A 1V 12mW 2GHz Receiver with 49dB Image Rejection in CMOS/SIMOX," *ISSCC Dig.*, 288–289 (2001).

25. S. Mathew, R. Krishnamurthy, M. Anders, R. Rios, K. Mistry, and K. Soumyanath, "Sub-500ps 64b ALUs in 0.18μm SOI/Bulk CMOS : Design & Scaling Trends," *ISSCC Dig.*, 318–319 (2001).

2

SOI CMOS Devices–Part I

Behaviors of SOI CMOS devices are quite different from those of the bulk ones. Understanding the unique behavior of the SOI CMOS devices is important for designing SOI CMOS VLSI circuits. In this chapter, fundamental SOI CMOS technology is described first, followed by the phenomena of SOI CMOS devices. The back gate bias effects of SOI CMOS devices are introduced. Then short and narrow channel effects are described, followed by mobility including the velocity overshoot phenomenon. Unique phenomena of SOI CMOS devices caused by their floating body structure may cause peculiar drain current characteristics--kink effects . In this chapter, floating body effects and subthreshold characteristics of SOI CMOS devices are analyzed. Also depicted are impact ionization, snapback, bipolar leakage, and bipolar history effects derived from the floating body structure. Because of the insulating buried oxide below the active thin-film region, SOI CMOS devices are susceptible to thermal effects–self-heating. In the final portion of this chapter, self-heating of SOI CMOS devices is presented, followed by the transient analysis of SOI CMOS devices.

2.1 BASIC SOI TECHNOLOGY

The fabrication process of SOI CMOS technology is similar to the bulk CMOS counterpart except for the starting silicon wafers. In this section, fundamental SOI CMOS technology in terms of SOI wafers and isolation technology is described. In addition, cross-section of an SOI CMOS technology is also explained.

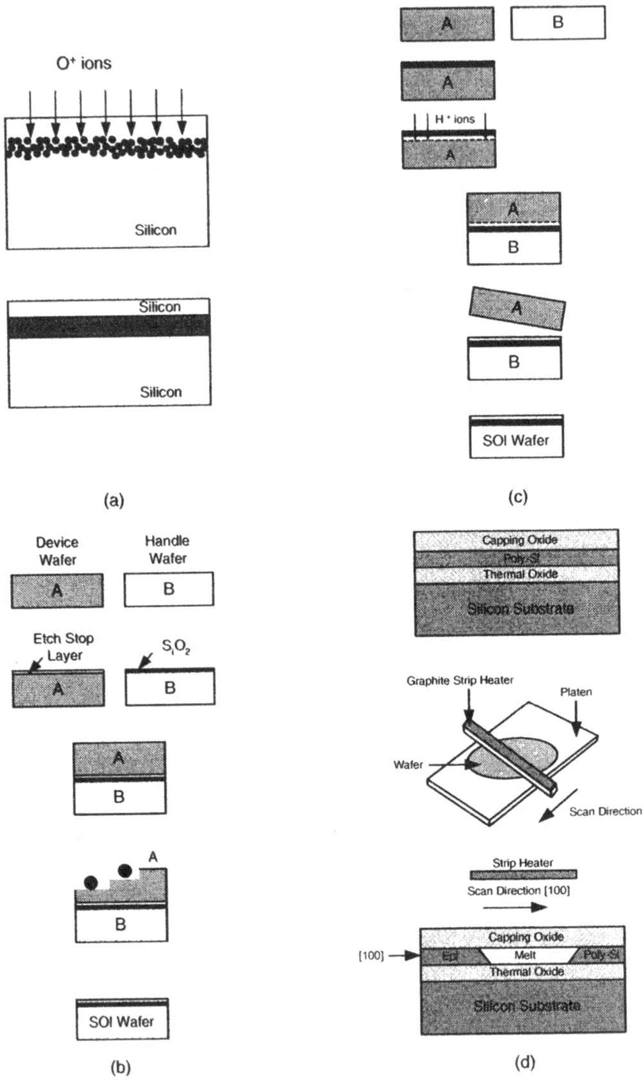

Fig. 2.1 SOI wafer technologies: (a)SIMOX–Synthesis by Implanted Oxygen (Adapted from Watanabe and Tooi [1]), (b)BESOI– Bond and Etch Back (Adapted from Godbey et al. [2]), (c) Smart-Cut (Adapted from Bruel et al. [3][4]), and (d) ZMR–Zone Melting Recrystallization (Adapted from Zavracky et al. [5].)

2.1.1 SOI Wafers

Figure 2.1 shows the simplified processing sequences of SOI wafer technologies. As shown in Fig. 2.1(a), in the synthesis by implanted oxygen (SIMOX) technology [1], high-dose oxygen ions (O^+) are implanted into the silicon wafer with the peak concentration at some distance below the surface. Via a high-temperature anneal, the oxygen ions in the silicon wafer react with silicon to form an oxide layer. At the top of the oxide layer, a layer of crystalline silicon layer is generated. If the thickness of the crystalline silicon layer above the oxide is not sufficient, an epitaxial layer of silicon can be grown on top of the silicon thin film by chemical vapor deposition (CVD) epitaxy. Then, a smooth surface of the silicon layer can be obtained by chemical mechanical polishing (CMP).

Figure 2.1(b) shows the simplified processing sequence of the bond and etch back (BESOI) wafer technology [2]. As shown in this figure, two silicon wafers, the device and handle wafer, are prepared. At the top of the device wafer, an epitaxial layer is grown as an etch stop layer. On top of the handle wafer, an oxide layer with very few silicon/oxide defects is formed by thermal oxidation. Then, the device and handle wafer are bonded together using van der Walls forces, reenforced by annealing. Thinning the device wafer by mechanical grinding and etching via the etch stop layer is used to remove the extra silicon layer at the top. followed by annealing and CMP to create the BESOI wafers with a clean and smooth surface.

Figure 2.1(c) shows the simplified processing sequence of Smart-Cut SOI wafer technology [3][4]. As shown in this figure, an oxide layer served as the buried oxide is grown thermally on top of silicon wafer A, followed by a hydrogen ion implant with a dose of $\sim 2 \times 10^{16} \sim 10^{17} cm^{-3}$. Silicon wafer B, which is used as a stiffener and as a substrate below the SOI buried oxide, is bonded with wafer A by hydrophilic bonding. At $400 \sim 600°C$, a layer of monocrystalline silicon is generated in the implanted wafer, which is bonded with wafer B by van der Walls forces. The rest of wafer A is split as shown. A 1100°C procedure is used to strengthen the chemical bonds between the monocrystalline silicon of wafers A and B. A surface polishing procedure is used to generate a smooth surface.

Figure 2.1(d) shows the simplified processing sequence of zone melting recrystallization (ZMR) SOI wafer technology [5]. As shown in this figure, a silicon substrate is thermally oxidized to grown an oxide layer of 1 μm, which is used as the buried oxide layer. The oxide at the edge of the wafer is etched off to serve as the seed during the recrystallization step. A layer of $0.75 \sim 3$ μm polysilicon is deposited on top of buried oxide by low-pressure chemical vapor deposition (LPCVD), followed by another layer of oxide deposited by low-temperature CVD, serving as the capping oxide to protect the wafer from contamination and to reduce the silicon thin film variation during the subsequent zone melting procedure. Then, a 2200°C graphite heater scans across the wafer back and forth at a speed of $0.1 \sim 1$mm/s from some distance above it, which is heated to 1200°C. Right under the heater, the polysilicon layer is melted, which is then cooled off to recrystallize and becomes a layer of single-crystalline silicon thin film after the heater moves away. Since the heater scan starts from the edge of the wafer, the substrate exposed to the polysilicon

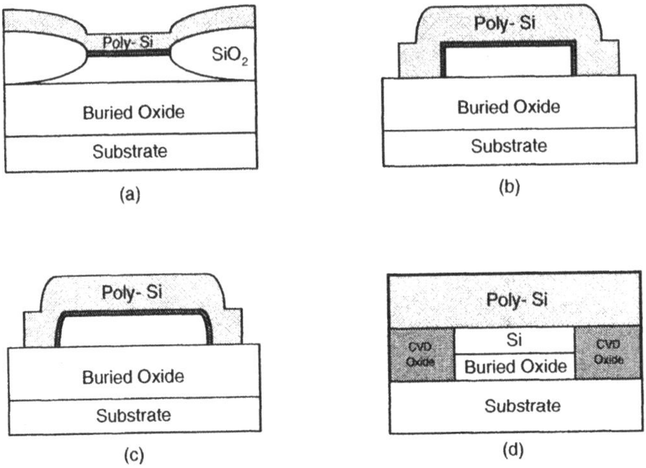

Fig. 2.2 Isolation technologies for SOI CMOS devices: (a) LOCOS, (b) Mesa (Adapted from Haond & Le Neel [6]), (c) Rounded-Edge Mesa (Adapted from Haond & Le Neel [7]), and (d) Shallow Trench Isolation (STI) (Adapted from Huang & Grula [8]).

layer at the edge becomes the seed during the melting and recrystallization procedure.

2.1.2 Shallow Trench Isolation

Figure 2.2 shows the isolation technologies for SOI CMOS devices: (a) LOCOS, (b) Mesa [6], (c) rounded-edge Mesa [7], and (d) shallow trench isolation (STI) [8]. As shown in Fig. 2.2(a), just as in bulk CMOS technology, LOCOS isolation based on the selective local oxidation techniques can also be used for SOI. In addition to LOCOS, mesa isolation as shown in Fig. 2.2(b) has also been frequently used in SOI technology. By masking the active region of the device, the field region of the thin film is etched off to form an island of silicon thin film. Mesa isolation is simple and has the capability of high density. However, due to the sharp edge of the mesa-isolated device, some polysilicon gate material may remain after the gate polysilicon etch procedure, which is not good for wiring. In addition, the sharp mesa sidewall corner may produce two-dimensional (2D) charge sharing effects, which lead to drawbacks in subthreshold characteristics. Figure 2.2(c) shows the rounded-edge mesa-isolated structure [7]. On top of the active region in the SOI wafer, a polysilicon layer with the thickness of the thin film is deposited. Anisotropic etching of the silicon and polysilicon layers are carried out until the buried oxide is exposed. The extra polysilicon layer and the thin film in the field region are totally etched off. Due to the predeposited polysilicon layer, the silicon island sidewall is in rounded-edge shape instead of steep shape. As shown in Fig. 2.2(d), as for sub-0.25 μm bulk CMOS technology, shallow trench isolation has also been used in an advanced SOI technology. As shown in this figure, the thin-film layer and the buried oxide in the field region are etched off and filled

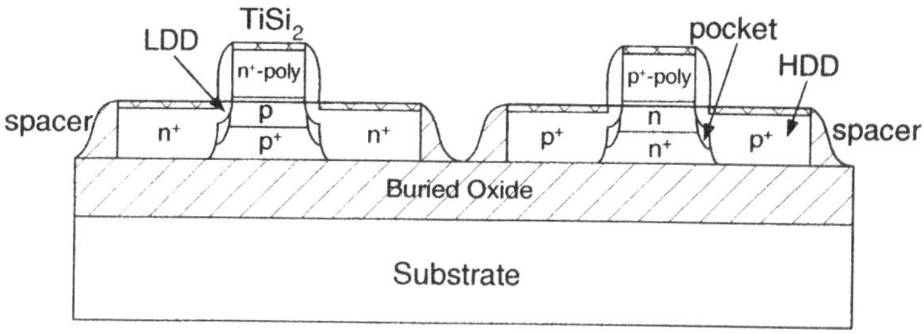

Fig. 2.3 Cross-section of a 0.13 μm PD SOI CMOS technology. (Adapted from Pindl et al.[11].)

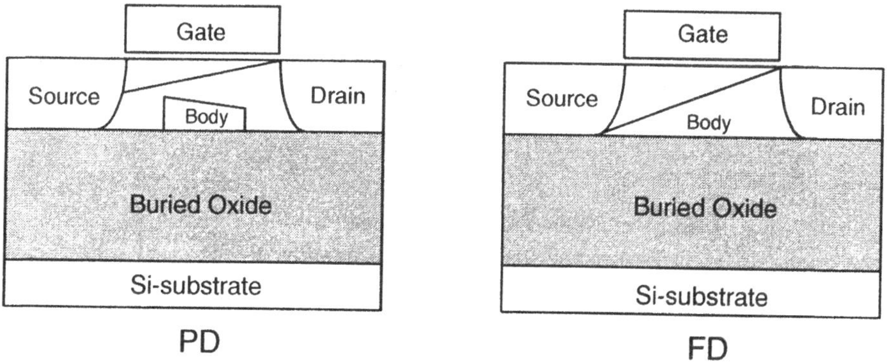

Fig. 2.4 Cross-section of PD and FD SOI CMOS devices.

with CVD oxide. The surplus CVD oxide on the top of the wafer is ground by CMP to form a smooth surface to reduce the formation of defects.

2.1.3 SOI Device Structure

Advanced SOI CMOS technologies with their feature size closed to 0.1 μm using copper interconnects and low-k dielectrics have been developed [9][10]. Figure 2.3 shows the cross-section of a 0.13 μm PD SOI CMOS technology with an advanced mesa isolation technique [11]. As shown in this figure, the SOI CMOS devices built on BESOI wafer have a thin film of 1900 Å, a buried oxide of 3900 Å, and a front gate oxide of 47 Å. A sidewall spacer and an arsenic/boron-type lightly doped drain (LDD) structure with a born/phosphorus pocket implant for the NMOS/PMOS device have been designed to reduce short channel effects. Titanium silicides have been used in the gate and the source/drain areas to reduce sheet resistance. As shown in this figure, except for the buried oxide structure, the device structure of the 0.13 μm PD SOI CMOS technology is similar to the bulk counterpart.

Depending on the condition of the thin film during operation, SOI CMOS devices are classified into two categories: (1) partially depleted (PD) and (2) fully depleted (FD) as shown in Fig. 2.4. For a PD device, its thin film is not fully depleted during

Item	PD	FD
Manufacturability	+	
Design Point (high V_T)	+	
Multiple V_T	+	
SCE	+	
Kink Effect		+
Body Contact	+	
Pass Gate Leakage	+	
History Dependence		+

Fig. 2.5 PD versus FD. (Adapted from Shahidi et al. [12].)

operation. For an FD device, the thickness of the thin film is small, hence during operation its thin film is fully depleted. As shown in Fig. 2.5, PD has advantages in manufacturability, design, multiple threshold voltages, and short channel effects. FD has advantages in history dependence and kink effects due to the floating body effect. Owing to the floating body, body contacts can be added to the PD device to eliminate the floating body effect and to provide a body control scheme for the low-voltage circuit designs.

2.2 BACK GATE BIAS EFFECTS

For bulk CMOS devices, back gate bias effect (body effect) is defined as the dependence of the threshold voltage on the back gate bias. For bulk CMOS devices, the back gate bias effect is caused by the depletion width under the gate oxide in the substrate. For a bulk NMOS device, when the back gate bias becomes more negative, the p-type depletion region under the gate widens. As a result, a larger gate voltage is needed to generate the inversion layer at the gate oxide interface–the threshold voltage increases. Owing to the buried oxide separating the active device region from the substrate, the threshold voltage of the SOI CMOS devices is less dependent on the back gate bias when compared to the bulk CMOS devices.

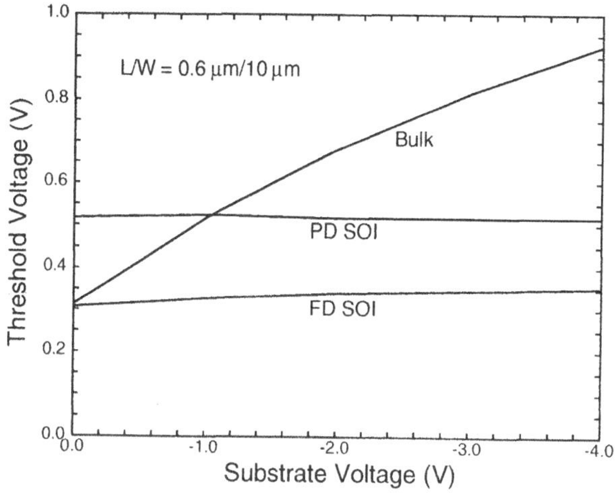

Fig. 2.6 Threshold voltage versus back gate bias of partially depleted (PD), fully depleted (FD) SOI and bulk NMOS devices. (Adapted from Eimori et al. [13].)

2.2.1 PD versus FD

Figure 2.6 shows the back gate bias effect–threshold voltage versus the back gate bias of partially depleted (PD), fully depleted (FD) SOI and bulk NMOS devices [13]. As shown in this figure, for the FD SOI NMOS device, the threshold voltage rises as the back gate bias voltage drops. Compared to the bulk, the back gate bias effect is much smaller. As for the PD SOI device, the back gate bias effect is even smaller as compared to the FD one owing to the more effective isolation of the neutral region from the electric field caused by the back gate bias. Both PD and FD SOI devices have less back gate bias effects as compared to the bulk devices–both PD and FD SOI devices are more suitable for low-voltage circuit applications. The reduced body effect provides a better performance for low-voltage circuits using SOI devices.

Back gate bias effects do not just determine the performance of a related circuit, back gate bias may also determine the operation of a specific SOI device. Under some situations, depending on the back gate bias, a PD device may become an FD device. Figure 2.7 shows the subthreshold characteristics of an SOI NMOS device with a channel length of 2 μm, a front gate oxide of 101 Å, a buried oxide of 3800 Å, and a thin film of 791 Å implanted with p-type dopants with a dose of $4 \times 10^{12}cm^{-2}$, biased at various back gate biases [14]. The device used for Fig. 2.7 is between PD and FD. As shown in this figure, depending on the back gate bias (V_{SUB}), this device may be FD or PD. Thus, its subthreshold slope may be different. At a negative back gate bias, the device tends to be PD. Under this situation, its threshold voltage is less influenced by the back gate bias (V_{SUB}). On the other hand, at a positive back gate bias (V_{SUB}), this device becomes FD, which means its thin film is fully depleted. At this time, the threshold voltage becomes more dependent on the change of the back

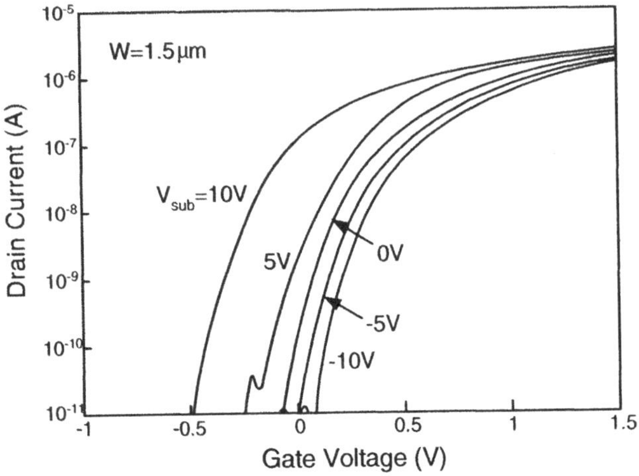

Fig. 2.7 Subthreshold characteristics of an SOI NMOS device with a channel length of 2 μm, a front gate oxide of 101 Å, a buried oxide of 3800 Å, and a thin film of 791 Å implanted with p-type dopants with a dose of 4×10^{12}cm^{-2}, biased at various back gate biases. (Adapted from Wang et al. [14].)

gate bias–back gate bias is more serious when the device works in the fully depleted regime.

When the back gate bias imposed on an FD SOI NMOS device is increased or decreased substantially, its back gate bias effect changes, which is similar to that of the one as shown in Fig. 2.8. For a back gate bias between -30 and 30 V (-30 V $<$ V_{SUB} $<$ 30 V), the trend of back-gate-bias-related threshold voltage is identical to the one of a typical FD device–the threshold voltage is linearly proportional to the back gate bias. At a very positive or negative back gate bias: $V_{SUB} > 30$ V, $V_{SUB} < -30$ V, the threshold voltage becomes insensitive to the change in the back gate bias–the back gate bias effect is reduced. For the back gate bias of less than -30 V ($V_{SUB} < -30$ V), holes start to accumulate at the bottom of the thin film. When the back gate bias of > 30 V ($V_{SUB} > 30$ V), at the bottom of the thin film, a back channel is generated due to strong inversion. In both cases, the generated carriers at the bottom of the thin film provide a shielding effect to isolate the front channel from the electric field in the buried oxide. Therefore, the back gate bias effect is reduced.

Generally speaking, for a FD device, the threshold voltage versus the back gate bias curve is linear in some region. The width of the linear region in the threshold voltage versus the back gate bias curve is dependent on the thickness and the doping density of the thin film. Figure 2.9 shows the threshold voltage versus the back gate bias of an SOI NMOS device with various thin-film thicknesses [16]. As indicated in this figure, several thin-film thicknesses have been assumed for the device. For the device with a thin film thickness larger than two times the depletion width in the substrate of a bulk MOS device doped with a certain doping density ($t_{si} > 2x_{d(max)}$) when

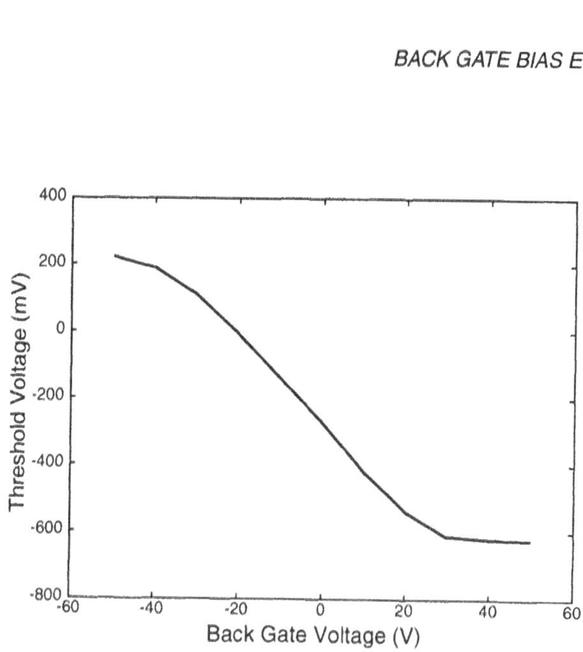

Fig. 2.8 Threshold voltage versus back gate bias of an SOI NMOS device with a channel length of 2 μm, a front gate oxide of 250 Å, a buried oxide of oxide of 20, 000 Å, and a thin film of 1800 Å. (Adapted from Haond & Tack [15].)

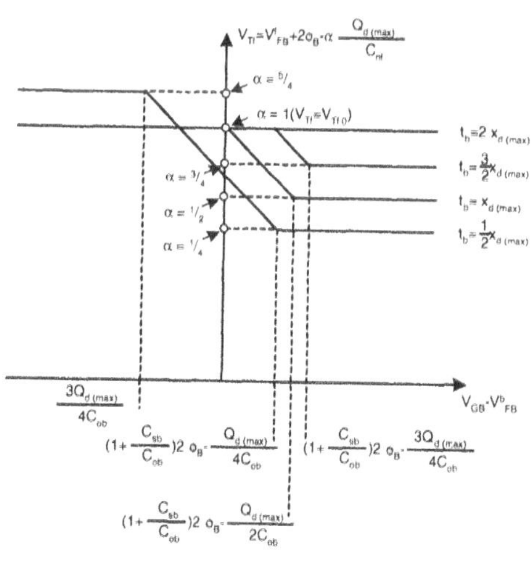

Fig. 2.9 Threshold voltage versus back gate bias of an SOI NMOS device with various thin-film thicknesses. (Adapted from Lim & Fossum [16].)

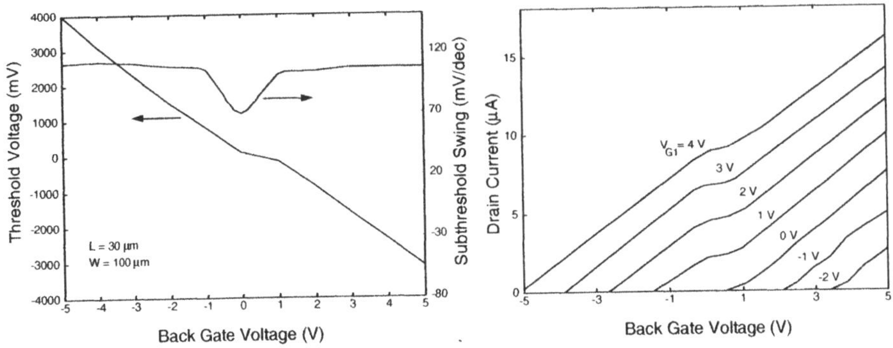

Fig. 2.10 Threshold voltage and drain current versus back gate bias of an SOI NMOS device with a thin film of 30 Å, a front gate oxide of 500 Å, and a buried oxide of 620 Å ,where comparable thicknesses have been intentionally selected for the front gate oxide and buried oxide to show the buried channel effect on body effect. (Adapted from Ernst et al. [17].) .

strong inversion is reached, there is no linear region in the threshold voltage versus gate bias curve no matter what the back gate bias (V_{SUB}) is, which indicates that the threshold voltage is independent of the back gate bias effect when $t_{si} > 2x_{d(max)}$. When the thin-film thickness is < $2x_{d(max)}$ ($t_{si} < 2x_{d(max)}$), within a certain range of the back gate bias (V_{SUB}), the thin film of the device can be maintained as FD. Hence, a linear region exists in the threshold voltage versus back gate bias curve. Under this situation, when the thin-film thickness becomes smaller, the FD situation can be reached more easily–a wider linear region in the threshold voltage versus the back gate bias curve is expected. The back gate bias effect can be very complicated when the back gate bias is large enough such that at the bottom of the thin film a back channel is formed.

Figure 2.10 shows the threshold voltage and the drain current versus the back gate bias of an SOI NMOS device with a front gate oxide of 500 Å, a thin film of 30 Å, and a buried oxide of 620 Å, where comparable thicknesses have been intentionally selected for the front gate oxide and buried oxide to show the buried channel effect on the body effect [17]. As shown in the figure, with the back gate bias < 0 V ($V_{SUB} < 0$ V) when the device turns on, it is with a front channel. When the back gate bias is > 1 V ($V_{SUB} > 1$ V), as the device turns on, it is with a back channel first, followed by the front channel. For the back gate voltage between 0 and 1 V (0 V < V_{SUB} < 1 V), it is the transition period. As indicated in this figure, at the front gate bias of $V_G = 4$ V, when the back gate bias is greater than -5 V ($V_{SUB} > -5$ V), the device turns on with a front channel. Along with the increase in the back gate bias, its threshold voltage becomes smaller. Thus the drain current increases. When the back gate bias is near 0 V ($V_{SUB} \cong 0$ V), inversion of electrons starts to exist at the back surface and the front channel does not vary with the change in the back gate bias (V_{SUB}). When the back gate bias is > 1 V ($V_{SUB} > 1$ V), the back channel is totally on. The drain current flowing through the front and the back channels starts to increase along with the increment in the back gate bias. At the front

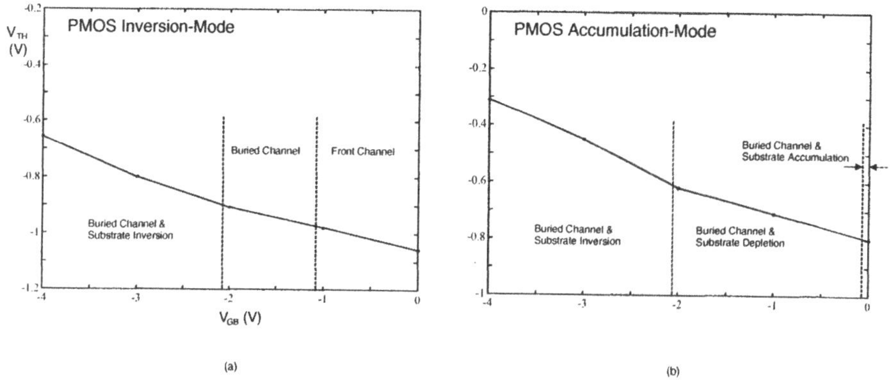

Fig. 2.11 Threshold voltage of inversion-mode and accumulation-mode SOI PMOS devices biased at the back gate bias from 0to $-$ 4 V. (Adapted from Kuo et al. [18].)

gate bias of -2 V ($V_G = -2$ V), only the back channel is on, which implies that no two-piece transition period appears as described before.

2.2.2 Inversion versus Accumulation

The back gate bias effects on the inversion-mode and accumulation-mode SOI devices are different. Figure 2.11 shows the threshold voltage of the inversion-mode and the accumulation-mode SOI PMOS devices biased at the back gate bias from 0 to -4 V [18]. Note that in the thin film it is $p^+ - n - p^+$ in the inversion-mode SOI PMOS device. In contrast; in the accumulation-mode SOI PMOS device, it is $p^+ - p - p^+$. As shown in this figure, for both the inversion-mode and the accumulation-mode, their threshold voltages drop as the back gate bias (V_{SUB}) increases. Under normal operation, the accumulation-mode device is on with the conduction via the buried channel. In contrast, in the inversion-mode device, depending on the back gate bias, the conduction is via the front channel or the back channel. Although only SOI PMOS devices are described here, the back gate bias effects for the SOI NMOS devices are similar.

2.3 SHORT CHANNEL EFFECTS

As for bulk, SOI CMOS devices also suffer from short channel effects (SCE). For low-voltage, low-power VLSI circuit applications using deep-submicron SOI CMOS devices, short channel effects are important. In this section, short channel effects of SOI CMOS devices are described. Both conventional bulk and SOI CMOS devices have similar short channel effects. When the channel length shrinks, the absolute value of the threshold voltage becomes smaller, which is due to the reduced controllability of the front gate over the depletion region from the increased charge sharing of the source/drain. Figure 2.12 shows the threshold voltage versus the channel length of

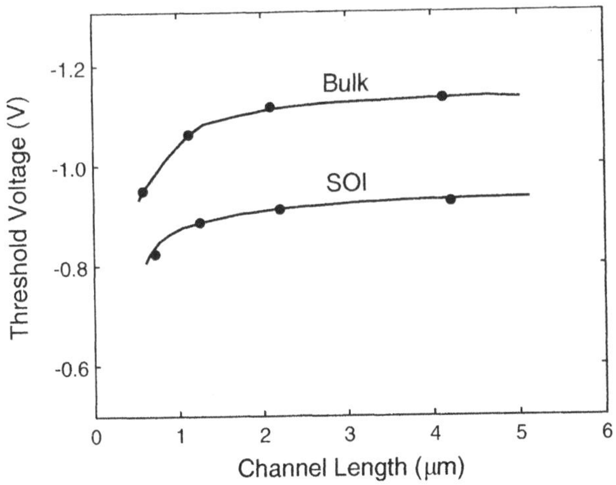

Fig. 2.12 Threshold voltage versus channel length of SOI and bulk PMOS devices. The SOI device has a front gate oxide of 150 Å, a thin film of 2200 Å, and a buried oxide of 4000 Å. The doping density of the n-type silicon is $8 \times 10^{15}\text{cm}^{-3}$. (Adapted from Hashimoto et al. [19].)

SOI and bulk PMOS devices with a front gate oxide of 150 Å, a thin film of 2200 Å (SOI), and a buried oxide of 4000 Å (SOI) [19]. As shown in this figure, the short channel effects of SOI MOS devices are lighter than those of the bulk ones, which is due to the thin-film structure of the SOI devices. In contrast to the bulk device, the front gate of the SOI device has better control over its active device region in the thin film. Therefore, the charge sharing effects from the source/drain region have been substantially reduced, and hence there is a smaller short channel effect.

2.3.1 Biasing Dependence

Short channel effect is also dependent on the back gate bias. Figure 2.13 shows the short channel effect of the FD SOI NMOS device with a front gate oxide of 92 Å, a buried oxide of 4000 Å, and a thin film of 800 Å, biased at the back gate bias of 0 and -5 V [20]. As shown in this figure at a negative back gate bias of -5 V, the threshold voltage is lifted upward as compared to the case at a back gate bias of 0 V. The extent of the upward shift when the back gate bias becomes negative is smaller for the device with a shorter channel length, which implies that SCE seems to improve. With a shorter channel, the controllability over the vertical direction of the channel region from the source/drain seems to be reduced at a more negative back gate bias, hence its back gate bias effect is smaller. Therefore, with a shorter channel, the change in the magnitude of the threshold voltage due to a change in the back gate bias is decreased.

In addition to back gate bias dependence, SCE can also be drain-bias dependent. Figure 2.14 shows the short channel effect of a FD SOI NMOS device with a front

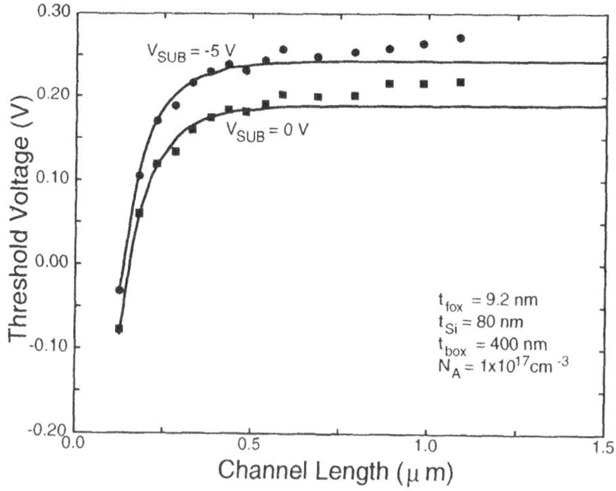

Fig. 2.13 Short channel effect of the FD SOI NMOS device with a front gate oxide of 92 Å, a buried oxide of 4000 Å, and a thin film of 800 Å, biased at the back gate bias of 0 and −5 V. (Adapted from Banna et al. [20].)

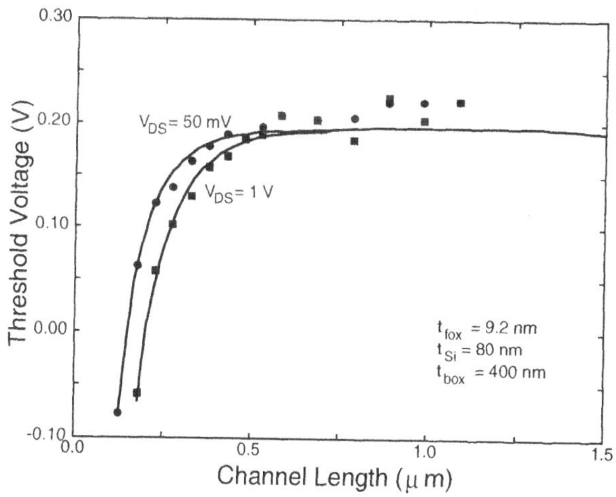

Fig. 2.14 Short channel effect of a FD SOI NMOS device with a front gate oxide of 92 Å, a buried oxide of 4000 Å, and a thin film of 800 Å, biased at $V_{DS} = 0.05$ and 1 V. (Adapted from Banna et al. [20].)

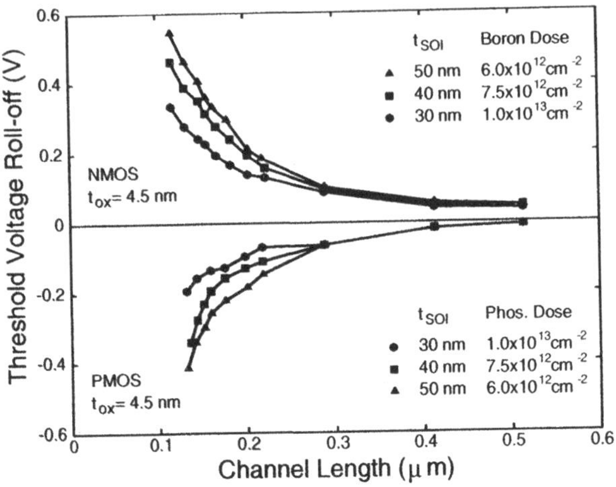

Fig. 2.15 Threshold voltage roll-off of FD SOI NMOS device with a front gate oxide of 45 Å and various thin-film thicknesses. (Adapted from Imai et al. [21].)

gate oxide of 92 Å, a buried oxide of 4000 Å, and a thin film of 800 Å, biased at $V_{DS} = 0.05$ and 1 V [20]. From this figure, when the drain voltage increases, the SCE becomes more serious. When the channel length is decreased, its threshold voltage becomes lower at a higher drain voltage due to the increased electric field in the lateral source/drain direction. This phenomenon becomes more noticeable when the channel length becomes shorter. Under this situation, the controllability of the front gate over the channel region is affected more. Hence, at a larger drain voltage, the lowering of the threshold voltage is more and the subthreshold current also increases due to the change in the internal potential distribution of an SOI device–drain-induced-barrier-lowering (DIBL). Compared to the back-gate-bias dependence, the drain voltage dependence has more effects on the short channel effects.

2.3.2 Structure Dependence

In addition to back gate and drain bias dependences, the SCE of an SOI MOS device can also be thin-film thickness dependent. Figure 2.15 shows the threshold voltage roll-off of the FD SOI NMOS device of a front gate oxide of 45 Å and with various thin-film thicknesses [21]. As shown in this figure, for the NMOS device, with a shorter channel length the threshold voltage moves toward the negative direction. In contrast, for the PMOS device, the threshold voltage moves toward the positive direction with a shorter channel length. For both NMOS and PMOS devices, when the thin-film thickness is reduced, their SCE becomes smaller since the controllability of the front gate over the active channel region is stronger and the source/drain has a less influence in the channel.

The short channel effect is also dependent on the thin-film doping density. Figure 2.16 shows the threshold voltage shift versus the thin-film thickness of an SOI NMOS

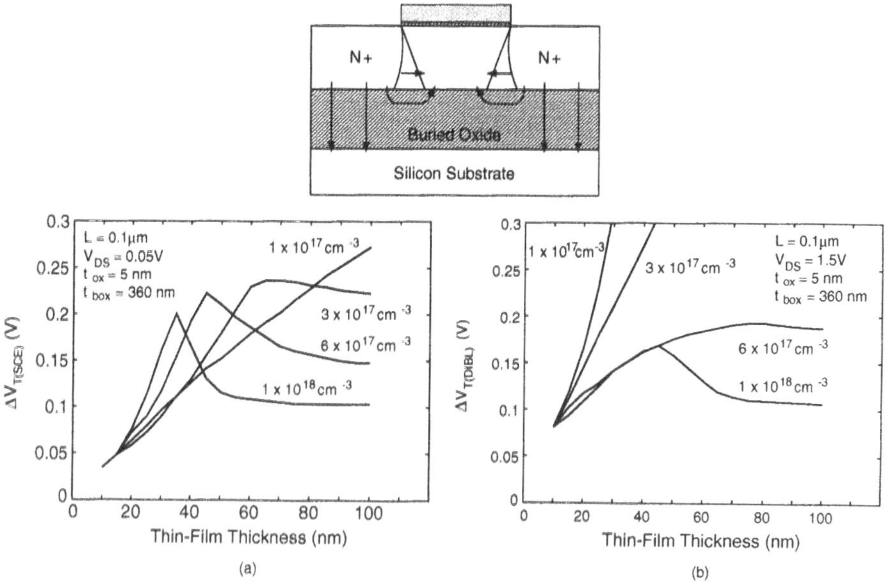

Fig. 2.16 Threshold voltage shift versus thin-film thickness of an SOI NMOS device with a front gate oxide of 50 Å and a buried oxide of 3600 Å, for various channel doping densities, biased at (a) $V_{DS} = 0.05$ V, and (b) 1.5 V. (Adapted from Su et al. [22].)

device with a front gate oxide of 50 Å and a buried oxide of 3600 Å for various channel doping densities, biased at (a) $V_{DS} = 0.05$ V, and (b) 1.5 V as shown in Fig. 2.16 [22]. At $V_{DS} = 0.05$ V, for each curve with a specific channel doping density, the threshold voltage shift reaches its peak at a certain thin-film thickness. If the thin-film thickness of this device exceeds this specific critical thin-film thickness, the device operates as a PD device–the PD regime. Below this specific thin-film thickness, it is in the FD regime. At this specific critical thickness, the threshold voltage shift does not show any noticeable sensitivity to the change in the thin-film thickness. In addition, with a lower thin-film doping density, the threshold voltage shift is larger. In the FD regime with the thin-film thickness below the critical thin-film thickness, similar trends can be observed. With a smaller thin-film thickness, the shift in threshold voltage is smaller. In the FD regime, with a lighter thin-film doping density, the SCE is smaller, which is opposite to that in the PD regime. At the boundary between the FD/PD regimes, the threshold voltage shift is the largest, which is due to the most serious influence of the electric field from the source/drain region via the buried oxide to the channel region. As shown in Fig. 2.16(b), at $V_{DS} = 1.5$ V, similarly, in the FD regime, with a thinner thin film, the SCE is less affected. At $V_{DS} = 1.5$ V, in both FD and PD regimes, the SCE can be reduced only at a higher thin-film doping density, which implies that only via increasing the thin-film doping density can DIBL be effectively lessened. For an SOI device with the thin-film thickness in the FD regime, in order to reduce SCE, the thin-film doping density should be kept low. In

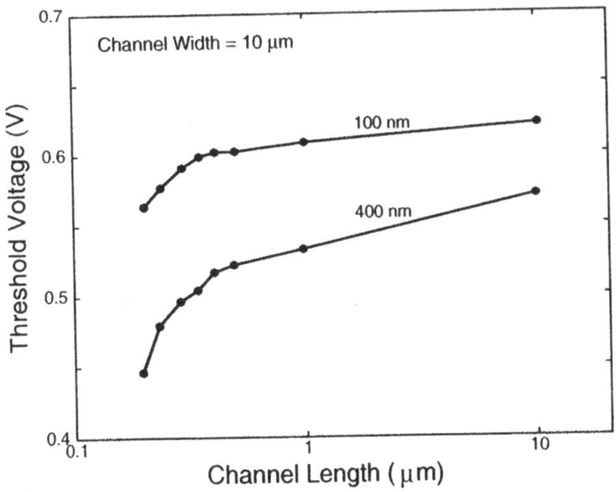

Fig. 2.17 Threshold voltage versus channel length of an SOI NMOS device with a front gate oxide of 60 Å and a thin film of 1000 Å, and a buried oxide of 1000 and 4000 Å. (Adapted from Lee et al. [23].)

contrast, for an SOI device with the thin-film thickness in the PD regime, the thin-film doping density should be raised appropriately to lessen the SCE. In order to minimize the SCE, SOI devices should be designed with their thickness not near the PD/FD boundary.

As described before, under some situations, the SCE becomes worse due to the influence from the source/drain via the buried oxide to the channel region– buried oxide thickness is also a factor in determining the SCE. Figure 2.17 shows the short channel effect of an SOI NMOS device with a front gate oxide of 60 Å and a thin film of 1000 Å, and a buried oxide of 1000 and 4000 Å [23]. As shown in this figure, with a thinner buried oxide, the SCE is indeed lessened. For a device with a thinner buried oxide, the threshold voltage is relatively higher– with a thinner buried oxide, the compressive stress is higher. Hence, during the thermal process in the fabrication, boron dopants in the thin film cannot diffuse easily. As a result, the doping density of the thin film is higher and its threshold voltage is higher. As the doping density of the thin film is raised, the SCE is reduced.

The thickness of the buried oxide can also affect the DIBL of an SOI device. Figure 2.18 shows the threshold shift versus the channel length of an FD SOI NMOS device with a front gate oxide of 40 Å, a thin film of 350 Å, and buried oxide thicknesses of 80 and 150 nm [24]. As described before, with a small channel length, DIBL lowers the magnitude of the threshold voltage of an SOI device biased by a larger V_{DS}. This DIBL-induced threshold voltage shrinkage is more noticeable for an SOI device with a smaller channel length. As shown in Fig. 2.18, when the thickness of the buried oxide gets smaller, the phenomenon of DIBL-induced threshold voltage shrinkage is reduced slightly. With a thinner buried oxide, the direct influence of the source/drain region via the buried oxide in the channel region is reduced accordingly.

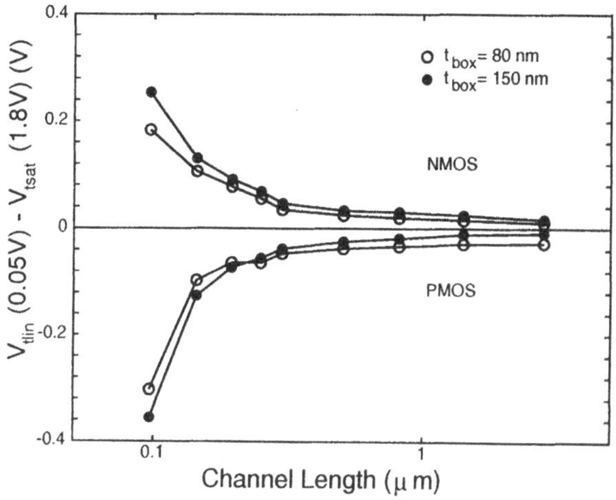

Fig. 2.18 Threshold voltage shift versus channel length of an FD SOI NMOS device with a front gate oxide of 40 Å, a thin film of 350 Å, and buried oxide thicknesses of 80 and 150 nm. (Adapted from Cao et al. [24].)

2.3.3 Processing Dependence

Various processing technologies may also affect the SCE of an SOI device. Figure 2.19 shows the threshold voltage versus the channel length of an FD SOI NMOS device with a front gate oxide of 50 Å, a thin film of 1000 Å, and a buried oxide of 4200 Å, using polysilicon and tantalum gates [25]. The use of a tantalum gate is to facilitate the adjustment of the threshold voltage of an SOI MOS device without raising the thin-film doping density substantially by taking advantage of the work function of tantalum. In addition, the low sheet-resistance of tantalum can help reduce the parasitic resistance of the gate. More important is that by adopting the tantalum gate, the SCE of the FD SOI NMOS device can be improved. By using the polysilicon gate, there exists a thin layer of the depletion region atop the front gate oxide due to the polysilicon depletion effect. As a result, the effective front gate oxide becomes thicker. Thus, the SCE has been amplified. With a thinner front gate oxide, this phenomenon becomes more noticeable. In contrast, by adopting a metal such as tantalum as the front gate material, there is no such problem, and therefore SCE is smaller.

In addition to tantalum, titanium nitride (TiN) can also be used to lessen the short channel effect for SOI CMOS devices. Figure 2.20 shows (a) the capacitance∼voltage characteristics of the FD SOI PMOS device with the front gate oxide of 40 Å and using TiN and polysilicon gates and (b) the threshold voltage versus the channel length of the FD SOI NMOS device with a front gate oxide of 40 Å and using the TiN [26]. As shown in this figure, by using the TiN gate, "polysilicon depletion effects" described before are also avoided. The SCE of the device with the TiN gate is smaller as compared to the poly-gate device. At certain biasing voltages, due to

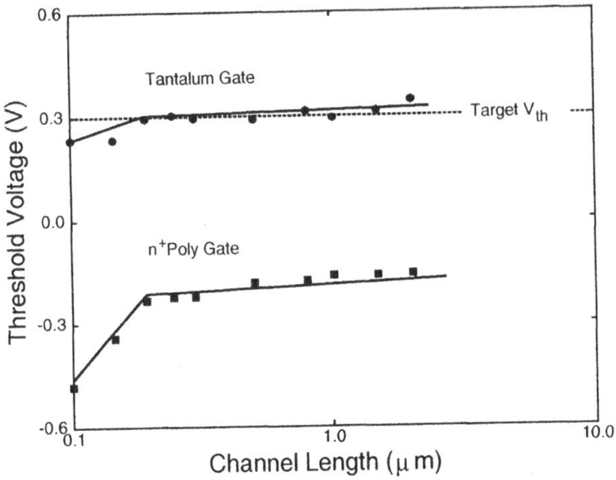

Fig. 2.19 Threshold voltage versus channel length of an FD SOI NMOS device with a front gate oxide of 50 Å, a thin film of 1000 Å, and a buried oxide of 4200 Å, using polysilicon and tantalum gates. (Adapted from Ushiki et al. [25].)

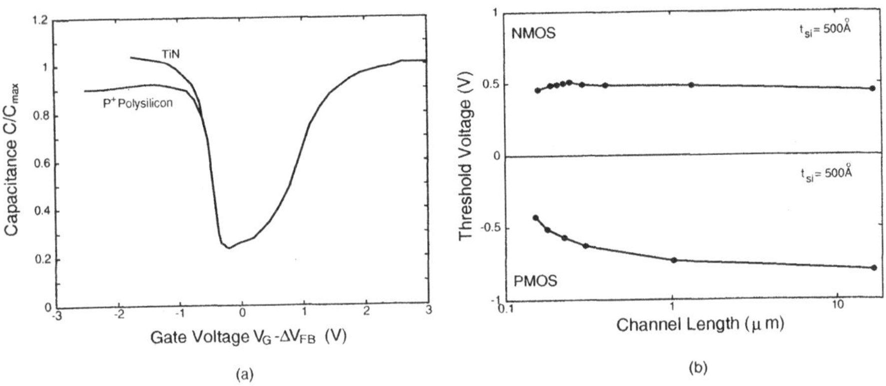

Fig. 2.20 (a) Capacitance~voltage characteristics of the FD SOI PMOS device with a front gate oxide of 40 Å, a thin film of 500 Å, and using TiN and polysilicon gates. (b) Threshold voltage versus channel length of the FD SOI NMOS device with a front gate oxide of 40 Å and using the TiN gate. (Adapted from Maiti et al. [26].)

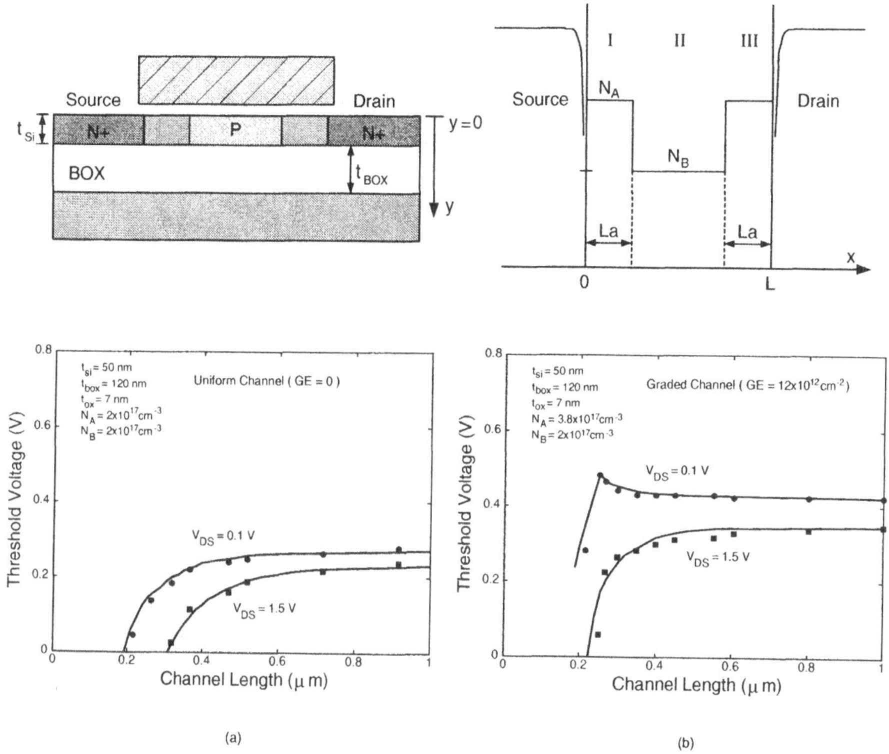

Fig. 2.21 Threshold voltage versus channel length of the FD SOI NMOS device with a front gate oxide of 70 Å, a thin film of 500 Å, a buried oxide of 1200 Å, and (a) a uniformly doped channel and (b) a graded channel. (From Adan et al. [27]. ©1999 IEEE.)

the polysilicon depletion effects, the capacitance of an SOI PMOS device with the polysilicon gate is reduced by 12% as compared to the titanium nitride gate device.

In addition to the nonuniform doping approach described above, the SCE of an FD SOI MOS device can also be reduced via a graded channel approach. Figure 2.21 shows the threshold voltage versus the channel length of the FD SOI NMOS device with a front gate oxide of 70 Å, a thin film of 500 Å, a buried oxide of 1200 Å, and (a) a uniformly doped channel and (b) a graded channel [27]. In the device with the uniformly doped channel, its doping density in the channel is $2 \times 10^{17} cm^{-3}$. In the device with a graded channel, in the center of the channel, the doping density is the same as for the device with the uniformly doped channel–$2 \times 10^{17} cm^{-3}$. Near the source/drain region, more highly doped regions with a length of 0.13 μm at each side have been generated via the gate-edge implant techniques. The graded channel approach for the SOI device is similar to the halo pocket implant for the bulk CMOS devices. As shown in Fig. 2.21, as compared to the case with the uniformly doped channel, with the gate-edge (GE) implanted graded channel, the SCE improves substantially, especially at a large drain voltage. For case with $V_{DS} = 0.1$ V, under certain situations, reverse short channel effect can occur– the threshold voltage of

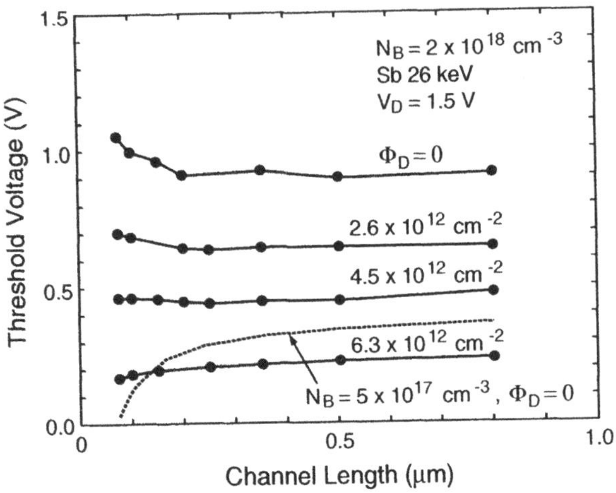

Fig. 2.22 Threshold voltage versus channel length of a PD SOI NMOS device with a front gate oxide of 40 Å and a nonuniformly doped thin film of 1000 Å with a counter-doped thin film using a n-type implant with various doses. (Adapted from Suzuki et al. [28].)

the SOI NMOS device increases when the channel length is scaled down because with a shorter channel, the graded channel occupies a more important portion of the overall channel. The average doping density in the channel also increases and thus the threshold voltage increases temporarily. When the channel becomes even shorter, the threshold voltage still falls off.

Short channel effects of PD SOI devices can be reduced by increasing the doping density of the thin film. However, a very high doping density of the thin film may lead to an excessive magnitude in the threshold voltage, which is not suitable for low-voltage operation. In addition, a highly doped thin film may also degrade the carrier mobility due to impurity scattering. In order to compromise the tradeoffs. A nonuniformly doped channel approach used for bulk devices has been adopted for the PD SOI devices– a nonuniformly doped thin film has been created for the PD SOI devices [28]. In a PD SOI device with a heavily doped thin film, by counter doping the upper portion of the thin film, an effective lowly doped portion under the front gate oxide can be created. As shown in Fig. 2.22, its threshold voltage can be appropriately adjusted by tuning the dose of the counter doping implant and its drain-induced-barrier-lowering (DIBL) can still be suppressed to within an acceptable level. In contrast, if the uniformly doped thin film is used for the PD SOI device, as shown in dashed lines, in order to have a reasonable threshold voltage, the doping density of the thin film cannot be too high, but on the other hand DIBL may not be acceptable. By using the nonuniformly doped approach, the threshold voltage of the PD SOI CMOS device can be appropriately controlled. Also shown in this figure, for the case without the counter doping ($\phi_D = 0$), the device may show the reverse short-channel effect, which is caused by the enhanced channel dopant diffusion from the source/drain implant.

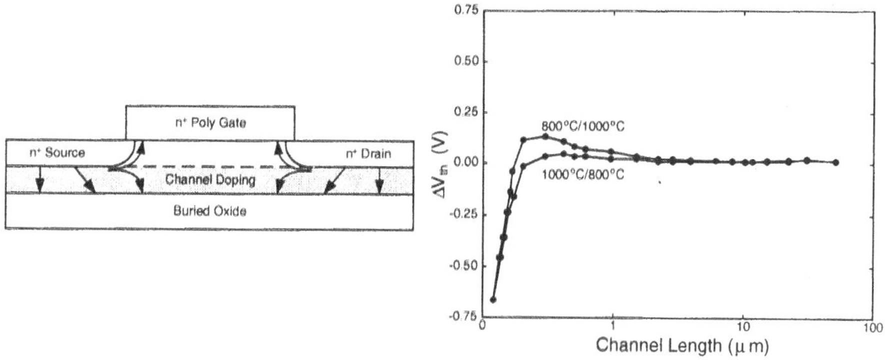

(a) (b)

Fig. 2.23 (a) Schematic of excess interstitial flux due to implant damage or arsenic deactivation causing enhanced channel dopant diffusion in an SOI device. (b) Threshold voltage change versus channel length of an SOI NMOS device having a 1000°C rapid thermal anneal and a 800°C furnace anneal and in an opposite order after the source/drain implant. (Adapted from Crowder et al. [29].)

The damage caused by the source/drain implant during the fabrication process of SOI MOS devices can also affect the SCE. As shown in Fig. 2.23(a), after the source/drain implant, the lattice in the thin-film region is damaged. Interstitials from source/drain diffuse to the bottom of the thin film [29]. If the thin film is thick enough, these interstitials may even affect the channel region. At the locations with interstitials, boron dopants diffuse more easily. Therefore, at the bottom of the channel region near source/drain the boron dopants tend to move toward the front gate oxide. As a result, the doping density in this region is increased. When the channel length is decreased, its threshold voltage increases on the contrary, which is the reverse short-channel effect. This phenomenon is reduced by adopting a 1000°C rapid thermal anneal followed by a 800°C furnace anneal after the source/drain implant. After the anneal step, the implant damage can be greatly reduced, thus the reverse short channel effect can be relieved. The reverse SCE effect caused by the defects in the lattice from the damage of the source/drain implant can be lessened by using a thinner thin film. As shown in Fig. 2.24, with a thinner thin film, the threshold voltage of the 2 μm device gets closer to that of the 12 μm, which implies the reverse SCE has been reduced [30]. With a thinner thin film, the interstitials from the source/drain are restricted more at the bottom of the source/drain, with less of a tendency to extend to the channel region, and thus less reverse SCE.

2.3.4 Subthreshold

Until now, short channel effects have been limited to the aspects related to threshold voltage. In fact, subthreshold characteristics should also be included as part of the

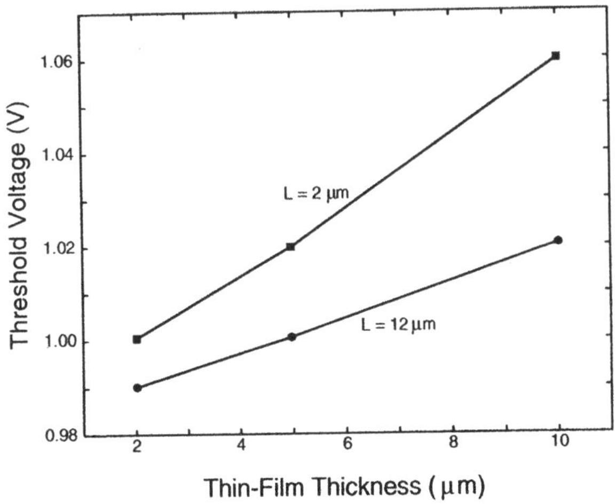

Fig. 2.24 Threshold voltage versus thin-film thickness of an SOI NMOS device with a front gate oxide of 190 Å and a channel length of 12 and 2 μm showing the reduced reverse SCE after anneal. (Adapted from Tsoukalas [30].)

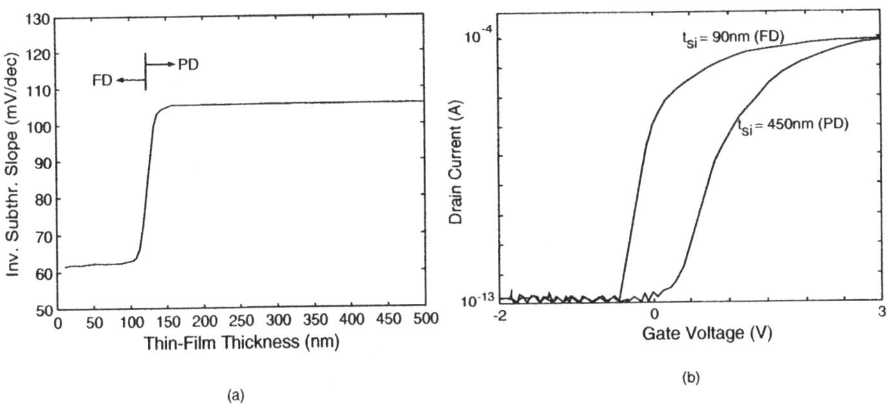

Fig. 2.25 (a) Inverse subthreshold slope versus thin-film thickness of an SOI NMOS device with a front gate oxide of 250 Å, a buried oxide of 1 μm, and a thin-film doping density of $8 \times 10^{16}\text{cm}^{-3}$. (b) Subthreshold drain current characteristics of the SOI NMOS device with a thin film of 900 and 4500 Å. (Adapted from Colinge [31].)

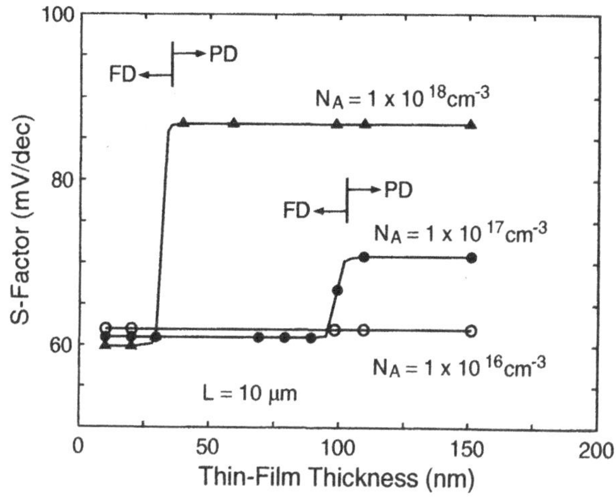

Fig. 2.26 The influence of the thin-film doping density in the inverse subthreshold slope for FD and PD SOI NMOS devices with a front gate oxide of 50 Å and a buried oxide of 5000 Å. (Adapted from Joachim et al. [32].)

short channel effects. Before describing the short channel effects on the subthreshold slope, the difference in the inverse subthreshold slope between PD and FD is discussed. Figure 2.25 shows (a) the inverse subthreshold slope versus the thin-film thickness of the SOI NMOS device with a front gate oxide of 250 Å, a buried oxide of 1 μm, and a thin-film doping density of $8 \times 10^{16}cm^{-3}$ and (b) the subthreshold drain current characteristics of the SOI NMOS device with a thin film of 900 and 4500 Å [31]. As shown in this figure, with a thick thin film, the device is in the PD regime and its inverse subthreshold slope is almost identical to the bulk one. With a thin thin film, the device enters the FD regime with its inverse subthreshold slope improved suddenly and close to the ideal value ($kT/q \times \ln 10$ mV/dec). In the FD regime, the channel is separated from the substrate by a thick buried oxide such that the controllability of the front gate over the channel is greatly enhanced.

Figure 2.26 shows the influence of the thin-film doping density in the inverse subthreshold slope for FD and PD SOI NMOS devices with a front gate oxide of 50 Å and a buried oxide of 5000 Å [32]. As shown in this figure, for the PD devices, with a more highly doped thin film, its subthreshold slope is worse, which is similar to bulk CMOS devices. As for the FD devices, although their subthreshold slope is less sensitive to the thin-film doping density, a higher thin-film doping density leads to a better subthreshold slope because the depletion width in the thin film is fixed for the FD devices. With a more highly doped thin film, a higher electric field makes the carriers closer to the surface. Therefore, the controllability of the front gate is improved–a better subthreshold slope.

The above discussion is for the long channel case. Now the short channel case is described. Figure 2.27 shows the inverse subthreshold slope versus the channel

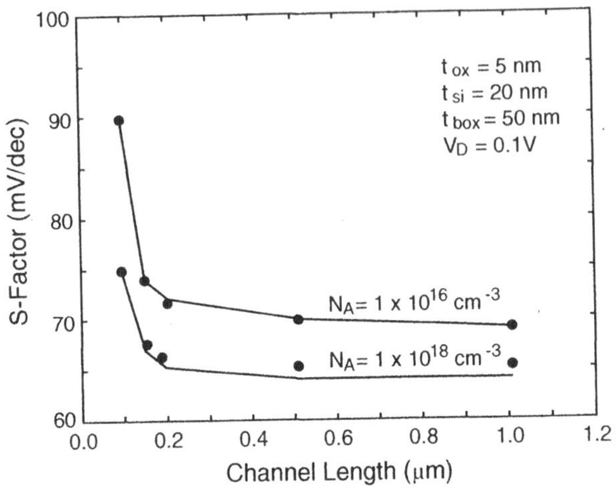

Fig. 2.27 Inverse subthreshold slope versus channel length of an FD SOI NMOS device with a front gate oxide of 50 Å, a thin film of 200 Å, and a buried oxide of 500 Å. (Adapted from Joachim et al. [32].)

length of an FD SOI NMOS device with a front gate oxide of 50 Å, a thin film of 200 Å, and a buried oxide of 500 Å [32]. As shown in this figure, when the channel length is decreased, its subthreshold slope is worsened due to the 2D effects from the source/drain region to decrease the front gate controllability over the channel region. When the thin-film doping density is raised, the subthreshold slope is improved. In addition, with a smaller channel length, the improvement in the subthreshold slope is more noticeable at a raised thin-film doping density.

The subthreshold slope of an FD SOI MOS device is also dependent on the thin-film thickness. Figure 2.28 shows the inverse subthreshold slope versus the channel length of an FD SOI NMOS device with a front gate oxide of 70 Å, a buried oxide of 3500 Å, and a thin film of 1000 and 500 Å with a doping density of $2 \times 10^{16}\text{cm}^{-3}$[33]. As shown in this figure, with a thinner thin film, the degradation of the subthreshold slope is reduced when the channel length is decreased due to the improved controllability of the front gate over the channel region.

Until now, short channel effects on the threshold voltage and the subthreshold slope has been discussed. In fact, both the threshold voltage and the subthreshold slope are reflected simultaneously on the device characteristics, especially on the standby leakage current for low-power application. Figure 2.29 shows the standby leakage current versus the drain voltage of an FD SOI NMOS device with a front gate oxide of 70 Å, a thin film of 500 Å, and a buried oxide of 1200 Å, with various channel lengths [34]. As shown in this figure, with a smaller channel length, the leakage current increases. Especially at an increased drain voltage, the leakage current may be worsened with a smaller channel length. When the channel length is decreased, both the threshold voltage and the subthreshold slope are worsened from 2D effects.

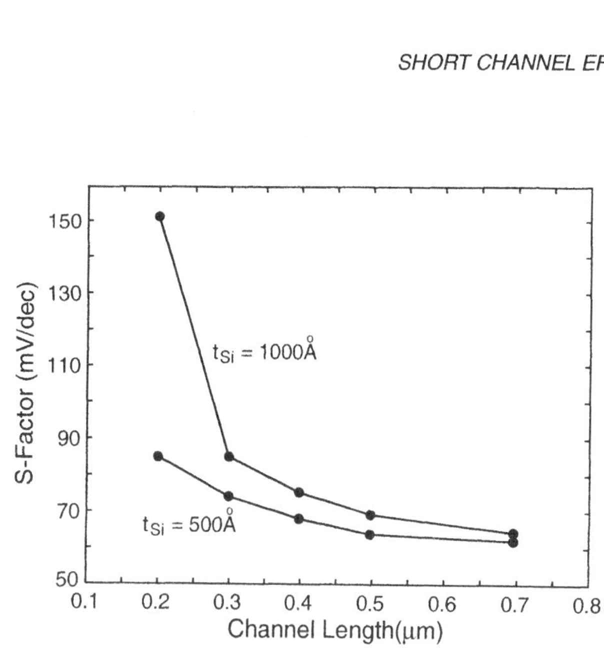

Fig. 2.28 Inverse subthreshold slope versus channel length of an FD SOI NMOS device with a front gate oxide of 70 Å, a buried oxide of 3500 Å, and a thin film of 1000 and 500 Å with a doping density of $2 \times 10^{16} cm^{-3}$. (Adapted from Yeh & Fossum [33].)

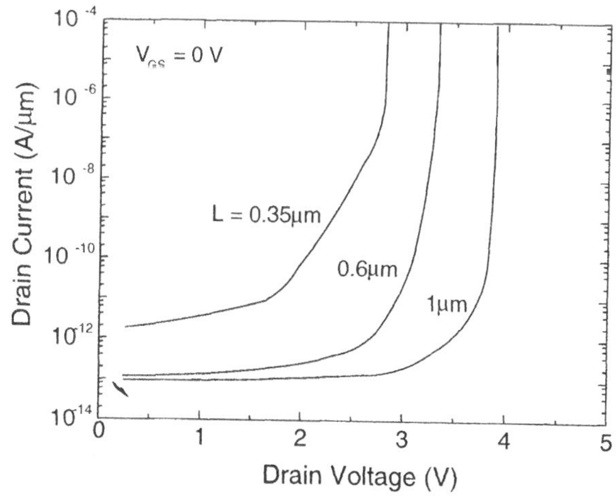

Fig. 2.29 Standby leakage current versus drain voltage of an FD SOI NMOS device with a front gate oxide of 70 Å, a thin film of 500 Å, and a buried oxide of 1200 Å, with various channel lengths. (Adapted from Adan et al. [34].)

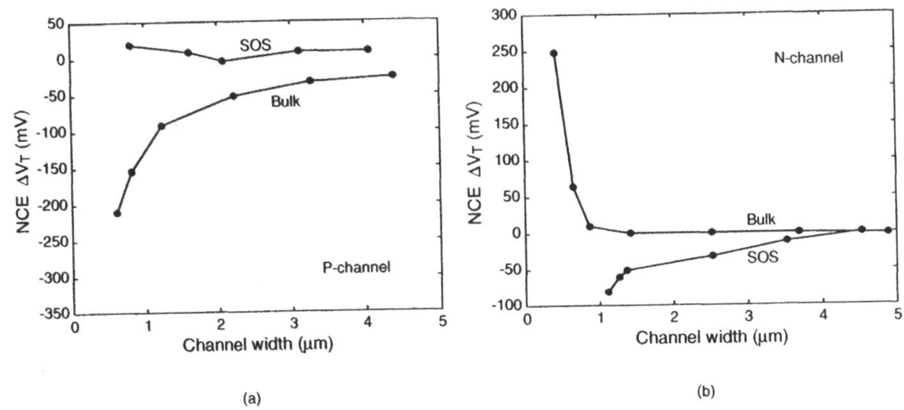

Fig. 2.30 Threshold voltage shift versus channel width of SOI (a) PMOS and (b) NMOS devices using mesa isolation. (Adapted from Brassington et al. [35].)

At a higher drain voltage, DIBL effects are more noticeable. Therefore, the extent in the increase of the standby leakage current increases quickly. To reduce the standby leakage current, a decreased thin-film thickness, an increased thin-film doping density, and adoption of a special channel doping profile can be used.

2.4 NARROW CHANNEL EFFECTS

In Section 2.3, short channel effects of SOI CMOS devices have been described. In addition to SCE, for low-voltage low-power VLSI circuit applications, narrow-channel effects (NCE) are also important. In this section, the NCE is discussed. For conventional bulk CMOS devices, narrow channel effects are due to the bird's beak and the channel stop implant in the field region if the LOCOS structure has been adopted. For deep-submicron bulk CMOS technology using shallow trench isolation, narrow channel effects have been substantially reduced as compared to the conventional bulk CMOS technology using LOCOS techniques. For SOI CMOS devices with mesa-isolation, due to the 2D sidewall effects, narrow channel effects can be serious. As shown in Fig. 2.30, for SOI CMOS devices, when the channel width is decreased, its threshold voltage is decreased–the narrow channel effect, which is different from NCE for the bulk CMOS devices [35]. For bulk CMOS devices fabricated by LOCOS techniques, the opposite trend on NCE is shown–when the channel width is decreased, the threshold voltage increases. From the figure, the trends on the NCE of the mesa-isolated SOI CMOS devices are similar to those of the shallow-trench-isolated bulk CMOS devices [36].

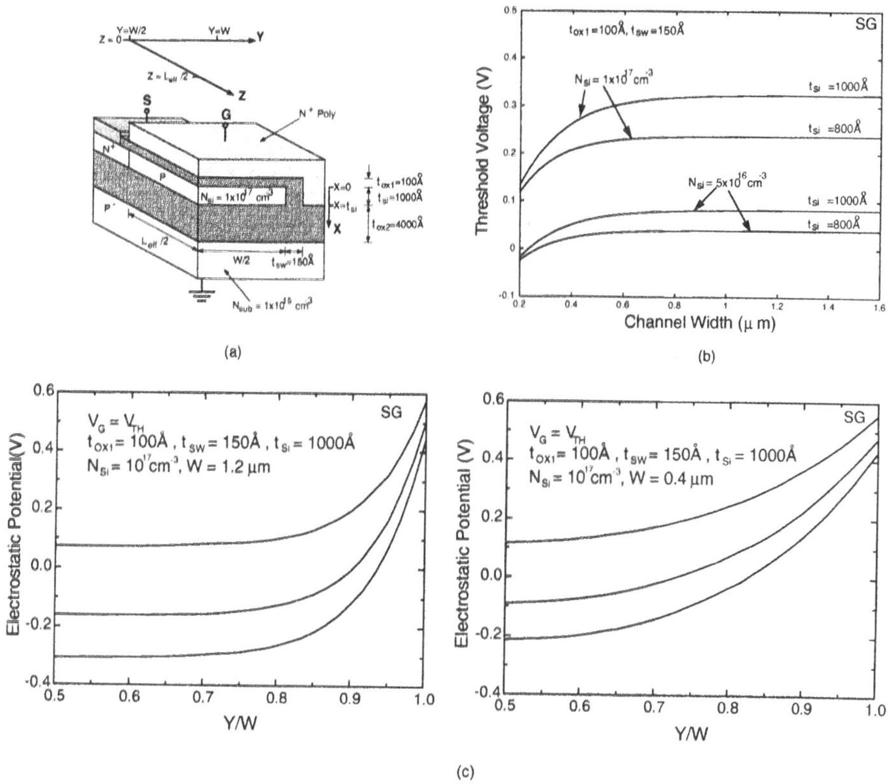

Fig. 2.31 (a) Cross-section and (b) threshold voltage versus channel width of a mesa-isolated SOI NMOS with a front gate oxide of 100 Å, a sidewall oxide of 150 Å, a thin film of 800 and 1000 Å, and a thin-film doping density of 1×10^{17} and $5 \times 10^{16} \text{cm}^{-3}$. (c) The electrostatic potential in the horizontal direction at one-quarter, two-quarters, three-quarters of the thin-film thickness in the thin film of the SOI device with channel widths of 0.4 and 1.2 μm. (Adapted from Su & Kuo [37].)

2.4.1 Structure Dependence

Narrow channel effects of SOI CMOS devices are also dependent on their internal parameters. Figure 2.31 shows (a) the cross-section and (b) the threshold voltage versus the channel width of a mesa-isolated SOI NMOS device with a front gate oxide of 100 Å, a sidewall oxide of 150 Å, and a thin-film doping density of 1×10^{17} and $5 \times 10^{16} \text{cm}^{-3}$ [37]. As shown in this figure, for an SOI device with a higher thin-film doping density, the NCE is more noticeable. When the channel width is decreased, its threshold voltage also decreases. If the thickness of the thin film can be decreased, NCE can be substantially improved. The internal mechanism of the NCE for a mesa-isolated SOI CMOS device can be understood by studying Fig. 2.31(c), which shows the electrostatic potential in the horizontal direction (channel width direction) at one-quarter, two-quarters, three-quarters of the thin-film thickness in the

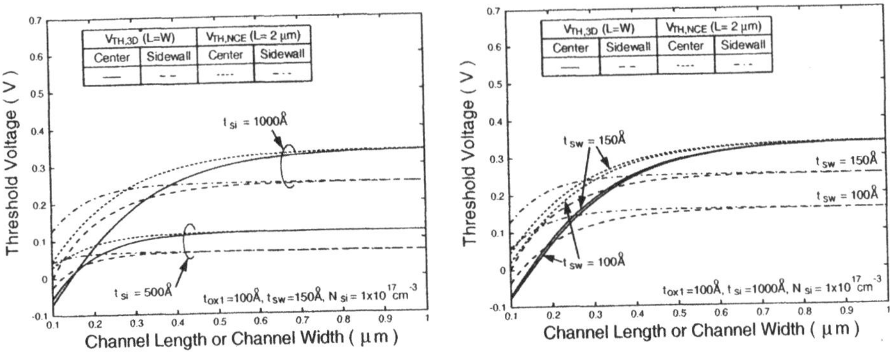

Fig. 2.32 Threshold voltage versus channel width or channel length of the center and the sidewall portions in the mesa-isolated SOI NMOS device with a front gate oxide of 100 Å, a silicon thin-film doping density of 10^{17}cm^{-3} with thin-film thicknesses of 1000 and 500 Å and sidewall oxide thicknesses of 100 and 150 Å. (Adapted from Su& Kuo [38].)

thin film of the SOI device with channel widths of 0.4 and 1.2 μm. As shown in this figure, for the device with a larger channel width (1.2 μm), due to the existence of the polysilicon sidewall gate, the sidewall gate could influence the thin-film internal region to raise the potential near the sidewall. Since the region with a rise in the potential is only restricted near the sidewall, its overall influence is small. For the device with a smaller channel width of 0.4 μm, the influence of the sidewall gate is extended to the overall channel region–the potential in the overall channel region has been raised. As a result, the threshold voltage is lowered. By decreasing the thin-film thickness, the influence of the sidewall gate can be reduced, which may result in the increased controllability of the front gate–NCE is reduced.

Since for small-geometry SOI CMOS devices both channel length and width are scaled down at the same time, SCE and NCE should be considered at the same time. For the mesa-isolated SOI MOS device, the trend on SCE and NCE is similar. Figure 2.32 shows the threshold voltage versus the channel width or the channel length of the center and the sidewall portions in the mesa-isolated SOI NMOS device with a front gate oxide of 100 Å, a silicon thin-film doping density of 10^{17}cm^{-3} with thin-film thicknesses of 1000 and 500 Å and sidewall oxide thicknesses of 100 and 150 Å [38]. As shown in the figure, the threshold voltage of the sidewall channel is more sensitive to the variation in the sidewall oxide thickness as compared to the center channel. However, the threshold voltage shift of the sidewall channel is smaller. The importance of the sidewall channel is reflected in the subthreshold current.

For a small-geometry SOI NMOS device with a small channel length, both SCE and NCE are additive–the drop in the threshold voltage is bigger than that for short- or narrow-channel devices when SCE or NCE is considered individually. Due to the complex three-dimensional (3D) effects in the device, the drop in the threshold voltage while considering a small-geometry device is smaller than the sum of the threshold voltage drops due to SCE and NCE individually considered– $\Delta V_{TH} < \Delta V_{TH(SCE)} + \Delta V_{TH(NCE)}$. As shown in the figure, decreasing the thin-film thickness can alleviate

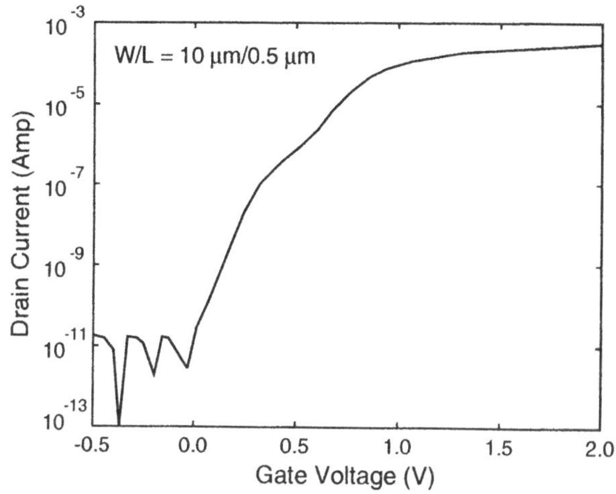

Fig. 2.33 Subthreshold characteristics of a mesa-isolated PD SOI NMOS device with a channel width of 10 μm. (Adapted from Tseng et al. [39].)

the 3D small-geometry effects. On the other hand, increasing the sidewall oxide thickness reduces NCE, thus the overall small-geometry effects can be improved.

2.4.2 Subthreshold

In addition to threshold voltage, narrow channel effect can also influence subthreshold characteristics of an SOI CMOS device. Figure 2.33 shows the subthreshold characteristics of a mesa-isolated PD SOI NMOS device with a channel width of 10 μm [39]. Due to the 3D effects near the sidewall of the device, the threshold voltage of the channel near the sidewall is smaller than that of the channel in the center. As a result, the sidewall channel turns on earlier. Thus, a current hump exists in the subthreshold characteristics, which causes an extra leakage current. The subthreshold current hump problem can be alleviated by a large angle tilt field implant to raise the doping density near the sidewall such that the threshold voltage of the sidewall channel can be raised. With a small channel width, the subthreshold current hump phenomenon can be eliminated, as described below.

The sidewall channel does not just change the threshold voltage and the subthreshold characteristics, but also the overall current characteristics. Figure 2.34 shows the drain current characteristics of an SOI NMOS device with a front gate oxide of 100 Å and a sidewall oxide of 150 Å with various channel widths [40]. Since the drain current is composed of the current in the center channel and the current in the sidewall channel, for a device with a smaller channel width, the relative portion of the sidewall channel in the overall drain current increases. From this figure, for the device with a large channel width, at a large gate voltage, the center channel dominates the large drain current regime. In the subthreshold region, there is a current hump as described

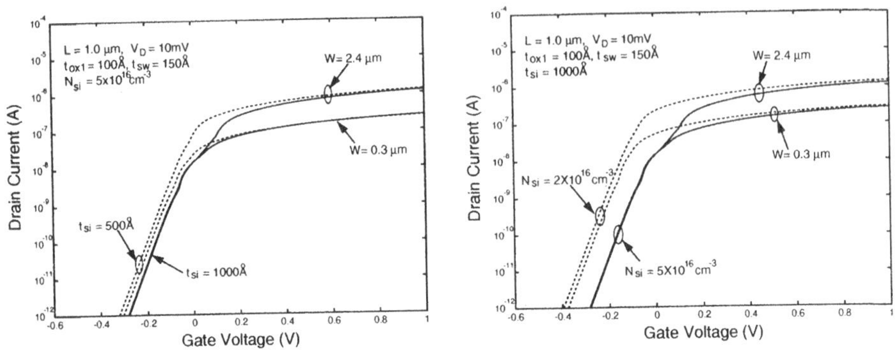

Fig. 2.34 Drain current characteristics of an SOI NMOS device with a front gate oxide of 100 Å and a sidewall oxide of 150 Å. (Adapted from Kuo & Su [40].)

before. With a small channel width, the importance of the sidewall channel becomes noticeable. In addition, the subthreshold current hump is substantially reduced. At a small gate voltage, because of the earlier turn-on of the sidewall channel for its smaller threshold voltage, the sidewall channel dominates the small drain current regime.

In fact, NCE also brings in advantages. As shown in Fig. 2.35, comparing the subthreshold drain current characteristics of the wide channel device with a channel width of 1.5 μm with that of a device with a channel width 0.3 μm, the subthreshold slope of the narrow channel is better than that of the wide channel [14]. At a large V_{DS}, compared to the wide channel device, the narrow channel demonstrates a better resistance to DIBL– its subthreshold slope does not increase much due to the increase in V_{DS}. Figure 2.35(b) also confirms the superiority of the narrow-channel SOI devices over the wide-channel ones in terms of subthreshold slope. At a smaller channel width, the degradation of the subthreshold slope due to the decreased channel length has been reduced substantially.

2.4.3 Back Gate Bias Dependence

If we combine the back gate bias effect, the NCE of SOI CMOS devices can be quite complicated. Figure 2.36 shows the threshold voltage versus the back gate bias of the FD SOI NMOS and PMOS devices with various channel widths [14]. With a smaller channel width, the back-gate bias effect becomes less noticeable because of the reduced controllability in the vertical direction from the back gate bias due to the 2D effect from the sidewall. As shown in this figure, when applying the back gate bias to the devices, the NCE of the FD SOI CMOS devices becomes complicated. Applying a sufficiently positive back gate bias, due to the body effect, the threshold voltage of the FD SOI NMOS device moves toward the negative direction. With a small channel width, the NCE offsets this negative shift in the threshold voltage of the FD SOI NMOS device due to the body effect. As a result, with a sufficiently

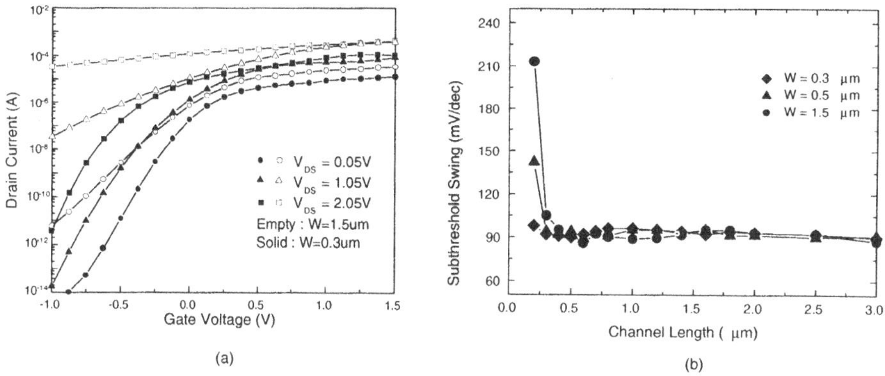

Fig. 2.35 (a) Subthreshold characteristics with an SOI NMOS device with channel widths of 0.3 and 1.5 μm, a front gate oxide of 101 Å, and a thin film of 779 Å. (b) Inverse subthreshold slope versus channel length of the SOI NMOS device with various channel widths. (Adapted from Hwang et al. [14].)

positive back gate bias, when the channel width is scaled down, the threshold voltage may move toward the positive direction (NMOS). On the contrary, when the back gate bias is negative enough, the threshold voltage of the FD SOI NMOS device with a small channel width moves toward the negative direction. When the channel width is shrunk, the NCE does lessen the back gate bias effect.

The NCE disturbed by the back gate bias effect can be further complicated by the interface charge effects. Generally speaking, the quality of the buried oxide is inferior to that of the front gate oxide in the SOI technology. At the interface between the buried oxide and the thin film, there exists a nonnegligible amount of traps or fixed positive charge, which function as if we were adding a positive back gate bias–the effective back gate bias has been increased. This complicates the NCE. As shown in Fig. 2.37, with a certain fixed interface charge, the NCE of the SOI CMOS devices can be reversed [14].

2.4.4 Isolation Dependence

In addition to mesa isolation, LOCOS device isolation techniques have also been used in SOI technology. The NCE of the LOCOS isolated SOI NMOS device in terms of the threshold voltage shift as shown in Fig. 2.38 is similar to that of mesa-isolated SOI NMOS devices [14], which is opposite to that of bulk CMOS devices [41]. Why do not LOCOS isolated SOI NMOS devices show the same trend in NCE as for the LOCOS isolated bulk NMOS devices? It is because the bird's beak of the LOCOS isolated SOI NMOS devices is less serious as compared to the LOCOS isolated bulk NMOS devices. In addition, during the fabrication process, no channel stop implant has been adopted for the LOCOS isolated SOI NMOS devices. During the source/drain implant, the generated interstitials may cause transient enhanced diffusion (TED) of the boron dopants to the buried oxide and the field oxide [43][44], which causes the

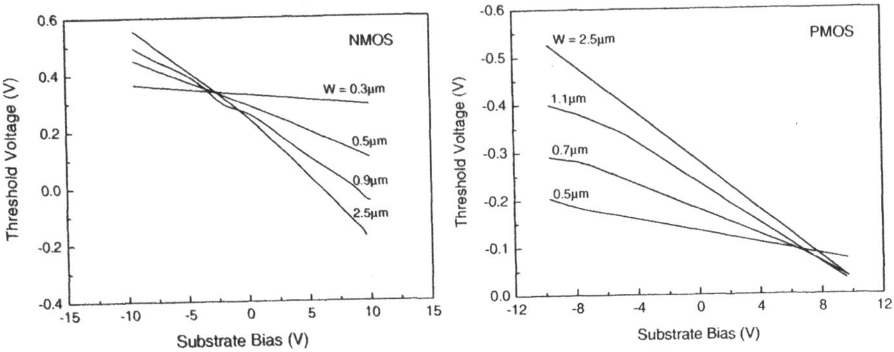

Fig. 2.36 Threshold voltage versus back gate bias of FD SOI NMOS and PMOS devices with various channel widths. (Adapted from Hwang et al. [14].)

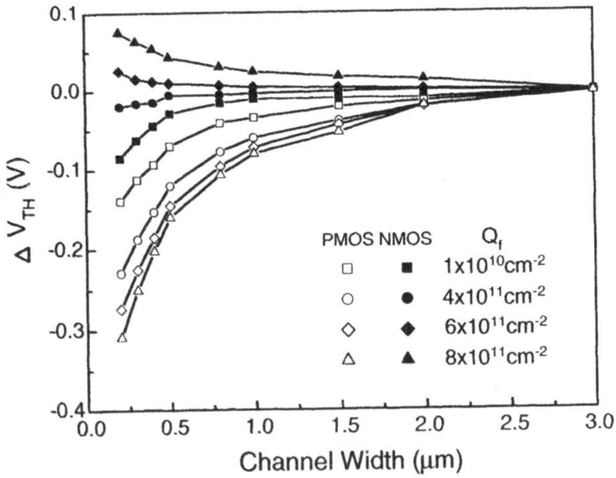

Fig. 2.37 Threshold voltage shift versus channel width of the SOI NMOS and PMOS devices with various fixed interface charge densities at the thin-film/ buried oxide interface. (Adapted from Hwang et al. [14].)

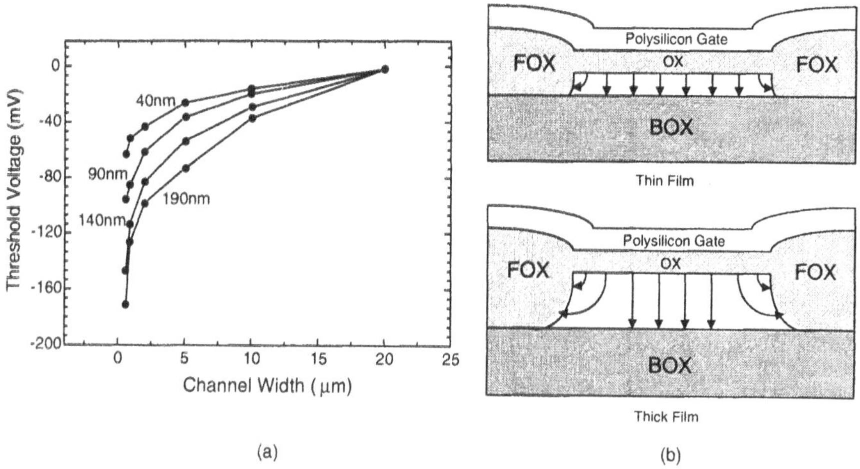

Fig. 2.38 (a) Threshold voltage shift versus channel width of SOI NMOS device with a buried oxide of 4000 Å and various thin-film thicknesses. (b) Cross-section of the LOCOS-isolated SOI NMOS devices for explaining the interstitial effects. (Adapted from Chang et al. [42].)

decrease in the doping density of the thin film near the field oxide. When the channel width is scaled down, its threshold voltage is decreased. As shown in Fig. 2.38, when the thin-film thickness is reduced, the NCE is improved.

Another LOCOS isolated SOI NCE phenomenon is related to compressive stress. As shown in Fig. 2.39, with a thicker buried oxide of 4000 Å, the trend on the NCE when the channel width is scaled down is similar to the one as shown in Fig. 2.38 [42]. When the buried oxide is reduced to 1000 Å, the NCE is reversed. This can be reasoned as follows. When the buried oxide is decreased below a certain value, the compressive stress is high, which is serious near the edge of the channel region at the interface between the thin film and the buried oxide. Because the compressive stress has the function of inhibiting the rediffusion of the boron dopants, the doping density of the thin film near the field oxide becomes higher than another area, which results in an increase in the threshold voltage at a reduced channel width. As long as the buried oxide is sufficiently thick, the compressive stress on the NCE does not occur.

2.5 MOBILITY

As for bulk CMOS devices, mobility is also important in determining the conduction current of SOI CMOS devices. In this section, the behavior referred to mobility in the SOI CMOS devices is described.

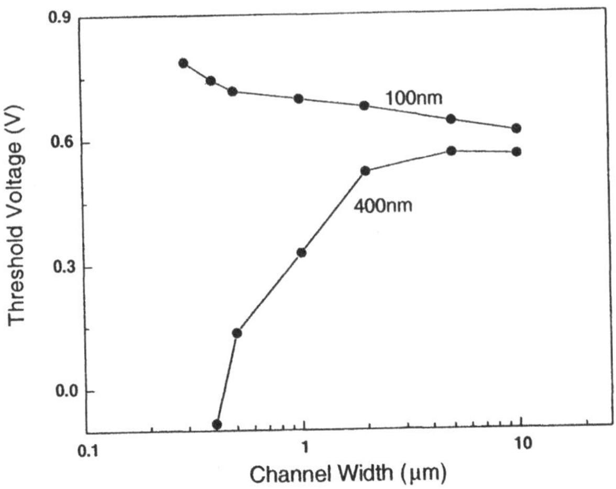

Fig. 2.39 Threshold voltage versus channel width of LOCOS isolated SOI NMOS device with a front gate oxide of 60 Å, a thin film of 1000 Å, and buried oxide thicknesses of 1000 and 4000 Å. (Adapted from Lee et al. [45].)

2.5.1 Vertical Field Dependence

Figure 2.40 shows the effective electron mobility versus the vertical electric field for several substrate concentrations [46]. As shown in this figure, when the vertical electric field is increased, mobility may increase initially, followed by a decrease, which can be reasoned as follows. When the vertical electric field is small, mobility is influenced by Coulomb scattering of the ionized impurities. In the low electric field region, a higher doping density leads to a decrease in mobility. When the vertical electric field is raised, the number of inverted mobile carriers increases, which gradually screens the function of the ionized impurity. Therefore, mobility is increased along with the increase in the vertical electric field in the low electric field regime. At the peak of the mobility, phonon scattering dominates. At this time, if the vertical electric field is raised further, the mobility drops quickly, which implies the high electric field regime, where the vertical electric field pushes the carriers to the interface between the thin film and the front gate oxide. Therefore, surface roughness scattering dominates. A further increase in the vertical electric field leads to a decrease in mobility. In the high electric field regime, the mobility is less dependent on the doping density. Instead, it is directly related to the vertical electric field and the surface roughness of the thin film.

Figure 2.40 shows that the effective mobility for the bulk device is also applicable for the SOI device. Figure 2.41 shows the effective mobility versus the vertical electric field in a fully depleted SOI NMOS device with (a) a thin-film thickness of 500 Å doped with various doping densities, and (b) thin-film thicknesses of 500 and 1500 Å with a doping density of $10^{15}cm^{-3}$ [47]. In addition, the mobility for a bulk device is also shown in Fig. 2.41(b). As shown in Fig. 2.41(a), similarly as for the bulk device, in the FD SOI NMOS device, when the vertical electric field increases, its effective

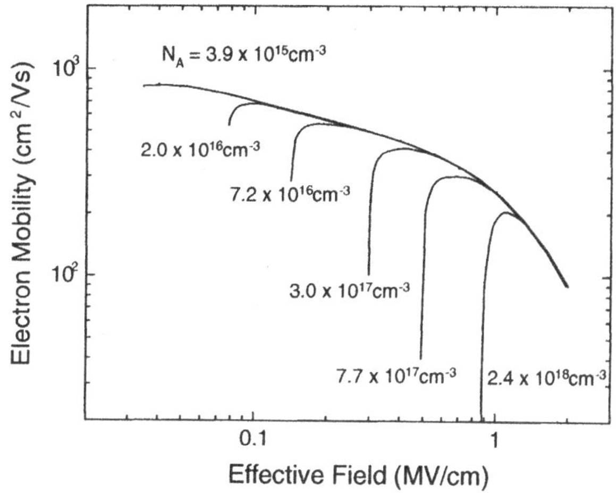

Fig. 2.40 Effective electron mobility versus the vertical electric field for several substrate concentrations. (Adapted from Sonoda et al.[46].)

mobility is degraded due to surface scattering. Only under a small vertical electric field is the mobility sensitive to the variation in the doping density of the thin film. As shown in Fig. 2.41(b), both bulk and SOI devices demonstrate about identical mobility characteristics. As indicated in this figure, a variation in the thin-film thickness does not lead to any significant difference in the mobility characteristics.

Along with the progress of this technology, the thickness of the thin film in an SOI device has been scaled down accordingly to improve its device characteristics. When the thickness of the thin film is below a certain value, the mobility may be affected. As shown in Fig. 2.42, when the thin-film thickness is > 100 Å, basically its mobility is not sensitive to the variations in the thin-film thickness [48]. On the other hand, when the thin-film thickness is < 100 Å, due to the substantial difference in the thermal expansion coefficients between the front gate/buried oxide and the single-crystal thin film, and also due to the very thin thin film, the accumulated stress cannot be relieved, which leads to a decrease in mobility. This behavior is especially noticeable for the SOI devices with a very thin thin film.

2.5.2 Lateral Field Dependence

Until now in this section, the influence of the vertical electric field in the mobility in an SOI device has been described. Now the effect of the lateral electric field on the SOI device is described. Fig. 2.43 shows electron and hole velocities versus the tangential electric field in the inversion layer of a bulk MOS device with various channel lengths [49]. As shown in this figure, along with the rise in the lateral electric field the electron drift velocity also increases. As shown in this log∼log plot, the electron velocity is linearly proportional to the lateral electric field when the lateral electric field is small, where the electron mobility is at a fixed value. When the lateral electric field rises

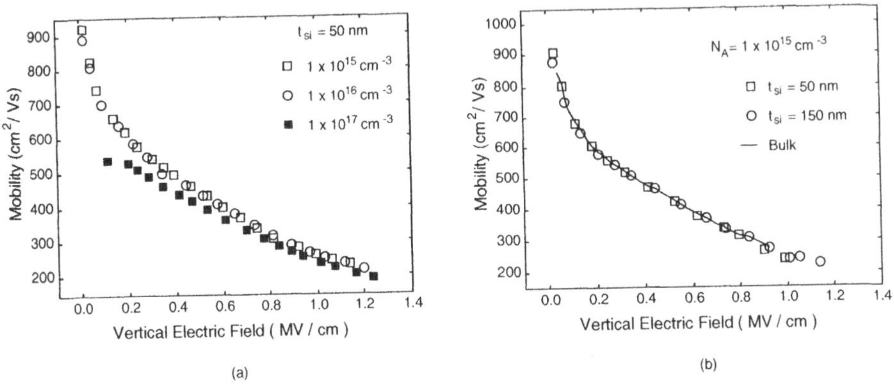

Fig. 2.41 Mobility versus vertical electric field in an FD SOI NMOS device (a) with a thin-film thickness of 500 Å doped with various doping densities, and (b) with thin-film thicknesses of 500 and 1500 Å and with a doping density of 10^{15}cm^{-3}. In addition, the mobility for a bulk device is also shown in (b). (Adapted from Wang et al. [47].)

further, the electron velocity is approaching its saturation velocity ($\cong 10^7\text{cm/s}$). In this regime, no matter how much the increase in the lateral electric field, the electron velocity becomes saturated, which indicates that the electron mobility drops at a high lateral electric field. This result is due to the fact that when carrying a high energy, it is easy for an electron to transfer the energy to the lattice. When the electron is traveling with a close to saturation velocity, most energy acquired from the electric field is transferred to the lattice, which leads to little change in the electron velocity and a decrease in the electron mobility. Compared to the electron, the hole has a similar behavior. Since the beginning hole mobility is much lower, the growth of the hole velocity with respect to the electric field is also at a slow pace until it reaches the hole saturation velocity. In contrast to the electron, the behavior of the hole velocity saturation as described above is less noticeable.

As described before, for normal operation, a balance between the traveling electron and the lattice is reached. In an advanced device with a very small channel length, not only is just the lateral electric field large, but the gradient of the electric field in the lateral direction is also large. Due to the large gradient in the lateral electric field, the electron energy accumulates quickly and cannot be transferred to the lattice quick enough to reach the energy balance. As a result, there may exist the phenomenon that the electron traveling velocity surpasses its saturation velocity, the so-called velocity overshoot. As shown in Fig. 2.44 in an SOI NMOS device with a front gate oxide of 40 Å, a thin film of 300 Å, a buried oxide of 800 Å, and a channel length of 500 Å, biased at a drain voltage of 2 V [50], in a certain region in the lateral channel, the electron traveling velocity may exceed its saturation velocity value– velocity overshoot.

The velocity overshoot phenomenon in the SOI device is different from that in the bulk counterpart. Figure 2.45 shows the electron velocity versus the gate over-drive voltage for (a) SOI and (b) bulk NMOS devices, biased at the drain voltage of 1 V

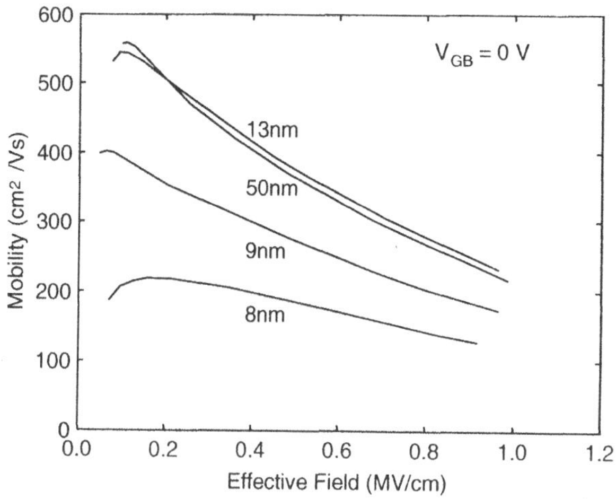

Fig. 2.42 Mobility versus vertical electric field in a FD SOI NMOS device with a front gate oxide of 170 Å and with various thin-film thicknesses. (Adapted from Choi et al. [48].)

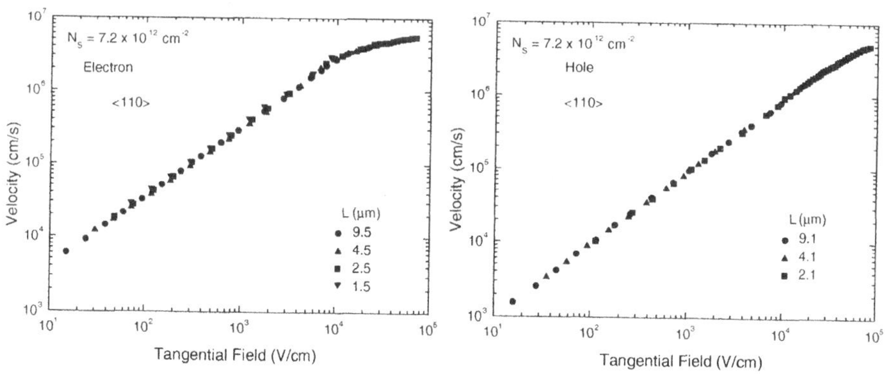

Fig. 2.43 Electron and hole velocities versus the tangential electrical field in the inversion layer of a bulk MOS device with various channel lengths. (Adapted from Tagaki et al. [49].)

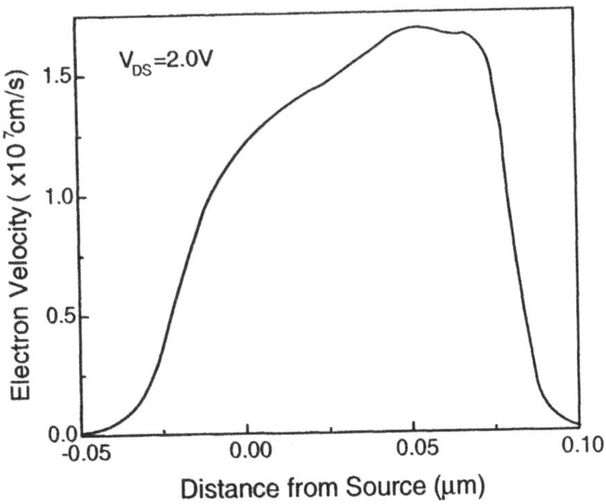

Fig. 2.44 Distribution of the electron velocity in the lateral channel region of an SOI NMOS device with a front gate oxide of 40 Å, a thin film of 300 Å, a buried oxide of 800 Å, and a channel length of 500 Å, biased at a drain voltage of 2 V. (Adapted from Ohuchi [50].)

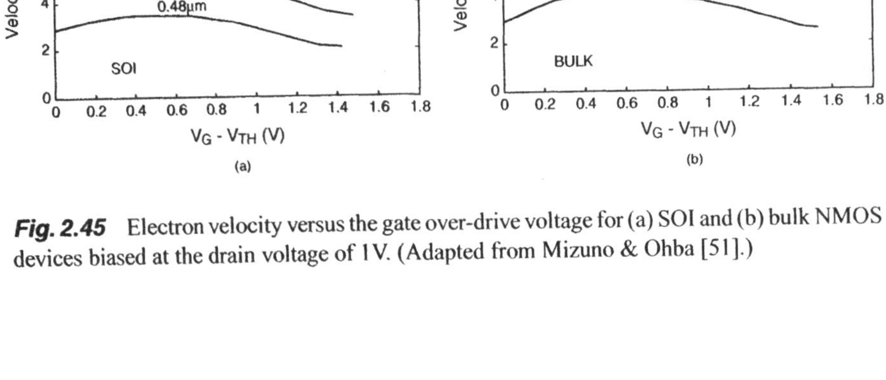

Fig. 2.45 Electron velocity versus the gate over-drive voltage for (a) SOI and (b) bulk NMOS devices biased at the drain voltage of 1 V. (Adapted from Mizuno & Ohba [51].)

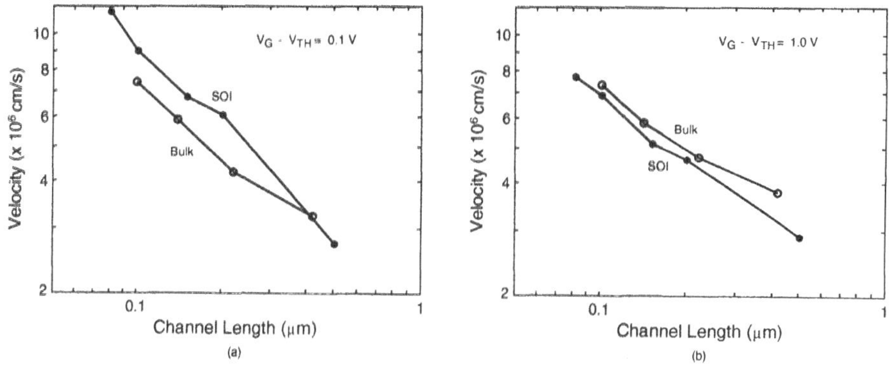

Fig. 2.46 Electron velocity versus channel length of SOI and bulk NMOS devices biased at the gate over-drive voltage of (a) 0.1V and (b)1V. (Adapted from Mizuno & Ohba[51].)

[51]. As shown in this figure, for the SOI device with a channel length of $< 0.1\ \mu m$, due to a large gradient in the lateral electric field, there exists velocity overshoot. In contrast, for the bulk device it is much more difficult to have velocity overshoot, which is due to the fact that in order to have an acceptable short channel behavior, the doping density of the bulk device is much higher than its SOI counterpart, which leads to a much smaller maximum mobility in the bulk device. Therefore, velocity overshoot is difficult to occur. In the SOI device, usually velocity overshoot occurs at a small gate voltage. At a small gate voltage, its vertical electric field is small, thus its surface scattering is small. Consequently, its mobility is high– it is more likely for the related electron velocity to exceed its saturation velocity value.

Figure 2.46 shows the electron velocity versus the channel length of SOI and bulk NMOS devices biased at the gate over-drive voltage of (a) 0.1 V and (b) 1 V [51]. As shown in Fig. 2.46(a), at a small gate over-drive voltage, the mobility is determined by Coulomb scattering. With a higher doping density, the low-field mobility of the device is lower. Under the situation that the doping density of the bulk is higher than that of the SOI, the electron velocity in the bulk is smaller than that in the SOI. From this figure, with a channel length of $< 0.1\ \mu m$, it is more likely for the SOI device to have velocity overshoot. As shown in Fig. 2.46, at a higher gate over-drive, surface scattering dominates. Thus, the electron velocity in the SOI and the bulk devices is about identical. Both devices are unlikely to have velocity overshoot.

2.6 FLOATING BODY EFFECTS

Due to the buried oxide structure, floating body effects of PD SOI MOS devices are important in determining their behavior. In this section, floating body effects of PD MOS devices are described.

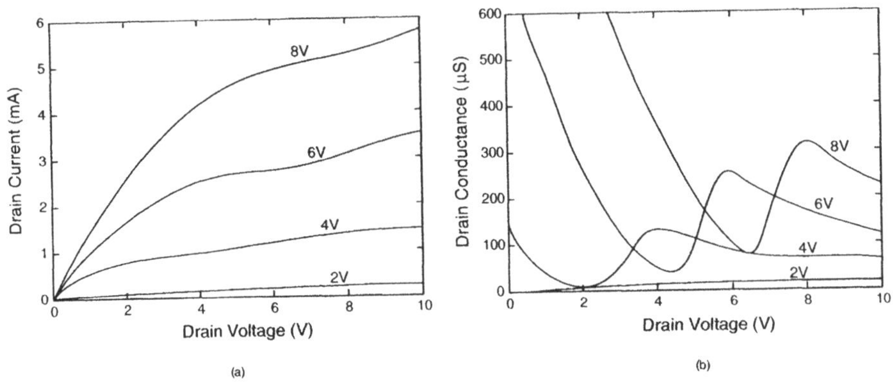

Fig. 2.47 (a) Drain current and (b) differential drain conductance versus drain voltage of an n-channel SOS MOSFET. (Adapted from Howes & Redman-White [52].)

2.6.1 Strong Inversion Kink Effects

Fig. 2.47 shows (a) the drain current and (b) the drain conductance versus the drain voltage of an n-channel SOS MOSFET [52]. As shown in Fig. 2.47(a), there are so-called kink effects– when the device is biased in the saturation region and the drain voltage exceeds a certain value, its drain current suddenly rises with a discontinuity in the drain current. As shown in this figure, in the strong inversion, with a smaller gate voltage, the drain current needed to trigger the kink effects is smaller. As shown in Fig. 2.47(b), kink effects worsen the differential drain conductance of the device, which may have a serious impact on the performance if the device is used for analog circuits.

Kink effects of an SOI device are strongly dependent on its operating speed. As shown in Fig. 2.48 [52], when the drain voltage exceeds a certain value, its drain admittance increases suddenly, which indicates kink effects. At a lower operating frequency, kink effects are more serious, which occurs at a smaller drain voltage. On the other hand, at a high operating frequency, kink effects become light and tends to disappear. Thus, the frequency-dependent kink effects affect the stability of the circuits using SOI devices. Kink effects are caused by the impact ionization of the PD SOI MOS device and the parasitic BJT effect. As shown in Fig. 2.49, at a large drain voltage, impact ionization caused by the traveling electrons with high energy in the high electric field region near the drain results in a large amount of electron/hole pairs, of which the electrons move toward the drain and the holes toward the floating body and thus accumulate at the buried oxide boundary near the source. Thus, the local body potential increases and the local threshold decreases, which triggers the sudden rise in the drain current. When the accumulation of holes and the related potential reach a certain extent, the source/body diode turns on. As a result, the body potential does not rise further. Instead, it maintains at $0.6 \sim 0.7$ V, which is the internal mechanism causing kink effects. From the analysis above, kink effects of a PD SOI NMOS device are related to the accumulation of holes in the body. Kink

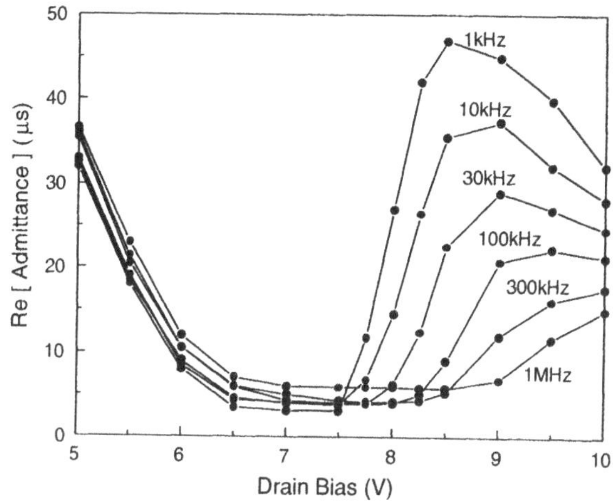

Fig. 2.48 Drain admittance versus drain voltage of the n-channel SOS MOSFET with its gate biased at 8 V. A DC voltage superimposed with a small AC signal for various frequencies is imposed at the drain voltage. (Adapted from Howes & Redman-White [52].)

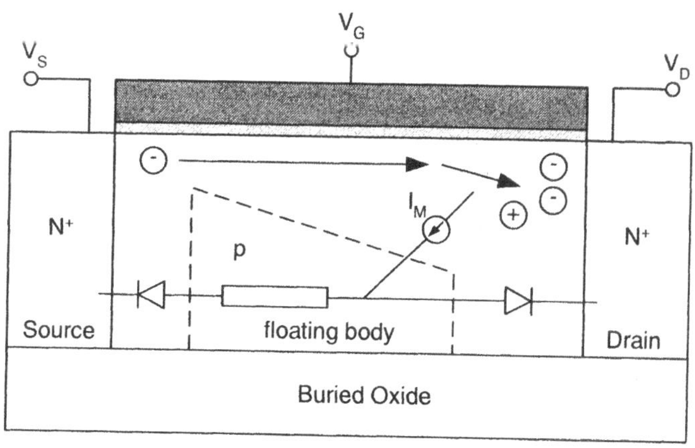

Fig. 2.49 Cross-section of an SOI NMOS device showing the current conduction path.

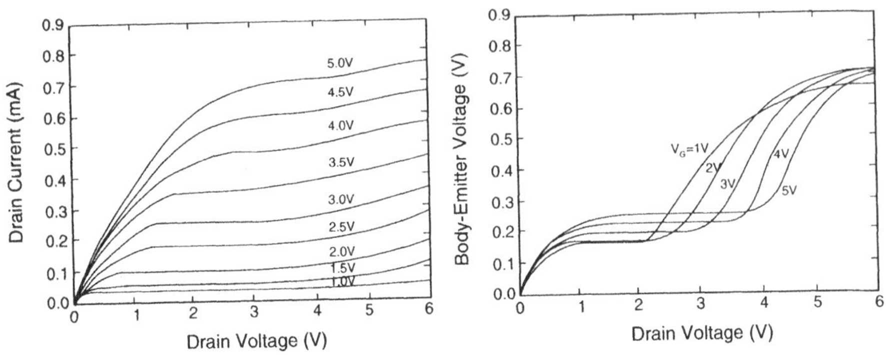

Fig. 2.50 Drain current and body-emitter voltage versus drain voltage of a PD SOI NMOS device. (Adapted from Chen & Kuo [53].)

effects are strongly correlated to the operating frequency of the device. Since impact ionization of holes is much smaller than that of electrons, SOI PMOS devices show much smaller kink effects as compared to NMOS devices.

The above analysis on kink effects can be confirmed by monitoring the body-emitter voltage of a PD SOI NMOS device as shown in Fig. 2.50. As shown in the figure, the body-emitter voltage is correlated to the drain current [53]. When the drain voltage rises to a certain value, impact ionization occurs. The generated holes accumulate in the floating body, thus its body potential rises and the local threshold voltage drops. Therefore, there is a sudden increase in the drain current–kink effects. After the body-emitter voltage rises to 0.7 V, it becomes saturated since the source/drain diode is turned on. From this figure, at a larger gate voltage, a larger drain voltage is needed to cause the rise in the body potential, thus its kink effects occur at a later time.

Kink effects are also determined by the structure of a PD SOI device. Figure 2.51 shows the triggering drain voltage at the onset of the kink effects versus the thin-film doping density of the PD SOI NMOS device at various oxide thicknesses [53]. As shown in this figure, when the doping density of the thin film is raised, its threshold voltage increases. Due to the fall of the saturation voltage (V_{DSAT}), at which the channel pinches off, and the large change of the electric field in the postsaturation region of the lateral conduction channel under the front gate, impact ionization increases and the triggering drain voltage to cause kink effects is lowered. On the other hand, when the front gate oxide becomes thinner, the change of the electric field in the postsaturation becomes larger. Despite the drop in the threshold voltage and the rise in V_{DSAT}, impact ionization still rises. Thus, the triggering drain voltage to cause kink effects also drops.

Kink effects are dependent on the channel length. Figure 2.52 shows the drain current versus the drain voltage of a PD SOI NMOS device with a front gate oxide of 250 Å, a buried oxide of 4500 Å, a thin film of 1000 Å doped with a doping density

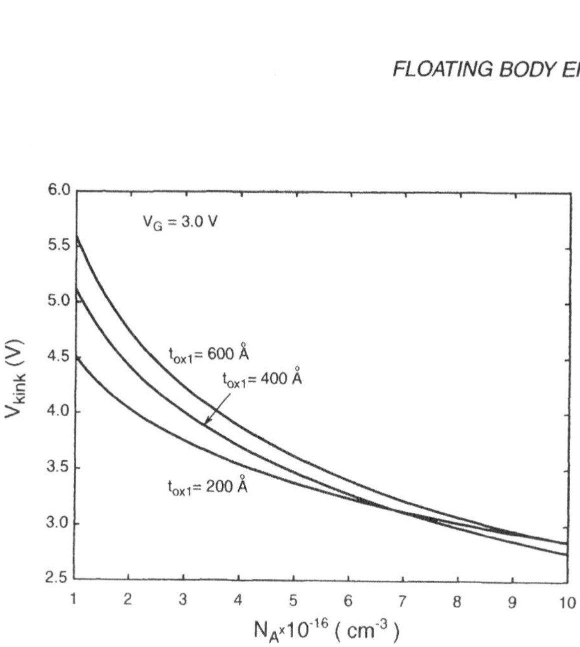

Fig. 2.51 The triggering drain voltage at the onset of the kink effects versus the thin-film doping density of the PD SOI NMOS device at various gate oxide thicknesses. (Adapted from Chen et al. [53].)

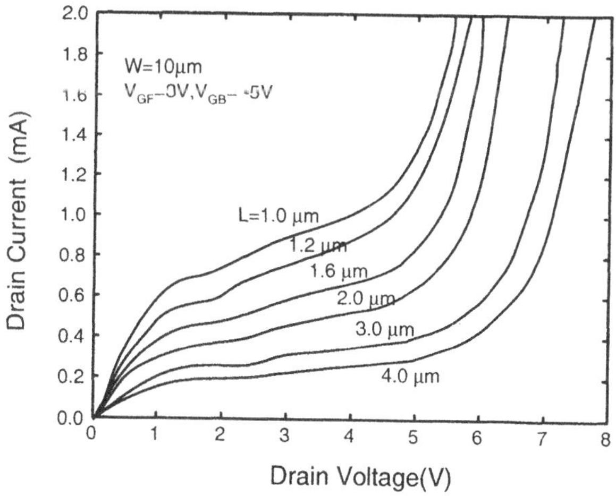

Fig. 2.52 Drain current versus drain voltage of a PD SOI NMOS device with a front gate oxide of 250 Å, a buried oxide of 4500 Å, a thin film of 1000 Å doped with a doping density of $1.2 \times 10^{17} cm^{-3}$ and various channel lengths, biased at a front gate voltage of 3 V and a back gate voltage of -5 V. (Adapted from Yu et al. [54].)

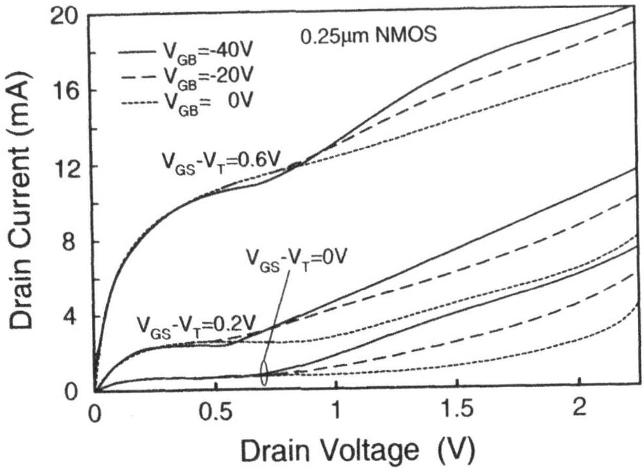

Fig. 2.53 Drain current characteristics of a 0.25 μm PD SOI NMOS device with a front gate oxide of 45 Å, a thin film of 400 Å, and a buried oxide of 3800 Å, biased at various back gate biases. (Adapted from Rozean et al. [55].)

of $1.2 \times 10^{17}\text{cm}^{-3}$ and various channel lengths, biased at a front gate voltage of 3 V and a back gate voltage of -5 V [54]. As shown in this figure, when the channel length becomes smaller, the electric field inside the device becomes larger. Thus, impact ionization becomes more serious. In addition, with a smaller channel length, the parasitic BJT at the bottom of the thin-film functions stronger, which provides a more noticeable amplification of the impact ionization. Thus, kink effects occur earlier and are more noticeable. As shown in this figure, the breakdown voltage drops substantially.

Back gate bias also determines kink effects of a PD SOI device. Figure 2.53 shows the drain current characteristics of a 0.25 μm PD SOI NMOS device with a front gate oxide of 45 Å, a thin film of 400 Å, and a buried oxide of 3800 Å, biased at various back gate biases [55]. As shown in this figure, at a more negative back gate bias, it is easier to have hole accumulation in the thin film– kink effects are more serious. Back gate bias effects on the kink effects are especially visible for SOI devices, which are defined at the boundary between FD and PD. As long as the back gate bias is sufficiently negative, even in an FD SOI device, hole accumulation in the thin film can still be caused to trigger kink effects.

As described before, kink effects caused by the floating body, may bring in bad influences to circuits made of PD SOI devices. To avoid the kink effects, the straightforward way is by connecting the body contact of the thin film to source or to ground. Once the body of the thin film is tied, instead of being accumulated in the thin film, holes are evacuated via the contact. Thus, kink effects disappear as shown in Fig. 2.54 [56]. Also shown in this figure, even under the floating body condition, it is difficult for PD SOI PMOS device to have kink effects due to the smaller impact ionization of holes and the worse function of the parasitic pnp bipolar device. Also shown in this

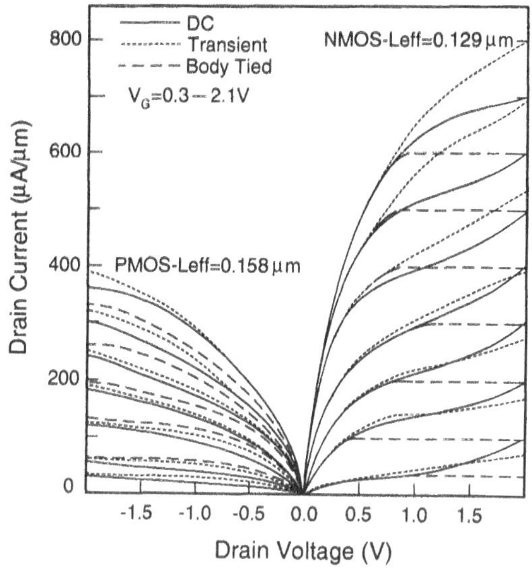

Fig. 2.54 Drain current characteristics of PD SOI NMOS and PMOS devices. (Adapted from Kraxenberger et al. [56].)

figure is the drain current measured via a transient approach, which requires only a short time for measurement. With the transient measurement, it is difficult for holes to be accumulated in the thin film and thus there are no kink effects. If we use the transient approach for measurement, the dissipated power cannot be accumulated, which implies no self-heating and that the measured current is higher.

2.6.2 Body Contact

Body contact has been used to resolve the kink effect problems. The body contact to a PD SOI NMOS device is usually via both sides of the channel width direction as shown in Fig. 2.55(a) [57]. However, the body contact approach also has its own limitations. When the channel length of the device is small, due to the very narrow neutral region from the body contact to the internal body, its equivalent body resistance can be very large. In addition, the cross-section of this neutral region is affected by V_{GS} and V_{DS}, hence this equivalent body resistance is bias dependent– not a constant. As shown in Fig. 2.55(b), with a channel width of 20 μm, with a smaller channel length, at a larger gate voltage, its body contact resistance becomes larger due to a smaller cross-section of the neutral region. The body contact resistance can be as high as 1 MΩ.

The influence of the body contact resistance in the device behavior can be observed from Fig. 2.56 [57]. As shown in the figure, with respect to various body contact resistances ranging from 100 kΩ to 10 MΩ, its drain current characteristics with the kink effects behavior can be quite different. Note that in the figure, a body contact resistance of 0 Ω implies an ideal body contact and a floating body implies the no body contact condition. From this figure, at the zero body contact resistance, kink

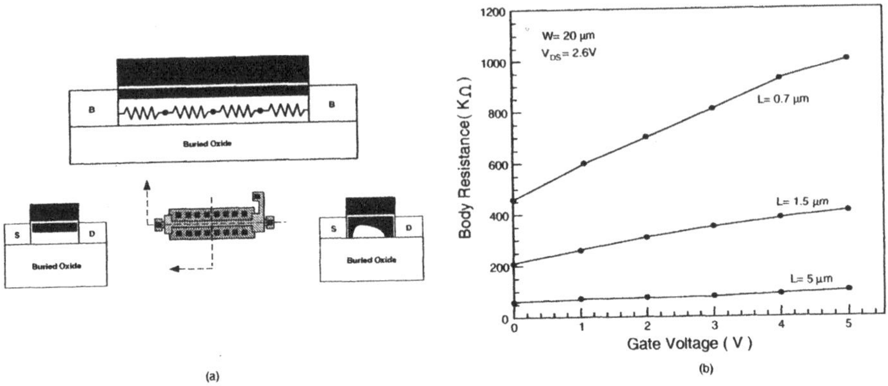

Fig. 2.55 (a) Cross-section of a PD SOI NMOS device at zero body bias and at a normal body bias. (b) Body contact resistance versus the gate voltage of the PD SOI NMOS device with various channel lengths. (Adapted from Edwards et al. [57].)

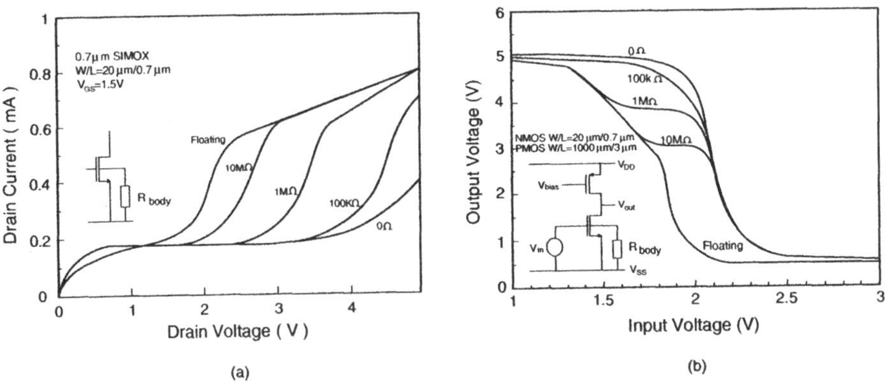

Fig. 2.56 (a) Drain current versus drain voltage of a PD SOI NMOS device with a channel length of 20 μm and a channel length of 0.7 μm and various body contact resistances, biased at $V_{GS} = 1.5$ V and (b) DC transfer curve of an amplifier using the PD SOI NMOS device with various body contact resistances. (Adapted from Edwards et al. [57].)

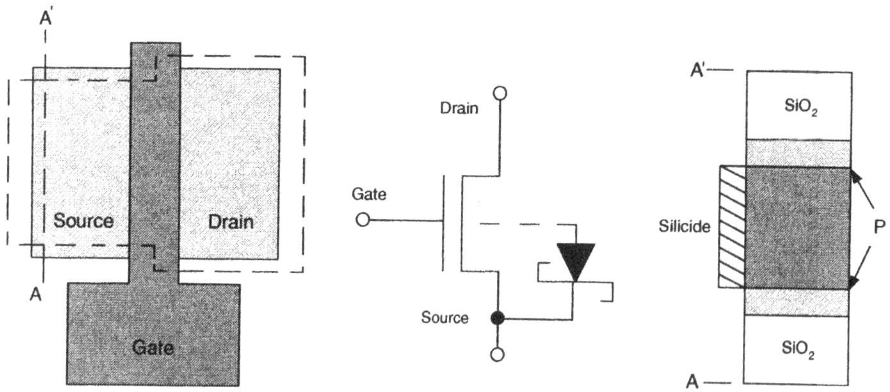

Fig. 2.57 A PD SOI NMOS device with Schottky body contact at the source. The dashed line indicates the n^+ implant mask. (Adapted from Sleight et al. [58].)

effects disappear completely. When the body contact resistance is > 100 kΩ, kink effects are noticeable. With a large body contact resistance, the current for carrying the holes out of the internal body is smaller than the hole current generated by the impact ionization. Thus, a substantial amount of holes are accumulated in the body to trigger the kink effects. Figure 2.56(b) shows the DC transfer curve of an amplifier using the PD SOI NMOS device described in Fig. 2.55 with various body contact resistances [57]. As shown in this figure, with a larger body contact resistance, the DC transfer curve becomes worse due to the worsened kink effects.

Adoption of the body contact approach may result in an increase in the layout and hence the increased complexity in interconnects. To lessen the kink effects, Schottky body contact technique, as shown in Fig. 2.57 [58], have been used. As shown in this figure, after the source/drain implant, silicide is formed on p-type and n^+ source regions simultaneously. In the region with silicide in contact with the p-type region, the Schottky diode is formed to provide a way to remove the holes from the body to the source. Usually, Schottky diode is also formed at the drain side for the PD SOI devices used as pass transistors and in circuits for devices without a fixed source/drain configuration. At this time, the implant mask for generating the symmetric source/drain Schottky body contact is shown in Fig. 2.57. As shown in Fig. 2.58(a), when the body-source Schottky diode is on, kink effects are noticeably reduced. Figure 2.58(b) shows the current characteristics of the source-body junction using the Schottky body contact approach and the conventional $p-n^+$ approach [58]. As shown in this figure, with the Schottky body contact approach, the body potential is decreased by 0.5 V, which is much smaller as compared to the conventional diode made of the $p-n^+$ junction. Thus kink effects can be substantially reduced. Note that with the Schottky contact approach, the reverse biased current is larger, which results in a larger leakage current in the source/drain region.

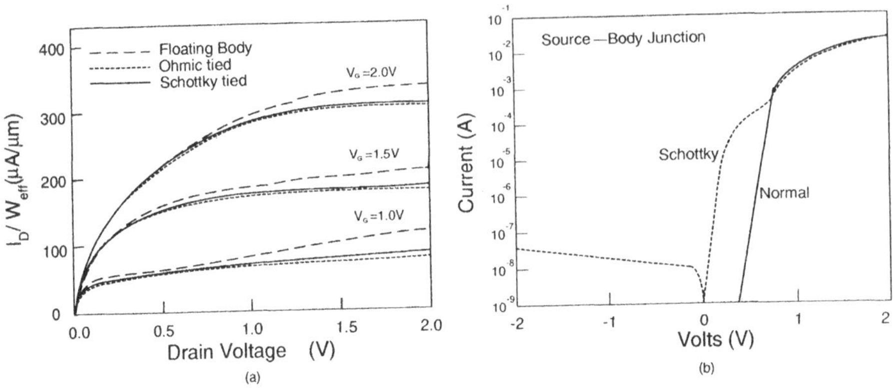

Fig. 2.58 (a) Drain current characteristics of a PD SOI NMOS device with a thin film of 1500 Å, a buried oxide of 2000 Å, and a channel length of 0.3 μm, with a floating body and a Schottky body contact and (b) the current characteristics of the source-body junction using the Schottky body contact approach and the conventional $p - n^+$ approach. (Adapted from Sleight et al. [58].)

2.6.3 Various Techniques to Reduce Kink Effects

The above techniques proposed to eliminate kink effects cannot effectively resolve the problems with the large-body contact resistance due to a large channel width. As shown in Fig. 2.59, by implanting germanium dopants to form SiGe in the source/drain region, Ge-implanted source/drain has been used to eliminate the kink effect [59]. As shown in Fig. 2.59(a), with the implanted Ge in the source/drain, bandgap narrowing occurs in the valence band. As a result, the potential barrier for the holes between the source and the body is reduced. Thus the accumulated holes at the bottom of the thin film can be easily expelled to the source and the kink effects are reduced. As shown in Fig. 2.59(c), with the Ge-implanted source/drain, kink effects are lessened. In addition, the function of the parasitic bipolar device is weakened. Therefore, the breakdown voltage of the device is improved. Since the whole source region can be used to remove holes, no body contact resistance problems exist. The Ge-implanted source/drain techniques to reduce kink effects and to improve the breakdown voltage of the PD SOI NMOS device can be further visualized by studying the lateral parasitic bipolar device, which uses the source/body/drain as the emitter/base/collector, as shown in Fig. 2.60 [60]. With the implanted-Ge source(emitter), bandgap narrowing occurs. The base current increases and the current gain of the parasitic bipolar device falls. Thus, the kink effects and the breakdown voltage become better. With a higher Ge dose, the improvement is more. In contrast, the change in the germanium implant energy does not show any substantial difference.

The Ge-implanted source/drain techniques can be helpful for the valence band, which means it is suitable only for NMOS. Although SOI PMOS devices have less

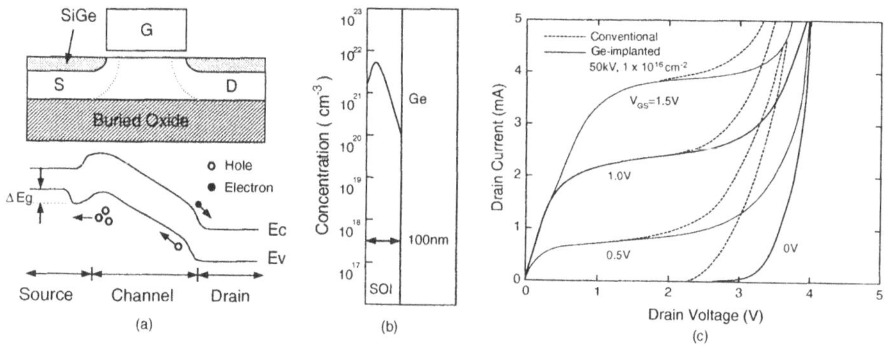

Fig. 2.59 (a) Structure and bandgap diagram of a PD SOI NMOS device with Ge-implanted source/drain. (b) Ge profile in the source/drain region. (c) Drain current characteristics of the Ge-implanted PD SOI NMOS device with a channel length of 0.15 μm and a channel width of 10 μm, a front gate oxide of 50 Å, and a thin film of 1000 Å. (Adapted from Yoshimi et al. [59].)

impact ionization and kink effects, under certain transient operations, extra carriers can also be accumulated in the thin film of the SOI PMOS device. Under this situation, a technique to remove the extra carriers in the thin films of both SOI NMOS and PMOS devices should be implemented. As shown in Fig. 2.61(a), by implanting argon ions, a large amount of recombination centers formed at the source/drain-body boundary increase the forward-bias current of the n^+/p and p^+/n diodes [61]. Thus, extra carriers in the body can be effectively released to reduce the kink/floating body effect.

Although the argon-implant technique can be used to resolve the floating body problems for both NMOS and PMOS devices at the same time, leakage current in the source/drain-body junction cannot be overlooked. There is another technique–bipolar embedded source structure (BESS) as shown in Fig. 2.62 can be used to lessen the floating body problems. As shown in Fig. 2.62(a) [62], the SOI device is still based on a silicon or argon implant to generate a large amount of recombination centers, which are not located at the source/body boundary as described in Fig. 2.61. Instead, they are located right under the source region. In addition, between the recombination center-scattered region and the body, there is another layer of n^- region. Thus, recombination center/n^- region/body forms the collector/base/emitter of another lateral bipolar device. Since the potential barrier of the n^-/p^- base-emitter junction is not high, once holes are being accumulated, due to the elevated body potential, holes enter the n^- region, most of which are recombined in the recombination center region. Therefore, any amount of holes accumulated in the body can be efficiently expelled. In addition, the junction between the n^- region of the drain and the body, which is reverse biased, can be used to effectively isolate the recombination center region to lower the leakage current. As shown in Fig. 2.62 [62], by using the BESS technique, kink effects are reduced and the breakdown voltage is improved.

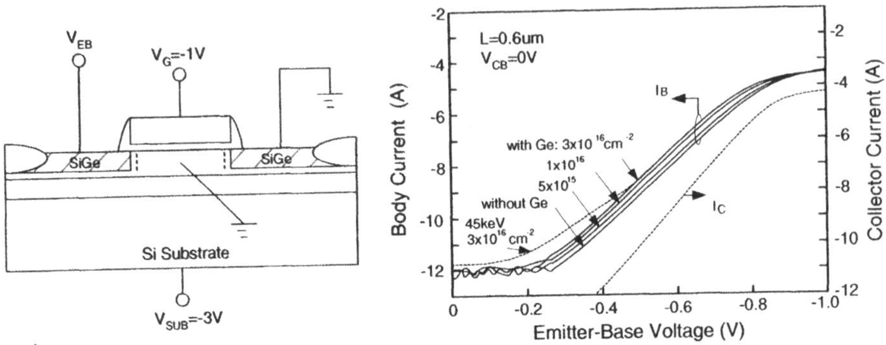

Fig. 2.60 Gummel plot of the parasitic bipolar device in the PD SOI NMOS device with and without the Ge-implanted source/drain. The width of the bipolar device is 6 μm. The dose of the germanium implant is $3 \times 10^{16} cm^{-3}$. Two implant energies of 25 and 45 KeV have been used. (Adapted from Nishiyama et al. [60].)

2.7 SUBTHRESHOLD BEHAVIOR

In Section 2.6, kink effects of PD SOI MOS devices biased in strong inversion have been studied. For PD SOI MOS devices biased in the subthreshold region, similar to strong-inversion kink effects, a sudden increase in the drain current owing to floating body and impact ionization can also occur. In this section, subthreshold behaviors of SOI CMOS devices are described.

2.7.1 FD versus PD

For the subthreshold drain current characteristics of the SOI devices biased at a small drain voltage ($V_D = 0.1$ V) as shown in Fig. 2.63, compared to the FD devices, the PD device has a worse subthreshold slope, which has been described before. In addition, at the large drain voltage ($V_D = 1.8$ V), the PD device has a better subthreshold slope as compared to the FD device. At a sufficiently large drain voltage, the subthreshold slope of the PD device is better than that biased at a small drain voltage, which is due to the fact that when the drain voltage is sufficiently large, impact ionization starts to occur in the device. The extra holes generated by impact ionization accumulate at the bottom of the thin film to cause the increase of the body potential, which rises until a balance between the source/body junction current and the hole current generated by impact ionization has been reached. This elevated body potential also raises the surface potential in the channel region to increase the diffusion current. On the other hand, a higher gate voltage also increases the hole current generated by impact ionization from a higher surface current. As a result, the body potential rises to a higher level to cause a further increase in the surface current. Under this operation, the subthreshold slope of the device is improved more at a higher drain voltage, which is the so-called subthreshold kink effects.

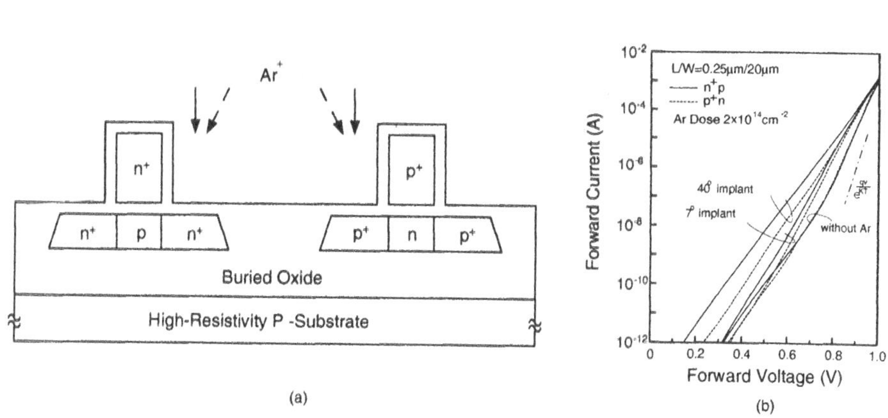

Fig. 2.61 (a) Cross-section of PD CMOS devices during an argon implant. Argon ions were implanted at two incident angles of 7° and 40° into the source/drain region [61]. (b) Forward current-voltage characteristics of n^+p junction (solid lines) and p^+n junctions (dashed lines). (Adapted from Ohno et al. [61].)

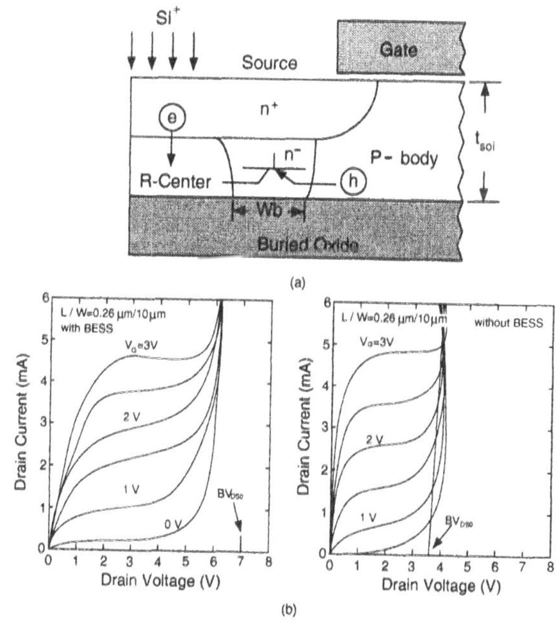

Fig. 2.62 (a) Cross-section of a PD SOI NMOS device using the bipolar embedded source structure (BESS) [62]. (b) Drain current characteristics of the PD SOI NMOS device with BESS and a channel length of 0.26 μm and a channel width of 10 μm, a thin film of 1500 Å and a buried oxide of 5000 Å. (From Horiuchi & Tamura [62]. ©1998 IEEE)

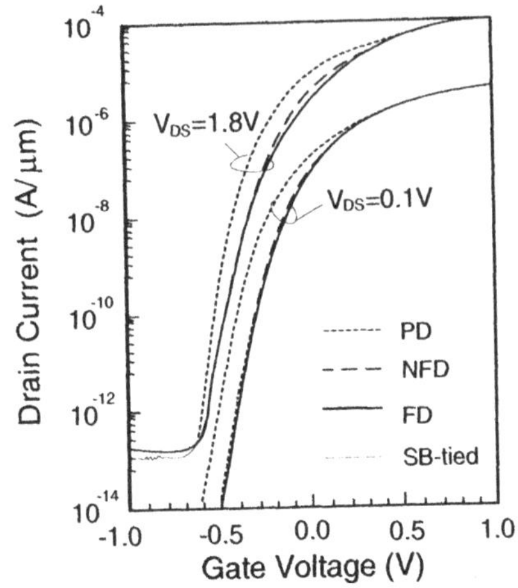

Fig. 2.63 Subthreshold drain current characteristics of PD and FD SOI NMOS devices biased at $V_D = 0.1$ V and 1.8 V. (Adapted from Tseng et al. [63].)

2.7.2 PD

Figure 2.64 shows the subthreshold drain current characteristics of the PD SOI NMOS device with a front gate oxide of 270 Å, a thin film of 2800 Å, a buried oxide of 3500 Å, a channel width of 8 μm and various channel lengths, biased at the drain voltage of 3 V [64]. As shown in this figure, with a smaller channel length, the subthreshold kink effects occur at a smaller gate voltage. This can be reasoned as being at the same drain voltage with a smaller channel length, and a larger internal electric field leading to more impact ionization, thus more kink effects. Therefore, the subthreshold slope of the device with a smaller channel length is steeper.

Figure 2.65 shows the drain voltage at the onset of the kink effects (V_{kink}) versus the gate voltage of a PD SOI NMOS device with a channel length of 3 μm for both subthreshold and strong inversion regions [64]. From this figure, the minimum V_{kink} occurs at the gate voltage of $V_G = V_T$. In the subthreshold region, an increase in the gate voltage leads to a decrease in V_{kink}. In contrast, in the strong inversion, an increase in the gate voltage causes an increase in V_{kink}. Although the mechanism of the kink effects for both subthreshold and strong inversion is the same, the trends are opposite. In the subthreshold region, impact ionization is controlled by the channel current. At a higher gate voltage, the higher channel current results in a larger impact ionization, which implies kink effects occur earlier. On the contrary, in the strong inversion region, impact ionization is controlled by the width of the postsaturation region in the device. When the gate voltage increases, the saturation voltage (V_{DSAT}) increases. Thus the postsaturation region becomes smaller and thus impact ionization effects become smaller. Therefore, kink effects occur later– V_{kink} rises.

As mentioned before, in the subthreshold region, a higher drain voltage brings in a better subthreshold slope, which is the case when drain induced barrier lowering (DIBL) is not serious. With DIBL, the subthreshold characteristics can be quite

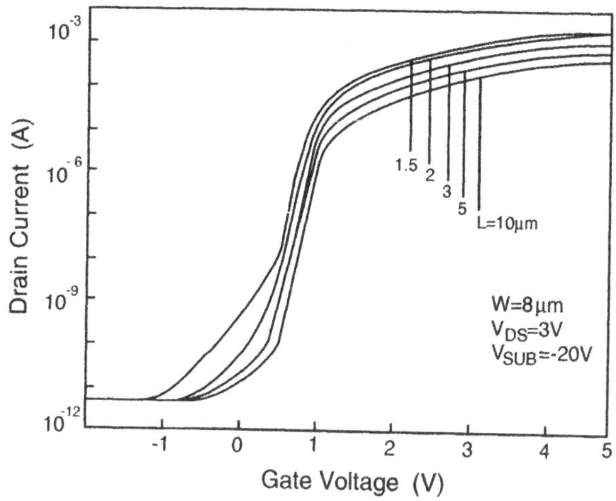

Fig. 2.64 Subthreshold drain current characteristics of the PD SOI NMOS device with a front gate oxide of 270 Å, a thin film of 2800 Å, a buried oxide of 3500 Å, a channel width of 8 μm and various channel lengths, biased at the drain voltage of 3 V. (Adapted from Davis [64].)

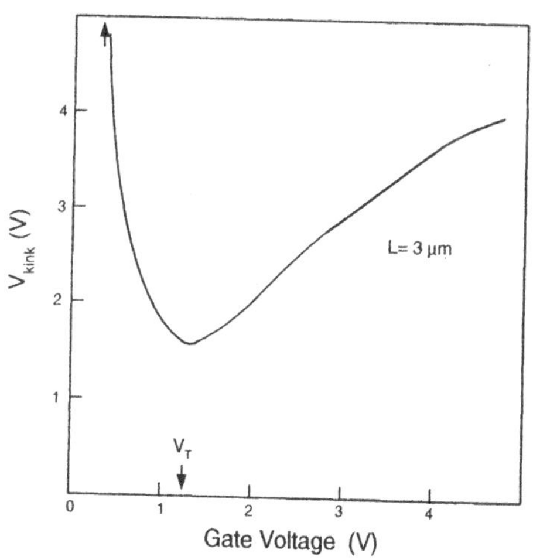

Fig. 2.65 Drain voltage at the onset of the kink effects (V_{kink}) versus the gate voltage of a PD SOI NMOS device with a channel length of 3 μm for both subthreshold and strong inversion regions. (Adapted from Davis et al. [64].)

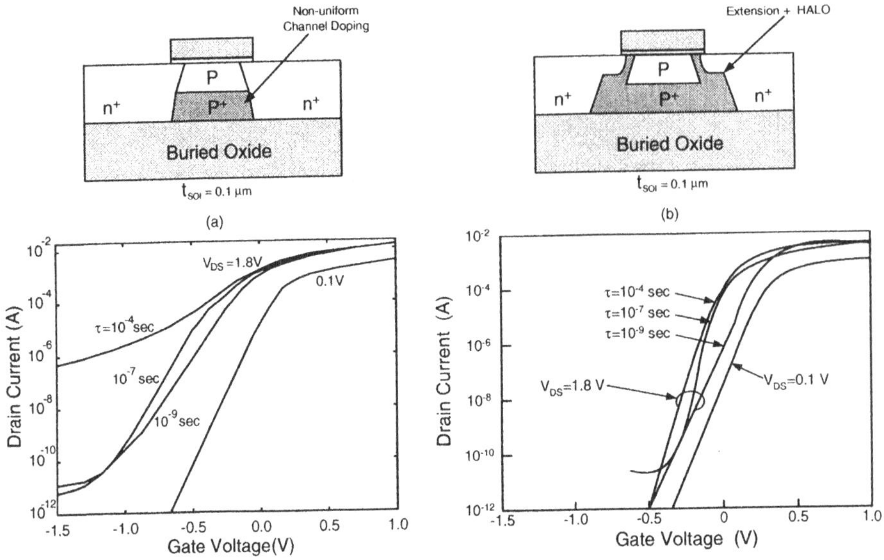

Fig. 2.66 Subthreshold drain current characteristics of the PD SOI NMOS device with a channel length of 0.1 μm, a thin film of 1000 Å, a front gate oxide of 50 Å, and with various carrier lifetimes (a) without and (b) with LDD and HALO structures. (From Shahidi et al. [65] ©1994 IEEE.)

different. Figure 2.66 shows the subthreshold drain current characteristics of the PD SOI NMOS device with a channel length of 0.1 μm, a thin film of 1000 Å, a front gate oxide of 50 Å, with various carrier lifetimes and (a) without and (b) with LDD and HALO structures [65]. Originally, a larger carrier lifetime could make the subthreshold slope steeper since its parasitic bipolar device functioned better and thus kink effects were larger. As shown in fig. 2.66(a), at a larger drain voltage, the subthreshold slope becomes worse due to a more serious DIBL, which dominates the subthreshold characteristics. As shown in Fig. 2.66(b), by adopting a similar to lightly doped drain (LDD) structure and HALO (the p-type doping density near source/drain is raised), DIBL has been inhibited and kink effects appear again. In addition, a larger lifetime leads to a steeper subthreshold slope. As a result, a better subthreshold slope at a higher drain voltage can occur again.

As described above, the subthreshold behavior of a PD SOI NMOS device is determined simultaneously by DIBL and kink effects. In addition to the special device structure described above, the bipolar embedded source structure (BESS) can also be used to improve the subthreshold characteristics. As shown in Fig. 2.67, for a typical PD SOI NMOS device, its threshold voltage drops with the increase in the drain voltage [66]. By using the BESS as described before, the kink effects are lessened. Therefore the trend on the drain voltage dependent threshold voltage can be improved.

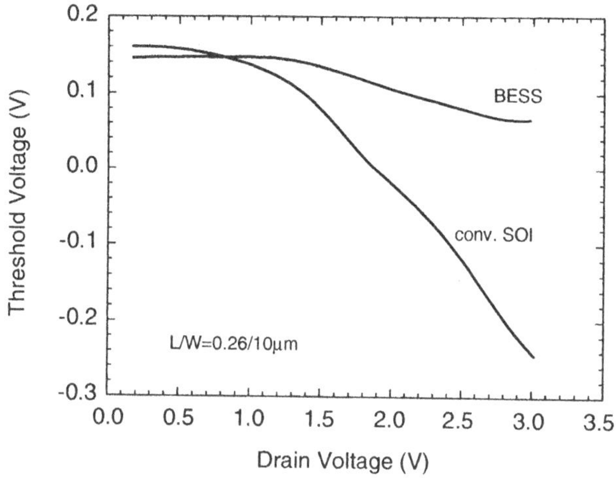

Fig. 2.67 Threshold voltage versus drain voltage characteristics of a PD SOI NMOS device with the bipolar embedded source structure (BESS). (Adapted from Horiuchi et al. [66].)

2.7.3 DIBL Dependence

The simultaneous function of DIBL and kink effects on an SOI device may bring in some interesting behaviors. Figure 2.68 shows the threshold voltage versus the channel length of an FD SOI NMOS device with a front gate oxide of 50 Å, a thin film of 500 Å and a buried oxide of 900 Å [67]. As shown in this figure, at the drain voltage of 0.1 V, kink effects do not occur yet. Its threshold voltage becomes smaller when the channel length is shrunk–short channel effect. At the drain voltage of 2 V, the whole threshold curve drops but the threshold voltage is not channel length dependent any more, which is different from traditional DIBL. Generally speaking, for an FD SOI NMOS device, when the drain voltage increases, due to kink effects, its subthreshold slope becomes steeper. From Fig. 2.69, for a channel length between 0.5 and 5 μm, this kink effect related subthreshold slope phenomenon becomes more noticeable when the channel length is shrunk [67]. When the channel length is further shrunk to 0.25 μm, this kink effect related subthreshold slope phenomenon becomes smaller. These phenomena are due to the simultaneous function of DIBL and the floating body effect.

As shown in Fig. 2.70(a) and (b), at the high drain voltage, impact ionization generated holes start to accumulate at the bottom of the thin film. Therefore, its body potential increases and its threshold voltage drops (kink effects). With a small channel length, although DIBL still lowers its threshold voltage, the potential barrier of the source/body junction shrinks due to DIBL. Hence, the accumulated holes can be released to the source–DIBL reduces kink effects and the drop in the threshold voltage is reduced. As shown in Fig. 2.70(c), the enhanced kink effects at an increased drain voltage make the overall threshold voltage curve shift downward, which is offset by

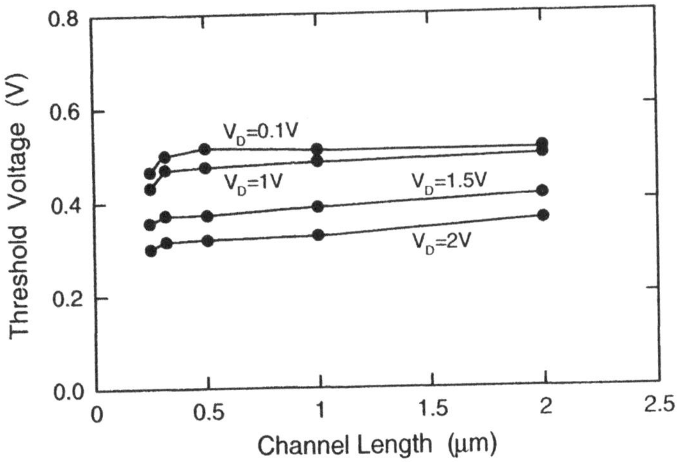

Fig. 2.68 Threshold voltage versus channel length of an FD SOI NMOS device with a front gate oxide of 50 Å, a thin film of 500 Å and a buried oxide of 900 Å. (Adapted from Tsuchiya et al. [67].)

the reduced kink effects due to DIBL [67]. Therefore, from the overall result, at an increased drain voltage, its threshold voltage is not sensitive to the channel length.

2.7.4 Latch/GIDL Behavior

When the drain voltage applied to an SOI NMOS device exceeds a certain value, its drain current increases substantially, which implies that the device cannot be turned off regardless of the gate voltage, as shown in Fig. 2.71(a) with the similar-to-latch behavior [68]. At a large drain voltage, the parasitic bipolar device at the bottom of the thin film is turned on and the current conduction is via the bottom of the thin film– the surface channel is not dominant any more. To resolve this problem, various techniques to lower the kink effects mentioned in this section can be used. For example, as shown in Fig. 2.71(b), by the adoption of the germanium implant into the source/drain region, the latch voltage has been raised at least 1 V.

There is another unique phenomenon for the SOI NMOS devices. When the device is turned off, its leakage current becomes larger and larger when the gate voltage becomes more negative. As shown in Fig. 2.72(a), when the gate voltage is less than -0.5 V, a more negative gate voltage leads to a larger drain current. In addition, this phenomenon is more noticeable at a higher drain voltage. It can be understood using the schematic as shown in Fig. 2.72(b). Generally speaking, for most SOI NMOS devices, there is overlap between the front gate and the n^+ drain, which results in a very thin depletion region with a very large electric field. When the difference between the drain voltage and the gate voltage ($V_{DS} - V_{GS}$) is sufficiently large, the electric field in this depletion region is high enough to make the

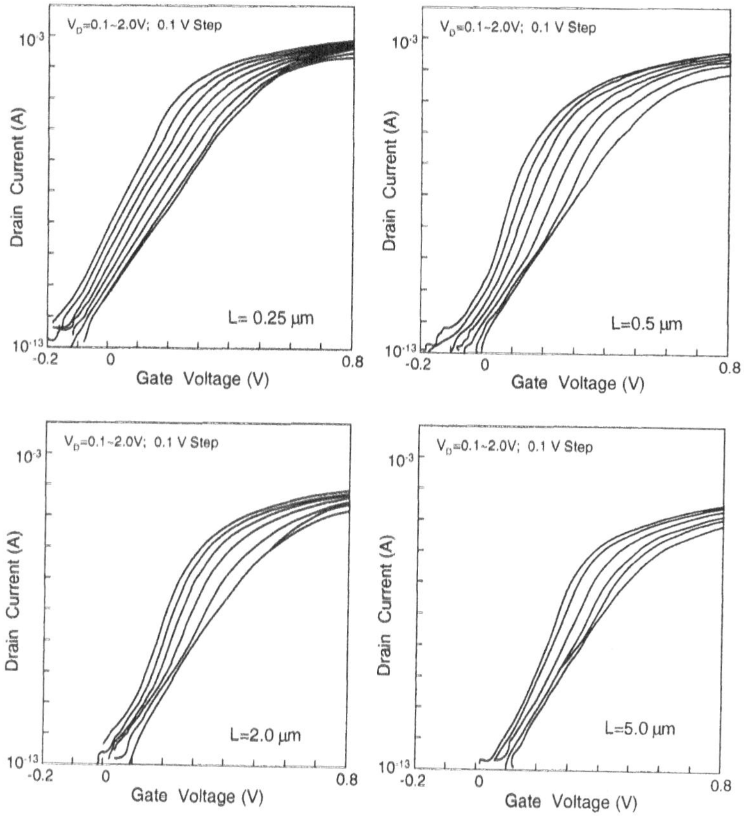

Fig. 2.69 Subthreshold characteristics of an FD SOI NMOS device with various channel lengths. (From Tsuchiya et al. [67]. ©1998 IEEE.)

electrons tunnel through the bandgap to generate electron/hole pairs flowing to the drain. As a result, the drain current rises. Some holes may flow toward the bottom of the thin film to turn on the parasitic bipolar device. Thus, the GIDL current is further amplified. As shown in Fig. 2.72(c), with a smaller channel length, the base width of the parasitic bipolar device becomes smaller. Therefore, the current gain of the parasitic bipolar device becomes better and GIDL becomes more noticeable. To reduce the GIDL phenomenon, the most straightforward method is to use the LDD structure such that the overlap is between the front gate and the n^- region. Under this situation, the electric field between the front gate and the drain is smaller and GIDL can be reduced. As shown in this figure, by adopting the LDD structure, GIDL has been inhibited.

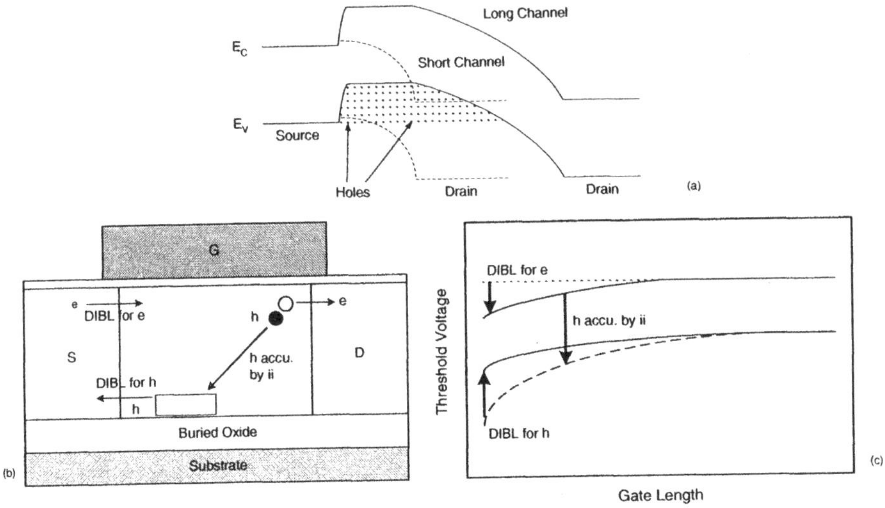

Fig. 2.70 (a) Energy band diagram near the bottom of the body of the FD SOI NMOS devices with a short channel and a long channel. (b) Three mechanisms of short channel effects in an FD SOI NMOS device. (c) Effects of the three mechanisms on the threshold voltage versus channel length curves. (From Tsuchiya et al. [67]. ©1998 IEEE.)

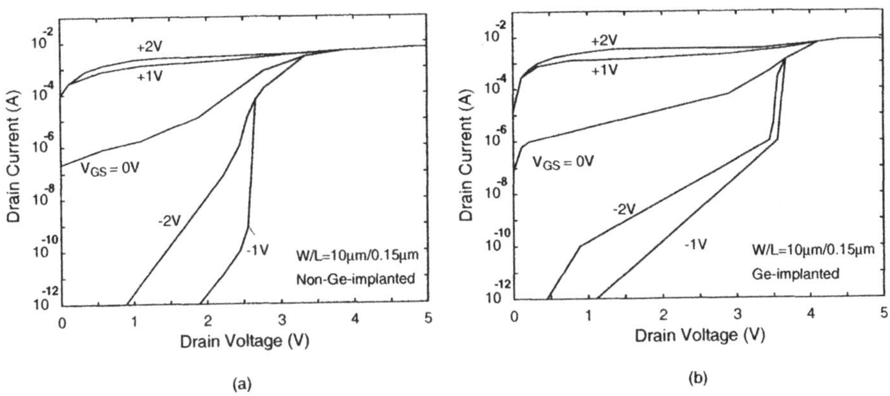

Fig. 2.71 Drain current versus drain voltage of an SOI NMOS device (a) with and (b) without a germanium implant and with a channel length of 0.15 μm. (From Yoshimi et al. [68]. ©1997 IEEE.)

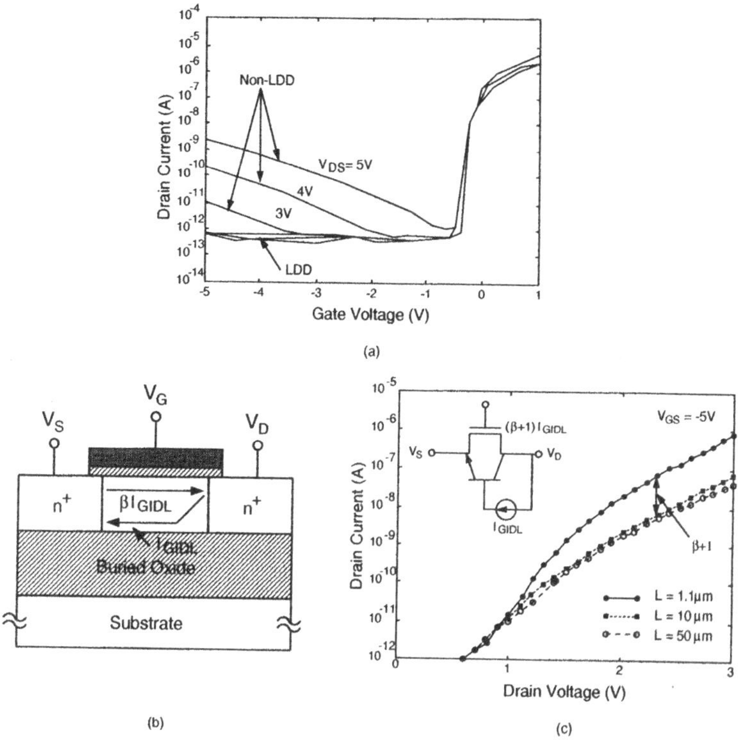

Fig. 2.72 (a) Gate induced drain leakage (GIDL) currents of non-LDD and LDD SOI NMOS devices. (b) Schematic of an SOI NMOS device current flow in a GIDL current model. (c) Off-state leakage current of long and short channel SOI NMOS devices. (From Chen et al. [69]. ©1992 IEEE.)

2.8 IMPACT IONIZATION

As for bulk devices, impact ionization is important in determining the performance of the SOI CMOS devices. In this section, impact ionization of PD SOI MOS devices is analyzed.

2.8.1 Basic Analysis

For PD SOI NMOS devices biased at a sufficiently large drain voltage, kink effects due to the floating body are unavoidable as described in Section 2.7. When biased at a large drain voltage, both PD and FD devices may demonstrate a rapid increase of the drain current when the drain voltage is increased. As shown in Fig. 2.73, the trend on the rapid increase of the drain current at an increased drain voltage is especially noticeable when the gate voltage is lowered [70]. This is due to the generated electron/hole pairs produced by the impact ionization of the traveling electrons under a high electric field. The newly generated electrons collide with the lattice to produce more electron/hole

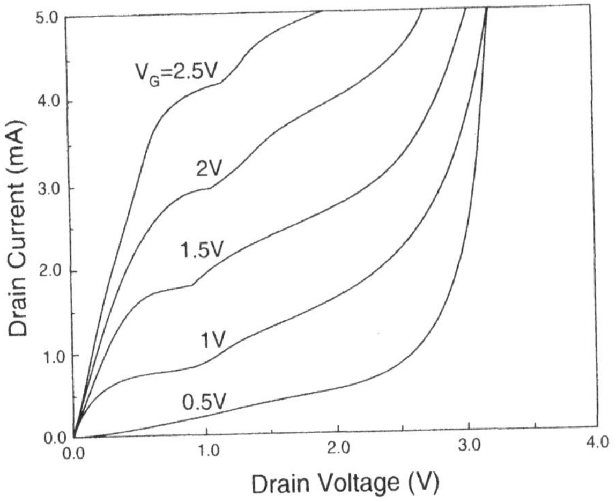

Fig. 2.73 Drain current versus drain voltage of the PD SOI NMOS device with a front gate oxide of 50 Å, a thin film of 750 Å, a buried oxide of 3600 Å, a channel width of 10 μm, and a channel length of 0.2 μm with its body floating. (Adapted from Suh & Fossum [70].)

pairs. The generation process repeats itself. As a result, the drain current is amplified quickly. When the gate voltage is lowered, the saturation voltage ($V_{DSAT} = V_G - V_T$) also drops, which leads to an expanded post-saturation region– the region with its potential between the drain voltage and the saturation voltage V_{DSAT} is enlarged. Therefore, impact ionization is even more serious and the rapid increase of the drain current is more noticeable. Furthermore, the parasitic lateral bipolar device in the thin film further amplifies the impact ionization effects of the SOI NMOS devices.

Figure 2.74 shows the drain current versus the drain voltage of the PD SOI NMOS device with a front gate oxide of 200 Å, a thin film of 3000 Å, a buried oxide of 4000 Å, a channel width of 2.4 μm, a channel length of 0.1 μm, and with (a) body tied and (b) body floating [70]. As shown in this figure, the connection of the body contact has a strong impact on the impact ionization behavior at a high drain voltage. With the body tied, the impact ionization behavior at a high drain voltage is reduced substantially as compared to the floating body case since with the body tied the parasitic bipolar device is not triggered to turn on any more.

Figure 2.75 shows the mechanism of the parasitic bipolar device action in a PD SOI NMOS device [71]. As shown in this figure, at a high drain voltage, the traveling electron enters the post-saturation region from the source and the channel. Due to impact ionization under a high electric field, a chain reaction of producing more electron/hole pairs has occurred. The generated electrons are attracted to move toward the drain. As a result, the drain current has been amplified. The generated holes move toward the bottom of the thin film and accumulate at locations close to the source/body boundary. As a result, the body potential is raised to turn on the parasitic bipolar device. At this time, a large amount of electrons are injected from the source

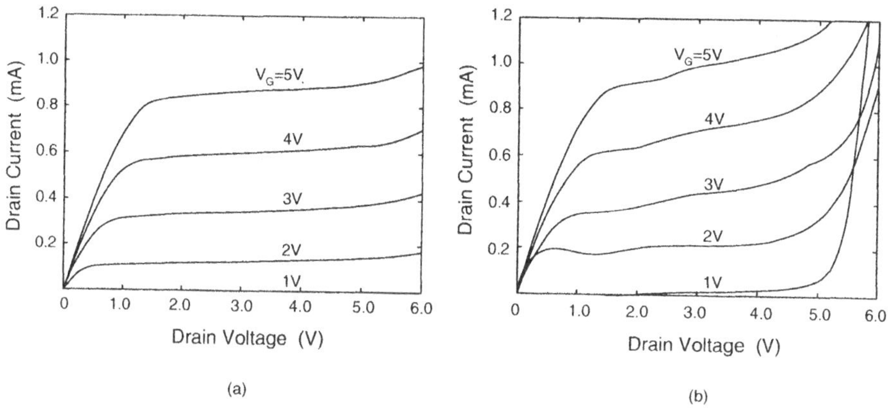

Fig. 2.74 Drain current versus drain voltage of the PD SOI NMOS device with a front gate oxide of 200 Å, a thin film of 3000 Å, a buried oxide of 4000 Å, a channel width of 2.4 μm, a channel length of 0.16 μm, and with (a) body tied and (b) body floating. (Adapted from Suh & Fossum [70].)

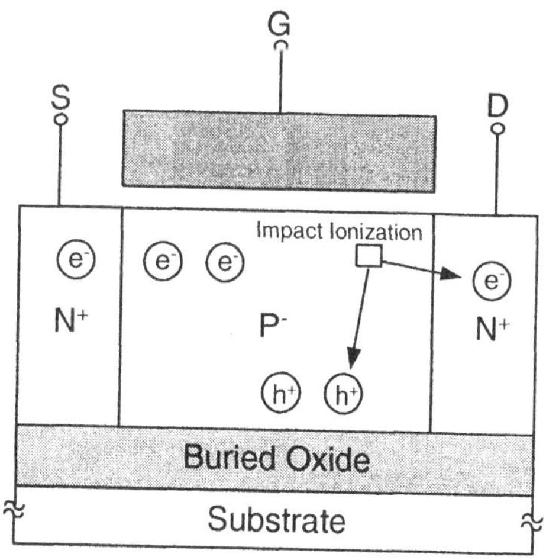

Fig. 2.75 Mechanism of the parasitic bipolar device action in a PD SOI NMOS device [71].

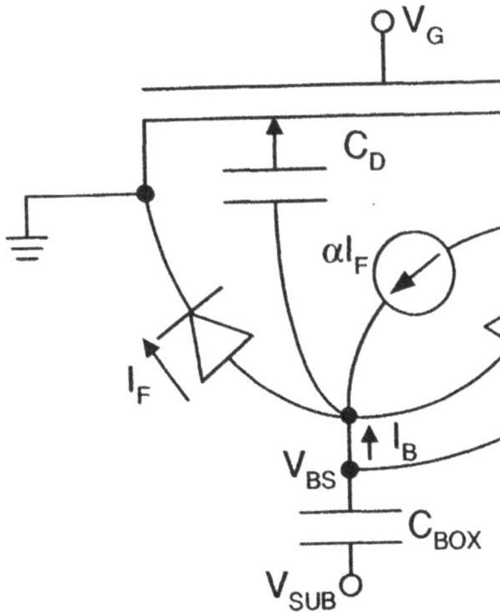

Fig. 2.76 Equivalent circuit of a PD SOI NMOS device including the impact ionization current and parasitic bipolar device effects. (Adapted from Adan et al. [72].)

(emitter) to the body (base) to reach drain (collector). On the other hand, the hole current functions as the role of the base current. Under this situation, the impact ionization occurring at the surface of the thin film has been amplified by the parasitic bipolar device at the bottom to cause a large impact on the final drain current.

Figure 2.76 shows the equivalent circuit of a PD SOI NMOS device including the impact ionization current and the parasitic bipolar device effects [72]. As shown in this figure, when impact ionization occurs, the hole current is charging the depletion capacitance (C_D) and the equivalent substrate capacitance (C_{BOX}) to raise the body potential (V_{BS}). As a result, the body/source junction is forward biased to trigger the parasitic bipolar device. The generated hole current from impact ionization becomes the triggering base current of the parasitic bipolar device. Via amplification by the parasitic bipolar device, a large amount of collector current flows out of the drain to cause a rapid increase of the drain current.

2.8.2 Body Current

From the above analysis, the rapid increase of the drain current for an SOI NMOS device biased at a high drain voltage is due to impact ionization generated electron/hole currents amplified by the parasitic bipolar device. Impact ionization generated hole current can be monitored via the body current (I_{sub}). As shown in Fig. 2.77, the body current is strongly correlated to the drain current. As shown in Fig. 2.77(a) [73], the body current reaches its maximum value at $V_G = V_T$. This result is due to the fact when the gate voltage is smaller than the threshold voltage ($V_G < V_T$), a small gate voltage leads to a small surface current, and hence a small impact ionization generated hole current. When the gate voltage is greater than the threshold voltage ($V_G > V_T$), the surface current is already sufficiently large. With a higher gate

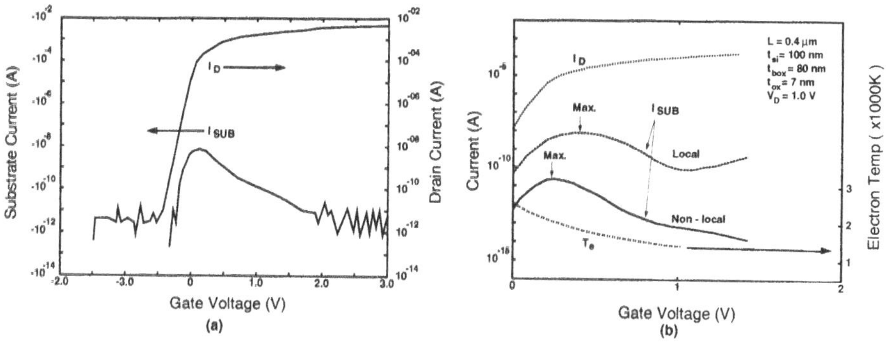

Fig. 2.77 Drain current and substrate current versus gate voltage of a PD SOI NMOS device with a front gate oxide of 70 Å, a thin film of 1000 Å, a buried oxide of 800 Å, and channel length of 0.4 μm. (a) Experimental result for the device biased at a drain voltage of 2.5 V and (b) 2D simulation results for the device biased at a drain voltage of 1 V. (Adapted from Omura & Izumi [73].)

voltage, the post-saturation region with a high electric field becomes smaller. Impact ionization effect is shrunk, and hence the hole current becomes smaller.

With its gate voltage biased at its threshold voltage its body current is at its maximum value, which is similar to the drain voltage equal to $V_G = V_T$ at the onset of kink effect. Impact ionization is closely related to the internal electric field. Figure 2.77(b) shows the body current based on the local and nonlocal effect model results. Local effect model means that the impact ionization of electrons is referred to the magnitude of the local electric field. Nonlocal effect model means in addition to the magnitude of the local electric field, the electric field gradient has also been considered, which is especially important for using the electron temperature to predict impact ionization. As shown in this figure, the local effect model overestimates the body current. For devices with a short channel, the nonlocal effect should be considered since the electric field in the channel may vary a lot. The acquired energy of the accelerated electrons in the channel with the diverse change in the electric field is smaller than that of the electrons traveling under a constant electric field. Since the traveling electrons are accelerated quickly, the electron energy may not be fully transferred to the lattice to produce impact ionization. Therefore, the actual effect when considering the nonlocal electric field effect is much smaller as compared to the case when the local effect model is used. This theory is similar to the one discussed in the velocity overshoot described before.

Figure 2.78 shows the drain current versus the drain voltage of the SOI NMOS device with a front gate oxide of 85 Å, a thin film of 420 Å doped with a density of $5 \times 10^{17}\text{cm}^{-3}$, and a buried oxide of 3000 Å, biased at a gate voltage of 2 V based on the conventional drift-diffusion (local) model and the energy balance (nonlocal) model [74]. As shown in the figure, the predicted drain current based on the nonlocal effect model (considering electron temperature) is quite different from that based on the conventional drift-diffusion model. As seen in this figure, at the high drain-voltage

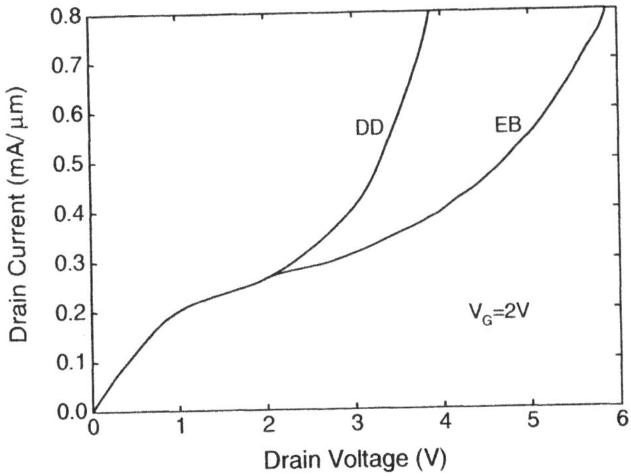

Fig. 2.78 Drain current versus drain voltage of the SOI NMOS device with a front gate oxide of 85 Å, a thin film of 420 Å doped with a density of $5 \times 10^{17}cm^{-3}$, and a buried oxide of 3000 Å, biased at a gate voltage of 2 V based on the conventional drift-diffusion (local) model and the energy balance (nonlocal) model. (Adapted from Apanovich [74].)

region with impact ionization, the predicted drain current based on the local effect model is much higher than that based on the nonlocal effect model.

2.8.3 Monitoring Techniques

Until now, the body current of an SOI NMOS device has been analyzed in terms of the gate voltage. The body current is also drain voltage dependent. Figure 2.79 shows the ratio of the body current to the drain current (I_{sub}/I_D) versus the inverse of the difference between the drain voltage and the saturation voltage [$1/(V_D - V_{DSAT})$] of the SOI NMOS device with a front gate oxide of 90 Å and the bulk NMOS device with a front gate oxide of 86 Å, with a channel length of 1.4 μm and a channel width of 9.5 μm [75]. As shown in the figure, in terms of log scale for the y axis, for both bulk and SOI devices, a linear relationship between $\ln(\frac{I_{SUB}}{I_D})$ and $1/(V_D - V_{DSAT})$ can be seen. In spite of differences in some parameters, the mechanism of impact ionization for both bulk and SOI devices is the same. Therefore, the analysis of impact ionization done for the bulk devices can be used for SOI devices.

Figure 2.80 shows the hole current due to impact ionization versus the drain voltage of the FD SOI NMOS device with a channel length of 1 μm, a thin film doping density of $10^{16}cm^{-3}$, and various thin-film thicknesses, biased at a gate voltage of threshold voltage ($V_G - V_T$). As shown in this figure, with a thinner thin film, impact ionization is more serious and the hole current is larger. The reasoning is as follows. When the thin film becomes thinner, the electric field in the post-saturation region increases much faster to a higher value, which leads to a stronger impact ionization effect.

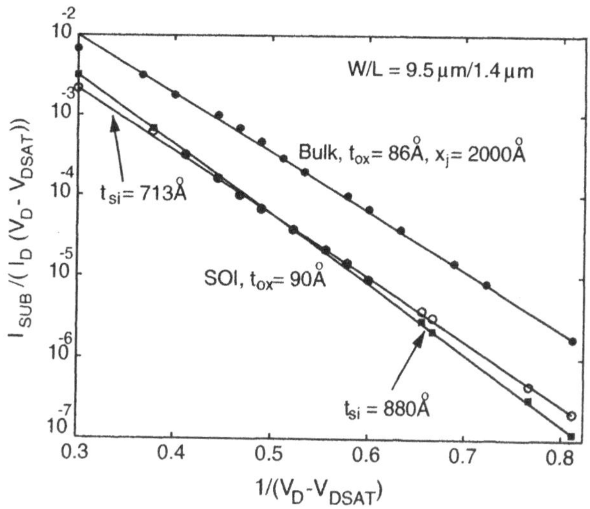

Fig. 2.79 Ratio of the body current to the drain current (I_{sub}/I_D) versus the inverse of the difference between the drain voltage and the saturation voltage [$1/(V_D - V_{DSAT})$] of the SOI NMOS device with a front gate oxide of 90 Å and the bulk NMOS device with a front gate oxide of 86 Å, with a channel length of 1.4 μm, and a channel width of 9.5 μm. (Adapted from Ma et al. [75].)

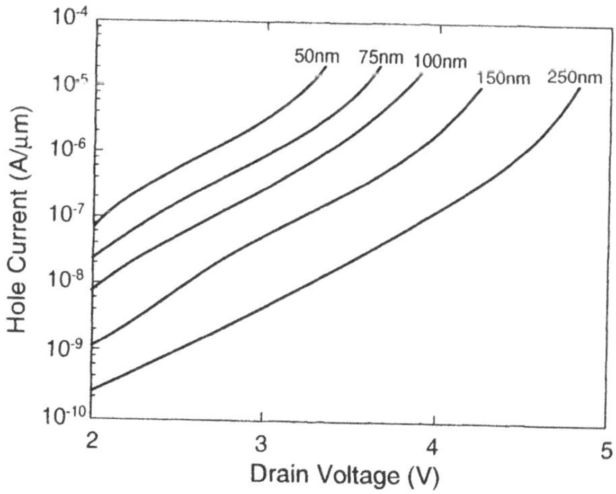

Fig. 2.80 Hole current due to impact ionization versus drain voltage of the FD SOI NMOS device with a channel length of 1 μm, a thin film doping density of 10^{16}cm^{-3}, and various thin-film thicknesses, biased at a gate voltage of the threshold voltage ($V_G = V_T$). (Adapted from Ver Ploeg [76].)

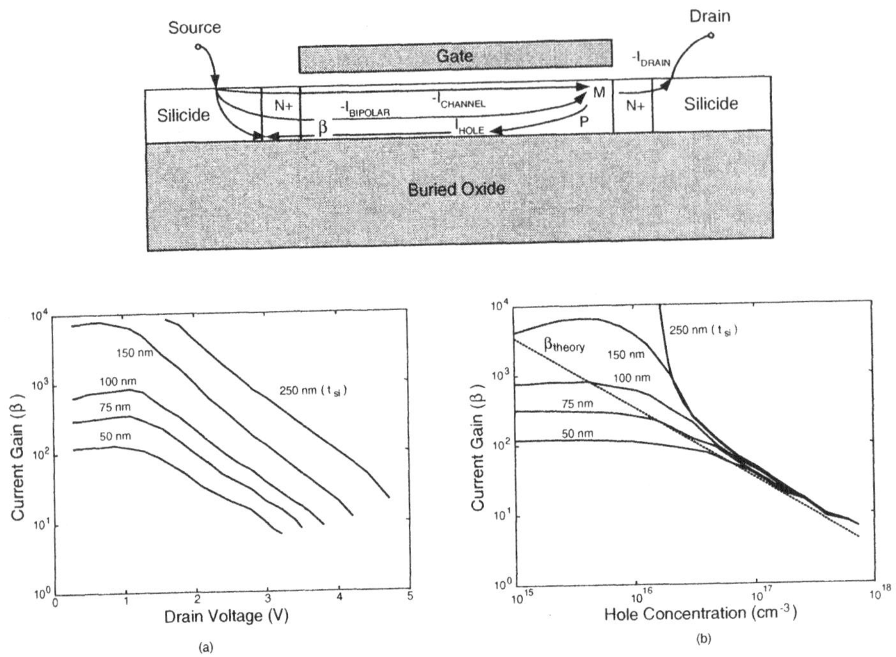

Fig. 2.81 Current gain of the parasitic bipolar device versus (a) drain voltage and (b) hole concentration of the FD SOI NMOS device with a channel length of 1 μm, a thin-film doping density of $10^{16}cm^{-3}$, and various thin-film thicknesses, biased at the gate voltage of the threshold voltage. (From Ver Ploeg [76]. ©1994 IEEE.)

As mentioned before, the parasitic bipolar device provides an amplification of impact ionization effects in an SOI NMOS device. The function of the parasitic bipolar device is discussed further. Figure 2.81 shows the current gain of the parasitic bipolar device versus (a) the drain voltage and (b) the hole concentration of the FD SOI NMOS device with a channel length of 1 μm, a thin-film doping density of $10^{16}cm^{-3}$, and various thin-film thicknesses, biased at the gate voltage of the threshold voltage [76]. As shown in Fig. 2.81(a), in the higher drain voltage regime, the current gain of the parasitic bipolar device drops when the drain voltage is increased. This result can be reasoned using Fig. 2.81(b). When the drain voltage is increased, impact ionization gets stronger and holes accumulate at the bottom of the thin film–the hole concentration increases. Thus, the current gain of the parasitic bipolar device is decreased. As shown in this figure, in the high hole concentration regime, the current gain of the parasitic bipolar device is insensitive to the variation in the thin-film thickness. Only in the low hole concentration regime does its current gain drop along with the decrease in the thin-film thickness.

Generally speaking, to improve the rapid increase in the drain current due to impact ionization in the SOI NMOS devices, the best way is via the adoption of the LDD structure. Figure 2.82 shows (a) the fabrication steps of the SOI MOS device with the gate-overlapped LDD structure, (b) the lateral electric field in the n^- region of

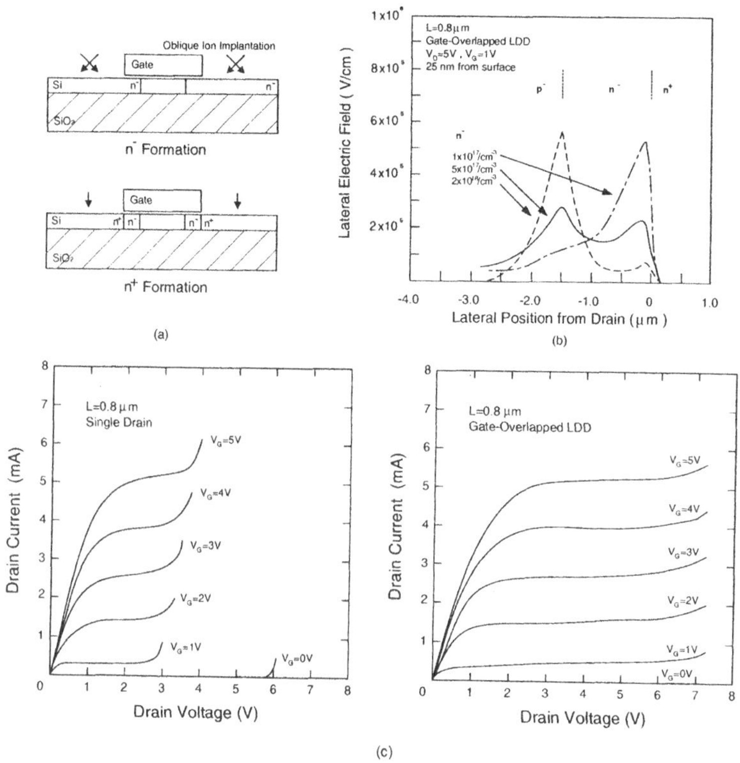

Fig. 2.82 (a) Fabrication steps of the SOI MOS device with gate-overlapped LDD structure. (b) Lateral electric field in the n^- region of the SOI MOS device with gate-overlapped LDD structure. (c) Drain current characteristics of the SOI NMOS device with a front gate oxide of 150 Å, a thin film of 800 Å, a buried oxide of 4600 Å, a channel width of 10 μm and a channel length of 0.8 μm, and the SOI NMOS device with and without gate overlapped LDD structures. The length of the n^- region is 0.15 μm and the doping density of the n^- region is $5 \times 10^{17} cm^{-3}$. (From Yamaguchi et al. [77]. ©1994 IEEE.)

the SOI MOS device with gate-overlapped LDD structure, and (c) the drain current characteristics of the SOI NMOS device with a front gate oxide of 150 Å, a thin film of 800 Å, a buried oxide of 4600 Å, a channel width of 10 μm and a channel length of 0.8 μm, and the SOI NMOS device with and without gate overlapped LDD structures. The length of the n^- region is 0.15μm and the doping density of the n^- region is $5 \times 10^{17} cm^{-3}$ [77]. As shown in Fig. 2.82, the SOI NMOS device with gate-overlapped LDD structure can be fabricated using the oblique ion implantation technique to generate an n^- region under the front gate between the n^+ source/drain and the body [77]. By properly selecting the doping density of the n^- region, the maximum electric field near the drain of the SOI NMOS device biased at a high drain voltage can be reduced substantially to lower impact ionization as shown in Fig. 2.82(b). As shown in Fig. 2.82(c), using the gate overlapped LDD structure,

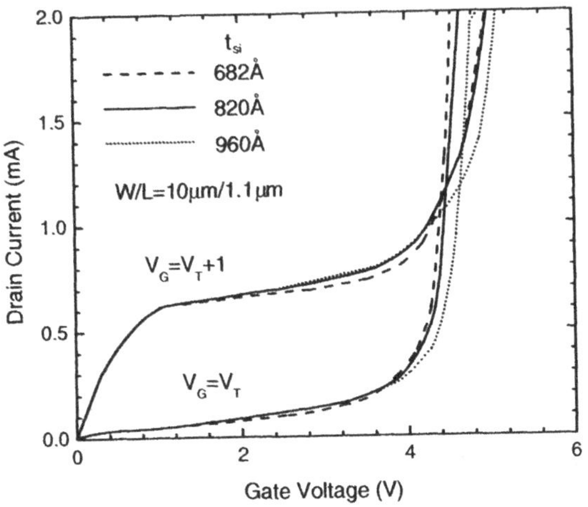

Fig. 2.83 Drain current characteristics of an SOI NMOS device with a channel width of 10 μm, a channel length of 1.1 μm, and with various thicknesses of the thin-film doped with a p-type density of 10^{17}cm^{-3}. (Adapted from Chen et al. [78].)

the rapid increase in the drain current due to impact ionization has been reduced effectively as compared to the SOI device using the conventional drain structure.

2.9 BREAKDOWN

As described in Section 2.8, when the drain voltage is sufficiently large, due to impact ionization effects amplified by the parasitic bipolar device, there is a rapid increase of the drain current. If the drain current increases further, breakdown of the device may occur. In this section, breakdown of the SOI devices is described.

2.9.1 Structure Dependence

Figure 2.83 shows the drain current characteristics of an SOI NMOS device with a channel width of 10 μm, a channel length of 1.1 μm, and with various thicknesses of the thin-film doped with a p-type density of 10^{17}cm^{-3} [78]. As shown in Fig. 2.83, the SOI NMOS device already shows similar-to-short-circuit behavior–breakdown at a high drain voltage, where the surface channel current triggers impact ionization. The hole current generated by impact ionization is amplified by the parasitic bipolar device to produce an even higher drain current. Then the elevated drain current produces a higher hole current via impact ionization. Thus a positive feedback is formed to cause the rapid increase of the drain current and breakdown occurs. In the following, the breakdown voltage of the SOI devices is analyzed in terms of the gate voltage and the device structure.

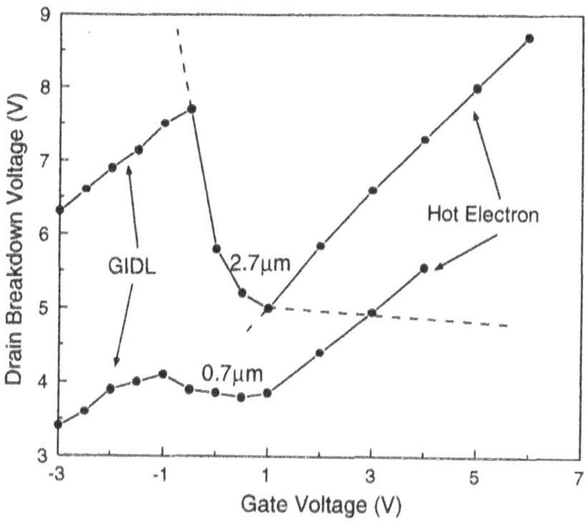

Fig. 2.84 Breakdown voltage versus gate voltage of an SOI NMOS device with a thin film doped with a p-type doping density of $10^{17}cm^{-3}$, and channel lengths of 2.7 and 0.7 μm. (Adapted from Chen et al. [78].)

Figure 2.84 shows the breakdown voltage versus the gate voltage of an SOI NMOS device with a thin film doped with a p-type doping density of $10^{17}cm^{-3}$, and channel lengths of 2.7 and 0.7 μm [78]. As shown in this figure, when the gate voltage is greater than the threshold voltage ($V_G > V_T$), a higher gate voltage leads to a larger breakdown voltage, since at a higher gate voltage, the saturation voltage (V_{DSAT}) is larger. The post-saturation region, which has a high electric field because of the location of its potential equal to the saturation voltage to the drain, is shrunk and hence impact ionization is reduced. Therefore, the breakdown voltage is increased. For the gate voltage smaller than the threshold voltage ($V_G < V_T$), the breakdown voltage drops along with the increase in the gate voltage because at this time the impact ionization is determined by the channel current. At a larger gate voltage, the channel current, which is made of diffusion current, is larger, which implies that a larger amount of electrons can be used to produce impact ionization. Therefore, the breakdown voltage drops. For the gate voltage equal to the threshold voltage, there is a local minimum for the breakdown voltage, which is correlated to the case for the body current discussed in Section 2.8. If the gate voltage is sufficiently negative, the breakdown voltage falls along with the decreased gate voltage because the channel current in this regime is caused by the GIDL current. A more negative gate voltage makes a larger potential difference between the gate and the drain, which raises the electric field in the depletion region under the gate/drain overlap. Thus the increased tunneling electron/hole pairs are multiplied further by the impact ionization and the parasitic bipolar device. As a result, the drain current increases sharply and the breakdown voltage is lowered. As shown in this figure, with a smaller channel length, the breakdown voltage is degraded due to an increased internal electric field and an enhanced current gain of the parasitic bipolar device.

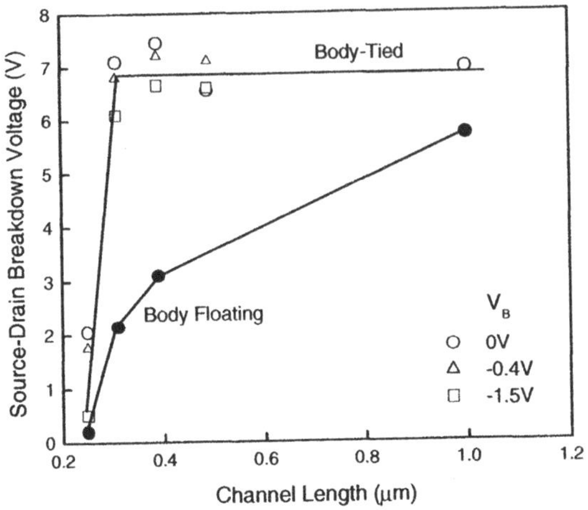

Fig. 2.85 Breakdown voltage versus channel length of the SOI NMOS device with a channel width of 10 μm, biased with body floating and body tied to various voltages. (Adapted from Maeda et al. [79].)

Further analysis of the breakdown voltage of an SOI NMOS device with respect to the floating body effect is shown in Fig. 2.85, which shows the breakdown voltage versus the channel length of the SOI NMOS device with a channel width of 10μm, biased with body floating and body tied to various voltages [79]. With its body tied (the internal body is connected to a fixed biasing voltage via the body contact), its breakdown voltage is degraded with the decreased channel length, which is noticeable for a channel length of $L < 0.3\mu m$. With body floating, its breakdown voltage is much smaller than that with the body tied because with the body tied, the body current is flowing out via the body contact and the holes do not accumulate in the internal body. Therefore, the parasitic bipolar device is not triggered and the breakdown voltage is better as compared to the body-floating case.

Figure 2.86 shows the influence of the thin-film thickness in the breakdown voltage of a PD SOI NMOS device with a front gate oxide of 60 Å and a buried oxide of 2000 Å. From this figure, when the channel length is large, the breakdown voltage of the device is not sensitive to the thickness of the thin film. When the channel length becomes smaller, since the device with a thinner thin film is more easily susceptible to the influence from the channel length, breakdown occurs earlier. On the contrary, with a thicker thin film, the device shows a higher breakdown voltage. When the thin film is thinner, the post-saturation region in the device is more influenced by the front gate. Therefore, the internal electric field is increased much faster to cause more impact ionization and the breakdown voltage is lowered. As shown in this figure, the breakdown voltage of the bulk device is much better than that of the SOI. Since the substrate of the bulk device is connected to the ground (for the NMOS device), the amplification of the parasitic bipolar device has been inhibited.

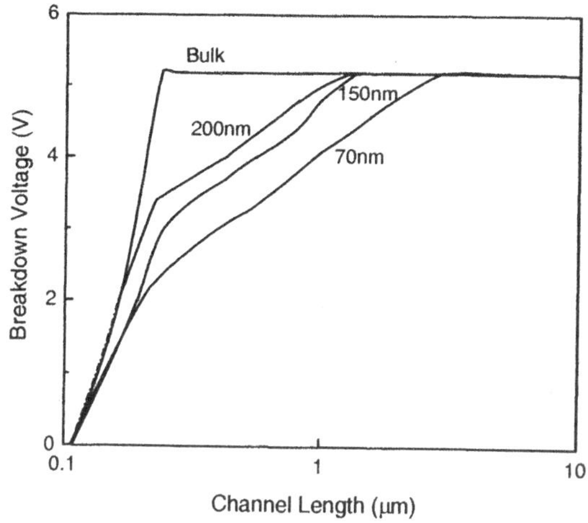

Fig. 2.86 Breakdown voltage versus channel length of a PD SOI NMOS device with a front gate oxide of 60 Å, a buried oxide of 2000 Å, and various thin-film thicknesses. (Adapted from Mistry et al. [80].)

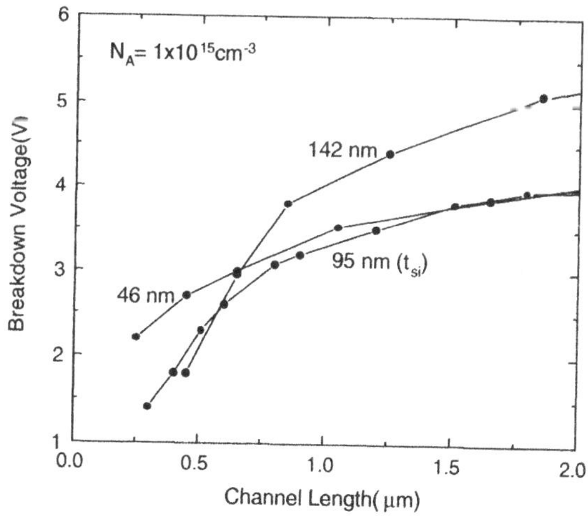

Fig. 2.87 Breakdown voltage versus channel length of the FD SOI NMOS device with a thin-film doped with a p-type doping density of 10^{15}cm^{-3} and various thin-film thicknesses. (Adapted from Kistler & Woo [81].)

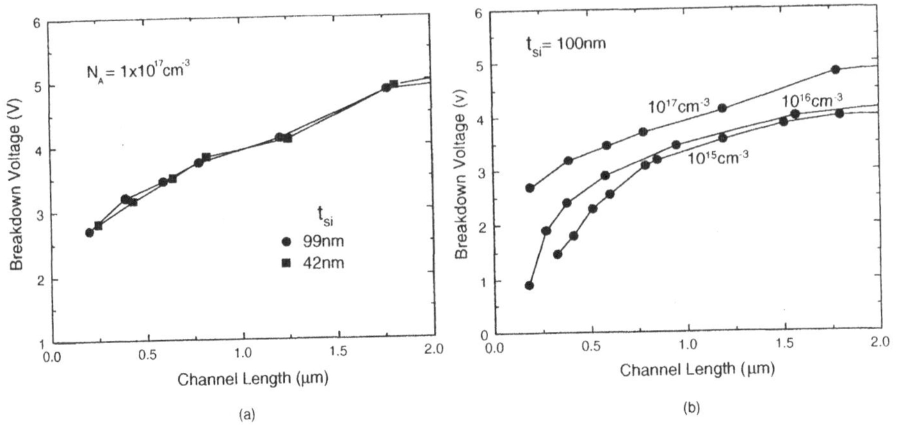

Fig. 2.88 Breakdown voltage versus channel length of the FD SOI NMOS device (a) with a thin film doped with a p-type doping density of $10^{17}\mathrm{cm}^{-3}$ and thin-film thicknesses of 990 and 420 Å and (b) with a 1000 Å thin film doped with p-type doping densities of 10^{15}, 10^{16}, and $10^{17}\mathrm{cm}^{-3}$. (Adapted from Kistler & Woo [81].)

Fig. 2.87 shows the breakdown voltage versus the channel length of the FD SOI NMOS device with a thin-film doped with a p-type doping density of $10^{15}\mathrm{cm}^{-3}$ and various thin-film thicknesses[81]. With a channel length of 2 μm, the breakdown voltage indeed becomes smaller with the decreased thin-film thickness. When the thin-film thickness is < 1000 Å, the breakdown voltage is not degraded with the decreased thin-film thickness. When the thin-film thickness is < 1000 Å, a thinner thin film leads to a worse parasitic bipolar device, which offsets the enhanced impact ionization. Thus in the regime with the thin-film thickness < 1000 Å for a channel length > 0.5 μm, the breakdown voltage is insensitive to the variation of the channel length. When the channel length is < 0.5 μm, the breakdown voltage is sensitive to the variation of the thin-film thickness for the thin film < 1000 Å. A thicker thin film leads to a smaller breakdown voltage due to the dominance of the punchthrough effect under the situation with a very small channel length. When the thin film becomes thinner, the controllability of the front gate becomes stronger, which inhibits the influence of the drain in the lateral direction. Thus, punchthrough becomes less and leakage current becomes smaller– the breakdown voltage is improved.

Figure 2.88 shows the breakdown voltage versus the channel length of the FD SOI NMOS device (a) with a thin film doped with a p-type doping density of $10^{17}\mathrm{cm}^{-3}$ and thin-film thicknesses of 990 and 420 Å and (b) with a 1000 Å thin film doped with p-type densities of 10^{15}, 10^{16}, and $10^{17}\mathrm{cm}^{-3}$ [81]. In contrast to Fig. 2.87, Fig. 2.88(a), with a higher doping density of $10^{17}\mathrm{cm}^{-3}$, a thin-film thickness < 1000 Å, and a channel length < 0.5 μm, has a breakdown voltage that is not as sensitive to the variation of the thin-film thickness any more. With a high thin-film doping density, punchthrough of the FD device has been effectively inhibited. As shown in Fig. 2.88(b), with a thin-film thickness of 1000 Å, a higher thin-film doping density leads to a higher breakdown voltage. In addition, the degradation of the

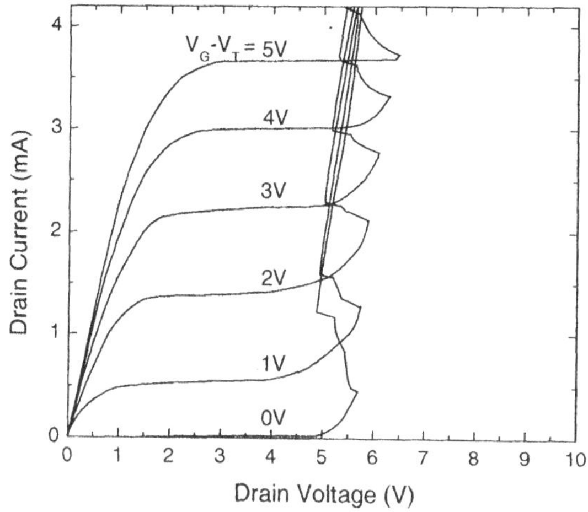

Fig. 2.89 Drain current characteristics of the FD SOI NMOS device with a front gate oxide of 120 Å, a thin film of 1550 Å doped with a p-type doping density of $6 \times 10^{15}\text{cm}^{-3}$, a buried oxide of 3700 Å, a channel width of 10 μm, and a channel length of 0.05 μm, showing bipolar-induced breakdown. (Adapted from Ver Ploeg et al. [82].)

breakdown voltage with respect to the shrunk channel length has been improved due to the fact that DIBL and punchthrough have been inhibited by the high thin-film doping density.

2.9.2 Bipolar Induced Effects

Figure 2.89 shows the drain current characteristics of the FD SOI NMOS device with a front gate oxide of 120 Å, a thin film of 1550 Å doped with a p-type doping density of $6 \times 10^{15}\text{cm}^{-3}$, a buried oxide of 3700 Å, a channel width of 10 μm, and a channel length of 0.05 μm [82]. As shown in this figure, there exists bipolar-induced snapback behavior due to the breakdown of the parasitic bipolar device. This result can be reasoned as follows. When the drain voltage exceeds 4.5 V, impact ionization starts to exist in the surface channel, which results in a gradual increase in the drain current. At this time, the hole current generated by impact ionization starts flowing to the body as the base current triggering the parasitic bipolar device. At the bottom of the thin film electrons are injected from the source (emitter) to the drain (collector). Since the base current is low, the current gain of the parasitic bipolar device is not high. If the drain voltage increases further to 5.5 V, the increased hole current generated by impact ionization of the surface current provides an increased base current to produce a much larger collector current. This collector current also results in impact ionization near the drain (collector). A portion of the hole current generated by bipolar's impact ionization is also used to supply its own base current. Via positive feedback, a large amount of current flows through the bottom of the thin film. Thus, the parasitic bipolar device is also close to breakdown. If the drain voltage

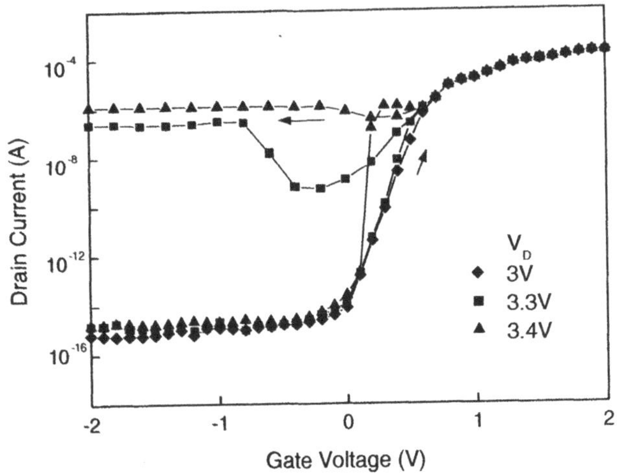

Fig. 2.90 Drain current versus gate voltage of the SOI NMOS device with a front gate oxide of 150 Å, a 4000 Å thin-film doped with a p-type doping density of $4 \times 10^{16}\text{cm}^{-3}$, and a channel length of 0.8 μm, biased at various drain voltages. The gate voltage is first changing in the positive direction and then in the negative direction. (Adapted from Duan et al. [83].)

is increased slightly further, the increased base current caused by impact ionization leads to an enhanced current gain. With this enhanced current gain, even the drain voltage becomes slightly smaller, the drain current becomes even higher. Even at a smaller drain voltage, the drain current still rises–the snapback phenomenon, which continues until the current gain becomes saturated. After that, the drain current can rise only by increasing the drain voltage– the drain current curve is back to a positive slope. Under this situation, it is compatible to the breakdown of the bipolar device with the base open (BV_{CEO}). At this time, the drain current is dominated by the bipolar device with the base current supplied by itself. No matter how negative a bias is imposed on the front gate, the device cannot be turned off– single-transistor latch phenomenon, that should be avoided. As shown in the figure, for this drain voltage > 5.5 V ($V_D > 5.5$ V), the device has only one stable state – latch. For the drain voltage between 5 and 5.5 V (5 V $< V_D <$ 5.5 V), the device is in one of two states– latch or normal depending on the operation procedure.

The single-transistor latch phenomenon with two stable states is further explained for the SOI NMOS device with a front gate oxide of 150 Å, a 4000 Å thin-film doped with a p-type doping density of $4 \times 10^{16}\text{cm}^{-3}$, and a channel length of 0.8 μm, biased at the drain voltages of 3, 3.3, and 3.4V, as shown in Fig. 2.90 [83]. The gate voltage is first changing in the positive direction and then in the negative direction. As shown in this figure, at the drain voltage > 3.3 V ($V_D >$ 3.3 V), if initially the gate voltage is very negative, the device is in the off state. When the gate voltage gradually becomes positive, the device gradually turns on its surface channel. The hole current generated by impact ionization, which is caused by the surface channel, triggers the parasitic bipolar device. After the turn-on of the parasitic bipolar device,

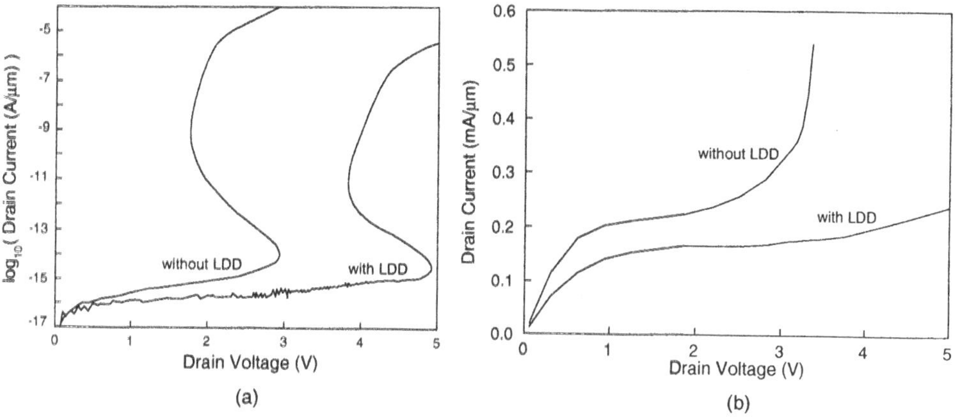

Fig. 2.91 Drain current characteristics of the FD SOI NMOS device with and without the LDD structure (a) at off-state with the gate voltage of -3 V and (b) at on-state with the gate voltage of $V_T + 2$ V. (Adapted from Choi & Fossum [84].)

due to impact ionization in the bipolar device, breakdown of the bipolar device occurs. At this time, if the gate voltage changes to negative to turn off the surface channel such that the supply of the hole current can be terminated, the parasitic bipolar device cannot be turned off since the needed base current is supplied by the impact ionization current in the base-collector junction. The device still has a large conducting current even with a very negative gate voltage– the latch state. Getting out of the latch state can be reached via lowering the drain voltage since the front gate has lost its control over the device during the latch state.

2.9.3 LDD

The breakdown voltage of an SOI NMOS device can be improved effectively by the adoption of the LDD structure. Figure 2.91 shows the drain current characteristics of the FD SOI NMOS device with and without the LDD structure (a) at off-state with the gate voltage of -3 V and (b) at on-state with the gate voltage of $V_T + 2$ V [84]. As shown in the figure, with the LDD structure, for both breakdown and snapback (single-transistor latch), the breakdown voltage has been improved for two reasons. First, at the drain end, the lightly doped region effectively lowers the maximum electric field to lessen impact ionization. Second, at the source end, the lightly doped region lowers the doping density of the emitter in the parasitic bipolar device such that its current gain is degraded. Due to the reduction in both impact ionization and the function of the parasitic bipolar device, the breakdown voltage is improved substantially.

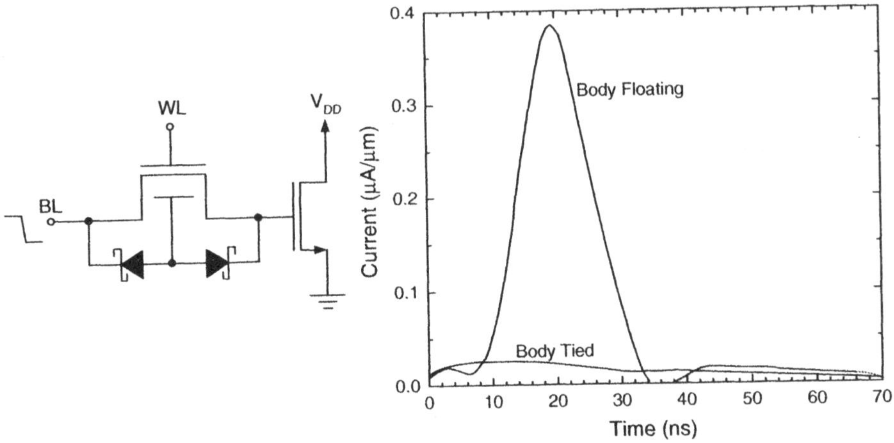

Fig. 2.92 Transient current of the parasitic bipolar device in the SOI pass transistor with its drain connected to the storage node in a DRAM cell and with its gate connected to the word line (WL), which is low, and with its source connected to the bit line (BL), which is switched from high to low, with its body floating and with its body tied to the source/drain via Schottky diodes. (Adapted from Sleight & Mistry [85].)

2.10 TRANSIENT-INDUCED LEAKAGE

In Section 2.9, the single-transistor latch phenomenon are discussed. Although the gate voltage is turned off, a substantial amount of current still exists in the SOI MOS device– the leakage current. In addition to the single-transistor latch, during the transient operation, the parasitic bipolar device in the SOI device may still be triggered to turn on to cause the temporary leakage current, which may cause errors for the dynamic circuits, consuming extra power. In this section, the phenomenon of the transient induced leakage current of the SOI devices is described. The effects on the circuit performance are described later.

Figure 2.92 shows the transient current of the parasitic bipolar device in the SOI pass transistor with its drain connected to the storage node in a DRAM cell and with its gate connected to the word line (WL), which is low, and with its source connected to the bit line (BL), which is switched from high to low, with its body floating and with its body tied to the source/drain via Schottky diodes [85]. First, both WL and BL are high and the storage node is pulled to high. At this time, the internal body potential of the SOI pass transistor is also high. Then, WL becomes low to turn off the pass transistor. The drain end should be separated from the source end. Any change at the source end should not affect the drain end. However, if the BL is pulled to low, there exists unexpected transient induced leakage current flowing from the drain, which may change the potential at the storage node. The existence of the transient related leakage current is correlated to the function of the parasitic bipolar device. When BL is pulled to low, a substantial potential drop between the body and the source occurs, which turns on the body/source junction to trigger the parasitic bipolar device. As a

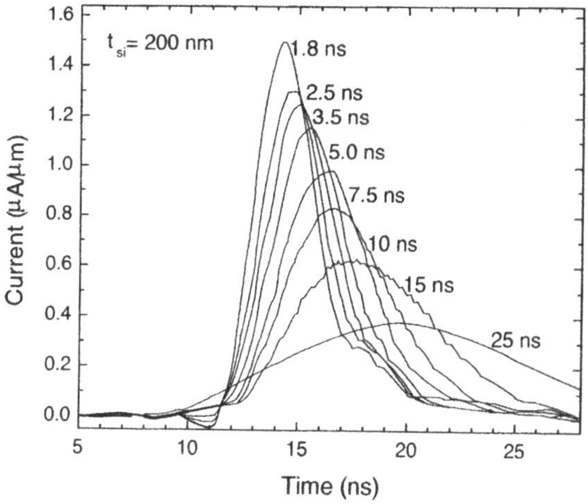

Fig. 2.93 Transient induced leakage current of the parasitic bipolar device in the SOI pass transistor used in the DRAM cell for various bitline fall times. (Adapted from Sleight & Mistry [85].)

result, a large amount of electrons are injected from the source into the body reaching the drain end to form the leakage current, which decays disappearing along with the decrease of the holes in the body and the decrease in the potential difference of the body/source junction.

The straightforward way to resolve this problem is to remove the floating body such that holes can also be expelled or supplied via the body contact. The SiGe source or Schottky-tied body described before can be used to inhibit this transient-induced leakage current effectively. As shown in Fig. 2.92, after adopting the body-tied structure, the transient-induced leakage current has been reduced greatly. In addition, the BL fall time also affects the transient-induced leakage current of the parasitic bipolar device in the SOI pass transistor used in the DRAM cell. As shown in Fig. 2.93, with a smaller BL fall time, the transient-induced peak leakage current is higher [85]. Among all cases with various BL fall times, the total area under each leakage current curve is about identical–the total charge flowing out of the drain is identical. With a larger BL fall time, its peak leakage current is smaller but the pulse width of the leakage current is wider. This result is probably due to the same amount of extra holes accumulated in the internal body, which can be used to trigger the turn-on of the parasitic bipolar device. Thus, the amount of the charge expelled from the internal body is also fixed.

Sometimes this kind of leakage current is not necessarily caused by the parasitic bipolar device. Figure 2.94 shows the drain and the body currents of the SOI NMOS pass transistor with a front gate oxide of 45 Å, a thin film of 1800 Å, and a buried oxide of 3700 Å, a channel width of 20 μm, and a channel length of 0.15 μm, biased at gate voltages of 0, -0.5, and -1 V [86]. As shown in the figure, for the gate voltage < -0.5 V ($V_G < -0.5$ V), when the body voltage is > 0.65 V ($V_{body} > 0.65$ V),

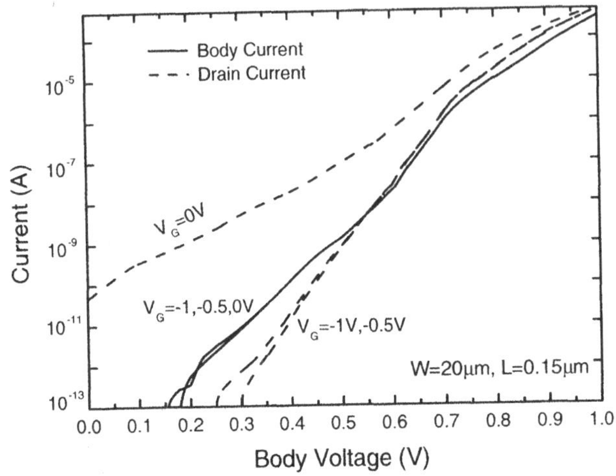

Fig. 2.94 Drain and body currents of the SOI NMOS pass transistor with a front gate oxide of 45 Å, a thin film of 1800 Å, a buried oxide of 3700 Å, a channel width of 20 μm, and a channel length of 0.15 μm, biased at gate voltages of 0, -0.5, and -1 V. (Adapted from Assaderaghi et al. [86].)

the current gain of the parasitic bipolar device is > 1. If the body voltage is < 0.65 V ($V_{body} < 0.65$ V), the current gain is < 1, which is correlated to the smaller current gain at a low base current for a traditional bipolar device. For the gate voltage of 0 V ($V_G = 0$ V), no similar behavior can be identified. Even at a very low body voltage there is still a large amount of leakage current, which is caused by DIBL. In addition, when the body potential drops, the threshold voltage rises and DIBL is lessened. Thus the leakage current is also reduced, but the amount of the reduction is smaller than that for the case with the gate voltage smaller than -0.5 V ($V_G < -0.5$ V).

Figure 2.95 shows the drain leakage current of the SOI NMOS pass transistor with a front gate oxide of 45 Å, a thin film of 1800 Å, a buried oxide of 3700 Å, a channel width of 20 μm, and a channel length of 0.15 μm, with its source driven by a step from high to low with a fall time of 200 ps, for various threshold voltages [86]. As shown in this figure, the leakage current is also dependent on the threshold voltage. For the threshold voltage < 0.4 V ($V_T < 0.4$ V), a smaller threshold voltage leads to a larger transient-induced leakage current with a wider pulse width because the DIBL induced leakage in the surface channel dominates the leakage current. For the threshold voltage > 0.4 V ($V_T > 0.4$ V), DIBL leakage is small. Under this situation, the transient drain current is not sensitive to the threshold voltage any more. The leakage current is mainly due to the conduction of the parasitic bipolar device.

Figure 2.96 shows the transient of the drain leakage current of the PD SOI NMOS pass transistor with a front gate oxide of 70 Å, a thin film of 1400 Å, a channel width of 10 μm, and a channel length of 0.3 μm, biased at the front voltage of 0V and the drain voltage of V_{DD}, after a voltage step from V_{DD} to 0 V is imposed at the source [87]. At a smaller V_{DD} of 1.5 V, the leakage current has a smaller peak and

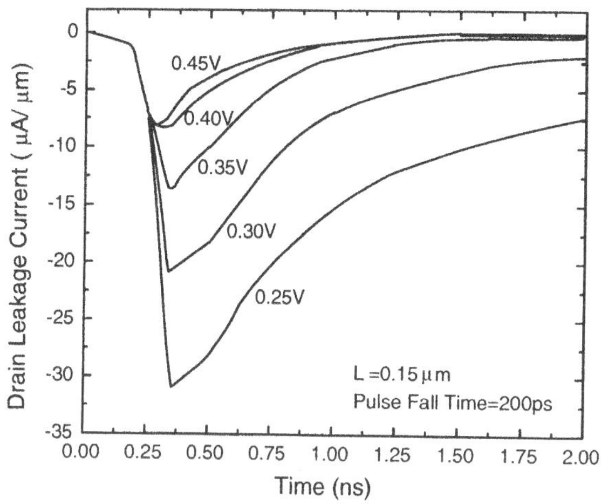

Fig. 2.95 Drain leakage current of the SOI NMOS pass transistor with a front gate oxide of 45 Å, a thin film of 1800 Å, a buried oxide of 3700 Å, a channel width of 20 μm, and a channel length of 0.15 μm, with its source driven by a step from high to low with a fall time of 200 ps, for various threshold voltages. (Adapted from Assaderaghi et al. [86].)

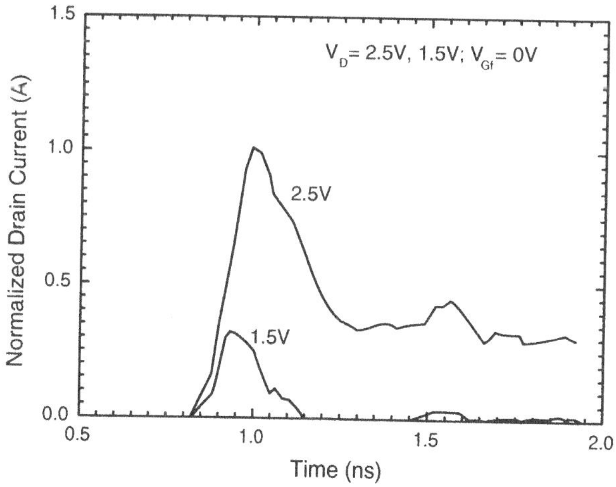

Fig. 2.96 Transient of the drain leakage current of the PD SOI NMOS pass transistor with a front gate oxide of 70 Å, a thin film of 1400 Å, a channel width of 10 μm, and a channel length of 0.3 μm, biased at the front gate voltage of 0 V and the drain voltage of V_{DD}, after a voltage step from V_{DD} to 0 V is imposed at the source. (Adapted from Pelella et al. [87].)

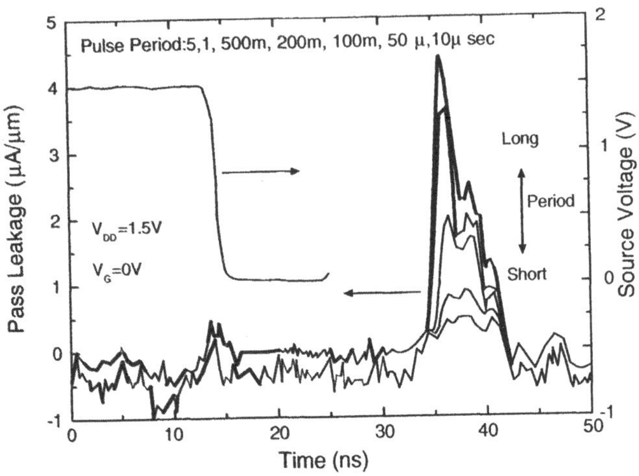

Fig. 2.97 Transient leakage current of the PD SOI NMOS pass transistor with a front gate oxide of 80 Å, a thin film of 1600 Å, and a buried oxide of 5000 Å, biased at the drain voltage of $V_{DD} = 1.5$ V, and the gate voltage of 0 V, with its source driven by periodic inverse pulses from 1.5 to 0 V with a pulse width of 100 ns, a rise/fall time of 2 ns, and various pulse periods from 10 μs to 5 s. (Adapted from Saraya & Hiramoto [88].)

a narrower width–the overall leakage current has been effectively improved. When V_{DD} is smaller, the maximum body-source voltage (V_{BS}) is also smaller, and hence the body-source current (I_{BS}) is smaller. Therefore, the maximum leakage current is lowered. In addition, at a smaller V_{DD}, when the imposed voltage at the source end falls from V_{DD} to 0 V, the amount of the extra holes in the internal body is reduced considerably since they have been recombined or expelled via the source. Thus the body potential drops quickly and the pulse of the leakage current is smaller as compared to the case with a larger V_{DD}.

Figure 2.97 shows the transient leakage current of the PD SOI NMOS pass transistor with a front gate oxide of 80 Å, a thin film of 1600 Å, and a buried oxide of 5000 Å, biased at the drain voltage of $V_{DD} = 1.5$ V, and the gate voltage of 0 V, with its source driven by periodic inverse pulses from 1.5 to 0 V with a pulse width of 100 ns, a rise/fall time of 2 ns and various pulse periods from 10 μs to 5 s [88]. As shown in this figure, the leakage current of the pass transistor is affected by the time between two consecutive inverse pulses. For the two consecutive pulses separated by 5 s, the two consecutive source pulldowns are independent. Under this situation, leakage current is identical to the analysis done before– it reaches a peak value then decays gradually. If the separation between two consecutive pulses is shrunk < 0.5 s, the peak of the leakage current is lower for a shorter pulse period because after the previous source pulldown, most holes in the internal body have been recombined. When the source of the device is back to high, the holes in the body are replenished via thermal generation. If before the replenishment is complete, another source pulldown is carried on, then the amount of the holes in the body is not sufficient. Thus,

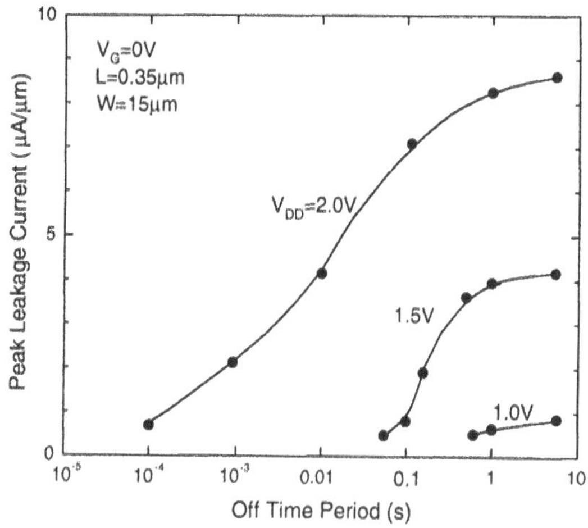

Fig. 2.98 Peak leakage current of the PD SOI NMOS pass transistor with a front gate oxide of 80 Å, a thin film of 1600 Å, a buried oxide of 5000 Å, a channel width of 15 μm, and a channel length of 0.35 μm, biased at the drain voltage of V_{DD} and the gate voltage of 0 V, with its source driven by periodic inverse pulses from V_{DD} to 0 V with a pulse width of 100 ns, the rise/fall time of 2 ns, versus pulse period. (Adapted from Saraya & Hiramoto [88].)

the leakage current is smaller. As long as the separation between two pulses is large enough, the holes in the body can be completely replenished and the leakage current is identical to the one described before.

As shown in Fig. 2.98, the peak of the leakage current is also dependent on the period of the periodic pulldown pulses imposed at the source of the pass transistor [88]. When the period of the source pulldown pulses is short, the separation between two consecutive pulldown pulses is small and the replenishment of holes in the body is not complete. Thus, the peak leakage current is low. As shown in this figure, with a lower V_{DD}, the peak leakage current is also smaller. For $V_{DD} < 1$ V, the leakage current almost disappears since the minimum body-source voltage to turn on the parasitic bipolar device is 0.7 V. When V_{DD} is < 1 V, it is difficult for body source to reach 0.7 V. Therefore, the leakage current is very small.

The transient-induced leakage current of SOI pass transistors may result in the loss of the charge at the storage node. Figure 2.99(a) shows a circuit, which is composed of a pass transistor and a storage node, to emulate the operation of a DRAM cell [89]. By turning on the pass transistor, the storage node is charged to high. When the word line switches to low, the pass transistor is supposed to turn off. After the BL is pulled down, the parasitic bipolar device is turned on to cause the leakage current. As a result, the charge at the storage node goes away and the storage node potential drops–the stored data is lost. As shown in Fig. 2.99(b), when the time for the BL to maintain high is short enough, the holes in the internal body cannot be replenished in time– it cannot trigger the turn-on of the parasitic bipolar device and thus has no extra leakage current. Under this situation, the storage node charge can be lost only via the

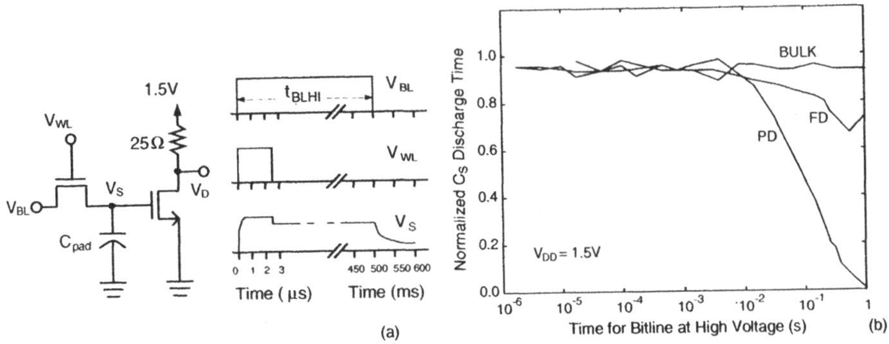

Fig. 2.99 (a) Circuit used to measure BL-induced transient effects in a DRAM cell due to the leakage of the SOI pass transistor. (b) Normalized time to discharge the storage node versus the time of the bitline pulse staying high (t_{BLHI}). (Adapted from Wei & Antoniadis [89].)

junction leakage and the discharge time is long. If the time for the BL to maintain high is long enough (> 0.01 s), the holes in the internal body can be replenished in time to trigger the turn-on of the parasitic bipolar device. As a result, the large leakage current worsens the discharge time of the storage node. With a longer time for the BL to maintain high the discharge time is shorter. As shown in this figure, the discharge time for the case using the FD pass transistor is longer than that using the PD since it is more difficult for the FD device to accumulate holes in the internal body. Consequently, the parasitic bipolar device is more difficult to turn on and the leakage current is smaller in FD. Since the body is usually tied to the ground, the bulk pass transistor does not have an extra leakage current due to the floating body as in the SOI devices. Thus the discharge of the bulk pass transistor is the best.

2.11 HISTORY EFFECTS

For PD SOI devices due to the floating body, device characteristics can be different depending on the initial condition, which are related to single-transistor latch or transient-induced results. In this section, behavior of the PD SOI devices with respect to various initial conditions is analyzed. First, the single-transistor latch is described.

Figure 2.100 shows the drain current versus gate voltage of the SOI NMOS device with a front gate oxide of 260 Å, a thin film doped with a p-type density of $10^{17}cm^{-3}$, biased at the drain voltages of 1, 4, and 5 V, and its gate voltage first changing from negative to positive, then back to negative [90]. As shown in this figure, at the drain voltage of 1 V, no breakdown of the parasitic BJT occurs. Under this condition, no matter what the scanning direction of the gate voltage is– from negative to positive or from positive to negative, the drain current behavior is consistent at a drain voltage of 5 V, considering the situation that the gate voltage is changing from negative to positive. Initially, the device is in the off state. When the gate voltage is close to 0 V, the subthreshold current starts to rise. Since the drain voltage of 5 V is

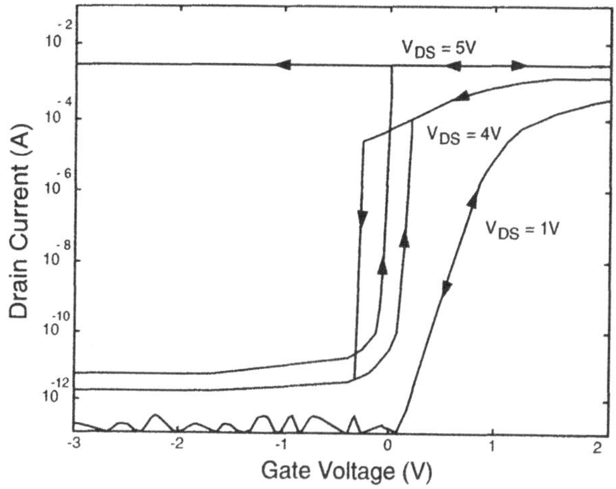

Fig. 2.100 Drain current versus gate voltage of the SOI NMOS device with a front gate oxide of 260 Å, a thin film doped with a p-type density of $10^{17}cm^{-3}$, biased at the drain voltages of 1, 4, and 5 V, and its gate voltage first changing from negative to positive, then back to negative. (Adapted from Chen et al. [90].)

sufficient to cause impact ionization, it generates a gradually increasing hole current. When the hole current is large enough to trigger the turn-on of the parasitic bipolar device, owing to a drain voltage of 5 V, which is sufficiently to cause breakdown of the parasitic bipolar device, at this time the drain current rises directly to reach its maximum value. If the gate voltage falls back to negative, no matter how negative the gate voltage is, the parasitic bipolar device cannot be turned off even though the surface channel is turned off. The parasitic bipolar device has already produced its own base current from the impact ionization of the collector current. The device maintains its high current state– single-transistor latch. Depending on the direction of scanning the gate voltage– from negative to positive or from positive to negative, the device characteristics can be totally different. At a drain voltage of 4 V, this case is between the two extremes. When the gate voltage is changing from negative to positive, its drain current curve is similar to the one with $V_{DS} = 5$ V with the initial impact ionization triggered by the subthreshold current. At a certain gate voltage, the parasitic bipolar device is turned on to cause breakdown, which is not as strong as compared to the case with $V_{DS} = 5$ V, where the hole current needed for the parasitic bipolar device is partially supported from the surface channel-related impact ionization. When the gate voltage is changing from the positive toward the negative direction, the surface channel is completely turned off, which decreases the supply of the hole current for the parasitic bipolar device. Finally, the hole current is less than the critical value, the bipolar device is turned off, and the whole device is off. Due to the function of the parasitic bipolar device, the whole curve ($V_{DS} = 4$ V) demonstrates the hysteresis behavior– turn-off is still more difficult than turn-on.

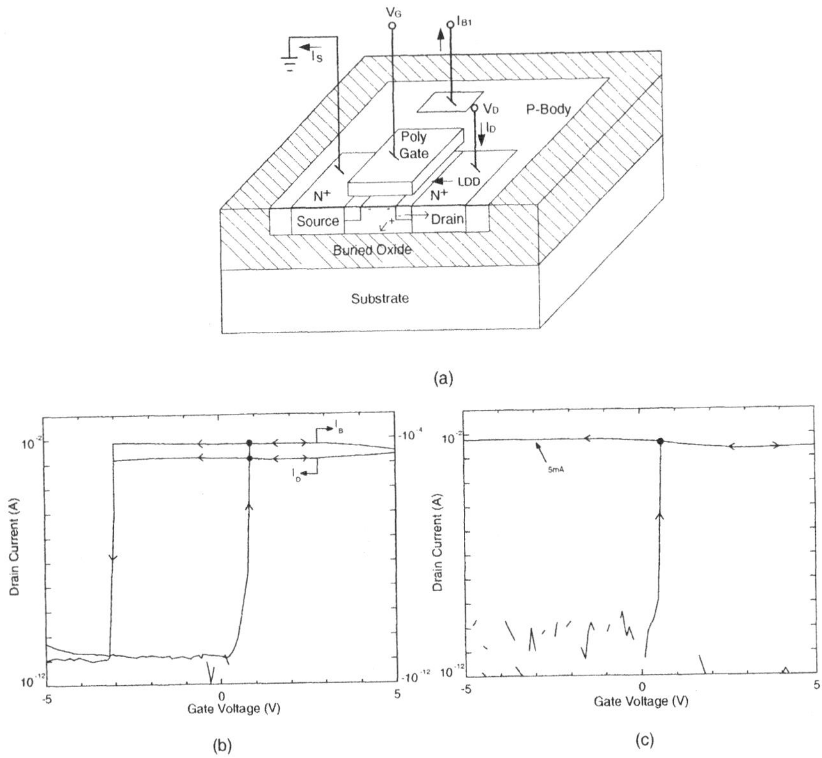

Fig. 2.101 (a) Cross-section of the SOI NMOS device with a front gate oxide of 200 Å, a 3000 Å thin film doped with a p-type density of 10^{17}cm^{-3}, a channel width of 10 μm and a channel length of 1.2 μm. (b) Drain current versus gate voltage of the device with its body tied to source, biased at the drain voltage of 7.1 V. (c) Drain current versus gate voltage of the device with its body floating, biased at the drain voltage of 7.1 V. (From Huang et al. [91]. ©1991 IEEE.)

To avoid the hysteresis for an SOI device caused by the single-transistor latch, the best way is to tie the body to the source such that hole current can flow out the internal body to avoid triggering the parasitic bipolar device. Note that the hole current caused by the single-transistor latch is large. If the resistance between the internal body and the body contact is not sufficiently small, the latch state still cannot be avoided. Figure 2.101 shows (a) the cross-section of the SOI NMOS device with a front gate oxide of 200 Å, a 3000 Å thin film doped with a p-type density of 10^{17}cm^{-3}, a channel width of 10 μm and a channel length of 1.2 μm and the subthreshold drain current characteristics at $V_D = 7.1$ V and (b) with its body tied to the source and (c) with its body floating [91]. As shown in Fig. 2.101, with the body tied to the source, the body contact resistance is still several MΩ for the arrangement of the body contact as shown in Fig. 2.101(a). Thus, the large voltage drop over the body contact resistance forward biases the body/source junction to turn on the parasitic bipolar device. As shown in Fig. 2.101(b), even with the body tied to the source, there is still a hysteresis. On the

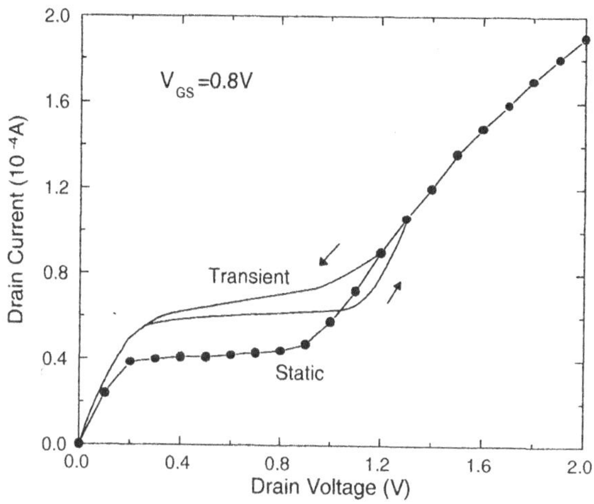

Fig. 2.102 Drain current versus drain voltage of the PD NMOS device with a front gate oxide of 70 Å, a 1000 Å thin film doped with a p-type density of 5×10^{17}cm^{-3}, a buried oxide of 4000 Å, a channel length of 1 μm and a channel width of 10 μm, biased at the gate voltage of 0.8 V, and with its drain voltage changing from negative to positive and from positive to negative with a scan time of 1 s. (Adapted from Perron et al. [92].)

other hand, when the gate voltage is changing from positive to negative, the parasitic bipolar device, which already has breakdown, cannot be turned off to terminate the progress of the latch state.

The hysteresis caused by various initial conditions of the transient measurement can be different. Figure 2.102 shows the drain current versus the drain voltage of the PD NMOS device with a front gate oxide of 70 Å, a 1000 Å thin film doped with a p-type density of 5×10^{17}cm^{-3}, a buried oxide of 4000 Å, a channel length of 1 μm and a channel width of 10 μm, biased at the gate voltage of 0.8 V, and with its drain voltage changing from negative to positive and from positive to negative with a scan time of 1 s [92]. As shown in this figure, for the curve measured by scanning the drain voltage from low to high with a scan time of 1 s, before the kink effects, the drain current is higher than its static value, which is due to the rise of the body potential caused by the coupling of the pullup drain voltage via the body-drain capacitance (C_{BD}) and the drop in the threshold voltage. Kink effects of the transient curve occur later as compared to the static effect because of the time needed for the hole current generated by impact ionization charging the floating body to raise the body potential. There exists a hysteresis in the drain current versus the drain voltage curves with the transient measurement.

Figure 2.103 shows the drain current versus the gate voltage of the PD SOI NMOS device described in Fig. 2.102, biased at the drain voltage of (a) 0.5 V and (b) 1 V, with its gate voltage scanning from 0 to 3 V and from 3 to 0 V with a scan time of 1 s [92]. At the drain voltage of 0.5 V [Fig. 2.103(a)], there are no kink or impact ionization

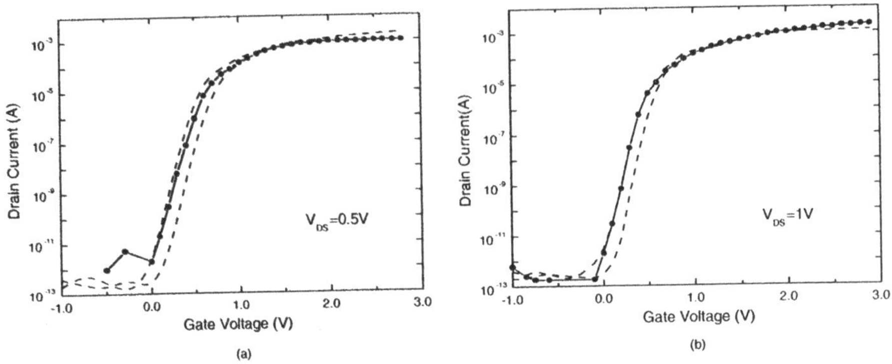

Fig. 2.103 Drain current versus gate voltage of the PD SOI NMOS device described in Fig. 2.102, biased at the drain voltage of (a) 0.5 V and (b) 1 V, with its gate voltage scanning from 0 to 3 V and from 3 to 0 V with a scan time of 1 s. The solid curves marked by circles are the static case. (Adapted from Perron et al. [92].)

effects. The V_G-pullup curve is higher than the static one and the V_G-pulldown curve is lower, which is due to change in the body potential caused by coupling of the gate voltage to the body via the body gate capacitance (C_{BG}). At a drain voltage of 1 V [Fig. 2.103(b)], the V_G-pulldown curve is farther from the static one and the V_G-pullup one is close to the static one. This occurs because during the V_G-pullup and V_G-pulldown processes, holes cannot accumulate in the internal body in time, which results in a lower body potential. Thus the whole hysteresis curve shifts downward as compared to the static one.

After analyzing the hysteresis behavior, the influence of the initial condition in the transient performance of the inverter circuit made of SOI CMOS devices is described. Figure 2.104 shows the gate delay versus time of an inverter using PD SOI CMOS devices with a front gate oxide of 45 Å, a thin film of 1500 Å, a buried oxide of 4000 Å, and a channel length of 0.145 μm, biased at $V_{DD} = 1.8$ V, with its input driven by a clock with a period of 1 ns and a 50% duty cycle. As shown in this figure, with a 1 GHz clock imposed at its input, its propagation delay gradually becomes stable. The output fall delay (TD_{fall}) and the output rise delay (TD_{rise}) are dominated by the NMOS and the PMOS devices, respectively. For the input clock initially at high (H-L), the output fall delay (TD_{fall}) is initially longer as compared to the L-H case with the input clock initially at low. Along with time, the TD_{fall} falls to a stable value. In contrast, the TD_{rise} for the input clock initially at low (L-H) is higher than its H-L case initially, and rises toward its stable value. This result can be reasoned as follows. In the case with the input initially at low (L-H) for the NMOS device, initially the gate voltage is low and the drain voltage is high. Thus, its body potential is higher as compared to the H-L case. So, for the L-H case, the NMOS threshold voltage is lower and its TD_{fall} is shorter. The TD_{rise} is similar. The TD_{rise} with the input clock initially low (L-H) is longer as compared to the H-L case. In the L-H case,

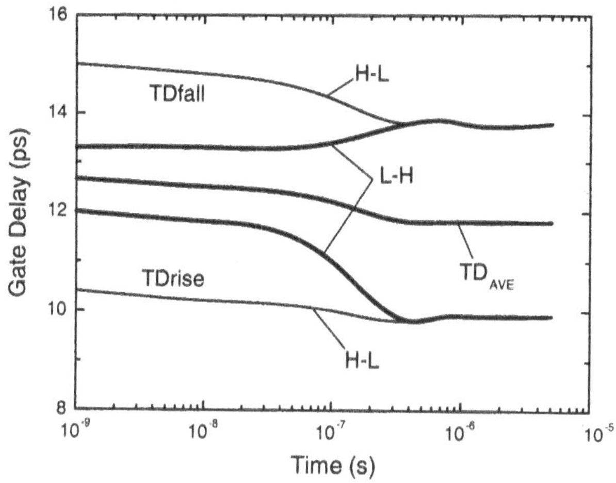

Fig. 2.104 Gate delay versus time of the inverter using PD SOI CMOS devices with a front gate oxide of 45 Å, a thin film of 1500 Å, a buried oxide of 4000 Å, and a channel length of 0.145 μm, biased at $V_{DD} = 1.8$ V, with its input driven by a clock with a period of 1 ns and a 50% duty cycle [93]. A 2:1 aspect ratio difference between the PMOS and the NMOS devices has been adopted. (Adapted from Pelella et al. [93].)

for the PMOS device, initially its gate voltage is low and its drain voltage is high, which implies a higher body potential. Consequently, the TD_{rise} is longer than the one for the H-L case. As time progresses, both H-L and L-H delays settle to the same stable state value. The average of the output rise and fall delay is identical regardless of the difference in the initial conditions.

2.12 SELF-HEATING

SOI devices are renown for their active thin-film region working on a silicon dioxide insulator, which is used to separate itself from the substrate. Since the silicon dioxide insulator is also a thermal insulator, the power consumed in the active device region cannot be dissipated easily. As a result, the lattice temperature of the thin film rises–the so-called self-heating.

2.12.1 Drain Current

Fig. 2.105 shows (a) the drain current and (b) the lattice temperature versus drain voltage of the SOI NMOS device with a front gate oxide of 70 Å, a thin film of 800 Å, a buried oxide of 4000 Å, a channel length of 0.2 μm and a channel width of 9.5 μm, using the model considering the energy balance equation and the lattice temperature and the data [94]. As shown in Fig. 2.105(a), at a high drain current (with a high gate voltage) in some high drain voltage region a higher drain voltage leads to a decrease

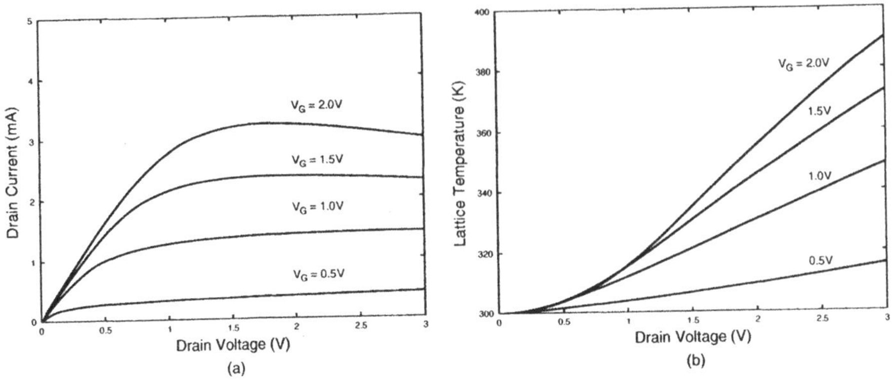

Fig. 2.105 (a) Drain current and (b) lattice temperature versus drain voltage of the SOI NMOS device with a front gate oxide of 70 Å, a thin film of 800 Å, a buried oxide of 4000 Å, a channel length of 0.2 μm and a channel width of 9.5 μm, considering the energy transport and the lattice temperature. (Adapted from Chen et al. [94].)

in the drain current, which is the so-called negative differential output resistance. At a higher drain voltage, the power consumption of the device is higher. Since the consumed power cannot be dissipated easily, the lattice temperature rises and the electron mobility falls. Consequently, the drain current decreases. Figure 2.105(b) shows the correlated lattice temperature in the device. From the figure, at a higher gate voltage and a higher drain voltage, its lattice temperature is higher– self-heating effect is noticeable. Self-heating of the NMOS devices is more noticeable as compared to the PMOS devices because of the higher electron mobility.

Self-heating is more noticeable for SOI devices operating at a low temperature. Figure 2.106 shows the drain current versus the drain voltage of an SOI NMOS device with a front gate oxide of 175 Å, a thin film of 800 Å, and a buried oxide of 3800 Å, operating at (a) 300 K and (b) 77 K [95]. As shown in this figure, the behavior of the negative differential output resistance in the high drain/gate voltage regime is more noticeable at 77 K. Since at the low temperature the electron mobility is higher and the drain current is larger. Therefore, power consumption and heat accumulation are higher. In addition, at low temperature, the mobility is more sensitive to the variation in the lattice temperature. At low temperature, the thermal resistance of the SOI device is increased because of the poorer thermal conductivity. As a result, it is more difficult for the SOI devices to dissipate power. Consequently, the self-heating effect is higher at a low temperature.

2.12.2 Thermal Resistance

Figure 2.107 shows (a) the thermal resistance versus temperature and (b) the thermal conductivity versus square root of temperature of the SOI NMOS device with a thin film of 800 Å and a buried oxide of 3800 Å [96]. As shown in this figure, at a higher temperature the thermal resistance of the SOI device is lower due to the increased

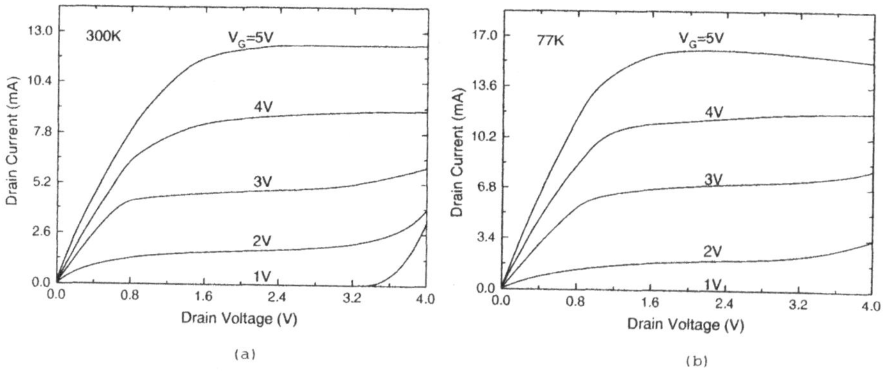

Fig. 2.106 Drain current versus drain voltage of an SOI NMOS device with a front gate oxide of 175 Å, a thin film of 800 Å, and a buried oxide of 3800 Å, operating at (a) 300 K and (b) 77 K. (Adapted from Jomaah et al. [95].)

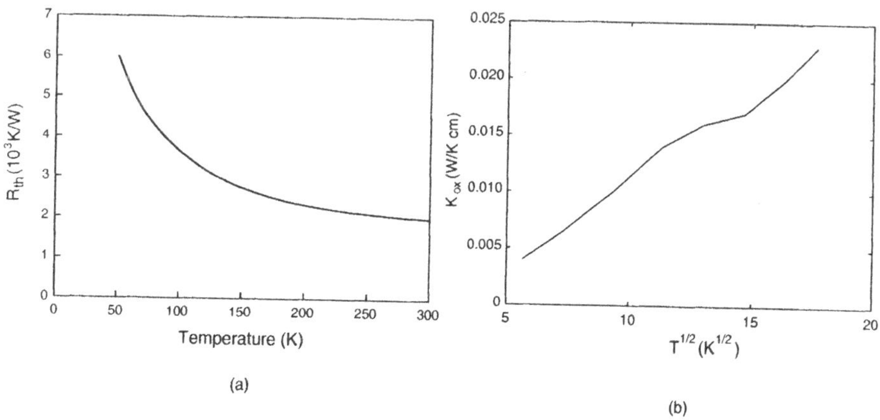

Fig. 2.107 (a)Thermal resistance versus temperature and (b) thermal conductivity versus square root of temperature of the SOI NMOS device with a thin film of 800 Å and a buried oxide of 3800 Å. (Adapted from Jomaah et al. [96].)

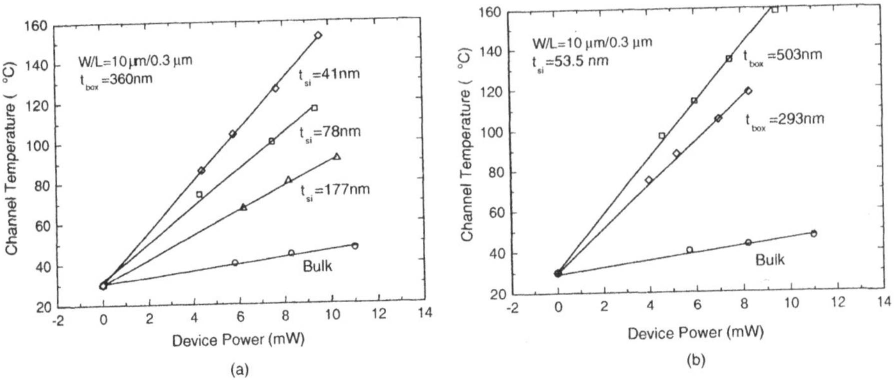

Fig. 2.108 Channel temperature versus power of the SOI NMOS device with a channel length of 0.3 μm, a channel width of 10 μm, (a) with a buried oxide of 3600 Å and various thin-film thicknesses, and (b) with a thin film of 535 Å and two buried-oxide thicknesses. (Adapted from Su et al. [97].)

oxide thermal conductivity. At a lower temperature, thermal resistance of the SOI device is higher. At low temperature, the slope of the thermal resistance versus the temperature curve is higher than that at room temperature.

Self-heating is strongly dependent on the heat dissipation capability of an SOI device, which is determined by the device structure. Figure 2.108 shows the channel temperature versus power of the SOI NMOS device with a channel length of 0.3 μm, a channel width of 10 μm, (a) with a buried oxide of 3600 Å and various thin-film thicknesses, and (b) with a thin film of 535 Å and two buried-oxide thicknesses [97]. As shown in Fig. 2.108(a), with a thinner thin film, the slope of the device temperature versus power curve becomes steeper, which implies a higher thermal resistance since with a reduced thin-film thickness it is more difficult for the heat generated in the channel region to dissipate via source/drain. As shown in Fig. 2.108(b), with a thicker buried oxide, the slope of the lattice temperature versus the power curve becomes steeper, which implies a higher thermal resistance. It is more difficult to dissipate heat through the thicker buried oxide to the substrate. In contrast, as shown in the figure, the channel temperature of the bulk devices without the isolation by the buried oxide is much smaller.

Thermal resistance is strongly determined by the layout of the SOI device. Figure 2.109 shows (a) the thermal resistance versus the channel length of the SOI NMOS device with a channel width of 20 μm and (b) the thermal resistance versus the channel width of the SOI NMOS device with a channel length of 10 μm [98]. As shown in the figure, when the channel length or the channel width becomes smaller, the thermal resistance between the SOI device and the environment is higher, which implies that it is more difficult for the device to dissipate the generated heat from the source/drain end and the gate contact. The thermal resistance between the SOI device and the environment is determined by the heat dissipation capability of the source, the drain, the gate, and the substrate.

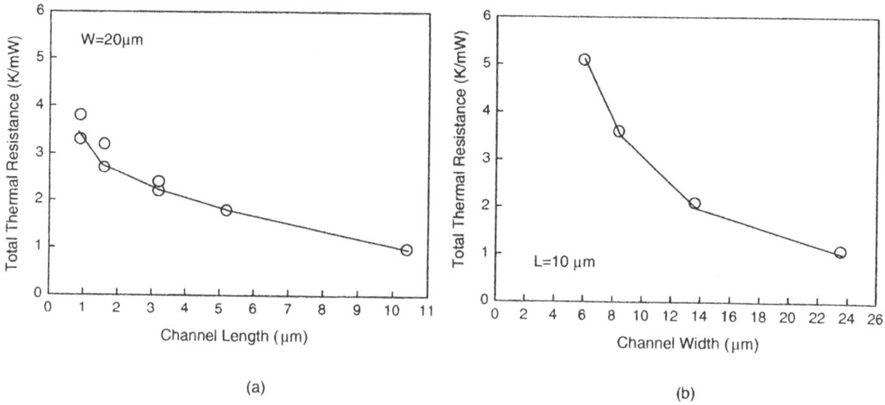

Fig. 2.109 (a) Thermal resistance versus channel length of the SOI NMOS device with a channel width of 20 μm. (b) Thermal resistance versus channel width of the SOI NMOS device with a channel length of 10 μm. (Adapted from Tenbroek et al. [98].)

2.12.3 Thermal Coupling

In addition to self-heating, thermal coupling between SOI devices is also important. Figure 2.110 shows the current mirror used to exemplify the thermal coupling behavior [99]. As shown in Fig. 2.110, the layout of the current mirror has been arranged as shown in layouts A and B. In layout A, both NMOS devices in the current mirror share the source region. So, heat can transfer between these two NMOS devices via the thin film directly. In layout B, both NMOS devices are separated by 20 μm. Between them there is a narrow and long polysilicon gate with the metal line connecting the sources, which can be used for heat transfer. As shown in Fig. 2.110, the measurement of thermal coupling is done by fixing the current in the reference device and changing the output voltage at the output device. Then the change in the gate voltage of the reference device is measured. As shown in Fig. 2.110(c), although the change in the gate voltage of the reference device is small, it is highly output voltage dependent, which can be reasoned as follows. When the output voltage is high, the power consumption of the output device is also high. Thus, the lattice temperature of the device increases due to the self-heating effect. If the two devices are close enough, the heat generated by the output device can be transferred to the reference device via the common metal line, the polysilicon gate, and the silicon thin film. As a result, the lattice temperature of the reference device rises–the so-called thermal coupling effect. Therefore, the electron mobility of the reference device falls. A larger gate voltage is needed to carry the same reference current. A higher output voltage leads to a higher gate voltage of the reference device. For layout A, thermal coupling is more noticeable– the slope of the gate voltage of the reference device to the output voltage is larger. In contrast, for the layout B, due to the relatively longer distance between two devices, the thermal coupling effect is small. Depending on the applications, thermal coupling may be good or bad for a circuit. For the current mirror, thermal coupling makes the lattice temperature in the two devices identical. As a result, the

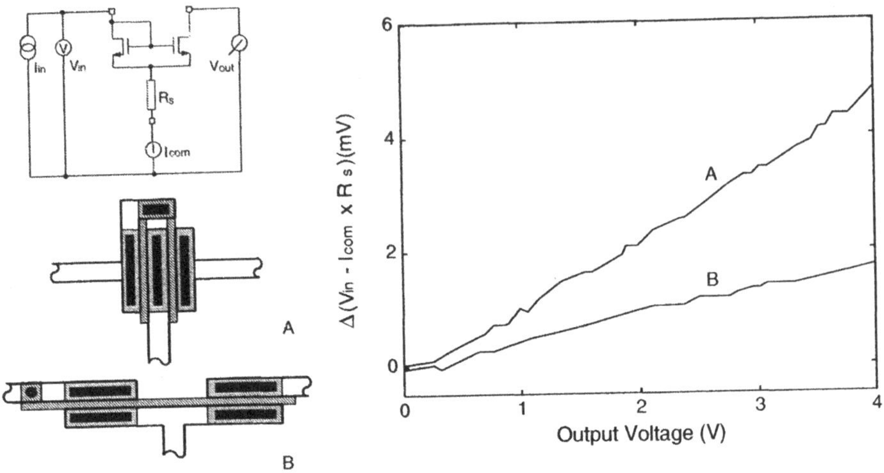

Fig. 2.110 (a) Layout of the current mirror for thermal coupling measurement of the SOI NMOS devices using mesa isolation, with a thin film of 1000 Å, a buried oxide of 3000 Å, a channel width of 20 μm, and a channel length of 2 μm. (b) Arrangement of the current mirror for measuring the input voltage versus output power dissipation. (c) Change in the gate voltage of the reference transistor versus output voltage (hence the output power dissipation). (From Tenbroek et al. [99]. ©1996 IEEE.)

properties of the two device are closer and the precision of the current mirror can be enhanced.

2.12.4 AC Behavior

Until now, analysis of thermal effects for SOI devices is for the steady-state cases. It takes time to reach thermal equilibrium starting from heat accumulation in the SOI MOS devices. Figure 2.111 shows the transient drain current of the SOI NMOS device with a front gate oxide of 90 Å, a thin film of 65 Å, and a buried oxide of 3600 Å, for various gate and drain voltages [100]. For the case with $V_G = 4$ V, $V_D = 4$ V and the case with $V_G = 4$ V, $V_D = 2$ V, when the device is turned on, its drain current rises to the maximum value. Along with the dissipated power and the accumulated heat in the device, its drain current falls with the rise in the device temperature, which is the self-heating effect, and finally settles down to a value at its thermal equilibrium. Note that at the thermal equilibrium the drain current for the case with $V_D = 4$ V and $V_G = 4$ V is lower than that for the case with $V_D = 2$ V and $V_G = 4$ V due to the negative resistance effect from self-heating. For the case with $V_G = 1$ V and $V_D = 1$ V, since the current is small, self-heating is slight. For the case with $V_G = 4$ V and $V_D = 3$ V, since it is a bulk device, with a good heat dissipation capability, no self-heating can be seen. From the cases with $V_G = 4$ V, the thermal time constant is $\sim$ 200 ns depending on the device structure. Since for

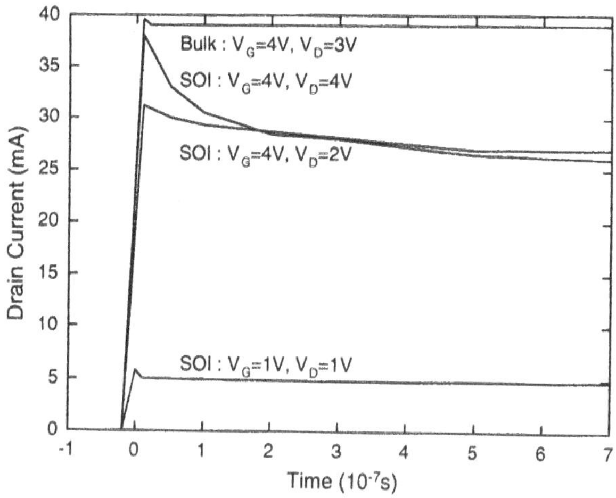

Fig. 2.111 Transient drain current of the SOI NMOS device with a front gate oxide of 90 Å, a thin film of 65 Å, and a buried oxide of 3600 Å, for various gate and drain voltages. (Adapted from Arora et al. [100].)

most logic gate circuits the process of the turn on/off takes only a few nanoseconds, transient thermal effects may not be significant.

At a high frequency, the self-heating effect on the negative resistance may be quite different. Figure 2.112 shows the small signal drain conductance of the PD SOI NMOS device with a front gate oxide of 250 Å, a thin film of 1780 Å, a buried oxide of 4150 Å, a channel width of 20 μm, and a channel length of 5 μm, biased in the saturation ($V_{DS} - V_{DSAT} = 1.8$ V), with the body tied to the source, plotted (a) versus drain current and (b) versus frequency [101]. As shown in Fig. 2.112(a), for a drain current > 1.3mA, its drain conductance is negative at the static state. If the scan frequency of the imposed small signal is higher, the negative resistance behavior may disappear since the heat cannot accumulate in time. When the scan frequency of the imposed small signal is > 1 MHz, the original behavior of the decreased drain conductance at the increased drain current disappears. For the frequency response of the drain conductance, as shown in Fig. 2.112(b), at the drain current of 1.57 mA, when the scan frequency is < 1 MHz, its drain conductance decreases quickly, which may be negative due to self-heating. With a scan frequency > 1 MHz, self-heating does not cause substantial effects on the device since heat accumulation does not occur in time. On the contrary, at a drain current of 0.86 mA, the drain conductance is not sensitive to the scan frequency because self-heating is not significant at the low drain current.

Self-heating effects on lowering the drain current of an SOI MOS device can be a cumbersome problem for device parameter extraction. With the biasing signal imposed at the high frequency, the self-heating behavior can be avoided. Figure 2.113 shows the drain current versus the drain voltage of an SOI NMOS device by pulse measurement with a 6 ns pulse width and periods of 1 ms and 1 μs [12]. As shown

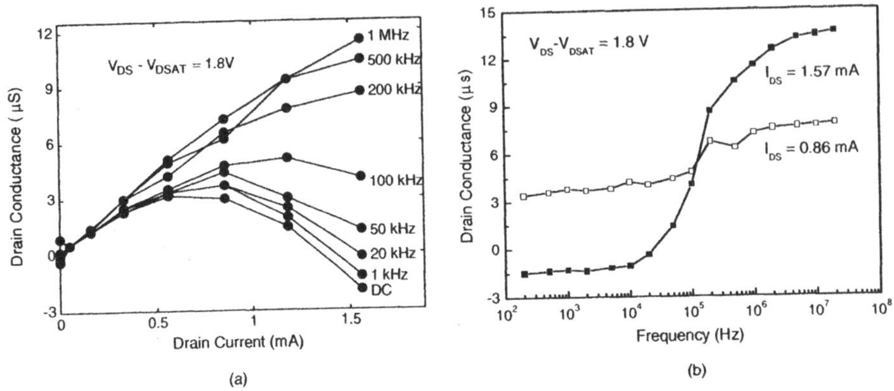

Fig. 2.112 Small signal drain conductance of the PD SOI NMOS device with a front gate oxide of 250 Å, a thin film of 1780 Å, a buried oxide of 4150 Å, a channel width of 20 μm, and a channel length of 5 μm, biased in the saturation ($V_{DS} - V_{DSAT} = 1.8$ V), with the body tied to the source, plotted (a) versus the drain current and (b) versus the frequency. (Adapted from Tenbroek et al. [101].)

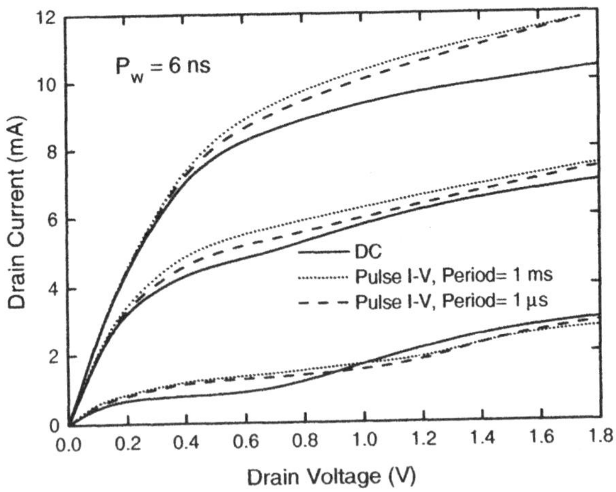

Fig. 2.113 Drain current versus drain voltage of an SOI NMOS device by pulse measurement with a 6 ns pulse width and periods of 1 ms and 1 μs. (Adapted from Shahidi et al. [12].)

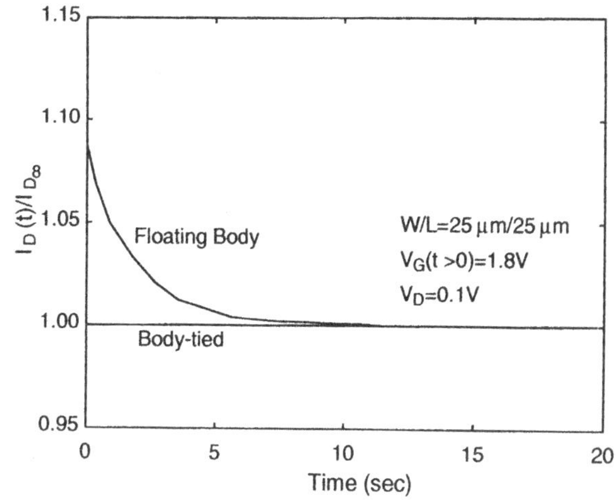

Fig. 2.114 Transient drain current of the PD SOI NMOS device with a front gate oxide of 105 Å, a 1000 Å thin film doped with a p-type density of $2.2 \times 10^{17} cm^{-3}$, a buried oxide of 4000 Å, channel width and length of 25 μm, biased at the drain voltage of 0.1 V, with a step voltage from 0 to 1.8 V imposed at the gate and with its body floating and tied. (Adapted from Shin et al. [102].)

in this figure, at the high gate voltage and the high drain current, the result obtained by pulse measurement is higher than its DC value. Imposing the biasing signal at the high frequency, self-heating cannot occur in time, and hence the lattice temperature does not rise. By using the pulse measurement, the drain current characteristics of an SOI device without the self-heating effects can be obtained. In addition, at the low gate voltage, using the pulse measurement, kink effects also disappear since the holes generated by impact ionization cannot accumulate in time with the biasing signal imposed at high frequency. At the low gate voltage and the low drain voltage, the drain current obtained by pulse measurement is higher than its DC value. This is not because of exclusion of self-heating–at the low drain current level self-heating is negligible. It occurs because by using the pulse measurement, the AC drain signal is coupled to the body via the drain-body capacitance (C_{DB}) such that the threshold voltage is lowered. This problem can be removed by connecting the body to the source.

2.13 TRANSIENT BEHAVIORS

Due to the floating body problems, the transient behaviors of PD SOI devices are more complicated. In this section, transient behaviors of SOI devices are described.

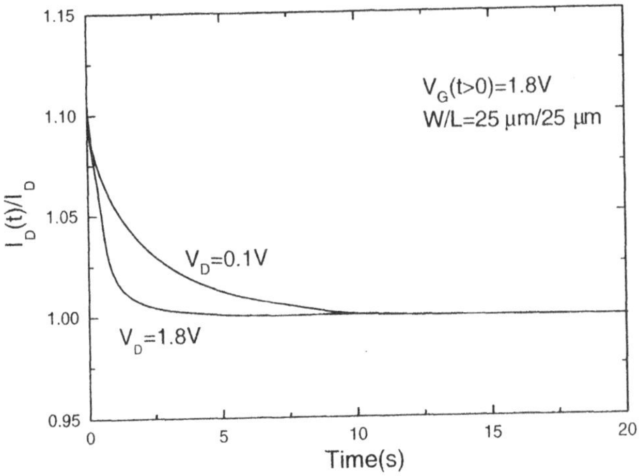

Fig. 2.115 Transient drain current of the PD SOI device with its body floating and with parameters as described in Fig. 2.114 except that two drain voltages (0.1 and 1.8 V) have been applied. (Adapted from Shin et al. [102].)

2.13.1 Floating-Body Induced

Figure 2.114 shows the transient drain current of the PD SOI NMOS device with a front gate oxide of 105 Å, a 1000 Å thin film doped with a p-type density of $2.2 \times 10^{17}cm^{-3}$, a buried oxide of 4000 Å, channel width and length of 25 μm, biased at the drain voltage of 0.1 V, with a step voltage from 0 to 1.8 V imposed at the gate with its body floating and tied [102]. Initially, when a step voltage is imposed at the gate, for the body floating case, the drain current overshoots to its maximum value. Then, it decays to a stable value. With its body tied, no such drain current overshoot can been seen. The existence of the drain current overshoot for the body floating case is attributed to the coupling of the switch high signal of the gate voltage to the body via the gate-body capacitance (C_{GB}). As a result, the body potential rises in a short time, which lowers the threshold voltage and results in the drain current overshoot. Along with time, the extra holes generated by the quick rise in the gate voltage, which can be removed in time, recombine gradually. Therefore, the body potential drops and the threshold voltage rises gradually. The drain current decreases with time until the extra holes in the body disappear.

The drain voltage affects the drain current overshoot and the speed of decay after the overshoot of the PD SOI device with its body floating. Figure 2.115 shows the transient drain current of the PD SOI device with its body floating and with parameters as described in Fig. 2.114 except that two drain voltages have been applied [102]. At the drain voltage of 1.8 V, the decay of the drain current after the overshoot is faster than that for the drain voltage of 0.1 V. At the higher drain voltage, the depletion region between the body and the drain is larger and the neural region at the bottom of the thin film becomes smaller. Thus, the extra holes in the body are much fewer.

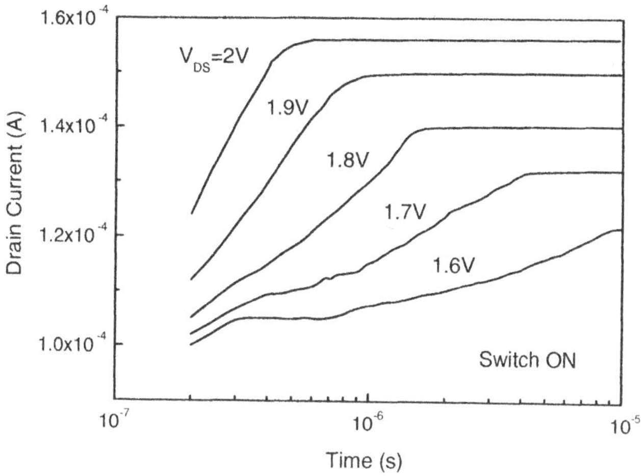

Fig. 2.116 Drain current during the turn-on transient of the PD SOI NMOS device with a front gate oxide of 70 Å, a 1000 Å thin film doped with a p-type density of $5 \times 10^{17} cm^{-3}$, a buried oxide of 4000 Å, a channel length of 1 μm, and a channel width of 10 μm, biased at various drain voltages, with its gate voltage switching from 0 to 0.8 V initially. (Adapted from Perron et al. [103].)

After turn-on of the device, the extra holes needed to be removed are less. Therefore, they are recombined more quickly to reach the equilibrium state earlier.

With impact ionization, the transient behavior of a PD SOI MOS device can be quite different. Figure 2.116 shows the drain current during the turn-on transient of the PD SOI NMOS device with a front gate oxide of 70 Å, a 1000 Å thin film doped with a p-type density of $5 \times 10^{17} cm^{-3}$, a buried oxide of 4000 Å, a channel length of 1 μm, and a channel width of 10 μm, biased at various drain voltages, with its gate voltage switching from 0 to 0.8 V initially [103]. As shown in this figure, the gate voltage switches from 0 to 0.8 V only if the effect of the coupling of the gate voltage signal that causes a sudden rise in the body voltage can be avoided. In addition, all of the applied drain voltages are > 1.6 V, which is large enough to cause kink effects. As shown in the figure, as opposite to Fig. 2.115 when the device turns on, its drain current rises gradually. The reason is that at the high drain voltage, the holes generated by impact ionization begin to accumulate in the bottom of the thin film. Therefore, along with the progress of time, the amount of the accumulated holes becomes large, and hence the body potential rises and the threshold voltage drops. Consequently, the drain current gradually increases. This trend continues until the body potential reaches 0.7V, where the body-source junction is forward biased. At this time, a stable state has been reached for the device. As shown in the figure, for a higher drain voltage, its drain current reaches its stable state earlier. At a higher drain voltage impact ionization is more serious to generate more holes, which results in a faster rise of the body voltage to 0.7 V.

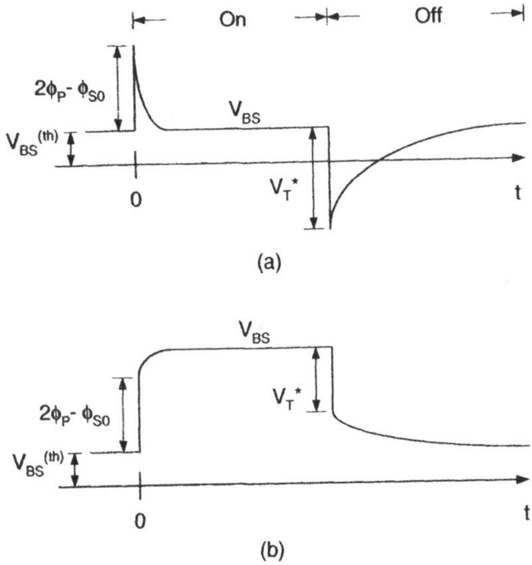

Fig. 2.117 Schematical transient of the body voltage of a PD SOI NMOS device during the turn-on and the turn-off voltage steps applied to the gate with its drain voltage biased (a) below the kink effect region and (b) well above the kink effect region. (Adapted from Perron et al. [104].)

Figure 2.117 shows the schematical transient of the body voltage during the turn-on and the turn-off voltage steps applied to the gate of a PD SOI NMOS device with its drain voltage biased (a) below the kink effect region and (b) well above the kink effect region [104]. As shown in Fig. 2.117, with a smaller drain voltage, kink effects do not occur. With a turn-on voltage step suddenly imposed at the gate, the body voltage jumps to a peak value due to signal coupling and then falls due to the recombination of the holes. On the contrary, as shown in fig. 2.117(b), with a larger drain voltage, after the turn-on voltage step at the gate, an up jump in the body voltage due to signal coupling can be seen. Then, the body voltage gradually grows due to the rise of the body voltage from the accumulation of holes generated by impact ionization until it reaches 0.7 V. Kink effects result in different transient behavior of the body voltage. Consequently, the transient drain current is different.

Figure 2.118 shows the drain current during the turn-off transient of the PD SOI NMOS device with a front gate oxide of 70 Å, a 1000 Å thin film doped with a p-type density of $5 \times 10^{17}\text{cm}^{-3}$, a buried oxide of 4000 Å, a channel width of 10 μm, and a channel length of 1 μm, biased at the drain voltage of (a) the drain voltage of 0.1 $\sim$ 1.2 V and (b) 1.3 and 1.5 V and with the gate voltage switching from 0.8 to 0.2 V [104]. As shown in Fig. 2.118, it is with its gate voltage switching from 0.8 to 0.2 V and with its drain voltage at 1.2 V without causing impact ionization. From Figs. 2.118(a) and 2.117, due to signal coupling from the gate to the body via the gate-body capacitance, at the beginning the body voltage drops substantially and an insufficient amount of holes exist in the thin film. Then via thermal generation, holes

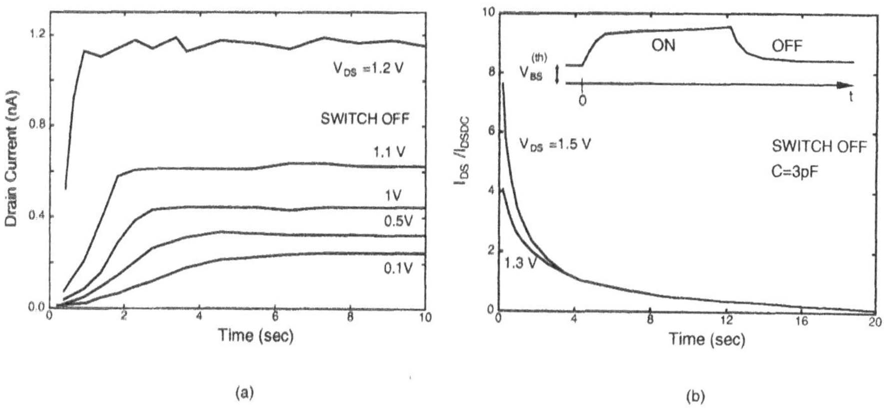

Fig. 2.118 Drain current during the turn-off transient of the PD SOI NMOS device with a front gate oxide of 70 Å, a 1000 Å thin film doped with a p-type density of $5 \times 10^{17}\text{cm}^{-3}$, a buried oxide of 4000 Å, a channel width of 10 μm, and a channel length of 1 μm, biased at (a) the drain voltage of 0.1 $\sim$ 1.2 V, and (b) the drain voltages of 1.3 and 1.5 V, and with its gate voltage switching from 0.8 to 0.2 V. (Adapted from Perron et al. [104].)

are replenished gradually in the thin film and the body potential. Hence, the drain current will rise accordingly. As shown in Fig. 2.118, with a larger drain voltage, the drain current reaches its stable value more quickly since at a larger drain voltage, the body/drain depletion region is wider and more thermal generation leads to a quicker replenishment of the holes. In addition, at a higher drain voltage, fewer holes need to be replenished in the smaller neutral region. Based on these reasons, at a higher drain voltage, the drain current reaches its stable value quicker.

As shown in Figs. 2.117 and 2.118(b), if the drain voltage is large enough to cause kink effects, when the turn-off voltage step is imposed at the gate due to the signal coupling, the body voltage has a mild down jump. Note that due to the large equivalent body capacitance, which is caused by the large amount of holes in the body, the down jump in the body voltage is mild with its magnitude determined by the ratio of the gate-body capacitance to the equivalent body capacitance. After the turn-off voltage step at the gate, the accumulated holes due to impact ionization cannot be removed immediately, which gradually disappear due to recombination. Hence, the body voltage and the drain current decrease gradually after the turn-off voltage step imposed at the gate.

Figure 2.119 shows the drain current transient of the PD SOI NMOS device with a channel length of 25 μm and a channel width of 25 μm, biased at the front gate voltage of 1 V, switching its back gate from weak to strong accumulation (generation effect) and vice versa (recombination effect) [105]. As shown in this figure, when the back gate switches low, due to signal coupling from the back gate to the body, the body voltage is pulled low and the holes in the thin film become insufficient. As time progress, the holes are replenished via thermal generation gradually and the body voltage rises steadily. Consequently, the drain current also rises accordingly. This

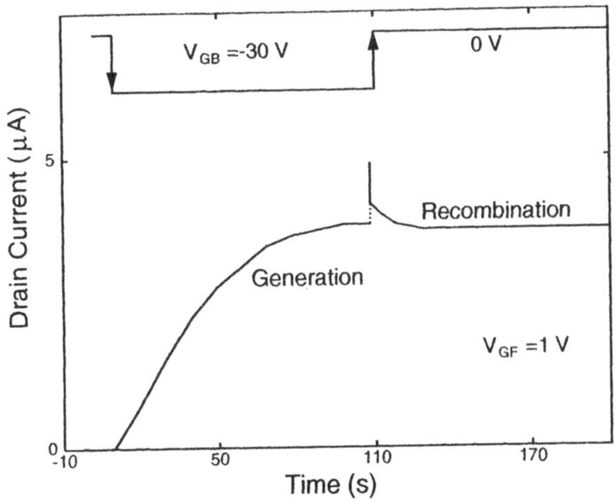

Fig. 2.119 Drain current transient of the PD SOI NMOS device with a channel length of 25 μm and a channel width of 25 μm, biased at the front gate voltage of 1 V, switching its back gate from weak to strong accumulation (generation effect) and vice versa (recombination effect). (Adapted from Munteanu et al. [105].)

trend continues until it reaches equilibrium. When the back gate switches high, due to signal coupling, the body voltage is raised high to a peak and too many holes exist in the thin-film. Then, via recombination the extra holes disappear gradually and the body voltage, and thus the drain current, fall to its steady state.

As described before, the device behavior under the transient operation is quite different from that at DC. Figure 2.120 shows the drain current characteristics of the PD SOI NMOS device with a front gate oxide of 70 Å, a 1400 Å thin film doped with a p-type density of $4 \times 10^{17}\text{cm}^{-3}$, biased at various drain voltages, using pulse measurement by imposing periodic pulses from low to high with a pulse width of 7 ns and periods of 1 ms and 10 μs [106]. As shown in this figure, kink effects are noticeable at DC. Using pulse measurement by imposing periodic pulses from low to high with a pulse width of 7 ns and a period of 1 ms on the gate, kink effects are not existent and the drain current is higher than its DC value. By using pulse measurement, the holes generated by impact ionization do not have enough time to accumulate in the thin film and thus there are no kink effects. In the high drain current regime, the higher drain current using pulse measurement is due to bypass of self-heating. By using pulse measurement, the heat consumed by the device cannot accumulate in time and the lattice temperature does not rise. Therefore, self-heating is avoided. In the low drain current regime, the drain current obtained by pulse measurement is still higher than its DC value due to the signal coupling effect. After the turn-on voltage step imposed at its gate, the gate voltage signal is coupled to the body, therefore the body voltage rises and its threshold voltage falls– a larger drain current.

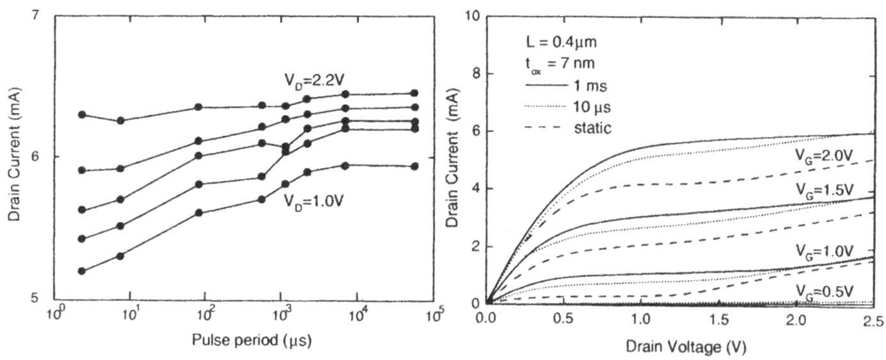

Fig. 2.120 Drain current characteristics of the PD SOI NMOS device with a front gate oxide of 70 Å, a 1400 Å thin film doped with a p-type density of $4 \times 10^{17}\text{cm}^{-3}$, biased at various drain voltages, using pulse measurement by imposing periodic pulses from low to high with a pulse width of 7 ns and periods of 1 ms and 10 μs. (Adapted from Jenkins et al. [106].)

2.13.2 History Effect

The influence of floating body in terms of the history effect on the current transient operation of a PD SOI device is important. Figure 2.121 shows the drain current during the transient of the PD SOI NMOS device with a front gate oxide of 90 Å, a 1150 Å thin film doped with a p-type density of $6 \times 10^{17}\text{cm}^{-3}$, a buried oxide of 3600 Å, a channel length of 0.2 μm, and a channel width of 50 μm, biased at the drain voltage of 1.5 V, with periodic pulses from 0 to 0.67 V imposed at the gate, (a) with a period of 1/50 s and a pulse width of 3 μs, (b) with a period of 1/50 s and pulse width of 100 ns, and (c) with a period of 1 μs and a pulse width of 100 ns [107]. The threshold voltage at $V_D = 1.5$ V is 0.47 V. As shown in Fig. 2.121(a), for the periodic pulses with a period of 1/50 s and a pulse width of 3 μs, the device will have plenty of time in accumulating holes generated by impact ionization and thus the body voltage rises to 0.7V and its drain current increases to 0.4mA at steady state. As shown in Fig. 2.121(b), for periodic pulses with a period of 1/50 s and a pulse width of 100 ns, before the device has enough time to accumulate holes, its gate voltage switches back to 0V. Therefore, the drain current cannot rise in time–it only rises to 0.03 mA. As shown in Fig. 2.121(c), for period pulses with a period of 1 μs and a pulse width of 100 ns, two consecutive pulses are much closer, separated by only 1 μs. Under this situation, the drain current could reach the maximum value (0.4 mA) as for Fig. 2.121(a), which can be reasoned as follows. During each short pulse width (100 ns) the amount of the accumulated holes is small, before they have enough time to be recombined the subsequent pulse has arrived. Thus the accumulated holes for each turn-on pulse are added up. Therefore, the body voltage rises along with each pulse until it reaches an equilibrium with the recombination rate or the body-source junction is forward biased. As long as the pulse frequency is sufficiently high, despite the short pulse width, holes still can be accumulated in the thin film such that the drain current can rise.

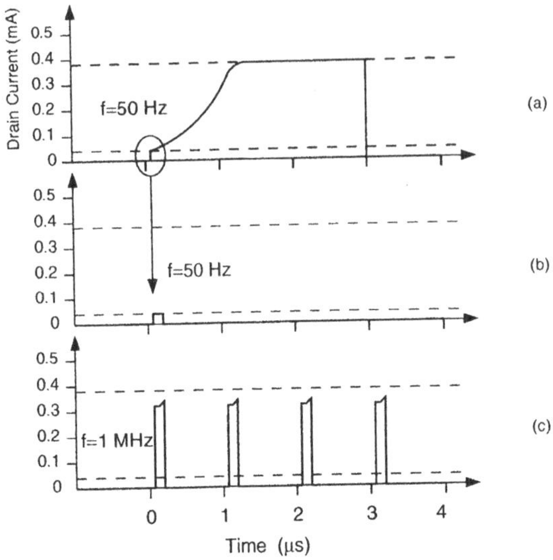

Fig. 2.121 Drain current during the transient of the PD SOI NMOS device with a front gate oxide of 90 Å, a 1150 Å thin film doped with a p-type density of $6 \times 10^{17}\text{cm}^{-3}$, a buried oxide of 3600 Å, a channel length of 0.2 μm, and a channel width of 50 μm, biased at the drain voltage of 1.5 V, with periodic pulses from 0 to 0.67 V imposed at the gate, (a) with a period of 1/50 s and a pulse width of 3 μs, (b) with a period of 1/50 s and a pulse width of 100 ns, and (c) with a period of 1 μs and a pulse width of 100 ns. The threshold voltage at $V_D = 1.5$ V is 0.47 V. (From Wei et al. [107]. ©1995 IEEE.)

Figure 2.122 shows the drain current normalized by its steady state value versus the frequency of the pulses from 0 to 0.87 V imposed at the gate of the PD SOI NMOS device with parameters described in Fig. 2.121, and biased at the drain voltage of 1.5 V, for various pulse widths [107]. As shown in the figure, if the pulse width is longer, during the turn-on pulse, the amount of the holes accumulated in the thin-film is larger and the body voltage becomes higher. Thus its drain current is higher. On the other hand, if the frequency of the pulses is high and the duration between consecutive pulses is short enough, the holes generated in the previous turn-on pulse can be preserved until the next turn-on pulse. Then the generated holes during each turn-on pulse can be added up and the body voltage can still be raised. Therefore, the drain current is increased even if the frequency is raised.

The transient-related history effect of PD SOI devices is important. This effect is different from what is described above. Impact ionization has been avoided by using a low drain voltage. Figure 2.123 shows the drain current during the transient of the PD SOI NMOS device with a front gate oxide of 700 Å, a 0.5 μm thin film doped with a p-type density of $5 \times 10^{16}\text{cm}^{-3}$, a buried oxide of 1 μm, biased at a drain voltage of 1 V, with its gate imposed by periodic pulses from 0 to 2 V with two consecutive turn-on periods separated by (a) 250 μs and (b) 20 μs[108]. As shown in this figure, since there is no impact ionization, when the gate voltage rises from 0 to 2 V, the drain

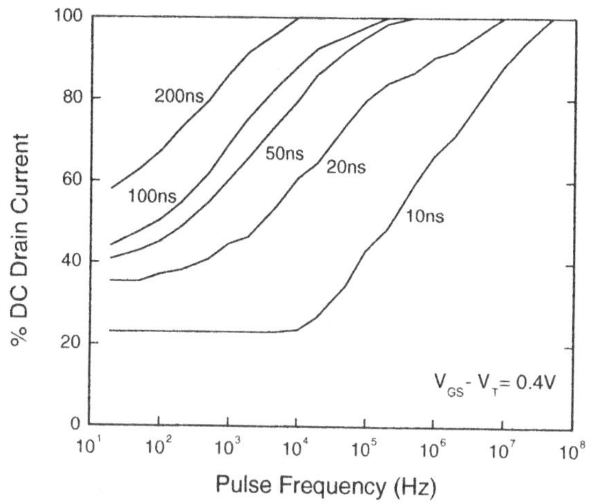

Fig. 2.122 Drain current normalized by its steady state value versus the frequency of the pulses from 0 to 0.87 V imposed at the gate of the PD SOI NMOS device with parameters described in Fig. 2.121, and biased at the drain voltage of 1.5 V, for various pulse widths. (Adapted from Wei et al. [107].)

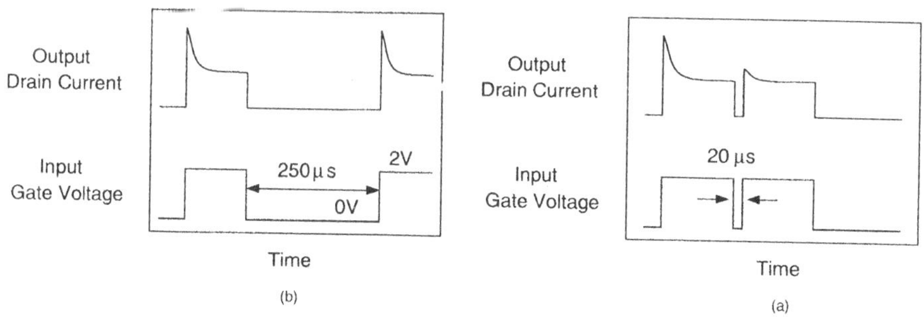

Fig. 2.123 Drain current during the transient of the PD SOI NMOS device with a front gate oxide of 700 Å, a 0.5 μm thin film doped with a p-type density of $5 \times 10^{16} cm^{-3}$, a buried oxide of 1 μm, biased at a drain voltage of 1 V, with its gate imposed by periodic pulses from 0 to 2 V with two consecutive turn-on periods separated by (a) 250 μs and (b) 20 μs. (Adapted from Kato & Taniguchi [108].)

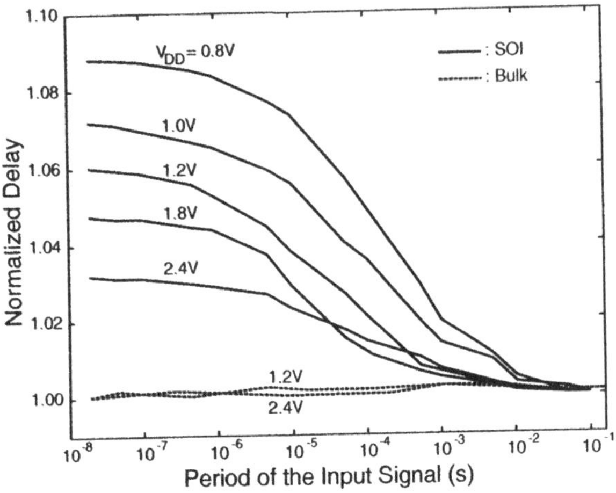

Fig. 2.124 Propagation delay versus period of the input pulses of a 28-stage 2-input NAND chain using SOI MOS devices with a front gate oxide of 45 Å, a thin film of 1400 Å, a buried oxide of 3600 Å, and a channel length of 0.3 μm, with its gate imposed by periodic pulses with a pulse width of 2 ns and for various supply voltages (V_{DD}). Propagation delay is normalized by the propagation delay of a single pulse. (Adapted from Assaderaghi et al. [109].)

current jumps to its maximum value and then decays to its steady-state value. This portion of reasoning has already been presented. If the turn-on periods of the two consecutive pulses are separated by an interval long enough as shown in Fig. 2.123(a), these two turn-on periods can be regarded as independent with no mutual influence. As shown in Fig. 2.123(b), if the turn-on periods of the two consecutive pulses are close, the peak drain current of the subsequent turn-on period is smaller than that of the previous one. During the first turn-on period, some holes in the thin film have been removed. They are supposed to be replenished during the turn-off period of the periodic pulses. If the interval between two consecutive turn-on periods is too short, before replenishment of holes can be completed the next turn-on period arrives. Since the amount of holes in the thin film is already small as compared to the previous period, the drain current overshoot during this turn-on period is smaller. When the frequency of the periodic pulses is higher, the drain current overshoot behavior is smaller, which is just opposite to the case described before with impact ionization.

Due to floating body, the drain current during transient of a PD device can be affected by the frequency of the periodic pulses imposed at the gate. If the drain voltage is high to have impact ionization in the device, the drain current during the transient may rise with increased frequency. If the drain voltage is not high enough to have impact ionization in the device, the drain current may fall with the increased frequency. Figure 2.124 shows the propagation delay versus period of the input pulses of a 28-stage 2-input NAND chain using SOI MOS devices with a front gate oxide of 45 Å, a thin film of 1400 Å, a buried oxide of 3600 Å, and a channel length of 0.3 μm, with its gate imposed by periodic pulses with a pulse width of 2ns and for various

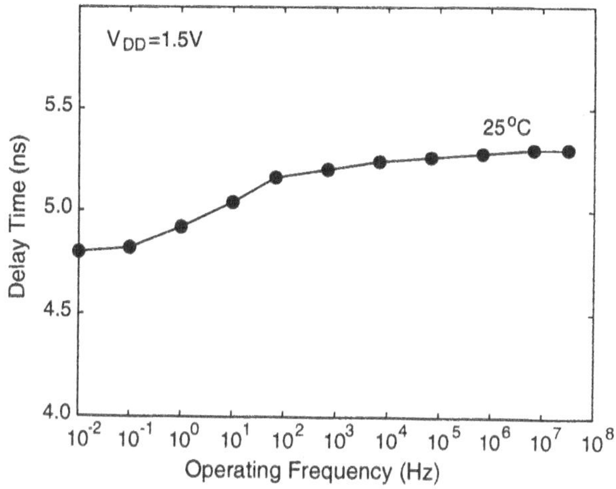

Fig. 2.125 Propagation delay time versus operating frequency of a 64-bit adder implemented by SOI CMOS devices, operating at $V_{DD} = 1.5$ V. (Adapted from Ueda et al. [110].)

supply voltages (V_{DD}) [109]. Propagation delay is normalized by the propagation delay of a single pulse. As shown in the figure, when the period of the input signal increases, the propagation delay is decreased, which implies that the average drain current in the device rises. Under this situation, the nonimpact-ionization transient analysis is appropriate for the logic circuit operation. In addition, at a lower V_{DD}, the propagation delay of the circuit is more sensitive to the change in the input signal frequency. At a V_{DD} of 0.8 V the impact ionization is slight and its propagation delay is the most frequency dependent. Also shown in the figure, for the circuit implemented by bulk devices, since they are without a floating body, no such frequency-dependent propagation delay can be seen. The frequency-dependent propagation delay indicates the instability of the logic circuits implemented by SOI devices with a floating body. To improve the stability of the SOI logic circuits, the most straightforward way is to use the body-tied structure.

In addition to the NAND gate, a 64-bit adder using carry look-ahead (CLA) has been used to exemplify the floating body effect of the SOI devices. Figure 2.125 shows the propagation delay versus the operating frequency of a 64-bit adder implemented by SOI CMOS devices, operating at $V_{DD} = 1.5$ V [110]. As shown in the figure, at a higher operating frequency, the drain current overshoot behavior is reduced and the drain current falls. As a result, the propagation delay of the adder increases, which correlates to the results for the NAND circuit described before.

2.14 SUMMARY

In this chapter, fundamental SOI CMOS technology has been described. The back gate bias effects, short and narrow channel effects, and mobility including the velocity overshoot phenomenon have been analyzed. Floating body effects and subthreshold characteristics of SOI CMOS devices have been analyzed. Also, the impact ionization, snapback, bipolar leakage, and bipolar history effects derived from floating body structure have been depicted. In the final portion of this chapter, self-heating of SOI CMOS devices has been presented, followed by the transient analysis of SOI CMOS devices. In Chapter 3, more behaviors of SOI CMOS devices will be described.

REFERENCES

1. M. Watanabe and A. Tooi, "Formation of SiO_2 Films by Oxygen-Ion Bombardment," *Jpn. J. Appl. Phys.*, **5**(8), 737–738 (1966).

2. D. J. Godbey, M. E. Twigg, H. L. Hughes, L. J. Palkuti, P. Leonov, and J. J. Wang, "Fabrication of Bond and Etch-Back Silicon on Insulator Using a Strained $Si_{0.7}Ge_{0.3}$ Layer as an Etch Stop," *J. Electrochem. Soc.*, **137**(10), 3219–3223 (1990).

3. M. Bruel, "Silicon on Insulator Material Technology," *Elec. Let.*, **31**(14), 1201–1202 (1995).

4. M. Bruel, B. Aspar, B. Charlet, C. Maleville, T. Poumeyrol, A. Soubie, A. J. Auberton-Herve, J. M. Lamure, T. Barge, F. Metral, and S. Trucchi, "Smart Cut: A Promising New SOI Material Technology," *SOI Conf. Dig.*, 178–179, (1995).

5. P. M. Zavracky, D.-P. Vu, and M. Batty, "Silicon-on-Insulator Wafers by Zone Melting Recrystallization," *Sol. St. Tech.*, **34**(4), 55–57 (1991).

6. M. Haond and O. Le Neel, "Lateral Isolation in SOI CMOS Technology," *Sol. St. Tech.*, **34**(7), 47–52 (1991).

7. M. Haond and O. Le Neel, "Rounded Edge MESA for Submicron SOI CMOS Process," *SOI Conf. Dig.*, 132–133 (1990).

8. C.-L. Huang and G. J. Grula, "Degradation Characteristics of STI and MESA-Isolated Thin-Film SOI CMOS," *IEEE Ele. Dev. Let.*, **18**(10), 474–476 (1997).

9. K. Sukegawa, M. Yamaji, K. Yoshie, K. Furumochi, T. Maruyama, H. Morioka, N. Naori T. Kubo, H. Kanata, M. Kai, S. Satoh, T. Izawa, and K. Kubota, "High-Performance 80-nm Gate Length SOI-CMOS Technology with Copper and Very-Low-K Interconnects," *Symp. VLSI Tech. Dig.*, 186–187 (2000).

10. P. Smeys, V. McGahay, I. Yang, J. Adkisson, K. Beyer, O. Bula, Z. Chen, B. Chu, J. Culp, S. Das, A. Eckert, L. Hadel, M. Hargrove, J. Herman, L. Lin, R. Mann, E. Maciejewski, S. Narasimha, P. O'Neil, S. Rauch, D. Ryan, J. Toomey, L. Tsou, P. Varekamp, P. Wachnik, T. Wagner, S. Wu, C. Yu, P. Agnello, J. Connolly, S. Crowder, C. Davis, R. Ferguson, A. Sekiguchi, L. Su, R. Goldblatt, and T. C. Che,

"A High Performance 0.13μm SOI CMOS Technology with Cu Interconnects and Low-k BEOL Dielectric," *Symp. VLSI Tech. Dig.*, 184–185 (2000).

11. S. Pindl, J. Berthold, T. Huttner, S. Reif, D. Schumann, and H. von Philipsborn, "A 130-nm Channel Length Partially Depleted SOI CMOS-Technology," *IEEE Trans. Ele. Dev.*, **46**(7), 1562–1566 (1999).

12. G. G. Shahidi, A. Ajmera, F. Assaderaghi, R. J. Bolam, H. Hovel, E. Leobandung, W. Rausch, D. Sadana, D. Schepis, L. F. Wagner, L. Wissel, K. Wu, and B. Davari, "Device and Circuit Design Issues in SOI Technology," *CICC Dig.*, 339–346 (1998).

13. T. Eimori, T. Oashi, F. Morishita, T. Iwamatsu, Y. Yamaguchi, F. Okuda, K. Shimomura, H. Shimano, N. Sakashita, K. Arimoto, Y. Inoue, S. Komori, M. Inuishi, T. Nishimura, and H. Miyoshi, "Approaches to Extra Low Voltage DRAM Operation by SOI-DRAM," *IEEE Trans. Elec. Dev.*, **45**(5), 1000–1008 (1998).

14. H. Wang, M. Chan, Y. Wang, and P. K. Ko, "The Behavior of Narrow- Width SOI MOSFET's with MESA Isolation," *IEEE Trans. Elec. Dev.*, **47**(3), 593–599 (2000).

15. M. Haond and M. Tack, "Rapid Electrical Measurements of Back Oxide and Silicon Film Thickness in an SOI CMOS Process," *IEEE Trans. Elec. Dev.*, **38**(3), 674–676 (1991).

16. H.-K. Lim and J. G. Fossum, "Threshold Voltage of Thin-Film Silicon-on-Insulator (SOI) MOSFET's," *IEEE Trans Elec. Dev.*, **30**(10), 1244–1251 (1983).

17. T. Ernst, D. Munteanu, S. Cristoloveanu, T. Ouisse, N. Hefyene, S. Horiguchi, Y. Ono, Y. Takahashi, and K. Murase, "Ultimately Thin SOI MOSFETs: Special Characteristics and Mechanisms," *SOI Conf. Dig.*, 92–93 (1999).

18. J.-H. Sim and J. B. Kuo, "An Analytical Back-Gate Bias Effect Model for Ultra-thin SOI CMOS Devices," *IEEE Trans. Elec. Dev.*, **40**(4), 755–765 (1993).

19. K. Hashimoto, T. I. Kamins, K. M. Cham, and S. Y. Chiang, "Characteristics of Submicrometer CMOS Transistors in Implanted-Buried-Oxide SOI Films," *IEDM Dig.*, 672–675 (1985).

20. S. R. Banna, P. C. H. Chan, P. K. Ko, C. T. Nguyen, and M. Chan, "Threshold Voltage Model for Deep-Submicrometer Fully Depleted SOI MOSFET's," *IEEE Trans. Elec. Dev.*, **42**(11), 1949–1955 (1995).

21. K. Imai, H. Onishi, K. Yamaguchi, K. Inoue, Y. Matsubara, A. Ono, and T. Horiuchi, "A 0.18μm Fully Depleted CMOS on 30nm Thick SOI for Sub-1.0V Operation," *Symp. VLSI Tech. Dig.*, 116–117 (1998).

22. L. T. Su, J. B. Jacobs, J. E. Chung, and D. A. Antoniadis, "Deep-Submicrometer Channel Design in Silicon-on-Insulator (SOI) MOSFET's," *IEEE Elec. Dev. Let.*, **15**(9), 366–369 (1994).

23. J.-W. Lee, H.-K. Kim, M.-R. Oh, and Y.-H. Koh, "Threshold Voltage Dependence of LOCOS-Isolated Thin-Film SOI NMOSFET on Buried Oxide Thickness," *IEEE Elec. Dev. Let.*, **20**(9), 478–480 (1999).

24. M. Cao, T. Kamins, P. V. Voorde, C. Diaz, and W. Greene, "0.18μm Fully-Depleted Silicon-on-Insulator MOSFET's," *IEEE Elec. Dev. Let.*, **18**(6), 251–253 (1997).

25. T. Ushiki, M.-C. Yu, Y. Hirano, H. Shimada, M. Morita, and T. Ohmi, "Reliable Tantalum-Gate Fully-Depleted-SOI MOSFET Technology Featuring Low-Temperature Processing," *IEEE Trans. Elec. Dev.*, **44**(9), 1467–1472 (1997).

26. B. Maiti, P. J. Tobin, C. Hobbs, R. I. Hegde, F. Huang, D. L. O'Meara, D. Jovanovic, M. Mendicino, J. Chen, D. Connelly, O. Adetutu, J. Mogab, J. Candelaria, and J. B. La, "PVD TiN Metal Gate MOSFETs on Bulk Silicon and Fully Depleted Silicon-On-Insulator (FDSOI) Substrates for Deep Sub-Quarter Micron CMOS Technology," *IEDM Dig.*, 781–784 (1998).

27. A. O. Adan, K. Higashi, and Y. Fukushima, "Analytical Threshold Voltage Model for Ultrathin SOI MOSFET's Including Short-Channel and Floating-Body Effects," *IEEE Trans. Elec. Dev.*, **46**(4), 729–737 (1999).

28. K. Suzuki, A. Satoh, and T. Sugii, "Counter Doping into Uniformly and Heavily Doped Channel Region of Sub 0.1μm SOI MOSFET's," *IEEE Elec. Dev. Let.*, **17**(1), 1–3 (1996).

29. S. W. Crowder, P. M. Rousseau, J. P. Snyder, J. A. Scott, P. B. Griffin, and J. D. Plummer, "The Effect of Source/Drain Processing on the Reverse Short Channel Effect of Deep Sub-Micron Bulk and SOI NMOSFETs," *IEDM Dig.*, 427–430 (1995).

30. D. Tsoukalas, C. Tsamis, D. N. Kouvatsos, P. Revva, and E. Tsoi, "Reduction of the Reverse Short Channel Effect in Thick SOI MOSFET's," *IEEE Elec. Dev. Let.*, **18**(3), 90–92 (1997).

31. J.-P. Colinge, "Subthreshold Slope of Thin-Film SOI MOSFET's," *IEEE Elec. Dev. Let.*, **7**(4), 244–246 (1986).

32. H.-O. Joachim, Y. Yamaguchi, K. Ishikawa, Y. Inoue, and T. Nishimura, "Simulation and Two-Dimensional Analytical Modeling of Subthreshold Slope in Ultrathin-Film SOI MOSFET's Down to 0.1μm Gate Length," *IEEE Trans. Elec. Dev.*, **40**(10), 1812–1817 (1993).

33. P. C. Yeh and J. G. Fossum, "Physical Subthreshold MOSFET Modeling Applied to Viable Design of Deep-Submicrometer Fully Depleted SOI Low-Voltage CMOS Technology," *IEEE Trans. Elec. Dev.*, **42**(9), 1605–1613 (1995).

34. A. O. Adan, K. Higashi, K. Niimi, and T. Ashida, "The OFF Leakage in SOI-MOS Transistors and the Impact on the Standby Current of ULSI's," *SOI Conf. Dig.*, 34–35 (1999).

35. M. P. Brassington, A. G. Lewis, and S. L. Partridge, "A Comparison of Fine-Dimension Silicon-On-Sapphire and Bulk-Silicon Complementary MOS Devices and Circuits," *IEEE Trans. Elec. Dev.*, **32**(9), 1858–1867 (1985).

36. S.-C. Lin, J. B. Kuo, K.-T. Huang, and S.-W. Sun, "A Closed-Form Back-Gate-Bias Related Inverse Narrow-Channel Effect Model for Deep-Submicron VLSI CMOS Devices Using Shallow Trench Isolation," *IEEE Trans. Elec. Dev.*, **47**(4), 725–733 (2000).

37. K.-W. Su and J. B. Kuo, "Analytical Threshold Voltage Formula Including Narrow-Channel Effects for VLSI Mesa-Isolated Fully Depleted Ultrathin Silicon-On-Insulator N-Channel Metal-Oxide-Silicon Devices," *Jpn. J. Appl. Phys.*, **34**(8A), 4010–4019 (1995).

38. K. W. Su and J. B. Kuo, "Analytical Threshold Voltage Model Considering Small-Geometry Effects for VLSI Mesa-Isolated Fully-Depleted Ultrathin SOI NMOS Devices using a Quasi-3D Approach," Chapter 6 of *CMOS VLSI Engineering: Silicon-On-Insulator*, Kluwer: Boston (1998).

39. Y. C. Tseng, J. Collett, T. O. Vu, J. S. Cable, and J. C. S. Woo, "Analysis of Edge Effects in the Mesa Isolated NMOS SOI," *SOI Conf. Dig.*, 90–91 (1996).

40. J. B. Kuo and K. W. Su, "Compact Current Model for Mesa-Isolated Fully-Depleted Ultrathin SOI NMOS Devices Considering Sidewall-Related Narrow Channel Effects," *SOI Conf. Dig.*, 84–85 (1997).

41. K. W. Su and J. B. Kuo, "Modeling Narrow-Channel Effect in VLSI Mesa-Isolated SOI MOS Devices using a Quasi-Two-Dimensional Approach," *Sol. St. Elec.*, **39**(9), 1321–1329 (1996).

42. C.-Y. Chang, S.-J. Chang, T.-S. Chao, S.-D. Wu, and T.-Y. Huang, "Reduced Reverse Narrow Channel Effect in Thin SOI NMOSFETs," *IEEE Elec. Dev. Let.*, **21**(9), 412–414 (2000).

43. C. S. Rafferty, H.-H. Vuong, S. A. Eshraghi, M. D. Giles, M. R. Pinto, and S. J. Hillenius, "Explanation of Reverse Short Channel Effect by Defect Gradients," *IEDM Dig.*, 311–314 (1993).

44. A. Ono, R. Ueno, and I. Sakai, "TED Control Technology for Suppression of Reverse Narrow Channel Effect in 0.1μm MOS Devices," *IEDM Dig.*, 227–230 (1997).

45. J.-W. Lee, H.-K. Kim, M.-R. Oh, and Y.-H. Koh, "Threshold Voltage Dependence of LOCOS-Isolated Thin-Film SOI NMOSFET on Buried Oxide Thickness," *IEEE Elec. Dev. Let.*, **20**(9), 478–480 (1999).

46. K. Sonoda, K. Taniguchi, and C. Hamaguchi, "Analytical Device Model for Sub-micrometer MOSFET's," *IEEE Trans. Elec. Dev.*, **38**(12), 2662–2668 (1991).

47. J. Wang, N. Kistler, J. Woo, and C. R. Viswanathan, "Mobility-Field Behavior of Fully Depleted SOI MOSFET's," *IEEE Elec. Dev. Let.*, **15**(4), 117–119 (1994).

48. J.-H. Choi, Y.-J. Park, and H.-S. Min, "Electron Mobility Behavior in Extremely Thin SOI MOSFET's," *IEEE Elec. Dev. Let.*, **16**(11), 527–529 (1995).

49. S. Takagi and A. Toriumi, "New Experimental Findings on Hot Carrier Transport under Velocity Saturation Regime in Si MOSFETs," *IEDM Dig.*, 711–714 (1992).

50. K. Ohuchi, R. Ohba, H. Nhyama, K. Nakajima, and T. Mizuno, "A High-Performance 0.05μm SOI MOS FET: Possibility of Velocity Overshoot," *Jpn. J. Appl. Phys.*, **35**(2B), 960–964 (1996).

51. T. Mizuno and R. Ohba, "Experimental Study of Carrier Velocity Overshoot in Sub-0.1μm Devices–Physical Limitation of MOS Structures–," *IEDM Dig.*, 109–112 (1996).

52. R. Howes and W. Redman-White, "A Small-Signal Model for the Frequency-Dependent Drain Admittance in Floating-Substrate MOSFET's," *IEEE J. Sol. St. Ckts.*, **27**(8), 1186–1193 (1992).

53. S. S. Chen and J. B. Kuo, "An Analytical CAD Kink Effect Model of Partially-Depleted SOI NMOS Devices Operating in Strong Inversion," *Sol. St. Elec.*, **41**(3), 447–458 (1997).

54. H.-K. Yu, J.-S. Lyu, S.-W. Kang, and C.-K. Kim, "A Physical Model of Floating Body Thin Film Silicon-On-Insulator nMOSFET with Parasitic Bipolar Transistor," *IEEE Trans. Elec. Dev.*, **41**(5), 726–733 (1994).

55. O. Rozean, J. Jomaah, J. Boussey, and C. Raynaud, "Impact of Floating-Body Effect on RF Performances of SOI MOSFET," *ESSDERC Dig.*, 204–207 (1999).

56. G. G. Shahidi, "Mainstreaming of the SOI Technology," *ESSDERC Dig.*, 3–10 (1999).

57. C. F. Edwards, W. Redman-White, B. M. Tenbroek, M. S. L. Lee, and M. J. Uren, "The Effect of Body Contact Series Resistance on SOI CMOS Amplifier Stages," *IEEE Trans. Elec. Dev.*, **44**(12), 2290–2294 (1997).

58. J. Sleight and K. Mistry, "A Compact Schottky Body Contact Technology for SOI Transistors," *IEDM Dig.*, 419–422 (1997).

59. M. Yoshimi, M. Terauchi, A. Nishiyama, O. Arisumi, A. Murakoshi, K. Matsuzawa, N. Shigyo, S. Takeno, M. Tomita, K. Suzuki, Y. Ushiku, and H. Tango, "Suppression of the Floating-Body Effect in SOI MOSFET's by the Bandgap Engineering Method Using a $Si_{1-x}Ge_x$ Source Structure," *IEEE Trans. Elec. Dev.*, **44**(3), 423–430 (1997).

60. A. Nishiyama, O. Arisumi, and M. Yoshimi, "Suppression of the Floating-Body Effect in Partially-Depleted SOI MOSFET's with SiGe Source Structure and Its Mechanism," *IEEE Trans. Elec. Dev.*, **44**(12), 2187–2192 (1997).

61. T. Ohno, M. Takahashi, Y. Kado, and T. Tsuchiya, "Suppression of Parasitic Bipolar Action in Ultra-Thin-Film Fully-Depleted CMOS/SIMOX Devices by Ar-Ion Implantation into Source/Drain Regions," *IEEE Trans. Elec. Dev.*, **45**(5), 1071–1075 (1998).

62. M. Horiuchi and M. Tamura, "BESS: A Source Structure that Fully Suppresses the Floating Body Effects in SOI CMOSFET's," *IEEE Trans. Elec. Dev.*, **45**(5), 1077–1083 (1998).

63. Y.-C. Tseng, W. M. Huang, D. J. Monk, P. Welch, J. M. Ford, and J. C. S. Woo, "AC Floating Body Effects and the Resultant Analog Circuit Issues in Submicron Floating Body and Body-Grounded SOI MOSFET's," *IEEE Trans. Elec. Dev.*, **46**(8), 1685–1692 (1999).

64. J. R. Davis, A. E. Glaccum, K. Reeson, and P. L. F. Hemment, "Improved Subthreshold Characteristics of n-Channel SOI Transistors," *IEEE Elec. Dev. Let.*, **7**(10), 570–572 (1986).

65. G. G. Shahidi, C. A. Anderson, B. A. Chappell, T. I. Chappell, J. H. Comfort, B. Davari, R. H. Dennard, R. L. Franch, P. A. McFarland, J. S. Neely, T. H. Ning, M. R. Polcari, and J. D. Warnock, "A Room Temperature $0.1\mu m$ CMOS on SOI," *IEEE Trans. Elec. Dev.*, **41**(12), 2405–2412 (1994).

66. M. Horiuchi, T. Sakata, and S. Kimura, "Suppression of Bit-Line-Induced Disturbance in SOI DRAM/SRAM Cells by Bipolar Embedded Source Structure (BESS)," *Symp. VLSI Tech Dig.*, 157–158 (1997).

67. T. Tsuchiya, Y. Sato, and M. Tomizawa, "Three Mechanisms Determining Short-Channel Effects in Fully Depleted SOI MOSFET's," *IEEE Trans. Elec. Dev.*, **45**(5), 1116–1121 (1998).

68. M. Yoshimi, M. Terauchi, A. Murakoshi, M. Takahashi, K. Matsuzawa, N. Shigyo, and Y. Ushiku, "Technology Trends of Silicon-On-Insulator–Its Advantages and Problems to be Solved," *IEDM Dig.*, 429–432 (1994).

69. J. Chen, F. Assaderaghi, P.-K. Ko, and C. Hu, "The Enhancement of Gate-Induced- Drain-Leakage (GIDL) Current in SOI MOSFET and Its impact on SOI Device Scaling," *SOI Conf. Dig.*, 84–85 (1992).

70. D. Suh and J. G. Fossum, "A Physical Charge-Based Model for Non-Fully Depleted SOI MOSFET's and Its Use in Assessing Floating-Body Effects in SOI CMOS Circuits," *IEEE Trans. Elec. Dev.*, **42**(4), 728–737 (1995).

71. K. Suma, T. Tsuruda, H. Hidaka, T. Eimori, T. Oashi, Y. Yamaguchi, T. Iwamatsu, M. Hirose, F. Morishita, K. Arimoto, K. Fujishima, Y. Inoue, T. Nishimura, and T. Yoshihara, "An SOI-DRAM with Wide Operating Voltage Range by CMOS/SIMOX Technology," *IEEE J. Sol. St. Ckts.*, **29**(11), 1323–1329 (1994).

72. A. O. Adan, K. Higashi, and Y. Fukushima, "Analytical Threshold Voltage Model for Ultrathin SOI MOSFET's Including Short-Channel and Floating-Body Effects," *IEEE Trans. Elec. Dev.*, **46**(4), 729–737 (1999).

73. Y. Omura and K. Izumi, "Physical Background of Substrate Current Characteristics and Hot-Carrier Immunity in Short-Channel Ultrathin-Film MOSFET's/SIMOX," *IEEE Trans. Elec. Dev.*, **41**(3), 352–358 (1994).

74. Y. Apanovich, P. Blakey, R. Cottle, E. Lyumkis, B. Polsky, and A. Shur, "Numerical Simulation of Ultra-Thin SOI Transistor Using Non-Isothermal Energy Balance Model," *SOI Conf. Dig.*, 33–34 (1994).

75. Z. J. Ma, H. J. Wann, M. Chan, J. C. King, Y. C. Cheng, P. K. Ko, and C. Hu, "Hot-Carrier Effects in Thin-Film Fully Depleted SOI MOSFET's," *IEEE Elec. Dev. Let.*, **15**(6), 218–220 (1994).

76. E. P. Ver Ploeg, C. T. Nguyen, S. S. Wong, and J. D. Plummer, "Parasitic Bipolar Gain in Fully Depleted n-Channel SOI MOSFET's," *IEEE Trans. Elec. Dev.*, **41**(6), 970–977 (1994).

77. Y. Yamaguchi, T. Iwamatsu, H.-O. Joachim, H. Oda, Y. Inoue, T. Nishimura, and K. Tsukamoto, "Source-to-Drain Breakdown Voltage Improvement in Ultrathin-Film SOI MOSFET's Using a Gate-Overlapped LDD Structure," *IEEE Trans. Elec. Dev.*, **41**(7), 1222–1226 (1994).

78. J. Chen, F. Assaderaghi, H.-J. Wann, P. Ko, and C. Hu, "An Accurate Model of Thin Film SOI MOSFET Breakdown Voltage," *IEDM Dig.*, 671–674 (1991).

79. S. Maeda, Y. Hirano, Y. Yamaguchi, T. Iwamatsu, T. Ipposhi, K. Ueda, K. Mashiko, S. Maegawa, H. Abe, and T. Nishimura, "Substrate-Bias Effect and Source-Drain Breakdown Characteristics in Body-Tied Short-Channel SOI MOSFET's," *IEEE Trans. Elec. Dev.*, **46**(1), 151–158 (1999).

80. K. Mistry. G. Grula, J. Sleight, L. Bair, R. Stephany, R. Flatley, and P. Skerry, "A 2.0V, 0.35μm Partially Depleted SOI-CMOS Technology," *IEDM Dig.*, 583–586 (1997).

81. N. Kistler and J. Woo, "Detailed Characterization and Analysis of the Breakdown Voltage in Fully Depleted SOI n-MOSFET's," *IEEE Trans. Elec. Dev.*, **41**(7), 1217–1221 (1994).

82. E. P. Ver Ploeg, T. Watanabe, N. A. Kistler, J. C. S. Woo, and J. D. Plummer, "Elimination of Bipolar-Induced Breakdown in Fully-Depleted SOI MOSFETs," *IEDM Dig.*, 337–340 (1992).

83. F. L. Duan, D. E. Ioannou, W. C. Jenkins, H. L. Hughes, and M. S. T. Liu, "Channel Coupling Imposed Tradeoffs on Hot Carrier Degradation and Single Transistor Latch-Up in SOI MOSFET's," *Symp. Reliability Phys.*, 194–202 (1998).

84. J.-Y. Choi and J. G. Fossum, "Analysis and Control of Floating-Body Bipolar Effects in Fully Depleted Submicrometer SOI MOSFET's," *IEEE Trans. Elec. Dev.*, **38**(6), 1384–1391 (1991).

85. J. W. Sleight and K. R. Mistry, "DC and Transient Characterization of a Compact Schottky Body Contact Technology for SOI Transistors," *IEEE Trans. Elec. Dev.*, **46**(7), 1451–1456 (1999).

86. F. Assaderaghi, G. Shahidi, L. Wagner, M. Hsieh, M. Pelella, S. Chu, R. Dennard, and B. Davari, "Accurate Measurement of Pass-Transistor Leakage Current in SOI MOSFETs," *SOI Conf. Dig.*, 66–67 (1996).

87. M. M. Pelella and J. G. Fossum, D. Suh, S. Krishnan, K. A. Jenkins, and M. J. Hargrove, "Low-Voltage Transient Bipolar Effect Induced by Dynamic Floating-Body Charging in Scaled PD/SOI MOSFET's," *IEEE Elec. Dev. Let.*, **17**(5), 196–198 (1996).

88. T. Saraya and T. Hiramoto, "Mechanisms of Dynamic Pass Leakage Current in Partially Depleted SOI MOSFETs," *SOI Conf. Dig.*, 84–85 (1999).

89. A. Wei and D. A. Antoniadis, "Measurement of Transient Effects in SOI DRAM / SRAM Access Transistors," *IEEE Elec. Dev. Let.*, **17**(5), 193–195 (1996).

90. C.-E. D. Chen, M. Matloubian, and R. Sundaresan, "Single-Transistor Latch in SOI MOSFET's," *IEEE Elec. Dev. Let.*, **9**(12), 636–638 (1988).

91. J. S. T. Huang, J. S. Kueng, and T. Fabian, "An Analytical Model for Snapback in n-Channel SOI MOSFET's, " *IEEE Trans. Elec. Dev.*, **38**(9), 2082–2091 (1991).

92. L. Perron, C. Hamaguchi, A. Lacaita, S. Maegawa, and Y. Yamaguchi, "Switching Characteristics and Static Parameter Extraction in PD SOI MOSFET's," *MIEL Dig.*, 293–296 (1997).

93. M. M. Pelella, C. T. Chuang, C. Tretz, B. W. Curran, and M. G. Rosenfield, "Hysteresis in Floating-Body PD/SOI CMOS Circuits," *Symp. VLSI Tech, Systems, and Applications*, 278–281 (1999).

94. Y.-G. Chen, S.-Y. Ma, J. B. Kuo, Z. Yu, and R. W. Dutton, "An Analytical Drain Current Model Considering Both Electron and Lattice Temperatures Simultaneously for Deep Submicron Ultrathin SOI NMOS Devices with Self-Heating," *IEEE Trans. Elec. Dev.*, **42**(5), 899–906 (1995).

95. J. Jomaah, F. Balestra, and G. Ghibaudo, "Self-Heating Effects in SOI MOSFET's Operated at Low Temperature," *SOI Conf. Dig.*, 82–83 (1993).

96. J. Jomaah, G. Ghibaudo, F. Balestra, and J. L. Pelloie, "Impact of Self-Heating Effects on the Design of SOI Devices Versus Temperature," *SOI Conf. Dig.*, 114–115 (1995).

97. L. T. Su, J. E. Chung, D. A. Antoniadis, K. E. Goodson, and M. I. Flik, "Measurement and Modeling of Self-Heating in SOI NMOSFET's," *IEEE Trans. Elec. Dev.*, **41**(1), 69–75 (1994).

98. B. M. Tenbroek, W. Redman-White, M. S. L. Lee, R. J. T. Bunyan, and M. J. Uren, "Characterization of Geometry Dependence of SOI MOSFET Thermal Resistance and Capacitance Parameters," *SOI Conf. Dig.*, 114–115 (1997).

99. B. M. Tenbroek, W. Redman-White, M. S. L. Lee, R. J. T. Bunyan, M. J. Uren, and K. M. Brunson, "Characterization of Layout Dependent Thermal Coupling in SOI CMOS Current Mirrors," *IEEE Trans. Elec. Dev.*, **43**(12), 2227–2232 (1996).

100. N. D. Arora, L. T. Su, B. S. Doyle, and D. A. Antoniadis, "Modeling the I-V Characteristics of Fully-Depleted SOI MOSFETs Including Self-Heating," *SOI Conf. Dig.*, 19–20 (1994).

101. B. M. Tenbroek, M. S. L. Lee, W. Redman-White, R. J. T. Bunyan, and M. J. Uren, "Self-Heating Effects in SOI MOSFET's and Their Measurement by Small Signal Conductance Techniques," *IEEE Trans. Elec. Dev.*, **43**(12), 2240–2248 (1996).

102. H. C. Shin, I.-S. Lim, M. Racanelli, W.-L. M. Huang, J. Foerstner, and B.-Y. Hwang, "Analysis of Floating Body Induced Transient Behaviors in Partially Depleted Thin Film SOI Devices," *IEEE Trans. Elec. Dev.*, **43**(2), 318–325 (1996).

103. L. Perron, C. Hamaguchi, A. Lacaita, S. Maegawa, and Y. Yamaguchi, "Dynamic Floating Body Effects in PD SOI MOSFET's Biased in the Kink Region," *ESSDERC Dig.*, 524–527 (1999).

104. L. M. Perron, C. Hamaguchi, A. L. Lacaita, S. Maegawa, and Y. Yamaguchi, "Switch-Off Behavior of Floating-Body PD SOI MOSFET's," *IEEE Trans. Elec. Dev.*, **45**(11), 2372–2375 (1998).

105. D. Munteanu, D. A. Weiser, S. Cristoloveanu, O. Faynot, J.-L. Pelloie, and J. G. Fossum, "Generation-Recombination Transient Effects in Partially Depleted SOI Transistors: Systematic Experiments and Simulations," *IEEE Trans. Elec. Dev.*, **45**(8), 1678–1683 (1998).

106. K. A. Jenkins, J. Y.-C. Sun, and J. Gautier, "History Dependence of Output Characteristics of Silicon-on-Insulator (SOI) MOSFET's," *IEEE Elec. Dev. Let.*, **17**(1), 7–9 (1996).

107. A. Wei, M. J. Sherony, and D. A. Antoniadis, "Transient Behavior of the Kink Effect in Partially-Depleted SOI MOSFET's," *IEEE Elec. Dev. Let.*, **16**(11), 494–496 (1995).

108. K. Kato and K. Taniguchi, "Numerical Analysis of Switching Characteristics in SOI MOSFET's," *IEEE Trans. Elec. Dev.*, **33**(1), 133–139 (1986).

109. F. Assaderaghi, G. G. Shahidi, M. Hargrove, K. Hathorn, H. Hovel, S. Kulkarni, W. Rausch, D. Sadana, D. Schepis, R. Schulz, D. Yee, J. Sun, R. Dennard, and B. Davari, "History Dependence of Non-Fully Depleted (NFD) Digital SOI Circuits," *Symp. VLSI Tech. Dig.*, 122–123 (1996).

110. K. Ueda, H. Morinaka, Y. Yamaguchi, T. Iwamatsu, I. J. Kim, Y. Inoue, K. Mashiko, and T. Sumi, "Floating-Body Effects on Propagation Delay in SOI/CMOS LSIs," *SOI Conf. Dig.*, 142–143 (1996).

Problems

1. For the body effect curve for the inversion-mode SOI PMOS device as shown in Fig. 2.11, derive the body effect dependent threshold voltage model from Poisson's equation. Hint: Ref. 2.18. Verify your result using MEDICI device simulation.

2. Summarize the back gate effects on the performance of PD and FD SOI NMOS devices in terms of narrow channel effect, short channel effect, DIBL, kink effects, and breakdown.

3. In Fig. 2.89, there exists the bipolar snapback in the drain current characteristics of the FD device. If the doping density of the thin film is increased 10 times, what will happen to the drain current characteristics? Use MEDICI device simulation to verify your reasoning.

3. Derive a boundary equation in terms of thin-film thickness thin-film doping density, front gate oxide, buried oxide thickness, and back gate bias, to determine if an SOI NMOS device is FD or PD biased at the threshold voltage.

4. How do you reduce the leakage current problem in a DRAM circuit as shown in Fig.2.99?

5. Why is self-heating of the SOI MOS devices more serious at a low temperature? Does it mean that SOI MOS devices are not suitable for low-temperature operation? What will happen to the floating body effect at low temperatures?

3

SOI CMOS Devices–Part II

In this chapter, more topics on SOI CMOS devices are presented. Starting from the hot carriers of the SOI CMOS devices, the accumulation-mode and double-gate SOI devices are analyzed. Then, dynamic-threshold (DTMOS) devices are introduced, followed by the scaling trends of SOI CMOS devices and the SOI single electron transistors (SET). Next, the temperature dependence of the SOI devices is analyzed. Finally, sensitivities and radiation effects of SOI CMOS devices are described.

3.1 HOT CARRIERS

For an SOI NMOS device biased at a high drain voltage, the electric field near the drain is large, where impact ionization may result in the rapid increase in the drain current and even breakdown. On the other hand, as shown in Fig. 3.1, the electrons and the holes with a high energy (hot carriers) generated by impact ionization may leave the thin film to enter the front or the buried oxide to cause interface traps. As a result, the properties of the device are changed [1]. As shown in Fig. 3.1, a portion of the hot electrons generated by impact ionization are collected by the drain. Another portion of the hot electrons goes toward the front oxide. The other portion of the hot electrons goes toward the buried oxide. The hot holes generated by impact ionization tend to move and accumulate in the bottom of the thin film near the source to trigger turn-on of the parasitic bipolar device. If the back gate bias is sufficiently negative, sometimes hot holes may be trapped in the buried oxide. In the following sections, hot carrier effects of the SOI MOS devices are analyzed.

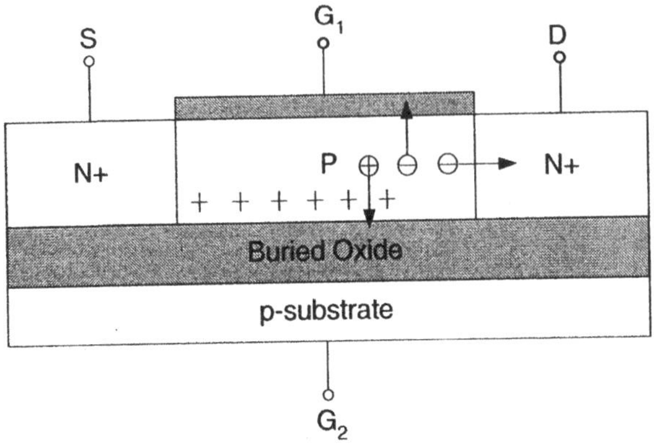

Fig. 3.1 Cross-section of an SOI NMOS device showing the front channel with impact ionization and hole accumulation in the back channel.

3.1.1 NMOS

Figure 3.2 shows the drain current characteristics of the SOI NMOS device with a front gate oxide of 250 Å, a thin film of 850 Å, and a buried oxide of 3600 Å, before and after a stress of 1500 s with the gate voltage of 1 V, the drain voltage of 8.5 V, the source voltage of 0 V, and the back gate voltage of 0 V in normal and reverse mode operations[2]. As shown in this figure, the normal mode operation is referred to the same biasing configuration after stress and the reverse mode operation is referred to the biasing configuration with the drain and the source reversed after stress. For the normal mode operation after stress, the drain current is reduced since the hot electrons are trapped in the front gate oxide to cause a positive shift of the threshold voltage. For the reverse mode operation, the decrease in the drain current is even larger. In addition, its breakdown voltage increases greatly. During the stress, a portion of the hot electrons generated by impact ionization cause a large amount of traps at the thin-film/buried oxide interface near the drain. Under the reverse mode operation, these interface traps become near the source, which can effectively recombine with the holes accumulating in this region, which are generated by impact ionization. Consequently, the turn-on of the parasitic bipolar device has been inhibited. Therefore, the amplification mechanism of the parasitic bipolar device does not function and the breakdown voltage is increased.

The behavior of the reverse mode operation can be also exemplified in the subthreshold characteristics of a PD SOI NMOS device. Figure 3.3 shows the subthreshold characteristics of the SOI NMOS device with a front gate oxide of 50 Å, a thin film of 500 Å, a buried oxide of 900 Å, a channel width of 10 μm and a channel length of 0.17 μm, before a 3-min stress at the drain voltage of 3 V and the gate voltage of 0.2 V, and after the stress in normal and reverse mode operation [3]. As shown in this figure, due to a short stress time of 3 min at a low drain voltage of 3 V, the subthreshold curve for the drain voltage of 0.1V after stress is about identical to the one before stress for both normal and reverse mode operations. For the subthreshold

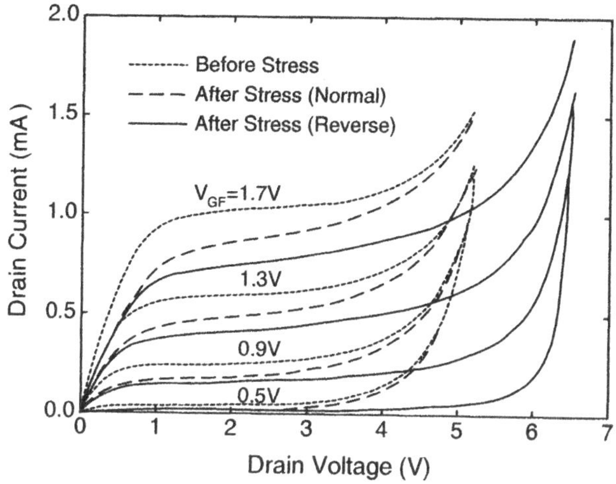

Fig. 3.2 Drain current characteristics of the SOI NMOS device with a front gate oxide of 250 Å, a thin film of 850 Å, and a buried oxide of 3600 Å, before and after a stress of 1500 s with the gate voltage of 1 V, the drain voltage of 6.5 V, the source voltage of 0 V, and the back gate voltage of 0 V, in normal and reverse mode operations. (Adapted from Zhang & Ma [2].)

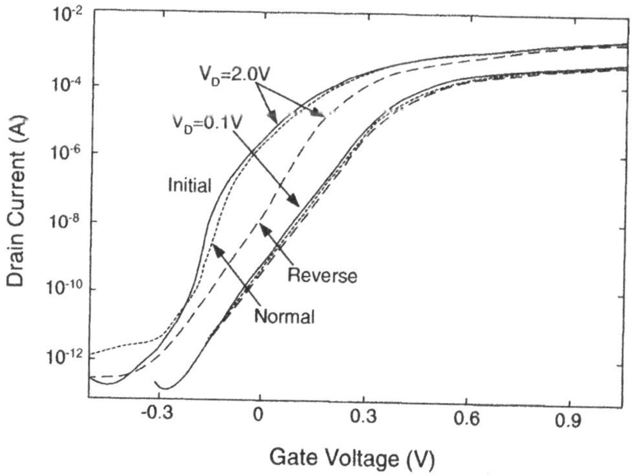

Fig. 3.3 Subthreshold characteristics of the SOI NMOS device with a front gate oxide of 50 Å, a thin film of 500 Å, a buried oxide of 900 Å, a channel width of 10 μm and a channel length of 0.17 μm, before a 3-min stress at the drain voltage of 3 V and the gate voltage of 0.2 V, and after the stress in normal and reverse mode operations. (Adapted from Tsuchiya et al. [3].)

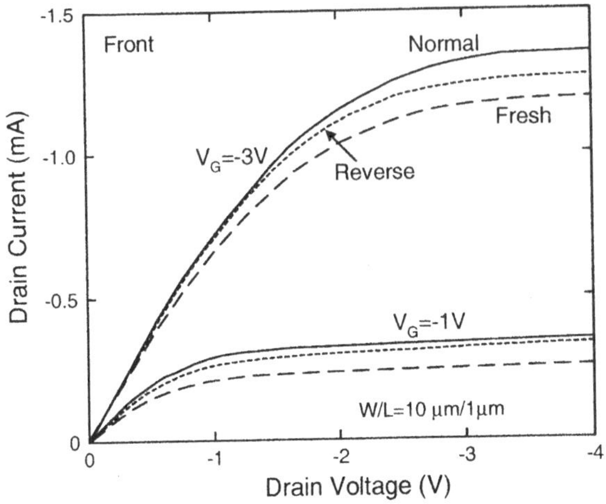

Fig. 3.4 Drain current versus drain voltage of the SOI PMOS device with a front gate oxide of 118 Å, a thin film of 916 Å, a channel width of 10 μm and a channel length of 1 μm, before and after hot-carrier stressing in normal and reverse mode operations. (Adapted from Chen et al. [4].)

behavior at the drain voltage of 2 V, the curve in the normal mode operation after stress is similar to the one before stress– kink effects still exist. In the reverse mode operation after stress, kink effects already disappear–in the reverse mode operation at the high drain voltage the drain current is decreased and the threshold voltage increases. This can be reasoned as follows. During the stress, some hot electrons cause some traps at the thin-film/buried oxide near the drain. In the reverse mode operation, these traps come near the source, which provide the effective recombination for the holes accumulating in this region. Thus the hole current related floating body effects and parasitic bipolar device effects have been inhibited effectively. Therefore, kink effects have been substantially lessened for the subsequent reverse mode operation.

3.1.2 PMOS

Hot-carrier effects of SOI NMOS devices have been described. In this subsection, hot-carrier effects of SOI PMOS devices are described. Figure 3.4 shows the drain current versus the drain voltage of an SOI PMOS device with a front gate oxide of 118 Å, a thin film of 916 Å, a channel width of 10 μm and a channel length of 1 μm, before and after hot-carrier stressing in normal and reverse mode operations [4]. As shown in Fig. 3.4, which is different from NMOS devices for both normal or reverse mode operations, the drain current after stress is always higher than the one before stress. It can be reasoned as follows. In the PMOS device, the influence of the hot electrons is still more significant. The hot-carrier generated electrons injected to the oxide attract more holes for channel conduction for the PMOS device since its threshold voltage has been reduced. Therefore, the drain current of the PMOS device

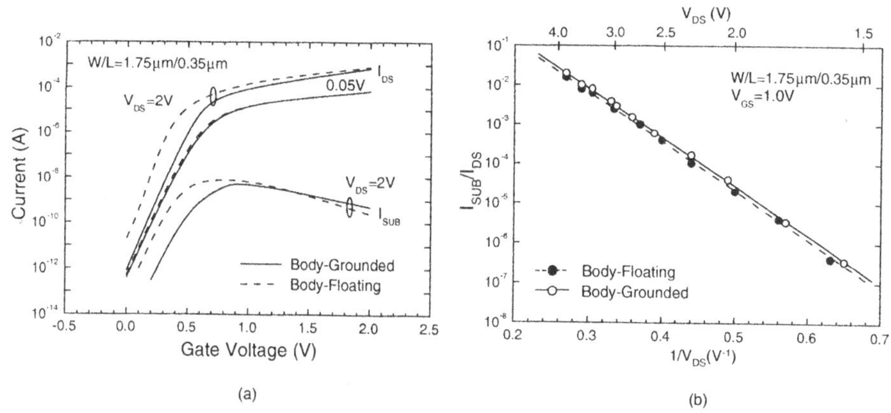

Fig. 3.5 (a) Drain and body currents versus gate voltage and (b) the ratio of the body current to the drain current versus the inverse of the drain voltage of the SOI NMOS device with a front gate oxide of 60 Å, a 2000 Å thin film doped with a p-type density of $2 \times 10^{17}\text{cm}^{-3}$, a buried oxide of 2000 Å, and a channel length of 0.35 μm. (Adapted from Sherony et al. [5].)

after stress is increased. Different from the NMOS device, in the normal operation after stress, the change in the drain current of the PMOS device is larger than that in the reverse mode operation.

3.1.3 Substrate Current

In addition to direct measurement of the device characteristics after stress, usually hot-carrier effects of SOI MOS devices can be assessed by their body current. Figure 3.5 shows (a) the drain and the body currents versus the gate voltage and (b) the ratio of the body current to the drain current versus the inverse of the drain voltage of the SOI NMOS device with a front gate oxide of 60 Å, a 2000 Å thin-film doped with a p-type density of $2 \times 10^{17}\text{cm}^{-3}$, a buried oxide of 2000 Å, and a channel length of 0.35 μm [5]. As shown in Fig. 3.5, at the high drain voltage, the electron/hole pairs are generated by impact ionization. The hot electrons are either absorbed by the drain or injected to the front/buried oxide. The hot holes are accumulated at the bottom of the thin film to form the body current. A high body current means a strong impact ionization, which implies a large amount of hot carriers generated and thus strong hot carrier effects. From Fig. 3.5, along with the increase in the gate voltage, the body current increases to its peak value, followed by falling. The key point for comparing the body current of an SOI MOS device is the peak body current. As shown in this figure, when the drain voltage increases, due to enhanced impact ionization, the body current of the SOI NMOS device also rises, which means stronger hot-carrier effects.

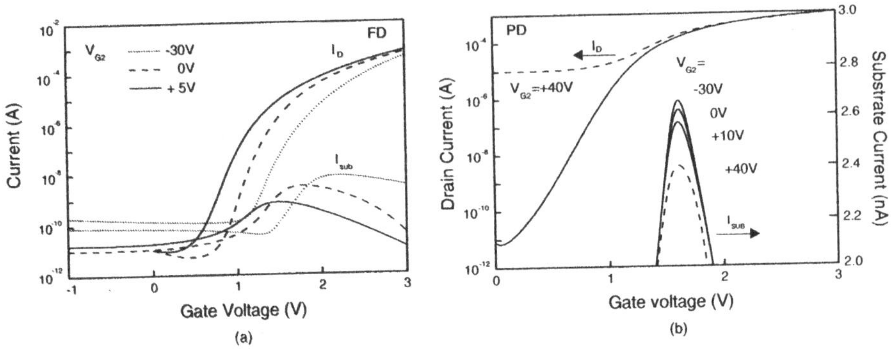

Fig. 3.6 Drain and body currents versus gate voltage (a) of the FD SOI NMOS device with a 1000 Å thin film doped with a p-type density of 10^{17}cm^{-3}, (b) of the PD SOI NMOS device with a 4000 Å thin film doped with a p-type density of 4×10^{16}cm^{-3}. Both FD and PD devices are with a front gate oxide of 150 Å, a buried oxide of 4000 Å, and a channel length of 0.8 μm, biased at various back gate biases. (Adapted from Duan et al. [6].)

3.1.4 Back Gate Bias

In addition to the drain voltage, the back gate bias effects on the hot-carrier behavior cannot be overlooked. Figure 3.6 shows the drain and the body currents versus the gate voltage (a) of the FD SOI NMOS device with a 1000 Å thin film doped with a p-type density of 10^{17}cm^{-3}, (b) of the PD SOI NMOS device with a 4000 Å thin-film doped with a p-type density of 4×10^{16}cm^{-3}. Both FD and PD devices are with a front gate oxide of 150 Å, a buried oxide of 4000 Å, and a channel length of 0.8 μm, biased at various back gate biases [6]. As shown in this figure, for the FD device, when the back gate bias becomes more negative, the peak value of the body current becomes higher, which implies more serious hot-carrier effects. In contrast, for the PD device, when the back gate bias becomes negative, the change in the peak body current is much smaller due to the shielding of the neutral region in the thin film.

3.1.5 Device Structure Dependence

In this subsection, the influence of the device structure in the hot-carrier effects of SOI NMOS devices is discussed. Figure 3.7 shows the body current versus the gate voltage of an SOI NMOS device with a front gate oxide of 90 Å, a thin film of 760 Å, and various channel lengths, biased at the drain voltage of 3 V [7]. As shown in this figure, when the channel length becomes smaller, the body current becomes larger, which implies more serious hot-carrier effects since the electric field in the device increases. In addition, the peak body current occurs at the gate voltage equal to its threshold voltage.

The thin-film thickness may have impact on the hot-carrier effects of an SOI MOS device. Figure 3.8 shows the degradation of the drain current and transconductance versus the thin-film thickness of the SOI NMOS device with a front gate oxide of 40 Å, a channel length of 0.1 μm, and a buried oxide of 800 Å, after a 10000-s

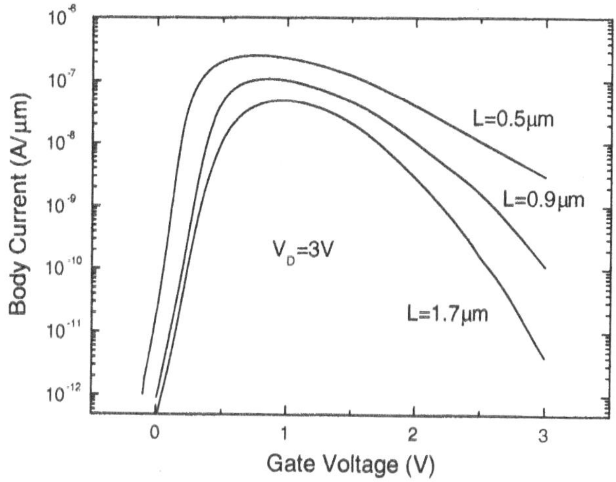

Fig. 3.7 Body current versus gate voltage of an SOI NMOS device with a front gate oxide of 90 Å, a thin film of 760 Å, and various channel lengths, biased at the drain voltage of 3 V. (Adapted from Wann et al. [7].)

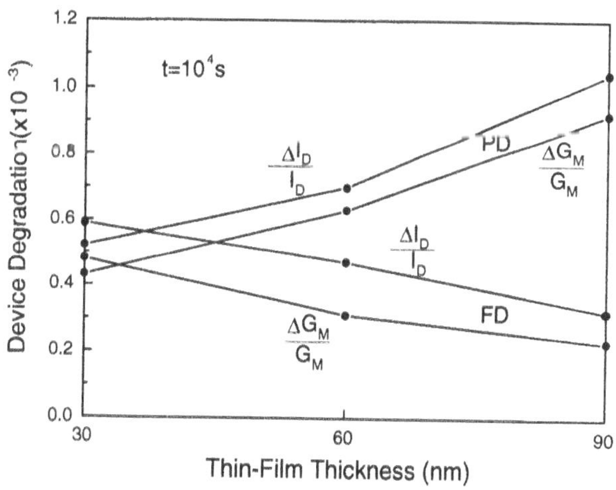

Fig. 3.8 Degradation of the drain current and transconductance versus the thin-film thickness of the SOI NMOS devices with a front gate oxide of 40 Å, a channel length of 0.1 μm, a buried oxide of 800 Å after a 10000-s stress at $V_G = V_T + 0.35$ V and $V_D = 1.5$ V, based on Monte Carlo simulation results. The PD device is with an N^+ polysilicon gate and a thin-film doping density of 10^{18}cm^{-3} and the FD device is with a mid-gate workfunction gate and a thin-film doping density of 10^{16}cm^{-3}. (Adapted from Hulfachor et al. [8]).

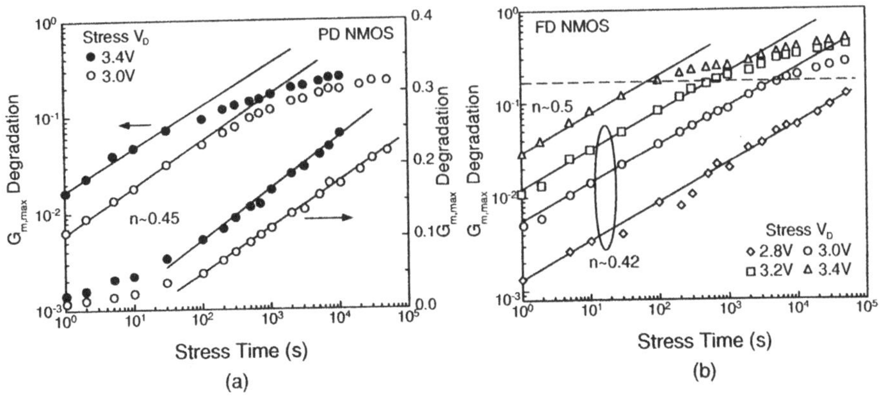

Fig. 3.9 Degradation of maximum transconductance versus stress time of (a) the PD SOI NMOS device with a front gate oxide of 45 Å, a buried oxide of 800 Å, a thin film of 1000 Å, and a channel length of 0.2 μm, stressed with the gate voltage at the threshold voltage and at the drain voltages of 3 and 3.4 V, and (b) the FD SOI NMOS device with a front gate oxide of 45 Å, a thin film of 400 Å, and a buried oxide of 3800 Å, stressed with the gate voltage at the threshold voltage and at various drain voltages. (From Renn et al. [9]. ©1998 IEEE.)

stress at $V_G = V_T + 0.35$ V and $V_D = 1.5$ V, based on Monte Carlo simulation results [8]. The PD device is with an N^+ polysilicon gate and a thin film doping density of 10^{18}cm^{-3} and the FD device is with a mid-gate workfunction gate and a thin-film doping density of 10^{16}cm^{-3}. As shown in this figure, for the PD device with a thicker thin film, hot-carrier effects are more noticeable and the degradation of the device is more serious. In contrast, for the FD device, a thicker thin film leads to an improvement in the device degradation, which can be reasoned as follows. For the PD device with a high thin film doping density, if the thin film becomes thicker, the vertical electrical field toward upward near the drain region becomes smaller, hence it is easier for hot electrons to enter the front oxide to affect the device performance. For the FD device with a low thin film doping density, if the thin film becomes thicker, the lateral electric field near the drain becomes smaller, hence the generation of hot carriers is reduced. Therefore, the device degradation becomes less.

3.1.6 Stress Time

Figure 3.9 shows the degradation of maximum transconductance versus the stress time of (a) the PD SOI NMOS device with a front gate oxide of 45 Å, a buried oxide of 800 Å, a thin film of 1000 Å, and channel length of 0.2 μm, stressed with the gate voltage at the threshold voltage and at the drain voltages of 3 and 3.4 V, and (b) the FD SOI NMOS device with a front gate oxide of 45 Å, a thin film of 400 Å, and a buried oxide of 3800 Å, stressed with the gate voltage at the threshold voltage and at various drain voltages [9]. As shown in this figure, for both PD and FD devices, with a higher drain voltage, the hot-carrier effect and thus the degradation are more serious. The degradation of the maximum transconductance with respect to the stress

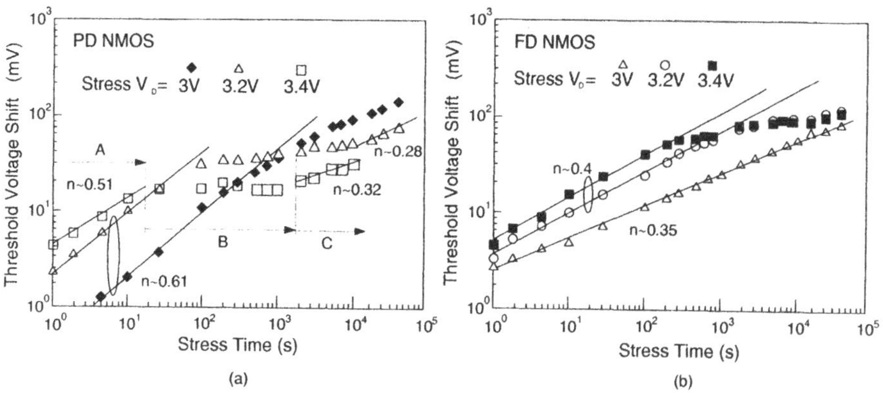

Fig. 3.10 Threshold voltage shift versus stress time of (a) the PD SOI NMOS device with a front gate oxide of 45 Å, a buried oxide of 800Å, a thin film of 1000 Å and a channel length of 0.2 μm, stressed with the gate voltage at the threshold voltage and at various drain voltages, and (b) the FD SOI NMOS device with a front gate oxide of 45 Å, a thin film of 400 Å, and a buried oxide of 3800 Å, stressed with the gate voltage at the threshold voltage and at various drain voltages. (From Renn et al. [9]. ©1998 IEEE.)

time is in terms of a power-law relationship. After reaching a certain stress time, the degradation tends to be saturated.

Figure 3.10 shows the threshold voltage shift versus the stress time of (a) the PD SOI NMOS device with a front gate oxide of 45 Å, a buried oxide of 800 Å, a thin film of 1000 Å and a channel length of 0.2 μm, stressed with the gate voltage at the threshold voltage and at various drain voltages, and (b) the FD SOI NMOS device with a front gate oxide of 45 Å, a thin film of 400 Å, and a buried oxide of 3800 Å, stressed with the gate voltage at the threshold voltage and at various drain voltages [9] As shown in this figure, the degradation of the threshold voltage for the PD device is more complicated as compared to the FD device and at a high drain voltage, the threshold voltage shift of the PD device with respect to the stress time has three stages. Compared to the FD device, stage B is unique for the PD device. In stages A and C, the degradation of the PD device is caused mainly by electron injection. In contrast, in stage B, due to the floating body and the parasitic bipolar device effect, which are unique for the PD device, holes may be trapped in the interface state to cause changes in the degradation of the threshold voltage.

Figure 3.11 shows the degradation of the drain current versus the stress time of the FD SOI NMOS device with a front gate oxide of 108 Å, a thin film of 700 Å doped with a p-type density of $5 \times 10^{16} cm^{-3}$, a buried oxide of 3600 Å, and a channel length of 0.8 μm, stressed with the drain voltage of 4.5 V and various gate voltages [10]. As shown in this figure, the trends on the drain current degradation are similar to those of the maximum transconductance degradation. Among all gate voltages, when the gate voltage is biased at the threshold voltage, the degradation of the drain current is the most, which is correlated to the situation of the maximum substrate current mentioned before. The worst degradation occurs at the gate voltage of the

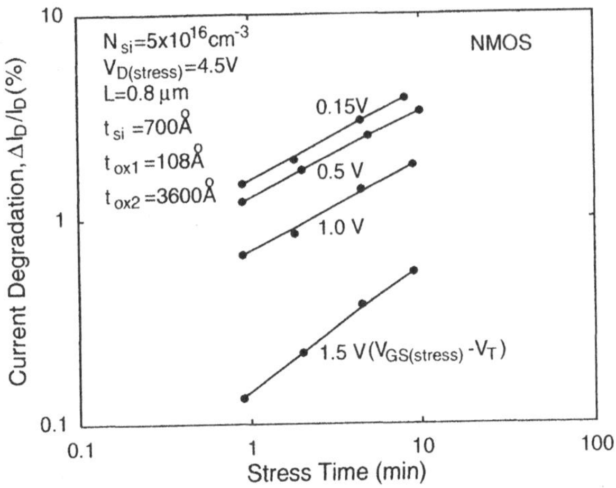

Fig. 3.11 Degradation of drain current versus stress time of the FD SOI NMOS device with a front gate oxide of 108 Å, a thin film of 700 Å doped with a p-type density of $5 \times 10^{16}\text{cm}^{-3}$, a buried oxide of 3600 Å, and a channel length of 0.8 μm, stressed with the drain voltage of 4.5 V and various gate voltages. (Adapted from Su et al. [10].)

threshold voltage ($V_G = V_T$), which is different from that for the bulk device, where the worst degradation occurs at the gate voltage of half of the power supply voltage ($V_G = \frac{1}{2}V_{DD}$). The difference in the occurrence of the maximum degradation between the SOI and bulk devices is due to the more serious parasitic bipolar device effects in the SOI device.

3.1.7 Isolation Structure

The isolation structure may affect the reliability of the SOI device. Figure 3.12 shows the degradation of the transconductance versus the stress time of an SOI NMOS device with a thin film of 1000 Å, a channel width of 10 μm, and a channel length of 0.25 μm, using LOCOS and shallow-trench isolation (STI), stressed at the gate voltage of the threshold voltage ($V_G = V_T$) and two drain voltages [11]. As shown in this figure, the SOI device using LOCOS has a worse degradation as compared to the SOI device using STI. In addition, the SOI device with LOCOS shows a two-stage degradation situation, which is different from the SOI device with STI. The reason is that for the SOI device with LOCOS, the physical stress at the channel edge is much higher than that for the device with STI. Consequently, it is easier to generate interface states to incur degradation.

Figure 3.13 shows the degradation of the transconductance versus the stress time of the SOI NMOS device with a thin film of 1000 Å, a channel width of 10μm, and a channel length of 0.25μm, for buried oxide thicknesses of 1000 and 4000 Å, stressed with the gate voltage of the threshold voltage ($V_G = V_T$) and two drain voltages [11]. As shown in the figure, for the SOI device with LOCOS, a thicker buried oxide leads

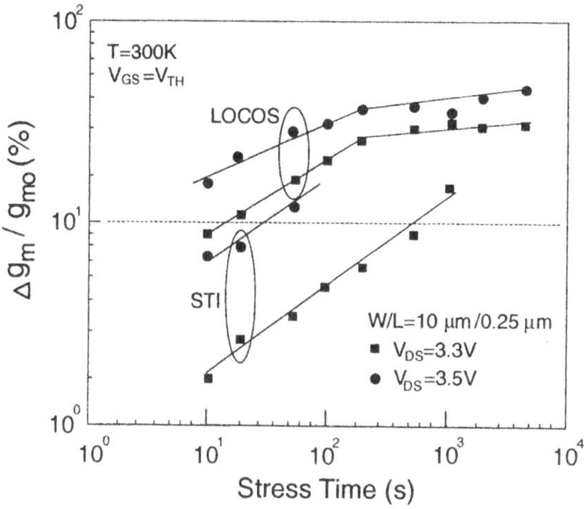

Fig. 3.12 Degradation of transconductance versus stress time of the SOI NMOS device with a thin film of 1000 Å, a channel width of 10 μm, and a channel length of 0.25 μm, using LOCOS and shallow-trench isolation (STI), stressed at the gate voltage of the threshold voltage ($V_G = V_T$) and two drain voltages. (Adapted from Lee et al. [11].)

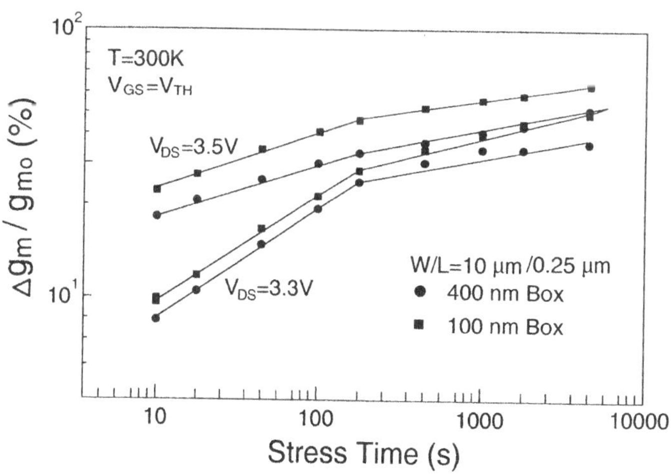

Fig. 3.13 Degradation of transconductance versus stress time of the SOI NMOS device with a thin film of 1000 Å, a channel width of 10 μm, and a channel length of 0.25 μm, for buried oxide thicknesses of 1000 and 4000 Å, stressed with the gate voltage of the threshold voltage ($V_G = V_T$) and two drain voltages. (Adapted from Lee et al. [11].)

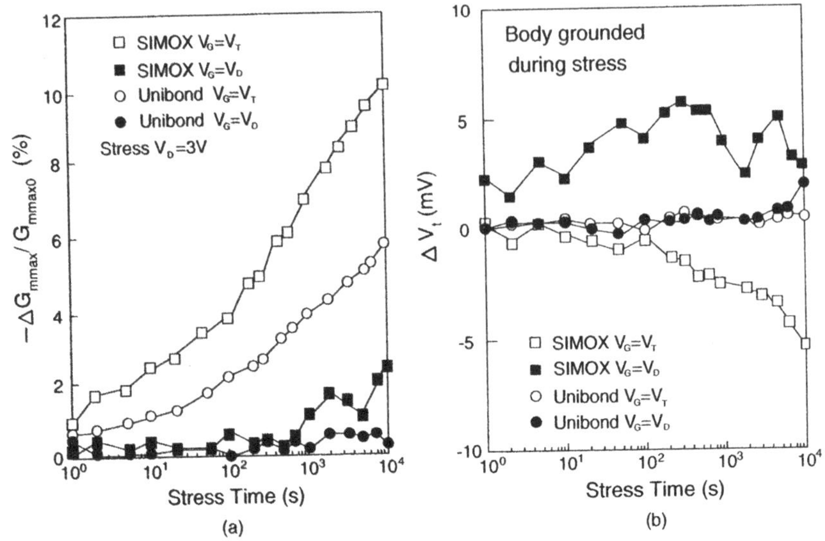

Fig. 3.14 (a) Degradation of transconductance and (b) threshold voltage shift versus stress time of the SOI NMOS device with a front gate oxide of 45 Å, a thin film of 400 Å, and a buried oxide of 3800 Å, based on SIMOX and Unibond wafers, stressed at the drain voltage of 3V and two gate voltages and the body grounded. (From Renn et al. [12]. ©1998 IEEE.)

to an improvement in the reliability. With a thicker buried oxide, there exists more space to relieve the physical stress generated by LOCOS. Therefore, the generation of the interface states is reduced and the reliability of the device is enhanced. Note that the physical stress is different from the electrical stress.

3.1.8 SOI Wafers

The quality of the SOI wafer, which is dependent on the manufacture method, directly affects the reliability of the SOI device. Figure 3.14 shows (a) the degradation of the transconductance and (b) the threshold voltage shift versus the stress time of the SOI NMOS device with a front gate oxide of 45 Å, a thin film of 400 Å, and a buried oxide of 3800 Å, based on SIMOX and Unibond wafers, stressed at the drain voltage of 3 V and two gate voltages and the body grounded [12]. Unibond wafers are referred to BESOI and Smart-Cut technologies using bonding to form SOI wafers. The SOI device based on the Unibond wafers shows a better performance in the degradation and the threshold voltage shift as compared to the SOI device based on SIMOX wafers. Since the buried oxide of the SOI device using the Unibond wafers is formed by thermal oxidation, its quality is better than that of the device using SIMOX technique. In addition, the defect density of the thin film in the device using Unibond wafers is smaller than that using the SIMOX technique. Accounting to both factors, the reliability of the Unibond SOI devices is better than that of the SIMOX ones.

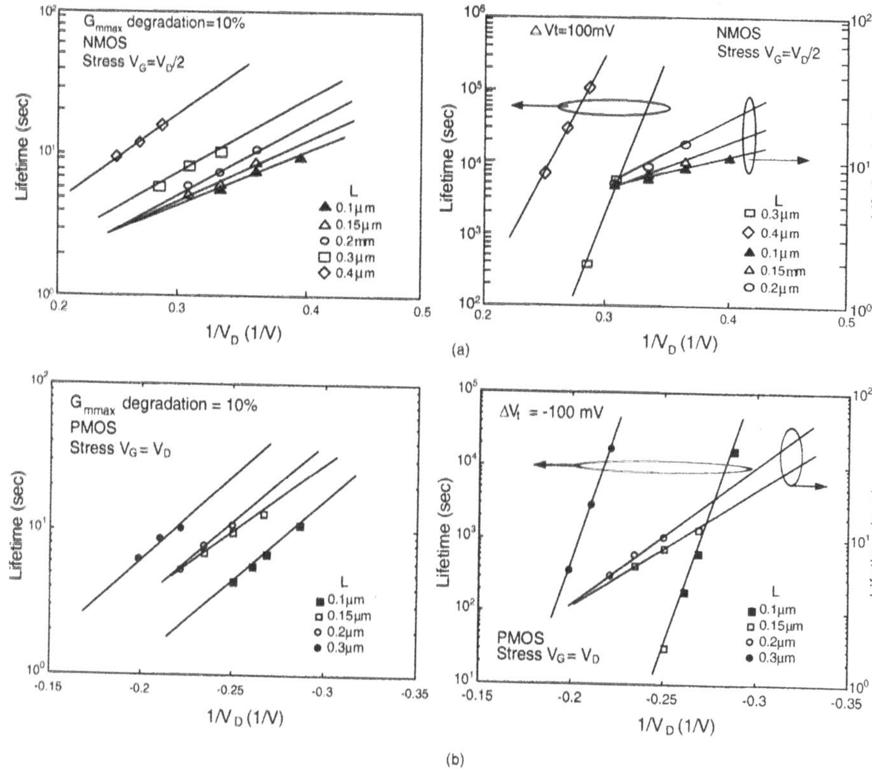

Fig. 3.15 Lifetime to reach the 10% transconductance degradation ($\Delta g_{mmax}/g_{mmax} = 10\%$) and the 100 mV threshold voltage shift versus the inverse of the drain voltage of the SOI device with a front gate oxide of 45 Å, a thin film of 1000 Å, a buried oxide of 800 Å, and various channel lengths, stressed with the gate voltage equal to one-half of the drain voltage ($V_G = V_D/2$) for (a) NMOS and (b) PMOS. (From Renn et al. [13]. ©1998 IEEE.) Note that for the PMOS device (b) the arrangement is identical to (a) except the gate voltage is equal to the drain voltage ($V_G = V_D$).

3.1.9 Lifetime

Figure 3.15 shows (a) the lifetime to reach the 10% transconductance degradation ($\Delta g_{mmax}/g_{mmax} = 10\%$) and the 100 mV threshold voltage shift versus the inverse of the drain voltage of the SOI device with a front gate oxide of 45 Å, a thin film of 1000 Å, a buried oxide of 800 Å, and various channel lengths, stressed with the gate voltage equal to one-half of the drain voltage ($V_G = V_D/2$) for (a) NMOS and (b) PMOS [13]. Note that for the PMOS (b) the arrangement is identical to (a) except the gate voltage is equal to the drain voltage ($V_G = V_D$). As shown in this figure, the two lifetimes have similar trends. With a larger drain voltage (a smaller gate voltage), or with a smaller channel length, the hot-carrier effects are stronger and the lifetime of the device is shorter.

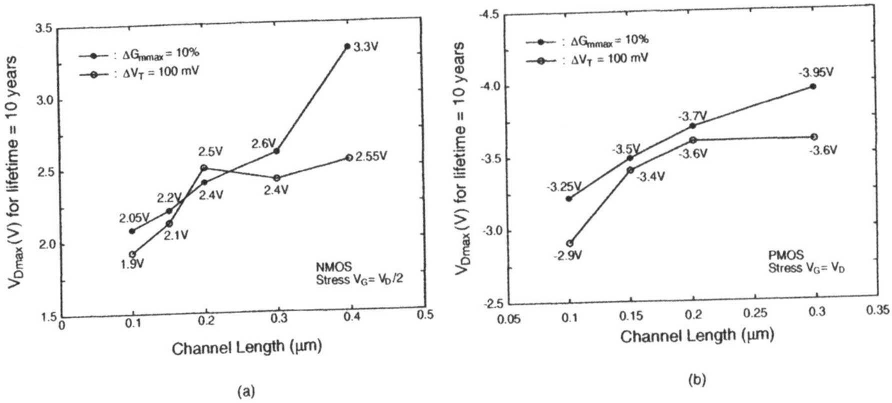

Fig. 3.16 Maximal drain voltage to obtain a lifetime of 10 years versus channel length of the SOI (a) NMOS and (b) PMOS devices with conditions and results extrapolated from Fig. 3.15. (Adapted from Renn et al. [13].)

Figure 3.16 shows the maximal drain voltage to obtain a lifetime of 10 years versus the channel length of the SOI (a) NMOS and (b) PMOS devices with conditions and results extrapolated from Fig. 3.15 [13]. As shown in this figure, the hot-carrier effect of the SOI NMOS device is more serious than that of the SOI PMOS device. In order to reach the standard with a lifetime of 10 years, the maximal drain voltage endurable by the NMOS device is lower than that by the PMOS device. When the channel length becomes smaller, in order to maintain a reasonable lifetime, its drain voltage needs to be scaled down accordingly.

When doing the lifetime extrapolation related analysis, the gate voltage is important. Figure 3.17 shows (a) the lifetime to reach the 10% transconductance degradation versus the inverse of the drain voltage and (b) the transconductance degradation versus the stress time of an SOI NMOS device with a front gate oxide of 105 Å, a thin film of 900 Å doped with a p-type density of $1.7 \times 10^{17}\text{cm}^{-3}$, and a buried oxide of 4000 Å, stressed with the gate voltage equal to the threshold voltage ($V_G = V_T$) and one-half of the drain voltage ($V_G = V_D/2$). Note for (b) only stressed with the gate voltage equal to the threshold voltage ($V_G = V_T$) is considered [14]. As shown in Fig. 3.17(a), for the case with the gate voltage equal to the one-half of the drain voltage ($V_G = V_D/2$), the trend on the lifetime is simple. In contrast, for the case with the gate voltage equal to the threshold voltage ($V_G = V_T$), a two-slope rise behavior can be identified. As shown in Fig. 3.17(b), the slopes of the curves can be classified into two cases: (a) the drain voltage of > 5 V ($V_D > 5$ V) and (b) the drain voltage < 5 V ($V_D < 5$ V). For case (a) with the drain voltage > 5 V ($V_D > 5$ V), the mechanism of the hot-carrier effect is dominated by the oxide charge trapping. For case (b) with the drain voltage < 5 V ($V_D < 5$ V), the hot-carrier effect is mainly determined by the generation of the interface state. Due to the different mechanisms in (a) and (b), their slopes are also different. Consequently, the lifetime versus the inverse of the drain voltage curve, as shown in Fig. 3.17(a), has been classified into two different slopes with the drain voltage of 5 V ($V_D = 5$ V) as the boundary.

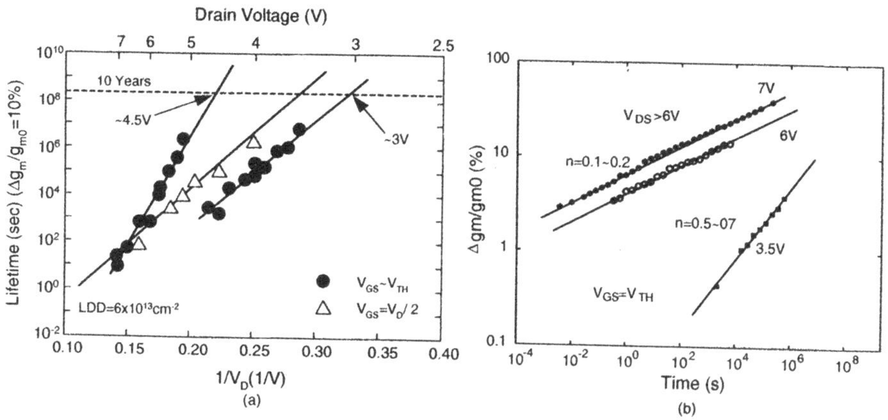

Fig. 3.17 (a) Lifetime to reach the 10% transconductance degradation versus the inverse of the drain voltage and (b) the transconductance degradation versus stress time of an SOI NMOS device with a front gate oxide of 105 Å, a thin film of 900 Å doped with a p-type density of $1.7 \times 10^{17} cm^{-3}$, and a buried oxide of 4000 Å, stressed with the gate voltage equal to the threshold voltage ($V_G = V_T$) and one-half of the drain voltage ($V_G = V_D/2$). Note that for (b) only stressed with the gate voltage equal to the threshold voltage ($V_G = V_T$) is considered. (From Wang-Ratkovic et al. [14]. ©1995 IEEE.)

The floating body effect of PD SOI devices may affect the hot carrier lifetime of the device. Figure 3.18 shows the hot carrier lifetime to reach the 10% transconductance degradation versus the inverse of the drain voltage of the PD SOI NMOS device with a front gate oxide of 70 Å, a thin film of 2000 Å, a buried oxide of 4000 Å, and a channel length of 0.3 μm, stressed at the gate voltage of 0.7 V, with its body floating and tied [15]. As shown in this figure, compared to the case with the body tied, the device with the body floating has a much smaller lifetime. In addition, for the drain voltage > 3 V ($V_D > 3$ V), the lifetime of the SOI device with the body floating suddenly becomes worse due to the turn-on of the parasitic bipolar device. As a result, the drain current increases quickly, which aggravates the hot-carrier effect.

3.2 ACCUMULATION-MODE DEVICES

For bulk devices, based on the reverse biased junction isolation, only inversion-mode devices are available. In the lateral channel direction of inversion-mode PMOS devices, its doping structure from the source via the substrate to the drain is $p^+ - n - p^+$. Based on buried oxide isolation, the accumulation-mode devices are available in SOI structures. Accumulation-mode SOI MOS devices are referred to the structure where the type of their thin-film doping is the same as that of the source/drain region. For the SOI accumulation-mode NMOS device, from the source via the thin-film body to the drain, it is with the $n^+ - n - n^+$ structure. For PMOS, it is $p^+ - p - p^+$. Due to the difference in the doping structure, the behaviors of the accumulation-

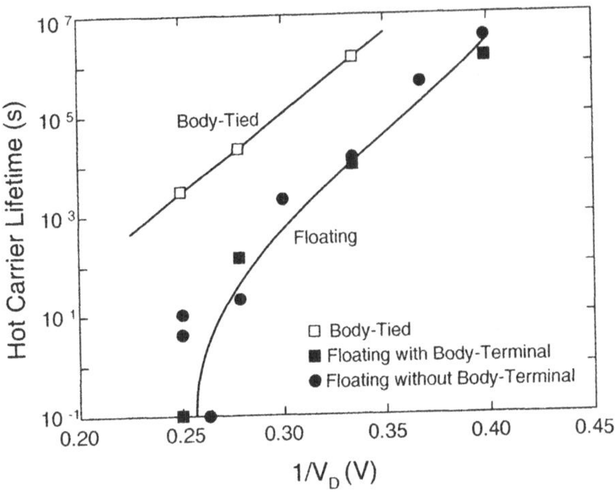

Fig. 3.18 Hot-carrier lifetime to reach the 10% transconductance degradation versus the inverse of the drain voltage of the PD SOI NMOS device with a front gate oxide of 70 Å, a thin film of 2000 Å, a buried oxide of 4000 Å, and a channel length of 0.3 μm, stressed at the gate voltage of 0.7 V, with its body floating and tied. (Adapted from Maeda et al. [15].)

mode SOI MOS devices are quite different from those of the inversion-mode. In this section, behaviors of the accumulation-mode SOI MOS devices are analyzed. Figure 3.19 shows the cross-section of an accumulation-mode FD SOI PMOS device with a recessed channel to reduce the thickness of the thin film in the channel region such that the controllability of the front gate can be enhanced. In the source/drain region, the thickness of the thin film is maintained to lower the parasitic resistance [16]. In the thin-film region, the doping structure is $p^+ - p - p^+$.

3.2.1 DC Behavior

In an accumulation-mode SOI MOS device, the doping density of the thin film is the same as the source/drain region. When the device is turned on, conduction current is via the buried channel or the front channel depending on the biasing condition of the front gate and the channel potential. Figure 3.20 shows the cross section of an accumulation-mode SOI PMOS device in the substrate direction with (a) both the upper and the lower portions of the thin-film depleted and (b) the lower portion depleted and the top surface accumulated with holes [17]. As shown in Fig. 3.20, when the device is turned on, if the front gate voltage is not negative enough, channel conduction can only be done via the existing buried channel. If the front gate voltage is sufficiently negative, as shown in Fig. 3.20(b), the surface of the thin-film can be accumulated with holes to form a surface channel for current conduction. Since both buried and surface channels exist in the device, depending on the biasing conditions, the operation of the device is classified into seven categories. In contrast, conventional inversion-mode MOS devices have only three operation regions– the subthreshold,

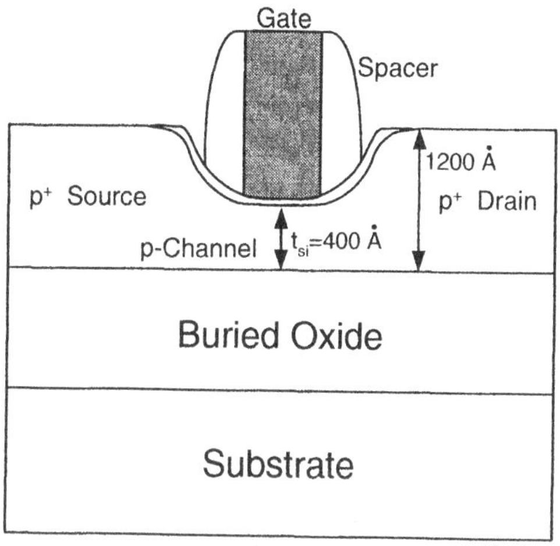

Fig. 3.19 Cross-section of an accumulation-mode FD SOI PMOS device with a recessed channel. (Adapted from Raynaud et al. [16].)

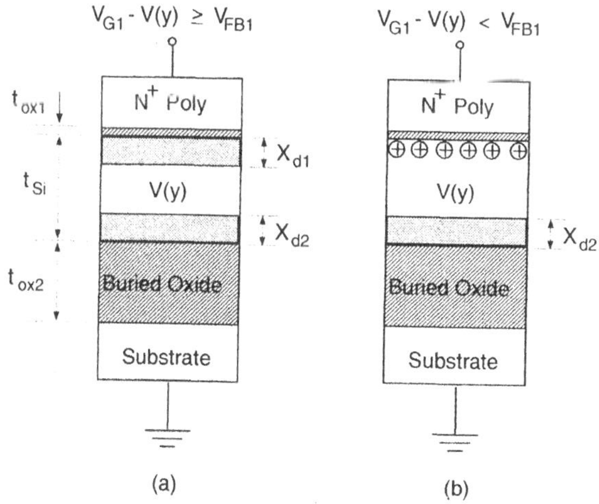

Fig. 3.20 Cross-section of an accumulation-mode SOI PMOS device in the substrate direction with (a) both the upper and the lower portions of the thin film depleted and (b) the lower portion depleted and the top surface accumulated with holes. (Adapted from Su & Kuo [17].)

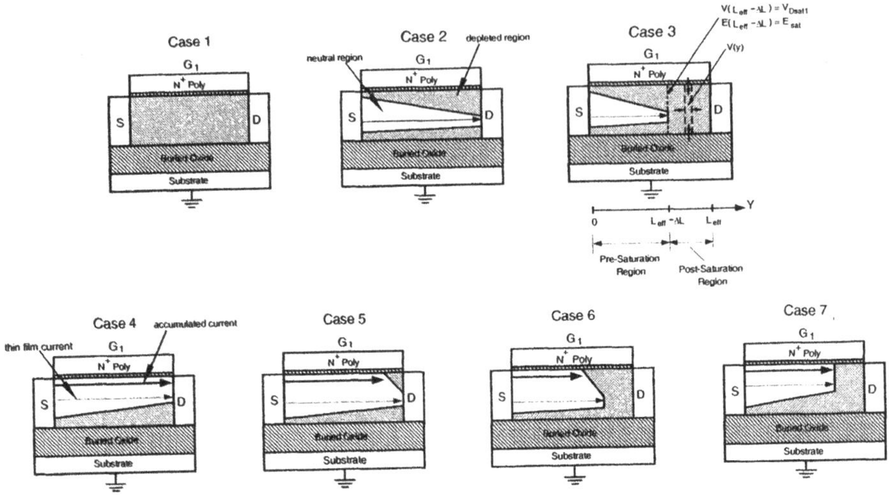

Fig. 3.21 Various operation conditions in the accumulation-mode SOI PMOS device biased at various conditions. Case 1: fully depleted. Case 2: depleted at upper and lower. Case 3: depleted at upper and lower and pinched off at center. Case 4: depleted at lower and accumulated at top. Case 5: depleted at lower and pinched off at top accumulated. Case 6: pinched at top accumulated and center central. Case 7: pinched off at the same lateral location at top accumulated and central neural. (Adapted from Su& Kuo [17].)

triode, and saturation regions. As shown in Fig. 3.21, Case 1 (fully depleted) implies the subthreshold region– the device is not turned on yet. Case 2 (depleted at upper and lower) implies that the device is turned on with only the buried channel for current conduction. If the drain voltage is increased, the buried channel near the drain may be pinched off as shown in Case 3. If the front gate voltage is enhanced, both the surface channel and the buried channel are on as shown in Case 4. If the drain voltage is increased further, the surface channel may first be pinched off as shown in Case 5. If the drain voltage is sufficiently large, both the surface channel and the buried channel are pinched off as shown in Case 6. In Case 7, due to a very large drain voltage, because of prepinch off velocity saturation, both the surface and the buried channels enter the post-saturation region at the same lateral location.

Depending on the biasing conditions, the accumulation-mode device may be operating in any one of the seven cases described. Figure 3.22 shows (a) the drain current versus the drain voltage in the strong inversion region and (b) the drain current versus the front gate voltage of the accumulated-mode SOI PMOS device with a front gate oxide of 125 Å, a 1000 Å thin-film doped with a p-type density of $2 \times 10^{16} cm^{-3}$, a buried oxide of 3500 Å, with a channel length of 0.55 μm, and channel width of 8 μm [17]. From the drain current characteristics, as shown in Fig. 3.22, the drain current behavior looks like a typical inversion-mode device. From turn-off to completely turn-on, the current conduction mechanism is complex. Once fully turned on, the device is mainly working in Case 7 with a high drain voltage or Case 4 with a low drain voltage.

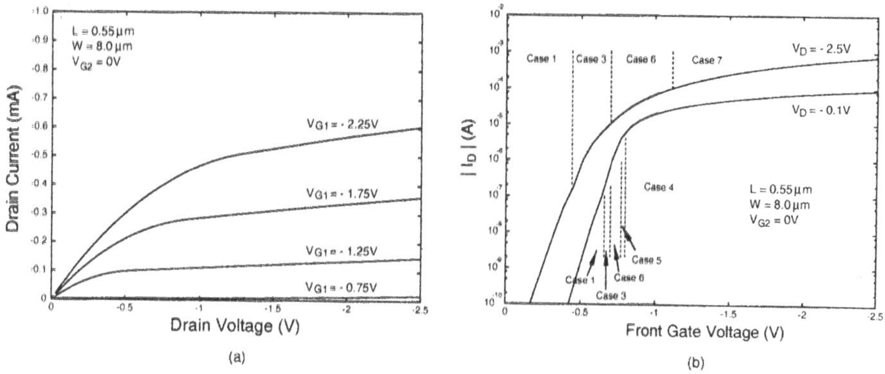

Fig. 3.22 (a) Drain current versus drain voltage and (b) drain current versus front gate voltage of the accumulation-mode SOI PMOS device with a front gate oxide of 125 Å, a 1000 Å thin film doped with a p-type density of $2 \times 10^{16}\text{cm}^{-3}$, a buried oxide of 3500 Å, with a channel length of 0.55 μm, and a channel width of 8 μm. (Adapted from Su & Kuo [17].)

3.2.2 AC Behavior

Figure 3.23 shows the gate-source capacitance (C_{GS}), the source-gate capacitance (C_{SG}), the gate-drain capacitance (C_{GD}), and the drain-gate capacitance (C_{DG}) versus the front gate voltage (V_{G1}) of (a) the accumulation-mode device and (b) the inversion-mode SOI PMOS devices with a front gate oxide of 125 Å, a thin film of 1000 Å, a buried oxide of 3500 Å doped with (a) a p-type density and (b) an n-type density of $4 \times 10^{16}\text{cm}^{-3}$, a channel length of 0.55 μm, and a channel width of 8 μm [18]. Compared to the inversion-mode device, the capacitance behavior of the accumulation-mode device is quite different. As shown in Fig. 3.23(a), after turn-on, the increase of its capacitances can be divided into two stages (first in Case 3, followed by Case 6). As for the inversion-mode device, after turn-on its capacitances increase quickly. The two-stage increase in the capacitances of the accumulation-mode devices is due to the existence of the buried channel. When the front gate voltage increases, the accumulation-mode SOI PMOS device turns on first with a buried channel. As shown in Fig. 3.24, under this situation, the equivalent capacitance of the device is determined by the front oxide capacitance (C_{ox1}) in series with the capacitance of the depletion region between the buried channel and the front oxide (C_{cap}). When the front gate voltage becomes more negative, its buried channel widens and the depletion region shrinks. Thus its depletion capacitance (C_{cap}) and hence its equivalent capacitance increase. This trend continues until the emergence of the surface channel, at which the equivalent capacitance reaches its maximum value. In contrast, for the inversion-mode device, when it turns on it is already with the surface channel--capacitances increase rapidly.

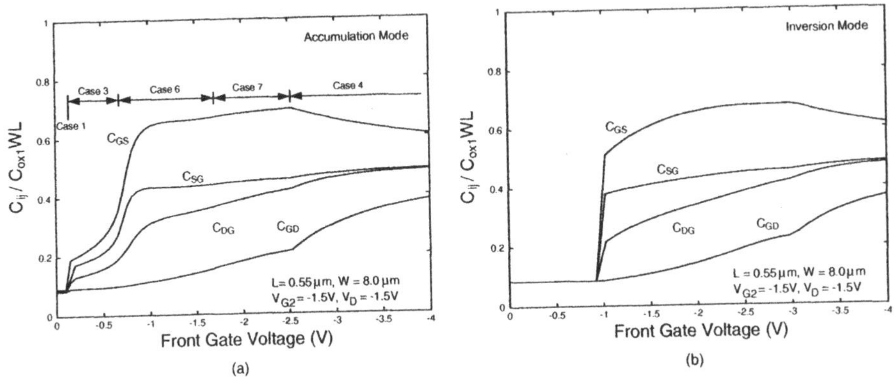

Fig. 3.23 Gate-source capacitance (C_{GS}), source-gate capacitance (C_{SG}), gate-drain capacitance (C_{GD}), and drain-gate capacitance (C_{DG}) versus the front gate voltage of (a) the accumulation-mode device and (b) the inversion-mode SOI PMOS device with a front gate oxide of 125 Å, a thin film of 1000 Å, a buried oxide of 3500 Å doped with (a) a p-type density and (b) an n-type density of $4 \times 10^{16}\text{cm}^{-3}$, a channel length of 0.55 μm, and a channel width of 8 μm. (Adapted from Su & Kuo [18].)

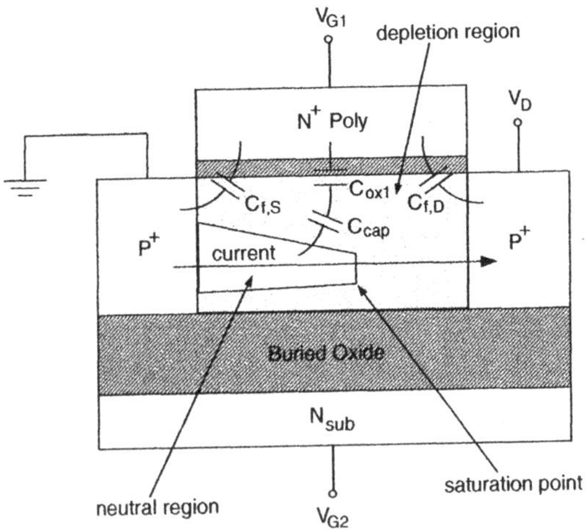

Fig. 3.24 Description of the internal capacitances of the accumulation-mode SOI PMOS device biased in Case 3. (Adapted from Su & Kuo [18].)

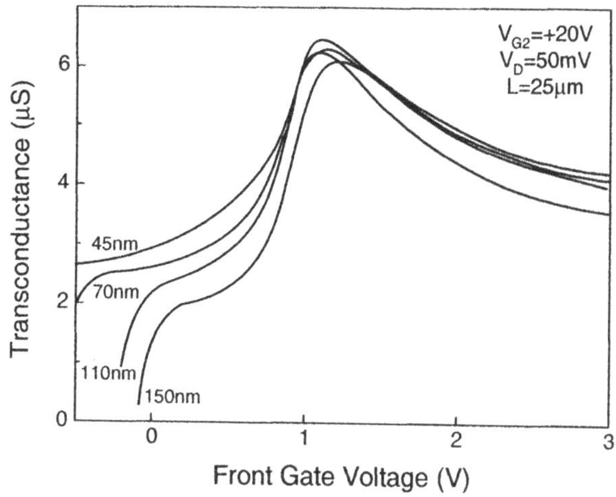

Fig. 3.25 Transconductance versus front gate voltage of the accumulation-mode SOI NMOS device with a front gate oxide of 175 Å, a thin film of various thicknesses from 45 to 150 nm, doped with an n-type density of $10^{16}cm^{-3}$, a buried oxide of 3800 Å, and a channel length of 25 μm, biased at the back gate voltage of 20V and the drain voltage of 50 mV. (Adapted from Faynot et al. [19].)

3.2.3 Thin-Film Thickness

In addition to capacitances, the transconductance of an accumulation-mode SOI MOS device also shows a two-stage increase behavior. Figure 3.25 shows the transconductance versus the front gate voltage of the accumulation mode SOI NMOS device with a front gate oxide of 175 Å, a thin film of various thicknesses from 45 to 150 nm, doped with an n-type density of $10^{16}cm^{-3}$, a buried oxide of 3800 Å, and a channel length of 25 μm, biased at the back gate voltage of 20V and the drain voltage of 50 mV [19]. As shown in this figure, after turn-on, due to the emergence of the buried channel, its transconductance rises quickly to a certain value. Then, along with the increase in the front gate voltage, its transconductance grows gradually due to the decrease in the width of the depletion region in the thin film as described before. The transconductance continues to grow to reach its peak when the front gate voltage reaches the flat-band voltage (V_{FB}), where the surface channel has appeared to provide another channel for current conduction. When the front gate voltage increases further, its transconductance drops due to a degraded mobility as its vertical electric field in the surface channel increases. In addition, the transconductance is sensitive to its thin-film thickness. From this figure, with a thinner thin film, when the device turns on its transconductance is larger. With a smaller thin-film thickness, its buried channel is closer to the front gate. Thus the controllability of its front gate is stronger and hence the transconductance is larger. In this study, the back gate bias is large (20 V), with a smaller thin-film thickness, the influence of the back gate bias is more serious and its threshold voltage is smaller.

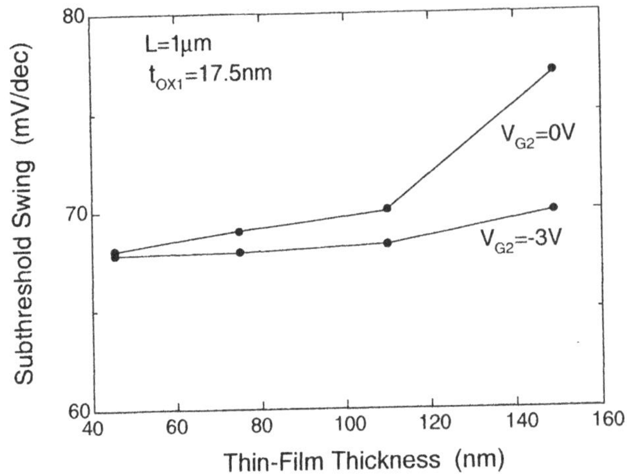

Fig. 3.26 Inverse subthreshold slope versus thin-film thickness of the accumulation-mode SOI NMOS device with its structure as described in Fig. 3.25 except for the channel length of 1 μm, biased at back gate biases of 0 and -3 V. (Adapted from Faynot et al. [19].)

When the thin-film thickness is changed, in addition to transconductance, its subthreshold slope is also affected. Figure 3.26 shows the inverse subthreshold slope versus the thin-film thickness of the accumulation-mode SOI NMOS device with its structure as described in Fig. 3.25 except for the channel length of 1 μm [19]. As shown in this figure, due to the existence of the buried channel, the channel of the accumulation-mode SOI MOS device is located farther from its front gate. Thus its subthreshold slope is worse than that of the inversion-mode device. In addition, when the thin-film thickness becomes larger, its buried channel is further away from the front gate. Hence, its subthreshold slope may be worsened further. From Fig. 3.26, with the back gate bias of -3 V, the degradation of the subthreshold slope due to an increased thin-film thickness has been effectively improved since with a negative back gate bias its buried channel has been pushed toward the front surface.

Owing to its thin-film doping structure (source-body-drain) of $n^+ - n - n^+$, the parasitic bipolar effect in the accumulation-mode SOI NMOS device in theory is negligible. Figure 3.27 shows the drain current versus the drain voltage of the accumulation-mode SOI NMOS device with its structure as described in Fig. 3.26 with various thin-film thicknesses, biased (a) at the front gate voltage of 2 V and the back gate voltage of 0 V and (b) at the front gate voltage of 0 V and the back gate voltage of 0 V [19]. As shown Fig. 3.27(a), no kink effects due to its $n^+ - n - n^+$ thin-film doping structure can be observed. However, as shown in Fig. 3.27(b), since the thin film has been depleted in the subthreshold region, behaviors similar to snapback and hysteresis still occur. As shown in this figure, the holding voltage, which is the drain voltage with the onset of hysteresis, is sensitive to the thin-film thickness. With a smaller thin-film thickness, its holding voltage improves substantially.

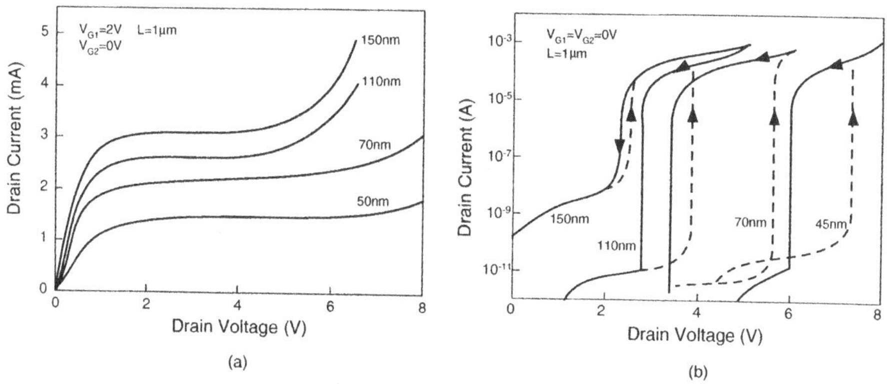

Fig. 3.27 Drain current versus drain voltage of the accumulation-mode SOI NMOS device with its structure as described in Fig. 3.26 with various thin-film thicknesses, biased (a) at the front gate voltage of 2 V and the back gate voltage of 0 V and (b) at the front gate voltage of 0 V and the back gate voltage of 0 V. (From Faynot et al. [19]. ©1995 IEEE.)

3.2.4 Accumulation versus Inversion

Performance comparison between accumulation-mode and inversion-mode SOI NMOS devices is discussed now. Figure 3.28 shows (a) the transconductance versus the gate voltage ($V_D = 0.1$ V), (b) the drain current versus the drain voltage (with both front and back gates grounded) of the accumulated-mode FD SOI NMOS device with a front gate oxide of 200 Å, a 1000 Å thin film doped with an n-type density of $2 \times 10^{16}\text{cm}^{-3}$, a buried oxide of 4000 Å, and a channel length of 0.8 μm [20]. Also shown in this figure are the results for a compatible inversion-mode device. For the inversion mode device, the thin-film doping density is $2 \times 10^{16}\text{cm}^{-3}$. As shown in Fig. 3.28, the transconductance of the accumulation-mode device is higher than that of the inversion-mode one owing to a higher carrier mobility in the buried channel as compared to the surface channel. In the accumulation-mode device, its vertical electric field is smaller, which leads to a larger surface mobility. Consequently, the transconductance of the accumulation-mode device is larger. Figure 3.28 shows that the latch phenomenon of the accumulation-mode device is more serious.

Figure 3.29 shows the cross-section of the mixed-mode SOI MOSFET [20]. At the top of the thin film, accumulation-mode structure ($n^+ - n - n^+$) is adopted. At the bottom, inversion-mode structure ($n^+ - p - n^+$) is used. Using the mixed-mode approach, the properties of the device are between the accumulation-mode device and the inversion-mode device as shown in Fig. 3.28. The structure of the mixed-mode SOI device is similar to the depletion-type device built on bulk.

Hot-carrier effects of accumulation-mode SOI MOS devices may be different from those of inversion-mode ones. Figure 3.30 shows the drain current degradation versus the stress gate voltage of both accumulation-mode and inversion-mode FD SOI NMOS devices, with a front gate oxide of 125 Å, a thin film of 1000 Å, a channel length of 0.8 μm, biased at the drain voltage of 2.5 V and $V_G - V_T = 2.8$ V, measured with source and drain reversed after a stress of 40,000 s at the drain voltage of 4 V [21].

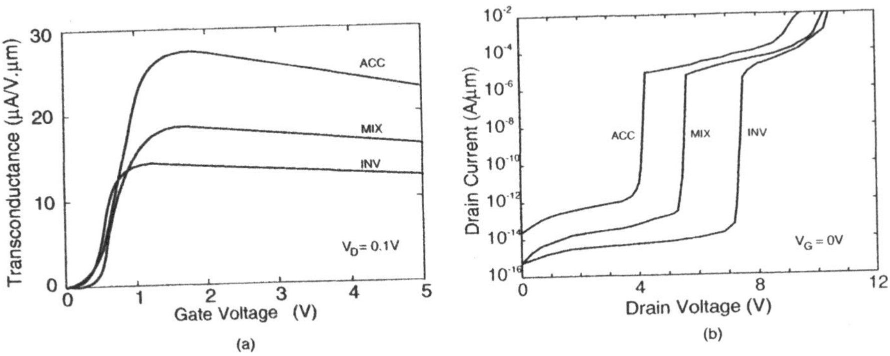

Fig. 3.28 (a) Transconductance versus gate voltage ($V_D = 0.1$ V), (b) Drain current versus drain voltage (with both front and back gates grounded) of the accumulation-mode FD SOI NMOS device with a front gate oxide of 200 Å, a 1000 Å thin film doped with an n-type density of 10^{15}cm^{-3}, a buried oxide of 4000 Å, and a channel length of 0.8 μm. For the inversion mode device, the thin-film doping density is $2 \times 10^{16}\text{cm}^{-3}$. (Adapted from Duan & Ioannou [20].)

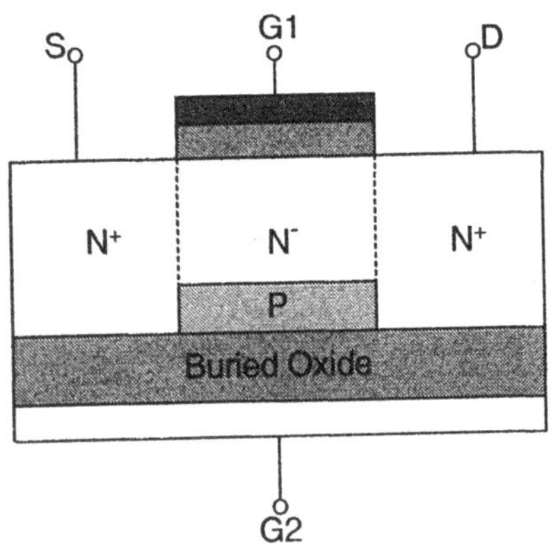

Fig. 3.29 Cross-section of the mixed-mode SOI MOSFET. (Adapted from Duan & Ioannou [20].)

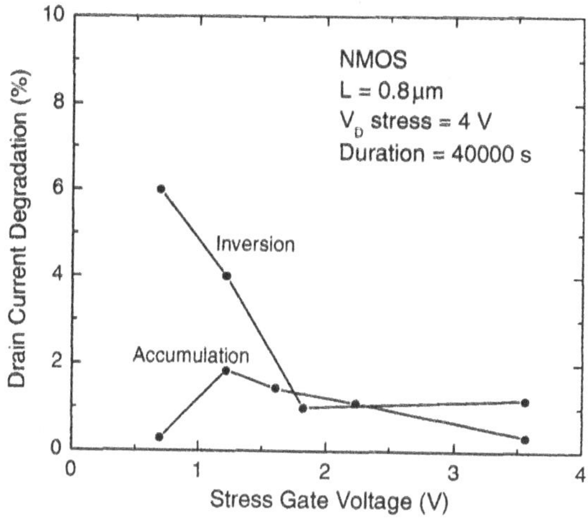

Fig. 3.30 Drain current degradation versus stress gate voltage of both accumulation-mode and inversion-mode FD SOI NMOS devices, with a front gate oxide of 125 Å, a thin film of 1000 Å, and a channel length of 0.8 μm, biased at the drain voltage of 2.5 V and $V_G - V_T =$ 2.8 V, measured with source and drain reversed after a stress of 40,000 s at the drain voltage of 4 V. (Adapted from Acovic et al. [21].)

As mentioned before, the accumulation-mode device may have a more serious latch behavior as compared to inversion-mode devices. From Fig. 3.30, the hot-carrier effect of the accumulation-mode devices is slightly better than that of the inversion-mode device, probably due to a smaller electric field in the accumulation-mode device.

3.3 DOUBLE GATE

Until now, all SOI devices described are based on front-gate technology. For SOI devices, their silicon thin-film region is surrounded by oxide. During the fabrication process, under the silicon thin film, in the buried oxide another gate-back gate can be arranged. For SOI MOS devices, with both the front gate and the back gate, they are double-gate SOI MOS devices[22]. In this section, double-gate SOI MOS devices are described.

As shown in Fig. 3.31, which is different from the single-gate device, for the double-gate device, at the bottom of the thin film there exists another polysilicon gate, that can be used to control the channel region with the front polysilicon gate at the same time [23]. Due to the simultaneous control of the top and bottom gates, the second-order effects of the double-gate devices are much better than those of the single-gate. Figure 3.32 shows the threshold voltage shift versus the channel length of double-gate and single-gate SOI NMOS devices with a front gate oxide of 10 Å, biased at the drain voltage of 1 V [23]. As shown in this figure, the double-gate

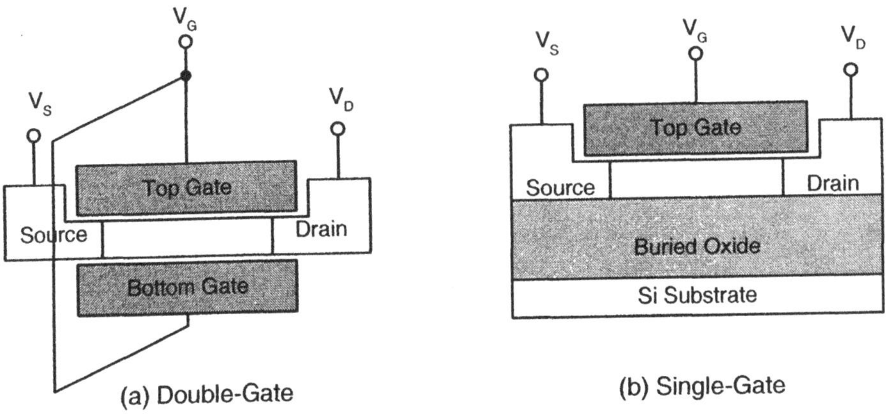

Fig. 3.31 Cross-section of double-gate and single-gate SOI MOS devices. (Adapted from Wong et al. [23] .)

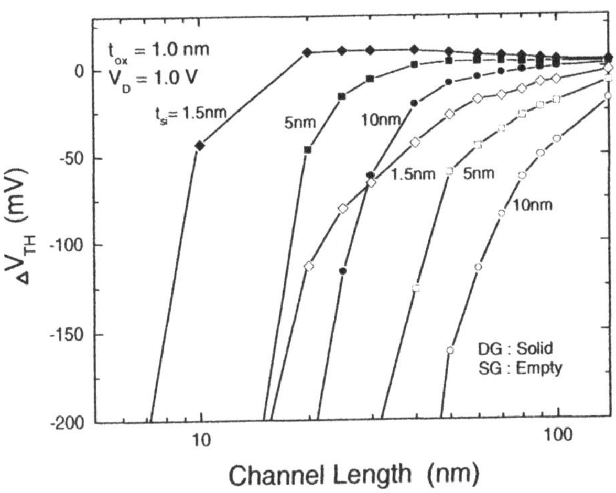

Fig. 3.32 Threshold voltage shift versus channel length of double-gate and single-gate SOI NMOS devices with a front gate oxide of 10 Å, biased at the drain voltage of 1 V. (Adapted from Wong et al. [23].)

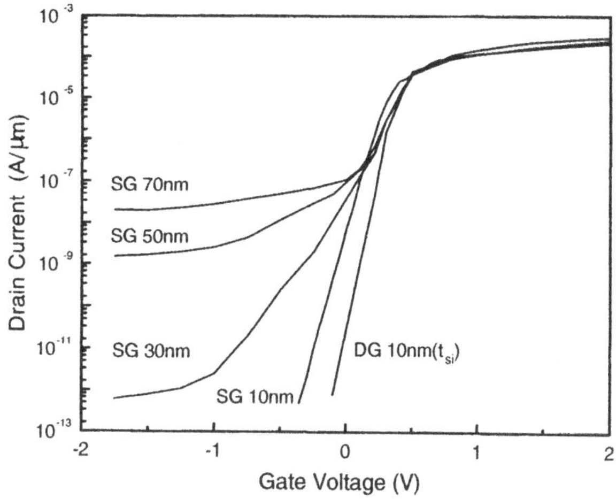

Fig. 3.33 Drain current versus gate voltage of double-gate and single-gate SOI NMOS devices with a front gate oxide of 30 Å, a buried oxide of 3800 Å for the single-gate device (a back gate oxide of 30 Å for the double-gate device), a thin film of 100 Å doped with a p-type density of $5 \times 10^{17} \mathrm{cm}^{-3}$, a channel length of 500 Å, and a $\mathrm{Si}_{0.3}\mathrm{Ge}_{0.7}$ mid-gap gate. (Adapted from Rauly et al. [24].)

device with a thin film of 100 Å has a better short channel effect as compared to the single-gate device with a thin film of 50 Å. By adopting the double-gate techniques, the requirement on the ultrathin film for deep sub-0.1 μm FD SOI MOS devices can be reduced.

In addition to the short channel effect, double-gate devices also have superior performance in the subthreshold characteristics. Figure 3.33 shows the drain current versus the gate voltage of double-gate and single-gate SOI NMOS devices with a front gate oxide of 30 Å, a buried oxide of 3800 Å for the single-gate devices (a back gate oxide of 30 Å for the double-gate device), a thin film of 100 Å doped with a p-type density of $5 \times 10^{17} \mathrm{cm}^{-3}$, a channel length of 500 Å, and a $\mathrm{Si}_{0.3}\mathrm{Ge}_{0.7}$ mid-gap gate [24]. As shown in this figure, the subthreshold slope of the double-gate device is better than that of the single-gate device. At the same off condition ($V_G = 0V$), the leakage current of the double-gate devices is smaller than that of the single-gate devices. Therefore, double-gate devices are more suitable for low-voltage circuits since their threshold voltage can be scaled down more, still meeting the leakage current requirement for digital VLSI circuit applications.

As described above, double-gate devices have superior short-channel threshold characteristics and a much lower leakage current during turn-off. Figure 3.34 shows the drain current versus the drain voltage of (a) double-gate and (b) single-gate SOI NMOS devices with a front gate oxide of 110 Å, a thin film of 600 Å, a back gate oxide of 110 Å for the double-gate device, a channel length of 0.33 μm, and a channel width of 10 μm [25]. Since at top and bottom surfaces of the thin film there exist

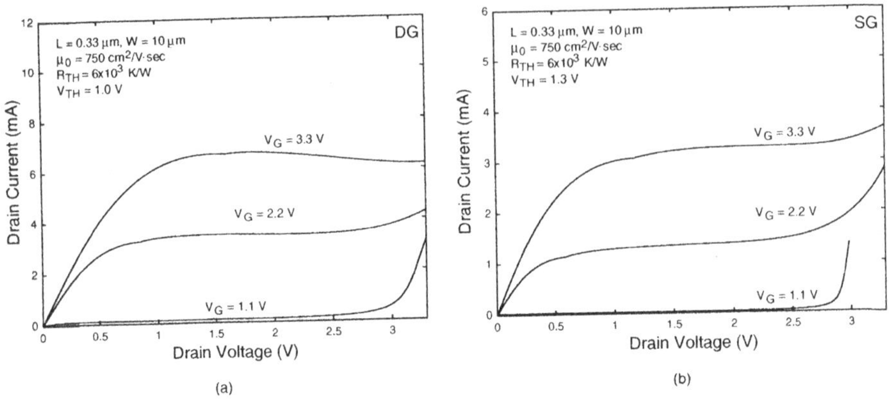

Fig. 3.34 Drain current versus drain voltage of (a) double-gate and (b) single-gate SOI NMOS devices with a front gate oxide of 110 Å, a thin film of 600 Å, a back gate oxide of 110 Å for the double-gate device, a channel length of 0.33 μm, and a channel width of 10 μm. (Adapted from Su & Kuo [25].)

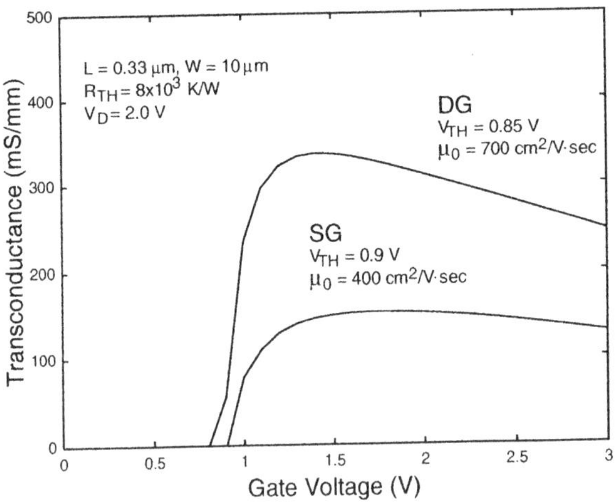

Fig. 3.35 Transconductance versus gate voltage of the double-gate and the single-gate SOI NMOS devices with their structures as described in Fig. 3.34, biased at the drain voltage of 2 V. (Adapted from Su & Kuo [25].)

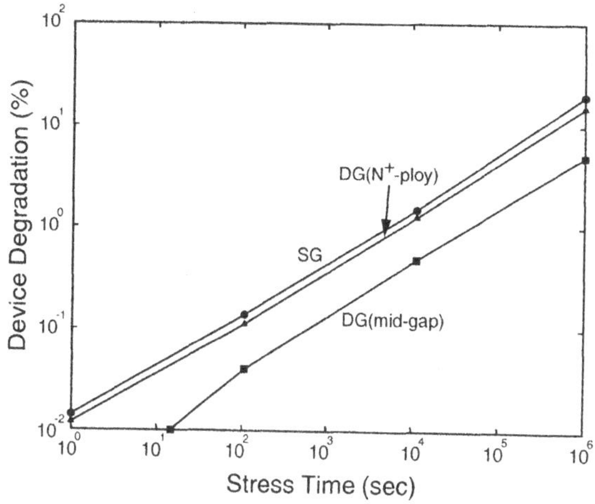

Fig. 3.36 Drain current degradation versus stress time of double-gate and single-gate SOI NMOS devices with a front gate oxide of 40 Å, a 250 Å thin film doped with a buried oxide of 1000 Å, a channel length of 0.08 μm, after hot-carrier stress at $V_{GS} = 1.2$ V and $V_{DS} = 0.05$ V [26]. The single-gate device has the gate made of a mid-gap material and the thin-film doping density of 10^{16}cm^{-3}. One double-gate device has the N^+ polysilicon gate and the thin-film doping density of 10^{18}cm^{-3}. The other double-gate device has the gate made of the mid-gap material and the thin-film doping density of 10^{16}cm^{-3} (Adapted from Williams et al. [26].)

channels, the conduction current of the double-gate device is much more than that of the single-gate device. By the same reasoning, the transconductance of the double-gate device is bigger than that of the single-gate device as shown in Fig. 3.35. Due to a higher conduction current in the double-gate device, the power consumption and the accumulated heat are higher. Thus, self-heating effect of the double-gate device is more noticeable than that of the single-gate. As a result, the negative output differential resistance as shown in Fig. 3.34(a) is more serious compared to the single-gate device. For the double-gate device, at a higher gate voltage, its transconductance falls due to self-heating effect as shown in Fig. 3.35 [25]. In contrast, the drop in the transconductance of the single-gate device due to self-heating effects is much less. Note that at a high drain voltage, the average lateral electric field in the post-saturation region of the double-gate device is higher than that in the single-gate device due to the function of top and bottom gates. Since the gradient of the electric field in the lateral direction is higher, the nonlocal impact ionization effect is more noticeable in the double-gate device. Despite the effects of self-heating and nonlocal impact ionization, the breakdown voltage of the double-gate devices is similar to that of the single-gate devices.

Figure 3.36 shows the drain current degradation versus stress time of double-gate and single-gate SOI NMOS device with a front gate oxide of 40 Å, a 250 Å thin film, a buried oxide of 1000 Å, and a channel length of 0.08 μm, after hot-carrier stress at $V_{GS} = 1.2$ V and $V_{DS} = 0.05$ V [26]. The single-gate device has the gate made

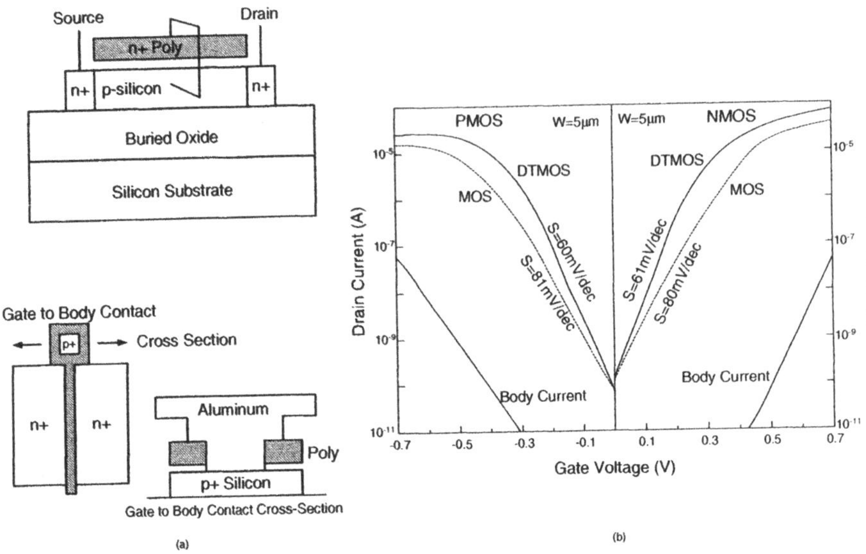

Fig. 3.37 (a) Cross-section and layout of an SOI NMOS device with body and gate tied together. (b) Subthreshold characteristics of SOI NMOS and PMOS devices operating in DTMOS configuration. (From Assaderaghi et al. [27]. ©1997 IEEE.)

of mid-gap material and the thin-film doping density of 10^{16}cm^{-3}. One double-gate device has the N^+ polysilicon gate and the thin-film doping density of 10^{18}cm^{-3}. The other double-gate device has the gate using the mid-gap material and a thin-film density of 10^{16}cm^{-3}. Note that due to the superior controllability of the gate over the thin film, even at a low thin-film doping density of 10^{16}cm^{-3}, DIBL effects can still be held for the double-gate device. From this figure, the double-gate devices with the polysilicon gate and the heavily doped thin film and the single-gate with the mid-gap gate and the lightly doped thin film have the similar drain current degradation. For the double-gate device using the gate made of a mid-gap material, its drain current degradation is smaller. The approach of the double-gate SOI MOS devices using a gate made of a mid-gap material and a thin film not highly doped provides a good hint for realizing deep sub-0.1 μm VLSI devices with an acceptable reliability.

3.4 DTMOS

Fig. 3.37(a) shows the cross section of the SOI dynamic threshold MOS (DTMOS) devices [27]. As shown in Fig. 3.37, by connecting the front gate to the neutral body, the controllability of the gate over the channel can be enhanced for the PD SOI MOS devices. Compared to the conventional body-tied device, the DTMOS device has a much larger turn-on current. In the subthreshold region, due to the better controllability of the front gate over the thin film, the subthreshold slope of the DTMOS device is also better. When the DTMOS is turned off with its gate voltage

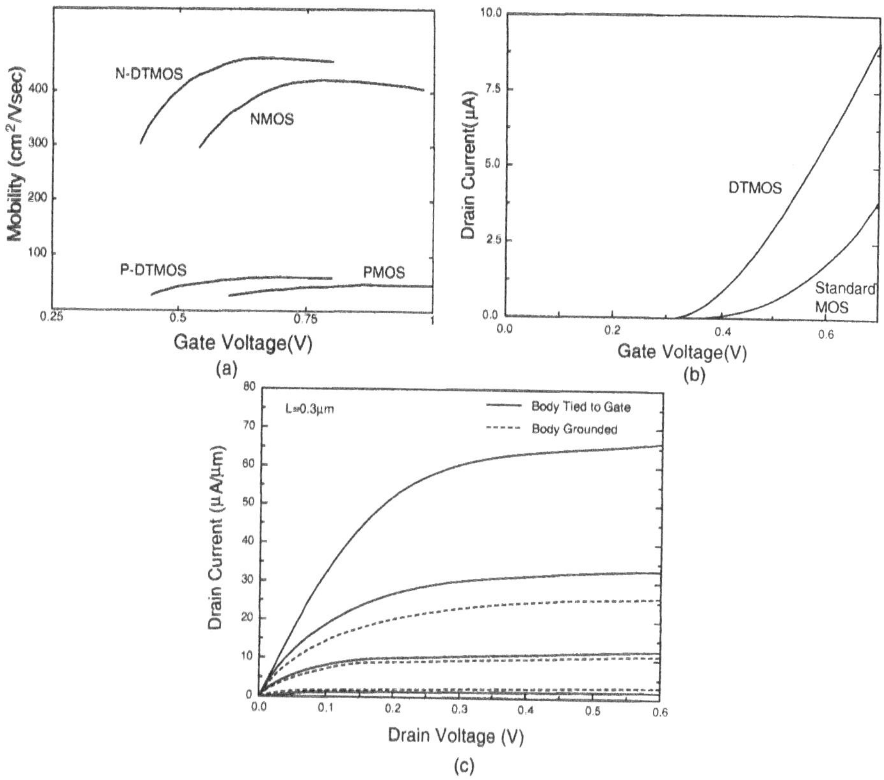

Fig. 3.38 (a) Electron and hole effective mobility versus gate voltage, (b) Drain current versus gate voltage, and (c) Drain current versus drain voltage of the SOI MOS devices in standard MOS configuration with its body grounded and DTMOS configuration. The SOI MOS device has a front gate oxide of 64 Å, a thin film of ~ 1500 Å doped with a density of $3 \times 10^{17} cm^{-3}$. (From Assaderaghi et al. [27]. ©1997 IEEE.)

of 0 V, the leakage current is identical to that of the conventional device without the DTMOS configuration. Since the body and the gate are connected together, there exist drain and gate leakage currents, especially when the power supply voltage is close to 0.7 V since body-source or body-drain diodes may be turned on to cause a substantially large leakage current. Therefore, the operating voltage of the DTMOS device is limited to < 0.7V.

3.4.1 Basic Performance

For DTMOS devices, the superior controllability of the gate over the thin film enhances the carriers available for current conduction. In addition, in the DTMOS devices, the vertical electric field in the thin film is smaller as compared to the conventional MOS device with its body grounded. As a result, the surface mobility of the DTMOS device is higher as shown in Fig. 3.38(a). From these two factors, the

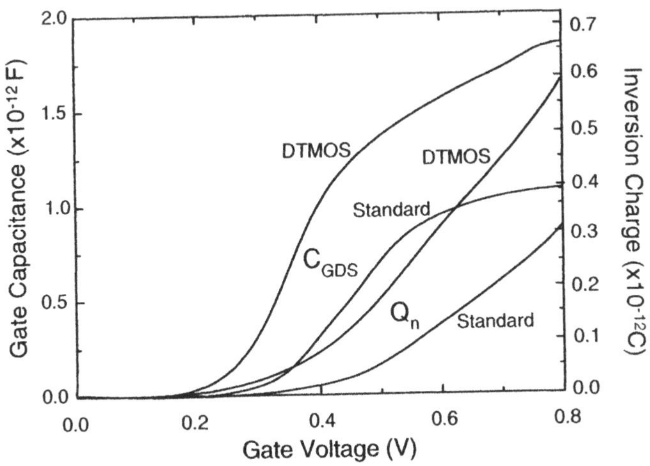

Fig. 3.39 Gate capacitance and inversion charge versus gate voltage of SOI NMOS device with a front gate oxide of 64 Å, a thin film of ~ 1500 Å doped with a p-type density of $3 \times 10^{17} cm^{-3}$, channel width and length of 20 μm, in standard MOS configuration with body grounded and in the DTMOS configuration. (Adapted from Assaderaghi et al. [27].)

conducting current of the DTMOS devices has been enhanced to some extent. As shown in Fig. 3.38(b) and (c), in terms of the drain current versus the gate voltage and the drain current versus drain voltage curves, DTMOS devices indeed provide superior properties.

Since the body and the gate are connected together, the effective load capacitance driven by the gate includes the front gate oxide, the equivalent capacitance accounting for the depletion region under the channel, and the junction capacitances of the body-source and the body-drain junctions. Therefore, the gate capacitance and the inversion charge of the SOI NMOS device in DTMOS configuration are larger than those of the device in the standard MOS configuration as shown in Fig. 3.39 [27]. Under this situation, the advantages of the DTMOS device on the enhanced current drive are offset by its increased capacitances.

Figure 3.40 shows the propagation delay time of the ring oscillator realized by SOI NMOS devices in DTMOS configuration, with its body grounded, and with its body floating [28]. Since the conducting current of the SOI NMOS device in DTMOS configuration is higher than that of the SOI NMOS device with its body grounded or tied, even with an increased gate capacitance, from overall speed of the ring oscillator, the DTMOS one is still the quickest. Also in this figure, for a supply voltage < 0.7 V, the speed performance has been degraded greatly for circuits using SOI MOS devices with body grounded or with body floating. In contrast, for the circuit using DTMOS devices, the speed performance is not degraded as much when using a supply voltage < 0.7 V. Consequently, DTMOS devices are especially suitable to integrate circuits using an ultra-low supply voltage.

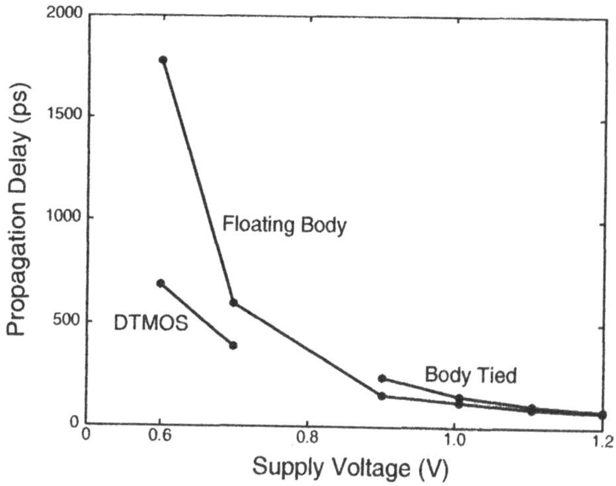

Fig. 3.40 Propagation delay time of the ring oscillator realized by SOI NMOS devices in DTMOS configuration, with its body grounded, and with its body floating. (Adapted from Ernst et al. [28].)

3.4.2 Second-Order Effects

Figure 3.41 shows (a) the inverse subthreshold slope and (b) the threshold voltage shift versus the drain voltage of the SOI NMOS device in DTMOS configuration, with body tied to ground and with body floating[28]. As shown in the figure, from the drain-voltage-dependent subthreshold slope and the drain-voltage-dependent threshold curves, the DTMOS devices show their advantages. The superiority of the DTMOS devices in terms of the second-order effects is derived from the decreased depletion region under the channel, which is due to the increased body potential when the device is turned on.

Figure 3.42 shows (a) the threshold voltage and (b) the inverse subthreshold slope versus the channel length of the SOI NMOS device with a nitride front gate oxide of 27 Å and a thin film of 1200 Å, in DTMOS configuration and in conventional configuration, biased at the drain voltage of 0.7 V [29]. As shown in Fig. 3.42(a), when the channel length is decreased, the decrease of the threshold voltage of the DTMOS device is less than that of the conventional device at the drain voltage of 0.7 V. As shown in Fig. 3.42(b), under most situations, the DTMOS device shows a better subthreshold slope. With a channel length of 0.12 μm, the inverse subthreshold slope of the DTMOS device is 68 mV/dec. The better subthreshold slope of the DTMOS device is due to the better controllability of the gate. In addition, the decrease in the depletion region lowers the influence of the source/drain region. Thus, the short channel effect of the DTMOS device is reduced as compared to the conventional one.

Figure 3.43 shows (a) the transconductance and (b) the output conductance versus the channel length of the SOI NMOS device with a nitride front gate oxide of 27 Å and a thin film of 1200 Å, in DTMOS configuration and in conventional configuration, biased at the drain voltage of 0.7 V [29]. As shown in Fig. 3.43(a), owing to the

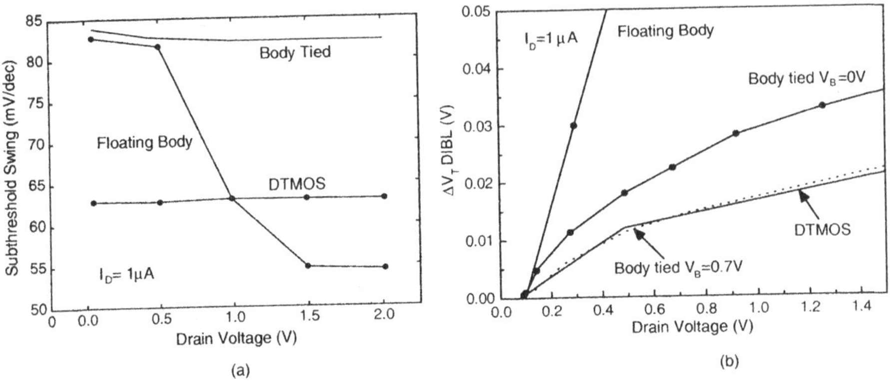

Fig. 3.41 (a) Inverse subthreshold slope and (b) threshold voltage shift versus drain voltage of the SOI NMOS device in DTMOS configuration, with body tied to ground and with body floating. (Adapted from Ernst et al. [28].)

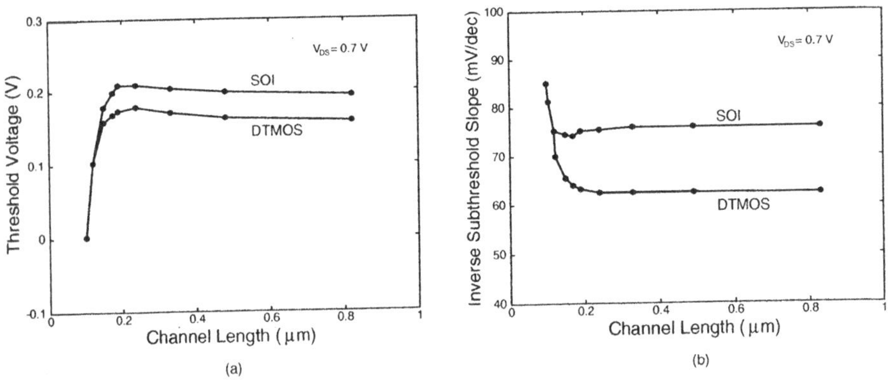

Fig. 3.42 (a) Threshold voltage (b) inverse subthreshold slope versus channel length of the SOI NMOS device with a nitride front gate oxide of 27 Å and a thin film of 1200 Å, in DTMOS configuration and in conventional configuration, biased at the drain voltage of 0.7 V. (Adapted from Tanaka et al. [29].)

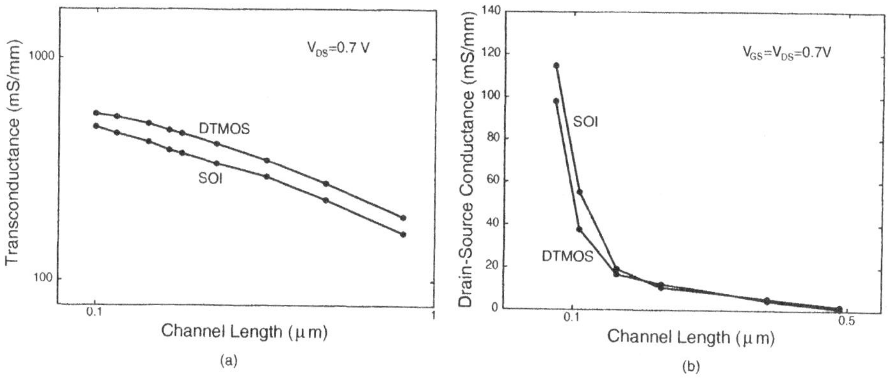

Fig. 3.43 (a) Transconductance and (b) output conductance versus channel length of the SOI NMOS device with a nitride front gate oxide of 27 Å and a thin film of 1200 Å, in DTMOS and conventional configurations, biased at the drain voltage of 0.7 V. (Adapted from Tanaka et al. [29].)

enhanced gate controllability for attracting more carriers, the transconductance of the DTMOS device is larger. Along with the shrinkage in the channel length, the transconductance has been increased steadily. As shown in Fig. 3.43(b), based on channel length modulation, the output conductance is degraded when the channel length is shrunk. The degradation of the output conductance at a shrunk channel length is less for the DTMOS device as compared to the conventional one. This result may be due to the thinner depletion region under the channel. Consequently, the shorter post-saturation region brings in reduced channel length modulation.

Figure 3.44 shows (a) the cut-off frequency (f_t) and (b) the maximum oscillation frequency (f_{max}) versus the channel length of the SOI NMOS device with a nitride front gate oxide of 27 Å and a thin film of 1200 Å, in DTMOS configuration and in conventional configuration, biased at the drain voltage of 0.7 V [29]. As shown in this figure, both DTMOS and conventional devices show a similar cut-off frequency (f_t). Although the current capability and the transconductance of the DTMOS device are larger, its gate capacitance is also larger due to the body tied to gate configuration. Therefore, the advantage of the transconductance has been offset by the drawback of the gate capacitance. Consequently, the cut-off frequency (f_t) of the DTMOS device is not improved. For the maximum oscillation frequency (f_{max}), the advantage of the DTMOS device is noticeable. In consideration of the maximum oscillation frequency, the output conductance is important. Due to the better output conductance, the maximum oscillation frequency of the DTMOS device is higher.

Figure 3.45 shows the off drain leakage current versus on drain current of the SOI NMOS device with a nitride front gate oxide of 27 Å and a thin film of 1200 Å in DTMOS and conventional configurations, biased at the drain voltage of 0.7 V [29]. As shown in this figure, the DTMOS device has a lower drain leakage current and a higher on drain current, which is especially suitable for high-speed, low-power circuits using a low power supply voltage.

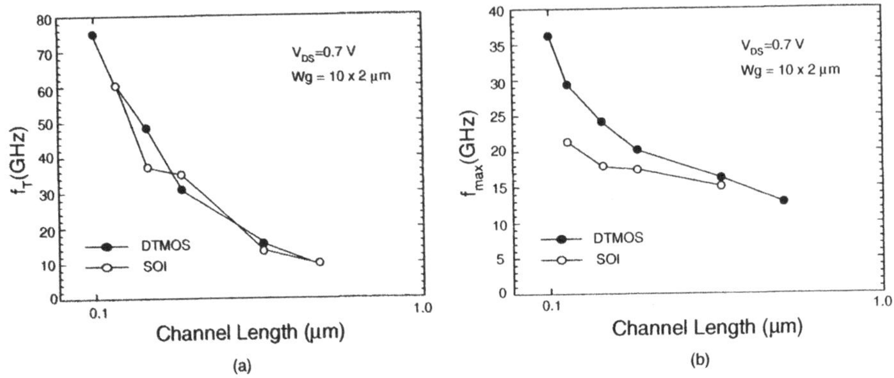

Fig. 3.44 (a) Cut-off frequency (f_t) and (b) maximum oscillation frequency (f_{max}) versus channel length of the SOI NMOS device with a nitride front gate oxide of 27 Å and a thin film of 1200 Å, in DTMOS and conventional configurations, biased at the drain voltage of 0.7 V. (Adapted from Tanaka et al. [29].)

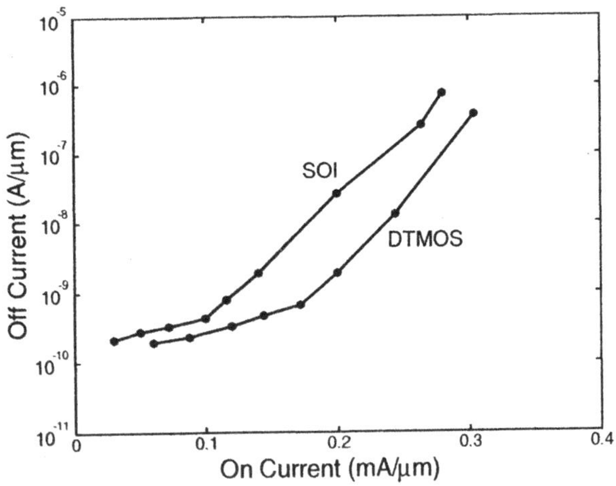

Fig. 3.45 Off drain leakage current versus on drain current of the SOI NMOS device with a nitride front gate oxide of 27 Å and a thin film of 1200 Å, in DTMOS and conventional configurations, biased at the drain voltage of 0.7 V. (Adapted from Tanaka et al. [29].)

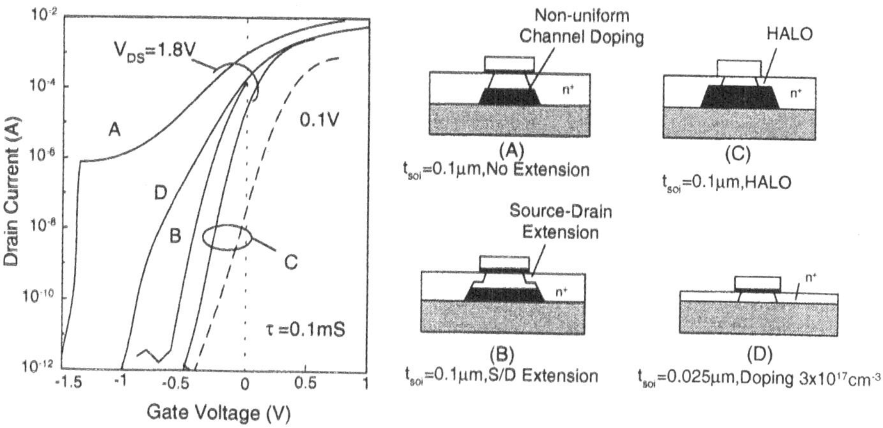

Fig. 3.46 Drain current versus gate voltage of an SOI NMOS device with various structures, a front gate oxide 42 Å, a channel length of 0.1 μm, and a thin-film thickness of 1000 Å for the PD devices (a)~(c) and 250 Å for the FD device (d), based on 2D device simulation results. (From Shahidi et al. [30]. ©1993 IEEE.)

3.5 SCALING TRENDS

SOI devices have been well recognized for their advantages in integrating sub-0.1 μm CMOS devices. When the channel length of SOI CMOS devices is shrunk below 0.1 μm, device structure becomes very critical. Figure 3.46 shows the drain current versus the gate voltage of an SOI NMOS device with various structures, a front gate oxide of 42 Å, a channel length of 0.1 μm, and a thin-film thickness of 1000 Å for the PD devices(a)~(c) and 250 Å for the FD device (d), based on 2D device simulation results [30]. As shown in this figure, for FD device (d), it is a fully depleted device with a thin-film doping density of $3 \times 10^{17}cm^{-3}$. At the high drain voltage of 1.8 V, the turn-off leakage current of device (d) is too large. The turn-off leakage current can be reduced by decreasing the thin-film thickness to < 250 Å or by increasing the thin-film doping density to over $3 \times 10^{17}cm^{-3}$. Via these two approaches, the sensitivity of the device performance to the thin-film thickness is increased. It is very difficult to fabricate devices having a very thin thin film with a tight accuracy control over the thin-film thickness. In addition, with a very thin thin film, the large source/drain parasitic resistance may degrade the performance of the device. These bottlenecks are yet to be overcome when developing sub-0.1 μm FD SOI CMOS devices.

The bottlenecks of developing deep sub-0.1 μm PD SOI CMOS devices are considered as follows. As described before, the most cumbersome problems with PD devices are their floating body and kink effects, which may cause instability of the devices. In general, second-order effects of the PD devices are more serious as compared to the FD devices. In order to improve the performance of the PD SOI devices in the sub-0.1 μm regime, much more complex thin-film doping profiles should be adopted. As shown in this figure, among three PD devices, devices (b) and (c) are the

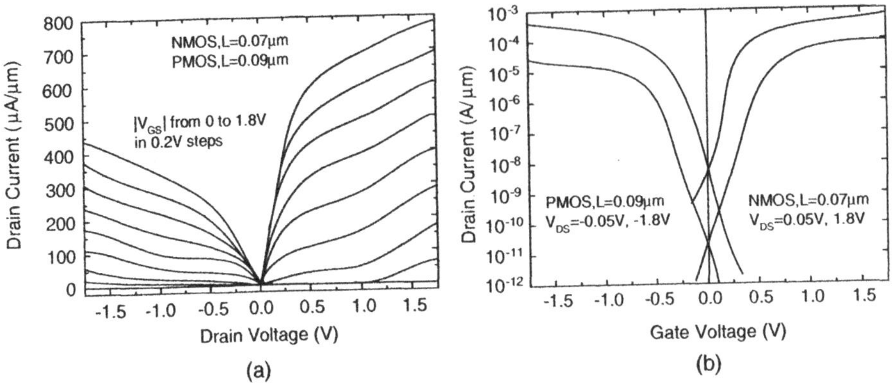

Fig. 3.47 Drain current (a) versus drain voltage and (b) versus gate voltage of the PD SOI CMOS devices with a front gate oxide of 35 Å. (Adapted from Assaderaghi et al. [31].)

best, where (1) deep source/drain junctions, (2) highly nonuniformly doped channel, (3) ultra-shallow source/drain extensions, and (4) HALO doping structure have been adopted. The adoption of the deep source/drain junctions (1) extending to the overall thin-film deep (1000 Å) can be used to effectively reduce the parasitic resistance. By the use of the highly nonuniformly doped channel (2), the thin film is divided into two layers. The thinner top layer is with a lower doping density. The thicker bottom layer is with a high doping density. Via this nonuniform doping approach, immunity of punchthrough at a high drain voltage can be improved and the active channel region is limited to the top layer to reduce short channel effects. In addition, the function of the parasitic bipolar device at the bottom of the thin film is degraded to improve the kink effects. While body contacts are required, the parasitic body contact resistance is small due to high doping density of the bottom layer. With the ultra-shallow source/drain extensions (3), which has a function similar to LDD, the properties of the device at a high drain voltage are improved. The influence of the source/drain in the channel region can be lowered to reduce the short channel effects. On the outskirt of the source/drain extension region, a local region with a higher doping density than that of the channel region is formed–the HALO doping structure to reduced DIBL effects (4) such that at a high drain voltage the leakage current is reduced. In addition, the adoption of silicide techniques to reduce parasitic resistance is necessary for realizing sub-0.1 μm SOI CMOS devices.

By using the techniques described above to reduce the short channel effects and the kink effects, the drain current versus the drain voltage and the drain current versus the gate voltage characteristics of the PD SOI CMOS devices with a front gate oxide of 35 Å are as shown in Fig. 3.47 [31]. Kink effects still cannot be avoided for the NMOS device with a channel length of 0.07 μm. For the PMOS device with a channel length of 0.09 μm, kink effects also exist. Due to the smaller impact ionization for the holes and the worse parasitic pnp bipolar device (as compared to npn), the kink effects of the PMOS device are much smaller.

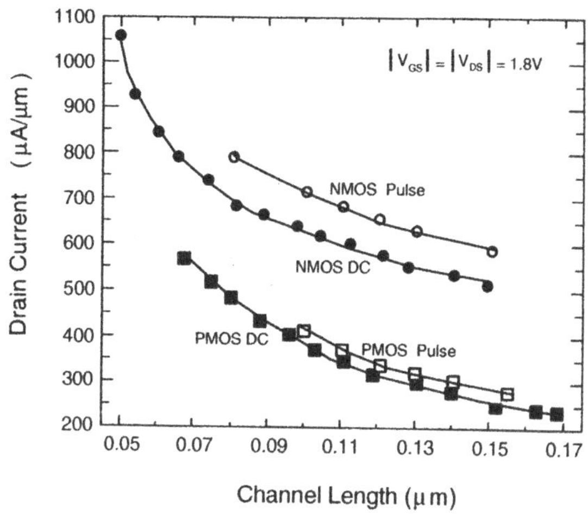

Fig. 3.48 Drain current versus channel length of the SOI CMOS devices described in Fig. 3.47 biased at the drain/gate voltage of 1.8 V for NMOS and −1.8 V for PMOS. (Adapted from Assaderaghi et al. [31].)

Figure 3.48 shows the drain current versus the channel length of the SOI CMOS devices described in fig. 3.47 biased at the drain/gate voltage of 1.8 V for NMOS and −1.8 V for PMOS [31]. As shown in this figure, along with the shrinkage of the channel length, the drain current of the PD SOI CMOS devices increases. Even with a channel length of 0.05 μm, this trend continues, which implies that the parasitic resistance is small. In addition, the devices may have velocity overshoot behavior because the drain current is not saturated due to velocity saturation.

Fig. 3.49 shows (a) the threshold voltage and (b) the off-state leakage current versus the channel length of the PD SOI CMOS devices[31]. As shown in this figure, at the low drain voltage, the short channel effect has been well controlled. At a high drain voltage, when the channel length is < 0.1 μm, the threshold voltage characteristics have been worsened due to DIBL, kink effects, and junction leakage. As shown in this figure, for the device with a channel length < 0.1 μm, the off-state leakage current rises quickly, which prevents the devices from being further scaled down.

Figure 3.50 shows (a) the propagation delay and (b) the power-delay product of the SOI CMOS inverter versus the channel length of the SOI NMOS device, for various power supply voltages [31]. Compared to the NMOS device, the channel length of the PMOS device is 0.022 μm longer. As shown in this figure, with a smaller channel length and at a larger power supply voltage, the speed of the inverter is faster. On the other hand, its power-delay product is reduced along with the shrinkage of the channel length. When the drain voltage is > 1.5 V and the channel length is < 0.08 μm, power-delay product may increase instead due to the high leakage current.

In general, for PD devices, leakage current from the parasitic bipolar device described before may occur. As shown in Fig. 3.51, the peak leakage current of the

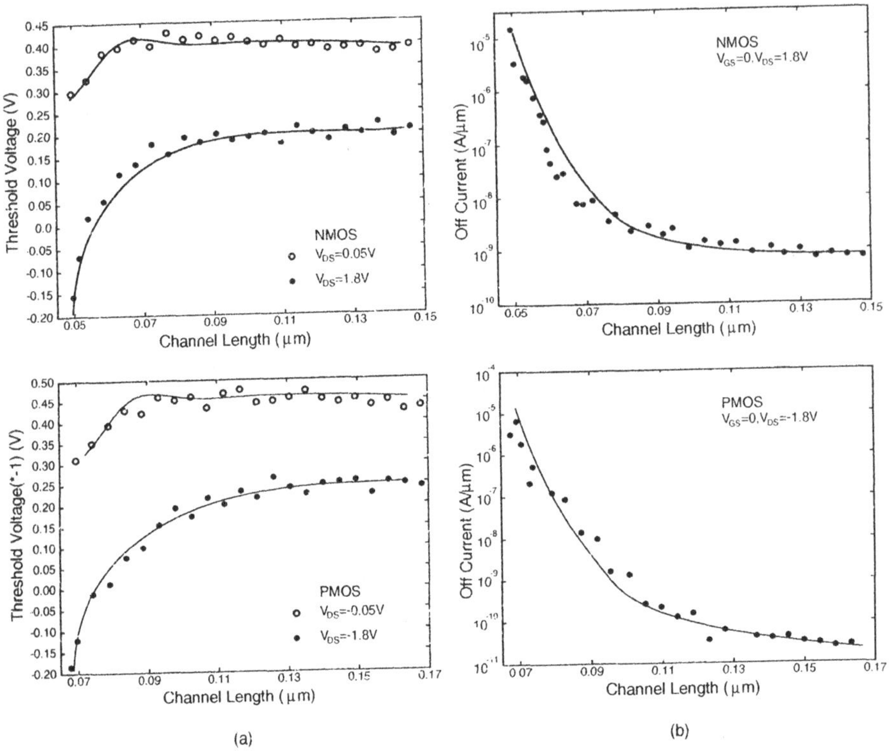

Fig. 3.49 (a) Threshold voltage (b) off-state leakage current versus channel length of the PD SOI CMOS devices. (From Assaderaghi et al. [31]. ©1997 IEEE.)

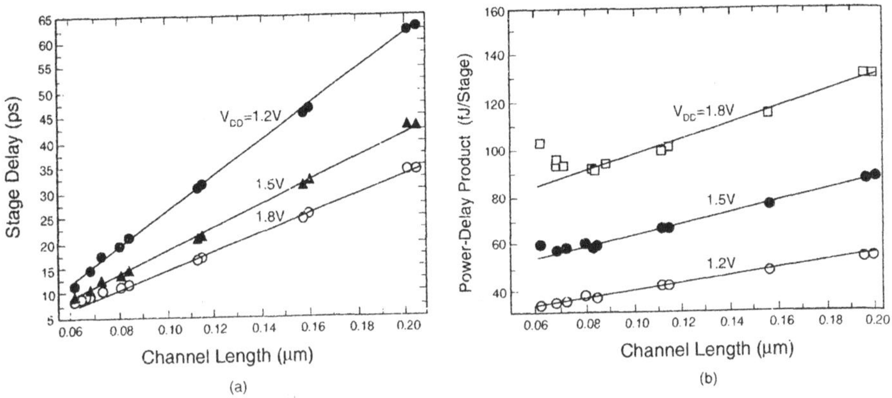

Fig. 3.50 (a) Propagation delay (b) power-delay product of the SOI CMOS inverter versus channel length of the SOI NMOS device, for various power supply voltages. Compared to the NMOS device, the channel length of the PMOS device is 0.022 μm longer. (Adapted from Assaderaghi et al. [31].)

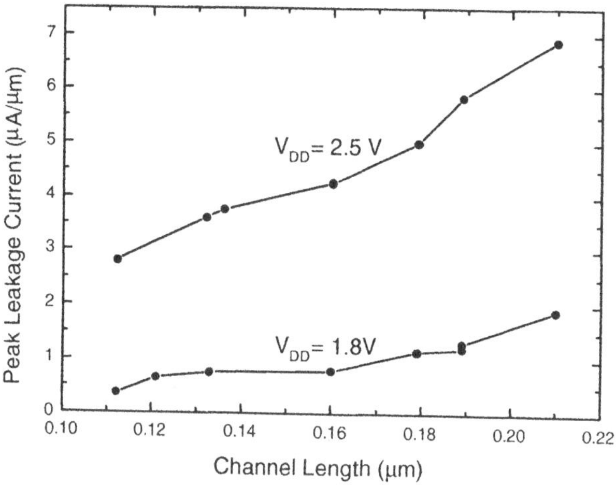

Fig. 3.51 Peak leakage current of the PD SOI NMOS pass transistor described in Fig. 3.50, biased at drain voltages of 2.5 and 1.8 V. (Adapted from Assaderaghi et al. [31].)

PD SOI pass transistor decreases with shrinkage of the channel length. Although the parasitic bipolar device functions more strongly with a decreased channel length, the gate-body capacitance drops substantially when the channel length is decreased. As a result, the holes in the thin film, which may trigger to generate the leakage from the parasitic bipolar device, are reduced to some extent. Consequently, its peak leakage current drops. This phenomenon occurs only when the channel length is small.

As shown in Fig. 3.52(a), along with the progress of time, size and performance of CMOS VLSI has been evolving continuously. Until now, the trend on the evolution has been steadily continuing. Soon the evolution of the CMOS VLSI will face two bottlenecks. First, in the sub-0.1 μm regime deep ultraviolet (DUV) light with a wavelength < 1930 Å should be adopted for lithography. With a shorter wavelength, the resolution of the lithography can increase and the implemented devices and circuits can be more refined. However, lithography with a shorter wavelength means more difficulties. As shown in Fig.3.52(b), when the device becomes smaller, the pace of the improvement for the bulk CMOS technology using aluminum interconnects has been slowing down, which implies conventional structure and material are not sufficient [32]. In order to have breakthroughs in these bottlenecks, the future CMOS VLSI technology will be based on Cu interconnect and low-k materials. In addition, SOI structures need to be adopted for the future CMOS VLSI.

Figure 3.53 shows (a) the propagation delay of the CMOS inverter versus channel length of the SOI devices used, at the supply voltage of 1.2 V and (b) the power versus the propagation delay of the CMOS inverter using SOI and bulk devices with channel lengths of 0.081 and 0.106 μm [33]. As shown in this figure, using the devices with various channel lengths, the speed performance of the SOI inverter provides a 33% advantage over the bulk one. When the channel length of the devices is < 0.1 μm,

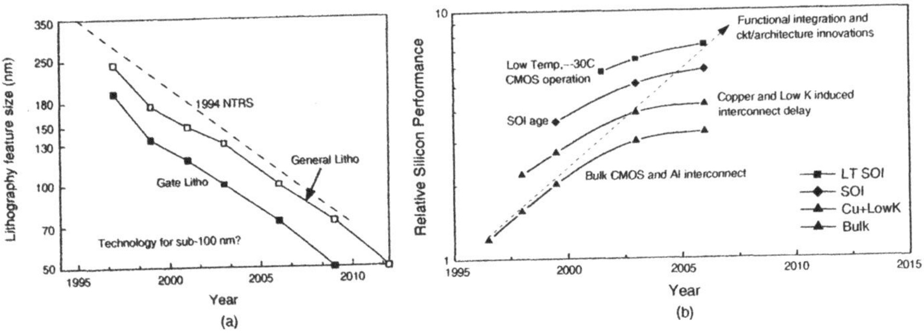

Fig. 3.52 (a) Lithography feature size and (b) relative silicon performance of CMOS VLSI. (Adapted from Davari [32].)

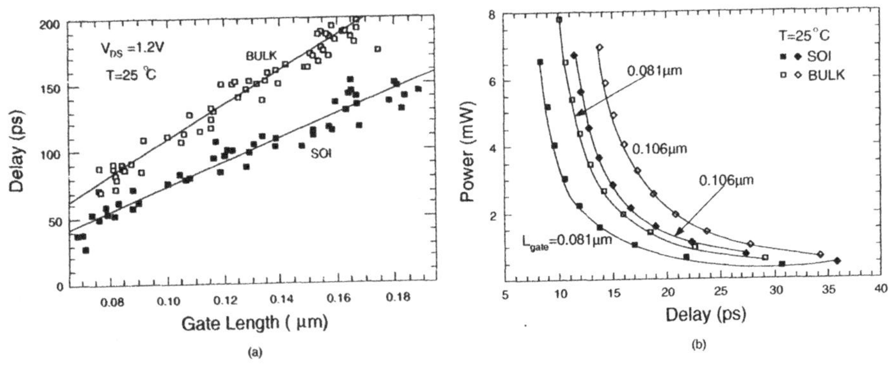

Fig. 3.53 (a) Propagation delay of the CMOS inverter versus channel length of the SOI devices used, at the supply voltage of 1.2 V. (b) Power versus propagation delay of the CMOS inverter using SOI and bulk devices with channel lengths of 0.081 and 0.106 μm. (From Leobandung et al. [33]. ©1998 IEEE.)

	L=0.25 μm	L=0.05 μm
$T_{ox} \approx L/50$	50 Å	10 Å
$X_{dmax} \approx L/5$	500 Å	100 Å
$N_B \propto X_{dmax}^{-2}$	2×10^{17} cm^{-3}	5×10^{18} cm^{-3}
$V_{CC} \propto T_{ox}$	2.5 V	0.25 V

Fig. 3.54 Extrapolation of the 0.25 μm MOS device design for realizing the 0.05 μm device based on bulk structure. (Adapted from Hu [34].)

at the same speed, the SOI inverter only consumes 50% of the needed power for the bulk one. From this figure, SOI devices indeed are suitable to be utilized for next-generation high-speed, low-power VLSI.

For future VLSI in the sub-0.1 μm regime, the adoption of the SOI devices are from the superior performance of the SOI CMOS devices as compared to the bulk ones. As shown in Fig. 3.54, based on down-scaling of a 0.25 μm CMOS device using the bulk structure, 0.05 μm CMOS devices are very difficult to implement [34]. As shown in this figure, when the channel length is scaled from 0.25 to 0.05 μm, in order to reduce short channel effects, the depletion depth under the channel should be shrunk accordingly. Consequently, the substrate doping density needs to increase from 2×10^{17} to $5 \times 10^{18}cm^{-3}$, which is too high from a practical manufacturing point of view. For the purpose of having an appropriate threshold voltage and an acceptable threshold slope, the front gate oxide needs to be shrunk from 50 to 10 Å, which may result in a large gate current due to electron tunneling. In order to avoid breakdown of the front gate oxide, the power supply voltage needs to be scaled down from 2.5 to 0.25 V, which is not reasonable for practical device designs. Although these difficulties for shrinking the bulk devices can be partially resolved via using nonuniform doping and low-k material, their effects are still limited basically.

PD SOI CMOS devices may face similar difficulties as bulk CMOS devices in the regime with a channel length < 0.05 μm. For the FD SOI CMOS devices, as long as the thin film is thin enough, these difficulties can be solved. When the thin film is very thin, the doping density of the thin film can be lowered and the front gate oxide does not have to be very thin. For the FD devices, how to the reduce the variation of the thin-film thickness and to reduce the parasitic resistance of the source/drain region are the main difficulties.

Recently, the FD SOI NMOS device with a channel length of 200 Å has been fabricated. As shown in Fig. 3.55, the FD SOI NMOS device has a thin film of 100Å doped with a p-type density of $10^{15}cm^{-3}$, a front gate oxide of 40 Å, a channel length of 20 nm, a 100 Å sidewall spacer, and a silicide source/drain region [35]. Figure 3.55(b) showed that although the parasitic resistance of the source/drain region is still large, from the drain current versus the gate voltage curves, the ratio of the

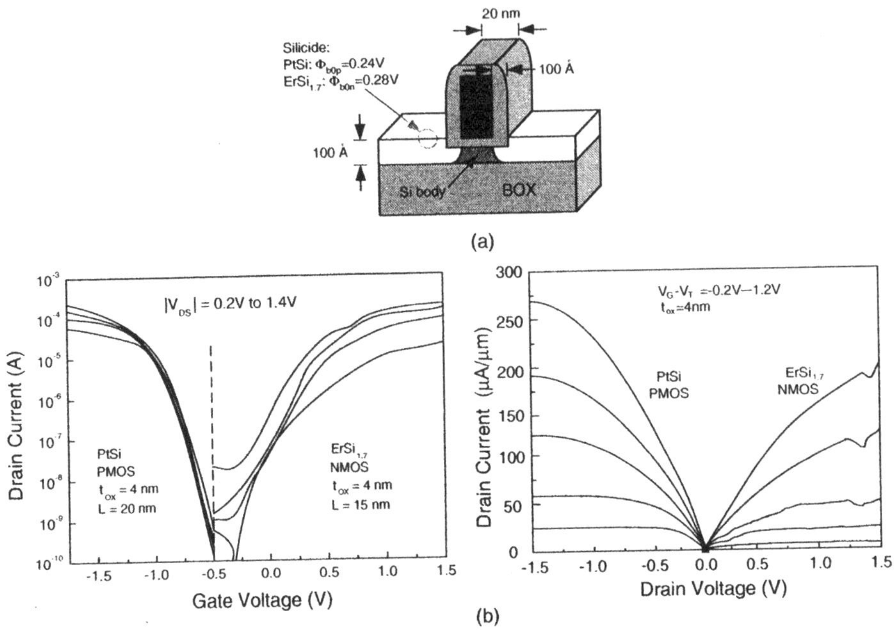

Fig. 3.55 (a) Cross-section of an FD SOI NMOS device with a thin film of 100 Å doped with a p-type density of $10^{15}\mathrm{cm}^{-3}$, a front gate oxide of 40 Å, a channel length of 20 nm, a 100 Å sidewall spacer, and a silicide source/drain region. (b) Drain current versus drain voltage and drain current versus gate voltage of the device. (From Kedzierski et al. [35]. ©2000 IEEE.)

on drain current to the off leakage current at the drain voltage of 1.4 V is 10^4 and its subthreshold swing is 150 mV/dec. From this figure, it shows that as long as the thin-film is sufficiently thin, the FD SOI NMOS with a channel length < 500 Å still can serve as an on/off switch effectively.

For sub-0.1 μm CMOS VLSI, in addition to scaling of device structure, scaling of supply voltage and threshold voltage is also important. Considering breakdown and reliability, power supply voltage is scaled down accordingly. Lowering the power supply voltage may provide advantages for reducing power consumption. Figure 3.56 shows the energy of a 101-stage ring oscillator versus the threshold voltage of the dual-gate SOI CMOS device used [36]. As shown in this figure, lowering the supply voltage with the threshold voltage accordingly can reduce power dissipation effectively. If the supply voltage/ threshold voltage is too low, power dissipation increases due to leakage current. The minimum power consumption occurs at the supply voltage < 1 V. Under such a low supply voltage, noise margin is small and hence the circuit can be interrupted by noises and high-energy particles. By using SOI CMOS devices, these problems can be avoided.

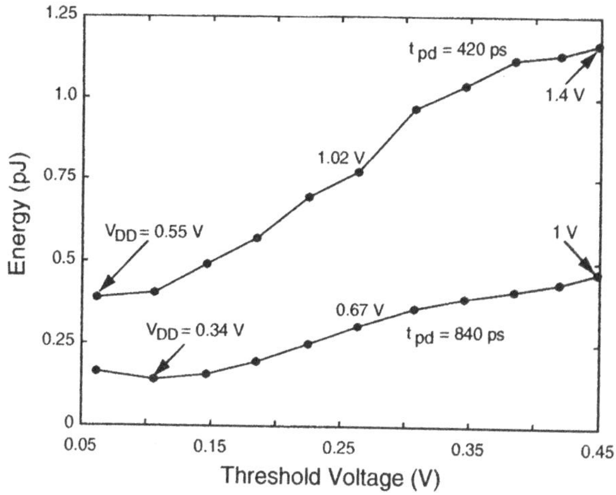

Fig. 3.56 Energy of a 101-stage ring oscillator versus threshold voltage of the dual-gate SOI CMOS device. (Adapted from Dancy & Chandrakasan [36].)

3.6 SINGLE ELECTRON TRANSISTORS (SET)

When both length and width of the active region in an NMOS device are very small (< 10 nm) there are quantum effects, which imply that in the device maybe only one electron is stored or transport. This is the so-called SET. The advantages of SET is its energy saving property, which means the operation of the device based on one electron, and high integration density (both the width and length of the active region < 10 nm), and high speed (electron transport via tunneling). Consequently, SET has been regarded as one possible candidate to replace the conventional MOS device structure. In an SET built on the SOI structure, the formation of the quantum dots made of silicon islands with a diameter < 10 nm is the key technique. Usually, in an SET it is very difficult to fabricate a very small silicon island by lithography. Instead, it is mostly formed by the specific properties of oxidation procedure.

Figure 3.57 shows the structure of an SET with its equivalent circuit using the pattern dependent oxidation (PADOX) technique [37]. The silicon island is separated from the source/drain region by a very thin oxide of several nanometers. Electrons can access the silicon island via tunneling through the oxide. Due to the small volume of the silicon island and its small equivalent capacitance, when a single electron enters it, the energy level of the quantum dot is changed to a large extent.

Figure 3.58 shows the conductance versus the gate voltage of the SET for various channel lengths at 40K [37]. As shown in this figure, the conductance versus the gate voltage curve is with an oscillating shape, which implies the quantum behavior is unique for the SET caused by Coulomb blockade effects. When the silicon island becomes large such that many electrons stored in the energy levels or when the operation temperature is too high such that the potential barrier of the quantum dot cannot function for the electron with a high energy, Coulomb blockade effects decease.

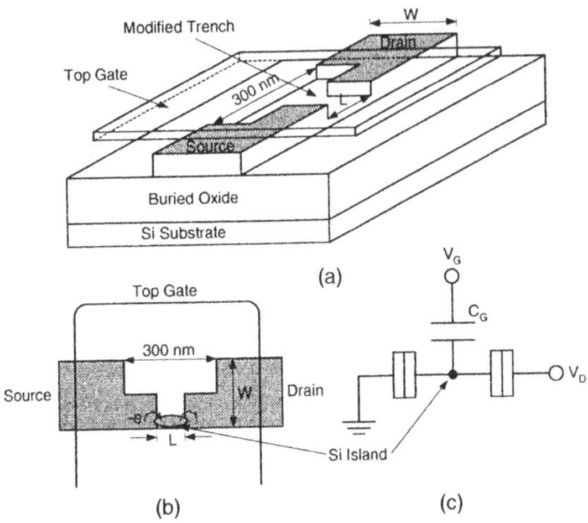

Fig. 3.57 Structure of an SET with its equivalent circuit. (From Ono et al. [37]. ©2000 IEEE.)

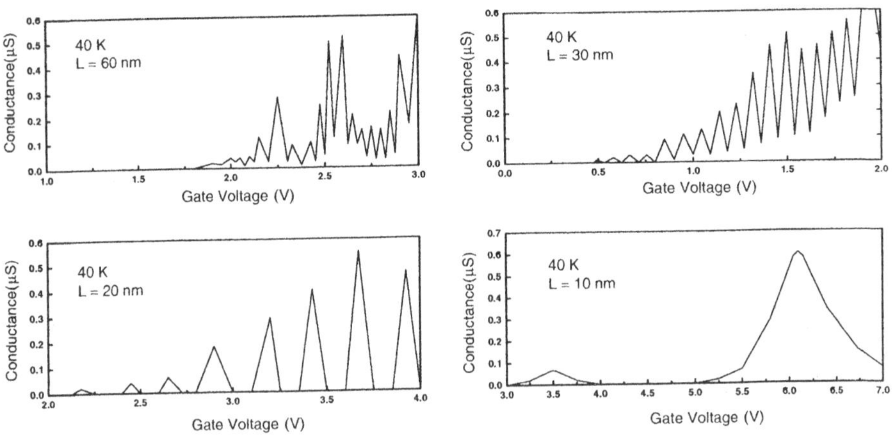

Fig. 3.58 Conductance versus gate voltage of the SET for various channel lengths operating at 40 K. (From Ono et al. [37]. ©2000 IEEE.)

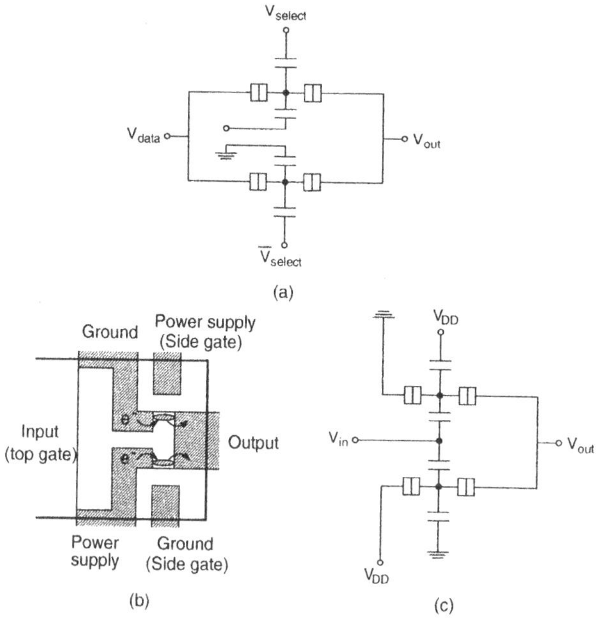

Fig. 3.59 CMOS-type inverter and its equivalent circuit using the SET structure with side gates. (From Ono et al. [37]. ©2000 IEEE.)

Figure 3.59 shows a CMOS-type inverter and its equivalent circuit using the SET structure with side gates [37]. As shown in the figure, for the side gate connected to the ground, its SET functions like a conventional PMOS device. For the side gate connected to V_{DD}, the SET acts like an NMOS device. If we connect the side gates to various biasing voltages to change the characteristics of the SET via imposing suitable voltages to the top gates, various circuits made of SETs can be accomplished.

Figure 3.60 shows a current switch using SETs controlled by charge injection into nano-crystal floating dots served as memory [38]. As shown in this figure, it is composed of the SET and the nano-crystal floating quantum dots. The SET has a silicon island between the source/drain region serving as the quantum dot. In addition, by depositing many crystal dots with a diameter of several nanometers on the top of the MOS channel, the single-electron quantum dots are formed, separated from the channel by a very thin oxide. As long as the gate voltage is sufficiently large, electrons may tunnel to the quantum dots and are confined in them. The nano-crystal floating dots serve as memory nodes of nonvolatile memory. Each time only one electron is stored. Combining the SETs with the nano-crystal floating dots results in an integrated device as shown in this figure. The stored electrons in the nano-crystal quantum dots at the top of the silicon island are used to change the properties of the SET. Various SETs with different properties programmed by their nano-crystal quantum dots, controlled by the operation voltages, can be used to implement circuits. Compared to the SET circuits programmed by the side gates described above, the SET

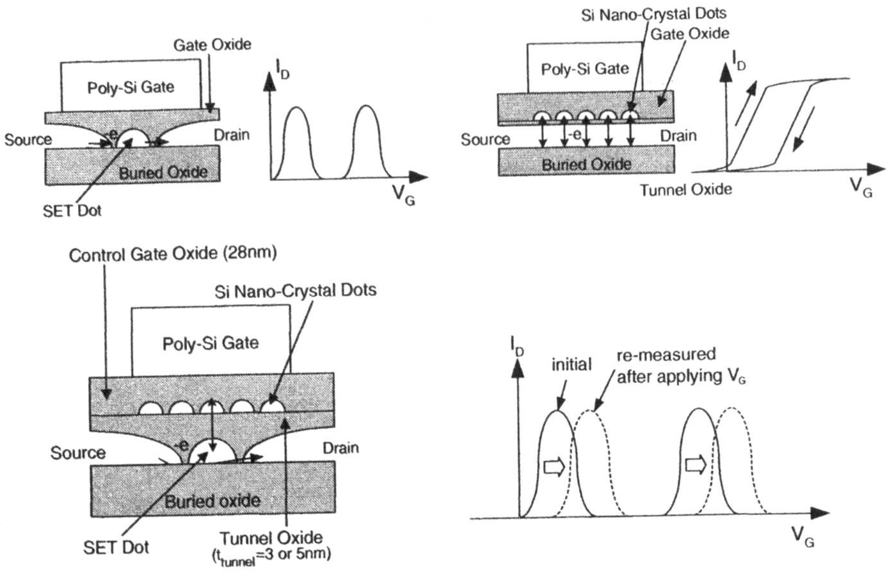

Fig. 3.60 Current switch using SETs controlled by charge injection into silicon nano-crystal floating dots served as memory. (From Takahashi et al. [38]. ©1999 IEEE.)

circuit controlled by nano-crystal floating dots provides an innovative approach for designing circuits, which save space and power.

Figure 3.61 shows the schematic of another SET [39]. As shown in this figure, the memory module at the top of the silicon island is not formed by deposition. Instead, the surface of the silicon island is processed with alkaline to form an undulated state. As a result, the potential of the surface changes substantially to form many quantum dots at the surface automatically. Due to many quantum dots at the surface, they may be linked to form a percolation channel, which has the characteristics of a SET. In addition, some independent quantum dots may serve as the function of the memory node. When the gate voltage is sufficiently large, via tunneling, some individual electrons may be stored as shown in the figure. Figure 3.62 shows the drain current versus the gate voltage of the SET biased at the drain voltage of 2 mV measured at room temperature [39]. As shown in this figure, at room temperature, Coulomb oscillations with the peak-to-valley current ratio (PVCR) of 100 in terms of a total capacitance of 0.4 aF with a 2.8 nm diameter of the SET island can be identified. As shown in Fig. 3.63(a), the SET has a memory function controlled by its gate voltage. By injecting electrons to the memory node with a π phase shift of Coulomb oscillation, the complementary SET is formed, which operates with an inverted input signal as compared to the conventional SET. As shown in Fig. 3.63(b), with both the conventional and the complementary SETs, various logic functions can be generated.

Until now the SETs described in this section are with only one silicon island. A SET having two silicon islands with mutual influences by tunneling has been developed [40]. Figure 3.64 shows the schematic of the double-island SET using the technology

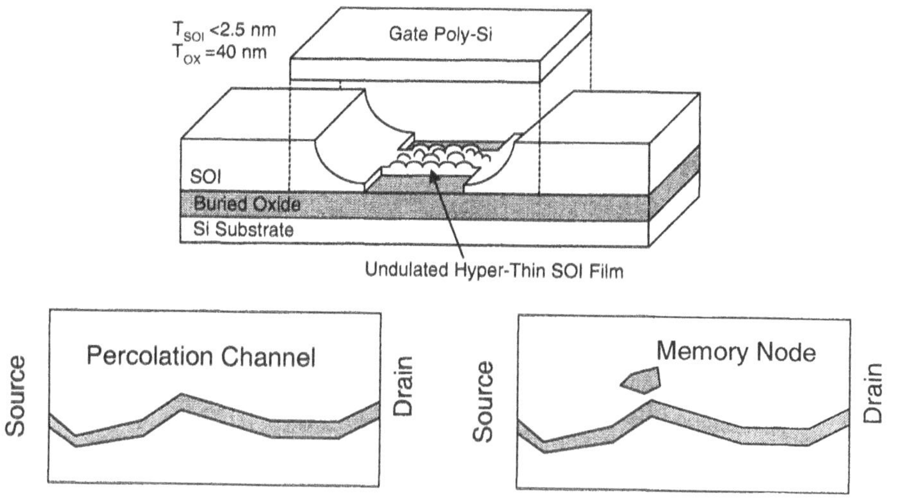

Fig. 3.61 Schematic of the fabricated SET with operation and memory function. (From Uchida et al. [39]. ©2000 IEEE.)

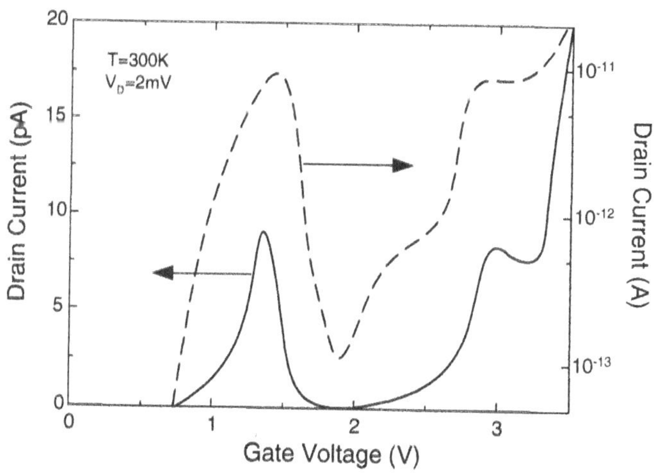

Fig. 3.62 Drain current versus gate voltage of the SET biased at the drain voltage of 2 mV measured at room temperature. (Adapted from Uchida et al. [39].)

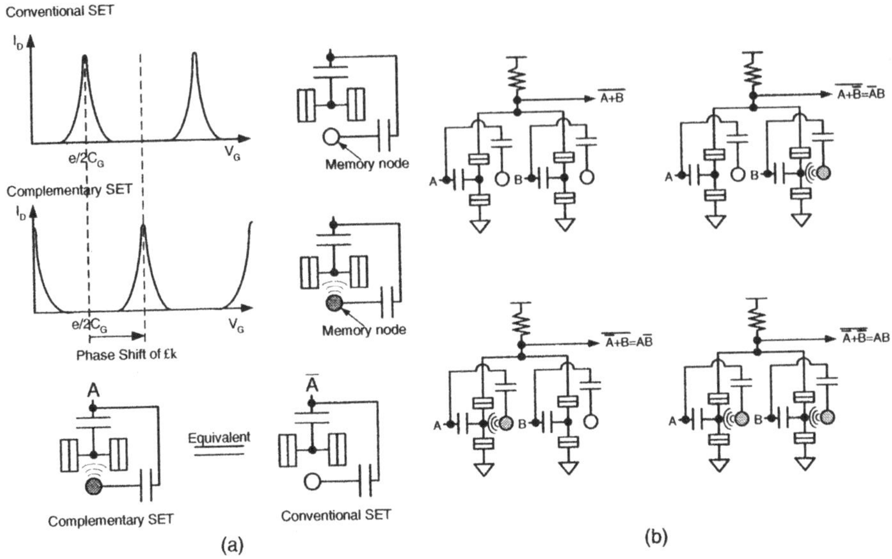

Fig. 3.63 (a) Conventional and complementary SETs. (b) Various logic functions realized using the conventional and the complementary SETs. (From Uchida et al. [39]. ©2000 IEEE.)

described before [40]. As shown in this figure, the SET device is with islands 1 and 2. Since islands 1 and 2 are close, mutual influence between the two islands is carried out via C_c by tunneling. As long as islands are sufficiently close, transfer of data does not have to be via the large-area source/drain. Instead, electrons can be transferred to other places via a series of silicon islands controlled by appropriate gate signals, as shown in Fig. 3.64(c). For the future SET circuits, signals are not transferred via the conventional interconnect lines. Via a series of tunneling capacitances, signals can be exchanged within a very short distance such that power and space can be saved to fully utilize the advantages of SET.

3.7 ELECTROSTATIC DISCHARGE (ESD)

During the down scaling of VLSI, the importance of electrical overstress (EOS) electrostatic discharge (ESD) phenomenon has been growing. In order to protect the internal circuits from the EOS ESD damages, protection circuits are implemented on chip with the circuits. How to built these protection circuits within a limited space with minimized side effects has been a challenge. During the evolution of the bulk CMOS VLSI technology in the past, a lot of precious ESD protection techniques have been accumulated. Unfortunately, most of them are not suitable for use in SOI technology for two reasons. First, since devices are built on the silicon thin film, the ESD techniques based on thick-field oxide devices are not feasible. Second, due to the existence of the buried oxide, the heat generated by the power consumed by

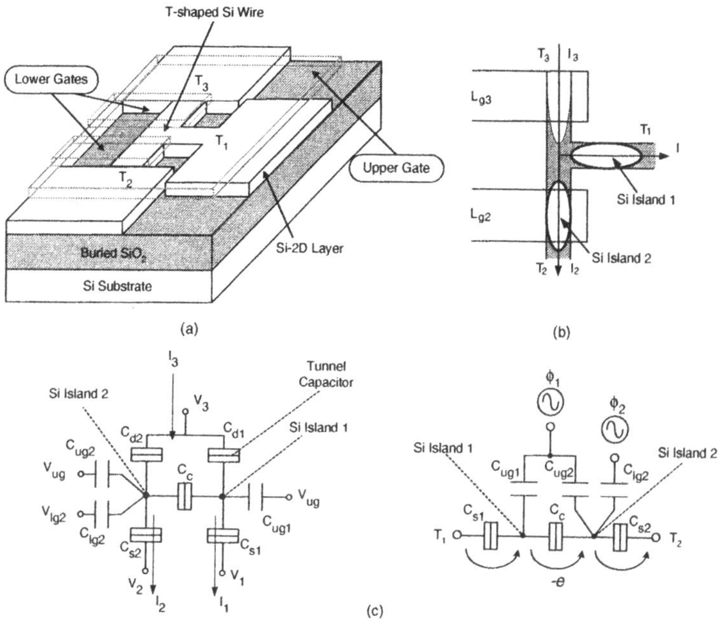

Fig. 3.64 (a) Schematic diagram of T-shaped silicon wire device. (b) Schematic of the double-island SET. (c) Single-electron transfer of the device with two gate controls. (From Fujiwara et al. [40]. ©1999 IEEE.)

the device cannot dissipate easily. Hence, the corresponding ESD devices at a high current have poor performance. Due to the existence of the buried oxide, the large-area parasitic diodes, bipolar devices, and silicon controlled rectifiers (SCR) frequently adopted in bulk technology cannot be implemented in SOI. Therefore, compared to bulk technology, the ESD devices for the SOI technology need to be reconsidered in many aspects. In this section, different ESD properties of SOI devices as compared to bulk devices are described first, followed by the techniques to improve the ESD properties in SOI technology. As for the ESD protection circuits, they are described in Chapter 4.

ESD phenomena can be divided into several categories. In order to effectively describe these ESD phenomena, various ESD testing standards such as the human body model (HBM), the machine model (MM), and the charged device model (CDM), etc., have been developed by the industry. For each testing method, the approach of imposing ESD to the wafer under measurement is different. This section is based on HBM with the equivalent circuit as shown in Fig. 3.65 [41]. In the HBM, the device under test is susceptible to a pulse with an effective pulse width of $\sim$ 150 ns.

At the emergence of ESD, the MOS device is susceptible to a high voltage for driving a high current. For HBM, the general requirement on ESD is the high sustained voltage of 2 kV. Therefore, the high-voltage characteristics of the device become the key points for studying ESD. Figure 3.66 shows the drain current versus the drain voltage of an MOS device in high- and low-current regimes [42]. For the

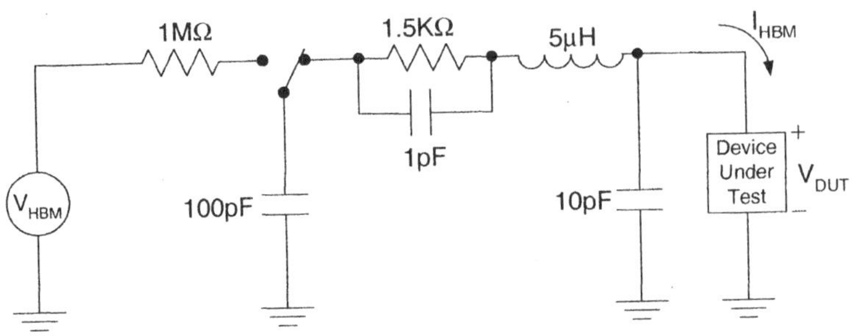

Fig. 3.65 Human body ESD model with a pulse width ~ 150 ns. (Adapted from Raha et al. [41].)

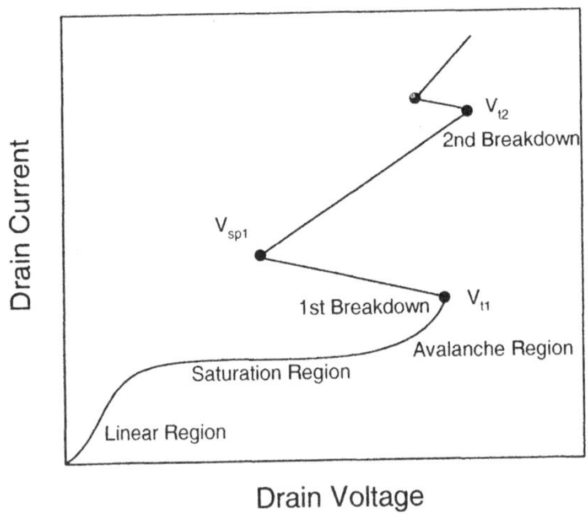

Fig. 3.66 Drain current versus drain voltage of an MOS device in higher- and low-current regimes. (Adapted from Volman [42].)

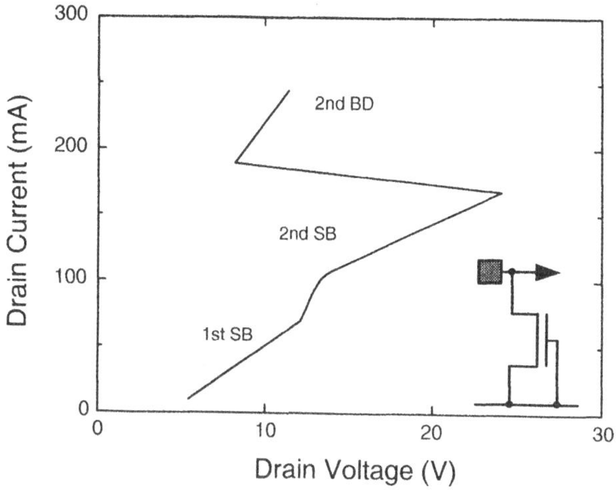

Fig. 3.67 The snapback characteristics of an SOI NMOS device with a front gate oxide of 190 Å, a thin film of 1800 Å doped with a p-type density of $10^{17}cm^3$, a channel length of 2 μm, and a channel width of 20 μm, measured by transmission-line pulsing techniques. (Adapted from Verhaege et al. [43].)

high-current behavior as shown in Fig. 3.66, the common condition is described as follows. By increasing the drain voltage gradually, the device changes from the linear region to the saturation region. Due to impact ionization the device enters the avalanche region, where the current rises quickly. If the drain voltage exceeds the threshold voltage V_{t1}, the parasitic bipolar device appears below the MOS device. The snapback after a large-scale impact ionization does not cause damages to the device. On the contrary, it provides a way to relieve the high current due to ESD. If the drain voltage surpasses the threshold voltage V_{t2}, the device cannot tolerate a large power. Therefore the high temperature may result in a high intrinsic carrier density surpassing the background doping density, which causes thermal runaway and second breakdown. Usually at this time the device already tends to be damaged. The starting point of the second breakdown is regarded as the upper limit of the device for resisting ESD. The properties of the snapback region determine the quality of operation of the device under a high voltage and a high current.

For some SOI NMOS devices, double snapback behavior may happen. Figure 3.67 shows the snapback characteristics of an SOI NMOS device with a front gate oxide of 190 Å, a thin film of 1800 Å doped with a p-type density of $10^{17}cm^{-3}$, a channel length of 2 μm, and a channel width of 20 μm, measured by transmission-line pulsing techniques [43]. As shown in this figure, before the second breakdown there are two staircase snapbacks. The first snapback occurs at the time when the parasitic bipolar device at the surface of the silicon thin film is triggered to turn on. Since the base-emitter potential barrier of the parasitic bipolar device at the surface is relatively low, it functions first. At the bottom of the thin film, the base of the parasitic bipolar device is made of the neutral region with the accumulated holes. Therefore its base-emitter

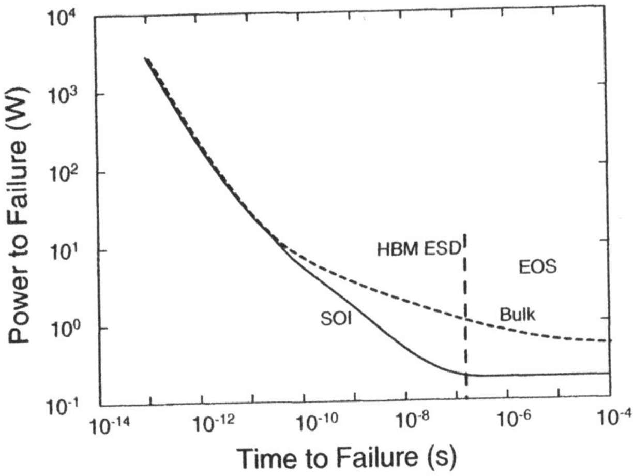

Fig. 3.68 Power to failure versus time to failure of an SOI NMOS device with a front gate oxide of 60 Å, a thin film of 500 Å doped with a p-type density of $4 \times 10^{17}\text{cm}^{-3}$, a buried oxide of 800 Å, a channel length of 0.4 μm, a channel width of 30 μm, and a contact-to-gate spacing of 2 μm. (Adapted from Ramaswamy et al. [44].)

potential barrier is higher. It turns on at a higher drain voltage. The parasitic bipolar devices in the different regions result in double-snapback phenomenon.

Figure 3.68 shows the power to failure versus the time to failure of an SOI NMOS device with a front gate oxide of 60 Å, a thin film of 500 Å doped with a p-type density of $4 \times 10^{17}\text{cm}^{-3}$, a buried oxide of 800 Å, a channel length of 0.4 μm, a channel width of 30 μm, and a contact-to-gate spacing of 2 μm [44]. As shown in this figure, when the power consumed by the device is higher, the time to failure is shorter since at a high power heat accumulates quickly. The time to reach the high temperature to have second breakdown is shorter. Thus, the time to failure is shorter. In comparison with the bulk device, when the time to failure is very short, the SOI device shows a similar power to failure. For a time to failure $>$ 10 ps, due to the poorer power dissipation capability, the power to failure of the SOI device is lower than that of the bulk one. At a longer time to failure, the difference in the power to failure between the SOI and the bulk devices gets larger. At the time to failure of 150 ns, which is the equivalent pulse width of HBM ESD, the HBM ESD of the SOI device is much worse than that of the bulk device.

Figure 3.69 shows the HBM failure voltage versus the channel length of the SOI and bulk NMOS devices with their gates grounded (a) under a positive ESD HBM stress and (b) under a negative ESD HBM stress [45]. As shown in Fig. 3.69(a), due to the poor power dissipation capability, it is easy for SOI devices to have thermal runaway and second breakdown. Thus the ESD failure voltage of the SOI device under a positive ESD HBM stress is lower than that of the bulk one. As shown in Fig. 3.69, the difference in the ESD failure voltage between the bulk and the SOI devices under a negative ESD HBM stress is even larger. Under a negative ESD HBM

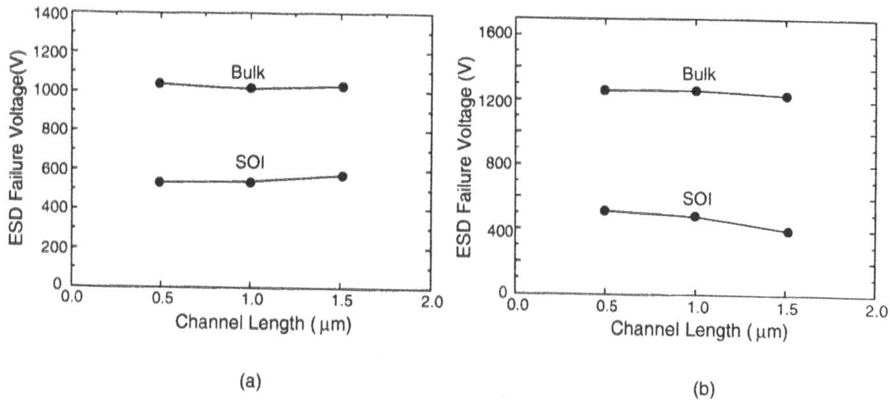

Fig. 3.69 HBM failure voltage versus channel length of the SOI and the bulk NMOS devices with their gates grounded (a) under a positive ESD HBM stress and (b) under a negative ESD HBM stress. (Adapted from Chan et al. [45].)

pulse, due to the turn-on of the large-area parasitic diodes between the drain/source and the substrate for providing a discharge path, the ESD failure voltage of the bulk device is higher than that under a positive ESD HBM stress. In contrast, the SOI device shows a totally opposite situation. Since between the source/drain and the substrate it is separated by the buried oxide and since most of the voltage drop is over the post-saturation region near the drain with more serious local heating, the failure voltage of the SOI device under a negative ESD HBM stress is worse than that under a positive one.

The structure of the SOI MOS devices may affect ESD. Figure 3.70 shows (a) the ESD failure voltage versus the thin-film thickness of an SOI NMOS device with a front gate oxide of 70 Å, a channel length of 1 μm, and a channel width of 20 μm under positive and negative ESD stresses and (b) the ESD failure voltage versus the channel width of an SOI NMOS device with a front gate oxide of 80 Å and a channel length of 1 μm, under positive and negative HBM ESD stresses [45]. As shown in this figure, when the thin-film thickness is thicker, thermal properties of the device improve since the heat can be distributed to a larger area. Thus, ESD improves. For both SOI and bulk devices, with a larger channel width, the ESD failure voltage rises owing to a larger discharge path. Compared to the bulk device, the improvement of the ESD failure voltage owing to a larger channel width for the SOI device is smaller, which implies the ESD problems of the SOI devices is much more serious.

As described above, the ESD failure voltage of the SOI device is > 50% smaller than that of the bulk. Figure 3.71 shows a CMOS output buffer using the through-oxide ESD protection scheme with the SOI PMOS device and the bulk NMOS device [45]. In the SOI chip, a portion of the buried oxide has been etched away to implement a bulk NMOS device for ESD protection purposes. But this approach is not practical and costly. Adding the bulk devices may complicate the processing techniques and may lower the advantages of the SOI circuits substantially.

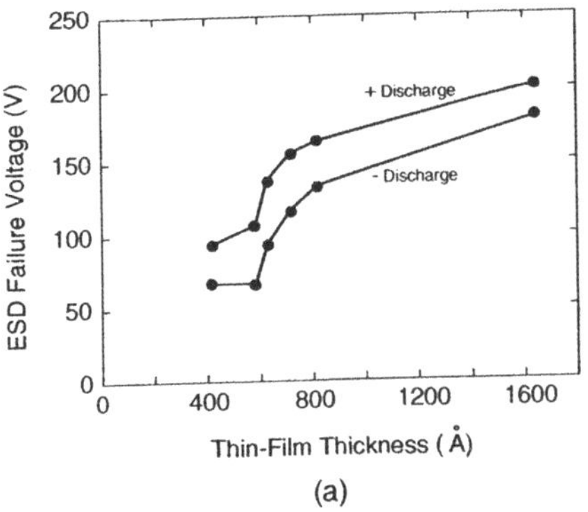

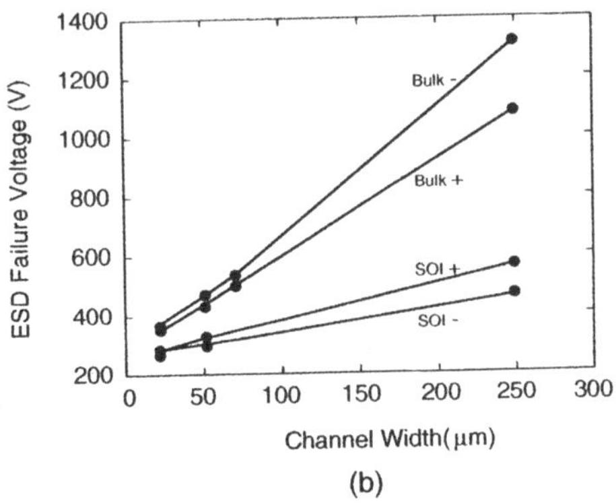

Fig. 3.70 (a) ESD failure voltage versus thin-film thickness of an SOI NMOS device with a front gate oxide of 70 Å, a channel length of 1 μm, and a channel width of 20 μm under positive and negative ESD stresses. (b) ESD failure voltage versus channel width of an SOI NMOS device with a front gate oxide of 80 Å and a channel length of 1 μm, under positive and negative HBM ESD stresses. (Adapted from Chan et al. [45].)

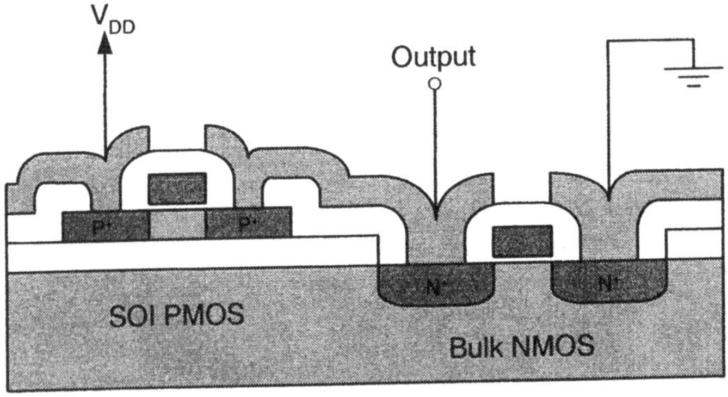

Fig. 3.71 CMOS output buffer using the through-oxide ESD protection scheme with the SOI PMOS device and the bulk NMOS device. (Adapted from Chan et al. [45].)

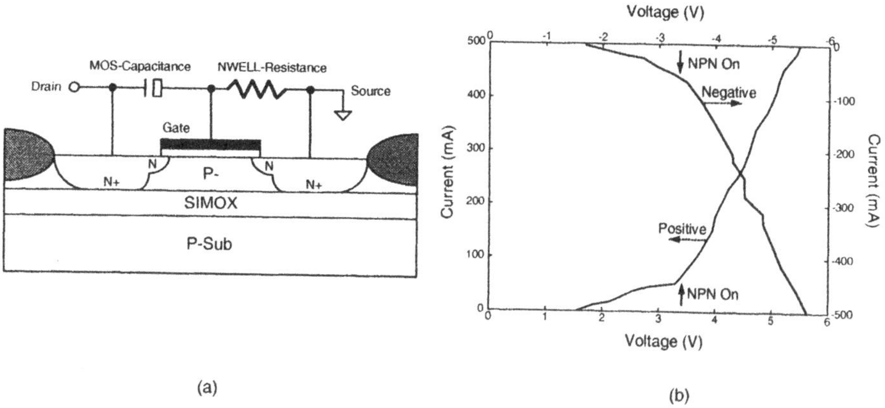

Fig. 3.72 (a) Gate-biased SOI NMOS ESD protection design. (b) Drain current versus drain voltage of the SOI NMOS device. (Adapted from Duvvury et al. [46].)

Although the ESD properties of the SOI devices are worse than those of the bulk devices, there are also advantages. Due to the simplicity of the ESD discharge path in the SOI devices, it is easy to control and improve their ESD behavior. In contrast, for bulk devices, in order to avoid triggering some mechanisms not related to the discharge path, the ESD protection device for the bulk device cannot be too small. In addition, the ESD protection device should be placed at some distance from other circuits with guard rings. From the scaling point of view, the design of ESD protection device for SOI technology has a lot of potentials to be developed.

The ESD failing voltage of the SOI device can be improved via inhibiting the triggering mechanism. Figure 3.72 shows the gate-biased SOI NMOS ESD protection design, whose NMOS gate is connected to a pad (the drain end) via a large capacitor made of a $10 \times 2\ \mu m$ MOS device and to the ground (source end) via a resistor with a high resistance for providing a DC path [46]. When there is no ESD pulse, the gate is connected to ground via the resistor. Once the pad receives a positive or negative ESD pulse, the signal coupling via the capacitor turns on the surface NMOS device, which results in impact ionization. The generated hole current triggers the turn-on of the parasitic bipolar device to provide discharge such that the ESD behaviors are improved. With a channel width of 400 μm, the ESD failure voltage of the SOI device using the gate-biased ESD protection design is over 2 kV, which meets the HBM industry standard.

In addition to the NMOS and parasitic bipolar devices, the diode can also be used in ESD protection circuits. Figure 3.73 shows the comparison of the SOI NMOS devices with the SOI diodes in terms of current to failure versus time to failure characteristics. As shown in this figure, at a high current the SOI diode has a better performance than the SOI NMOS device [47].

Figure 3.74(a) shows an SOI lubistor (lateral unipolar transistor) based on an SOI diode structure, in which the gate is connected to the cathode or V_{DD} [48]. As shown in Fig. 3.74(b), with a channel length of 1.2 μm, the ESD robustness reaches

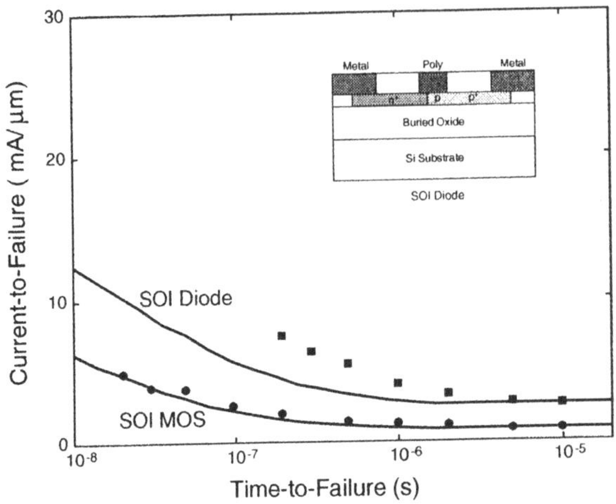

Fig. 3.73 Current to failure versus time to failure of the SOI NMOS device with a channel length of 0.5 μm, a channel width of 50 μm, and the SOI diode on the top of a buried oxide of 4000 Å. (Adapted from Raha et al. [47].)

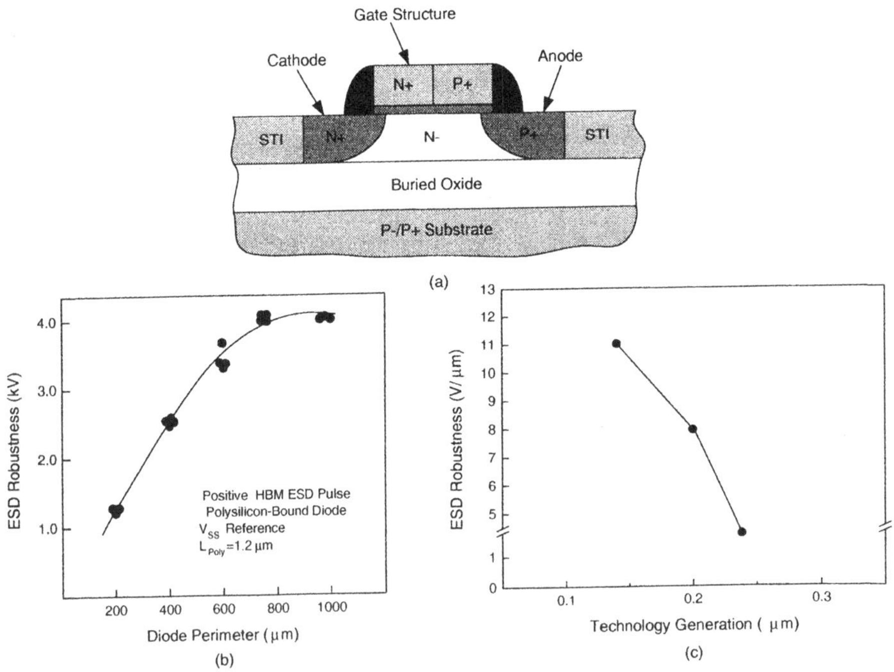

Fig. 3.74 (a) Cross-section of an SOI lateral gated diode with the abrupt implant drain structure. (b) ESD robustness versus perimeter of the polysilicon gated diode with a channel length of 1.2 μm under positive HBM ESD pulses. (c) ESD robustness versus technology. (Adapted from Voldman et al. [48].)

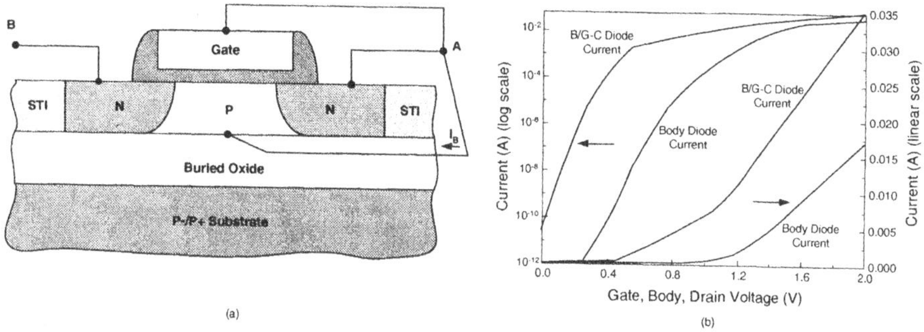

Fig. 3.75 (a) Cross-section of the body/gate-coupled (B/G-C) DTMOS SOI diode as the ESD protection device. (b) IV characteristics of the SOI MOS device in body source and B/G-C diode configurations. (Adapted from Voldman et al. [48].)

$5\text{V}/\mu\text{m}$ for a channel width of < 800 μm with its maximum ESD failure voltage of 4 kV. The ESD properties of the SOI lubistor improve with technology scaling as shown in Fig. 3.74(c). When the channel length is scaled down, the diode series resistance is reduced and consequently discharge is easier. Thus heat is not easier to accumulate. As shown in this figure, the SOI lubistor may be a good candidate for the ESD protection device in the future advanced SOI VLSI technology.

In addition to lubistor, following the DTMOS approach, Fig. 3.75(a) shows the cross section of the body/gate-coupled (B/G-C) DTMOS diode and the ESD protection device[48]. As shown in this figure, as an ESD protection device, both the diode and the surface MOS become the discharge path simultaneously. In addition, with the DTMOS configuration, when the drain is susceptible to an ESD pulse, the body voltage rises. The lowering in the threshold voltage of the MOS device turns on the MOS device for facilitating discharge. Using the B/G-C diode, the ESD robustness reaches $18.8\ \text{V}/\mu\text{m}$, which is two times of the lubistor's value. As shown in Fig. 3.75, the triggering voltage of the B/G-C diode is < 0.5 V, which is especially valuable for the future VLSI environment using a supply voltage < 1 V.

3.8 TEMPERATURE DEPENDENCE

In this section, temperature-dependent performance of the SOI MOS devices is described. Figure 3.76 shows the drain current versus the gate voltage of the SOI NMOS device with a front gate oxide of 250 Å, a thin film of 700 Å doped with a p-type density of 10^{17}cm^{-3}, a buried oxide of 3860 Å, a channel length of 25 μm, and a channel width of 50 μm, biased (a) in the strong inversion region and (b) in the subthreshold region, with body floating, at the drain voltage of 50 mV and the substrate bias of -20 V, operating at various temperatures [49]. As shown in this figure, in the strong inversion, at a higher temperature, the drain current is lowered due to a reduced mobility. In addition, at a higher temperature its threshold voltage becomes smaller. Due to these two contradicting trends simultaneously functioning,

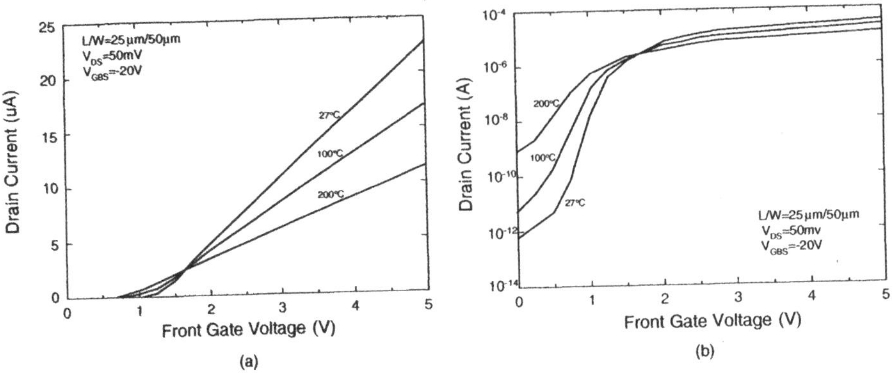

Fig. 3.76 Drain current versus gate voltage of the SOI NMOS device with a front gate oxide of 250 Å, a thin film of 700 Å doped with a p-type density of 10^{17}cm^{-3}, a buried oxide of 3860 Å, a channel length of 25 μm, and a channel width of 50 μm, biased (a) in the strong inversion region and (b) in the subthreshold region, with body floating, at the drain voltage of 50 mV, and the substrate bias of -20 V, operating at various temperatures. (Adapted from Jeon & Burk et al. [49].)

there exists the zero temperature coefficient (ZTC) point, where the drain current is almost independent of the change in the temperature. In the subthreshold region, the subthreshold slope becomes worse for the increase in the temperature. In addition, an increased temperature raises the thermal generation/recombination current. These two effects result in a quick rise in the off-leakage current at an elevated temperature.

Figure 3.77 shows the threshold voltage versus the temperature of FD and PD SOI and bulk NMOS devices [50]. As shown in Fig. 3.77, the temperature effects of the PD SOI and bulk devices are similar. The temperature effects of the FD device are much smaller than the other two. For PD and bulk devices, when the temperature is changed, the depletion region under the channel changes accordingly, thus the corresponding threshold voltage varies. As for the FD device, no such effects exist at all because the thin film is always fully depleted regardless of the temperature. At a high temperature ($> 220°$C), the temperature effects of the FD SOI device increase due to the increase in the intrinsic carrier concentration– the thin film tends to have a neutral region just as in a PD device.

Figure 3.78 shows the off leakage current versus the temperature of the bulk device and of the accumulation-mode SOI PMOS devices: (1) SOI1 with a front gate oxide of 150 Å, a thin film of 900 Å doped with a p-type density of $4 \times 10^{16}\text{cm}^{-3}$, a buried oxide of 4000 Å, a channel width of 20 μm, and a channel length of 5 μm, (2) SOI2 with a front gate oxide of 550 Å, a thin film of 1000 Å doped with a p-type density of 10^{16}cm^{-3}, a buried oxide of 4000 Å, a channel width of 3 μm, and a channel length of 3 μm, biased at the gate voltage of 0 V and the drain voltage of -3 V [51]. As shown in Fig. 3.78, the rise in the off-leakage current of the bulk device when the temperature rises is more when compared to the SOI device. The leakage current in the bulk is mainly made of the reverse-biased diode current between the source/drain and the substrate. In addition, the diffusion current is proportional

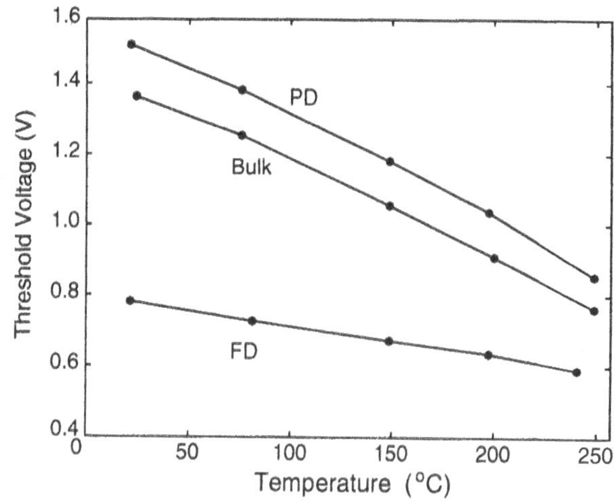

Fig. 3.77 Threshold voltage versus temperature of FD and PD SOI and bulk NMOS devices. (Adapted from Groeseneken et al. [50].)

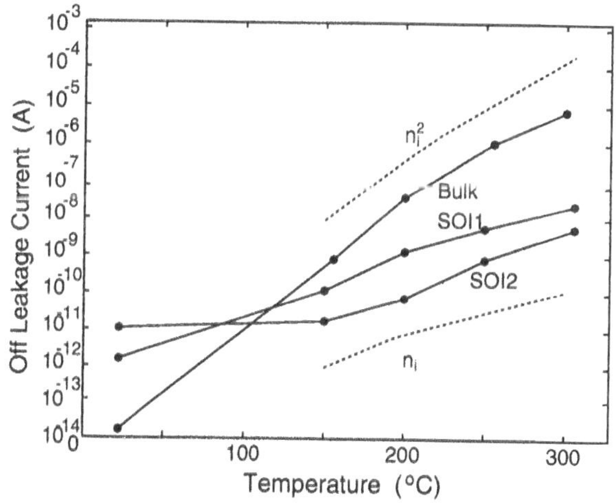

Fig. 3.78 Off-leakage current versus temperature of the bulk device and the accumulation-mode SOI PMOS devices: (1) SOI1 with a front gate oxide of 150 Å, a thin film of 900 Å doped with a p-type density of $4 \times 10^{16}\text{cm}^{-3}$, a buried oxide of 4000 Å, a channel width of 20 μm, and a channel length of 5 μm, and (2) SOI2 with a front gate oxide of 550 Å, a thin film of 1000 Å doped with a p-type density of 10^{16}cm^{-3}, a buried oxide of 4000 Å, a channel width of 3 μm, and a channel length of 3 μm, biased at the gate voltage of 0 V and the drain voltage of -3 V. (Adapted from Flandre et al. [51].)

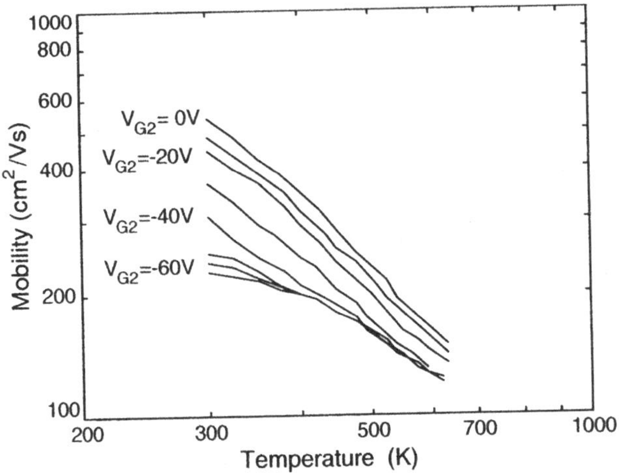

Fig. 3.79 Mobility versus temperature of the SOI MOS device for various front gate voltages. (Adapted from Reichert et al. [52].)

to the square of the intrinsic carrier density (n_i^2). Thus the leakage current of the bulk device is more affected by the temperature. In contrast, in the SOI device, the source/drain-substrate diode does not exist. The leakage current is mainly from the thermal-generation current in the thin film, which is proportional to the intrinsic carrier density (n_i). Thus, a smaller temperature dependence is observed for the SOI device. The superiority of the SOI device in terms of temperature dependence is exemplified at the temperature $> 100°C$. At a lower temperature, the leakage current of the bulk device is still lower than that of the SOI device probably due to the higher defect density in the thin film of the SOI device.

Mobility is a key factor in determining the drain current of an SOI device. Mobility is dependent on the temperature. Figure 3.79 shows the mobility versus the temperature of the SOI MOS device [52]. As shown in this figure, when the temperature rises, the mobility drops due to increased phonon scattering. When the back gate voltage is more negative, the increased vertical electric field leads to a lower mobility. At a more negative back gate voltage, the dependence of the mobility with respect to the temperature tends to decrease.

Figure 3.80 shows the drain current versus the temperature of PD and FD SOI NMOS devices with a channel width of 50 μm and a channel length of 0.6 μm, biased at the drain voltage of 2.5 V and the gate voltage of 2.5 V [53]. As shown in this figure, at a higher temperature, the drain current of both FD and PD devices drops. At the elevated temperature, both the threshold voltage and the mobility drop. At a high gate voltage, the influence of the mobility is more important. Thus, the drain current drops along with the increase in the temperature.

Figure 3.81 shows (a) g_m/I_D versus the scaled drain current of the SOI NMOS device with a front gate oxide of 300 Å, a thin film of 800 Å, doped with a p-type density of $10^{17}cm^{-3}$, a buried oxide of 4000 Å, and a channel length/width of 20 μm,

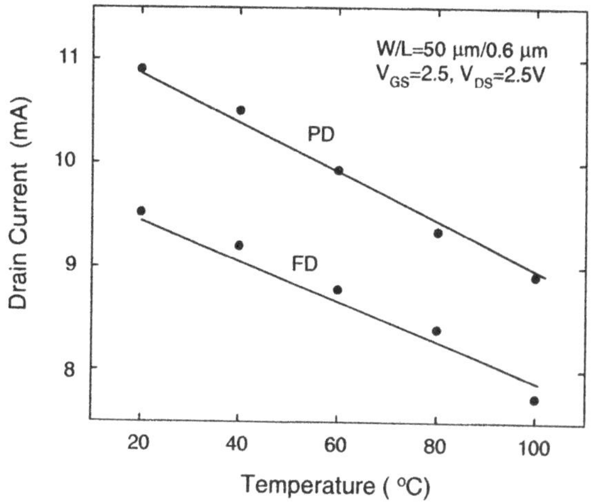

Fig. 3.80 Drain current versus temperature of PD and FD SOI NMOS devices with a channel width of 50 μm and a channel length of 0.6 μm, biased at the drain voltage of 2.5 V and the gate voltage of 2.5 V. (Adapted from Jin et al. [53].)

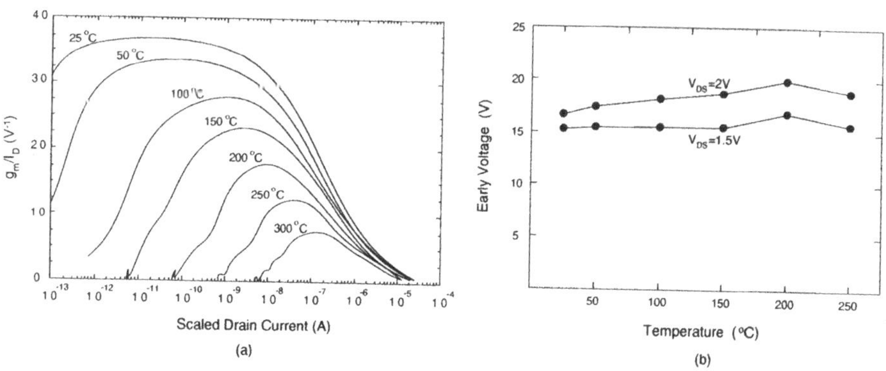

Fig. 3.81 (a) g_m/I_D versus scaled drain current of the SOI NMOS device with a front gate oxide of 300 Å, a thin film of 800 Å, doped with a p-type density of 10^{17}cm^{-3}, a buried oxide of 4000 Å, and a channel length/width of 20 μm, for various temperatures. (b) Early voltage versus temperature of the SOI NMOS device with a channel width of 50 μm and a channel length of 2 μm, biased at the drain current of 50 μA. (Adapted from Eggermont et al. [54].)

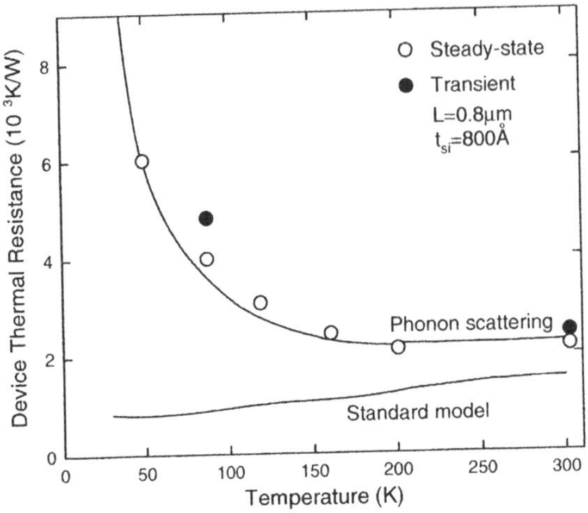

Fig. 3.82 Thermal resistance versus temperature of the SOI NMOS device with a thin film of 800 Å and a buried oxide of 3800 Å. (Adapted from Asheghi et al. [55].)

for various temperatures and (b) Early voltage versus temperature of the SOI NMOS device with a channel width of 50 μm and a channel length of 2 μm, biased at the drain current of 50 μA [54]. From Fig. 3.81, g_m/I_D has a maximum value in the subthreshold region. After the peak of the drain current in the strong inversion region, g_m/I_D becomes smaller. When the drain current is small, the leakage current occupies a substantially large portion of the drain current. Thus g_m/I_D becomes small. At a higher temperature, the maximum value of g_m/I_D falls. Due to the large increase in the leakage current, the location of the maximum g_m/I_D point tends to shift toward the higher drain region. As for the Early voltage, at a fixed drain current, the Early voltage of an FD device is almost independent of the temperature– the output resistance is almost fixed, which is helpful for analog circuit designs. In contrast, at the temperature exceeding 150°C, the Early voltage of a bulk device decreases quickly due to the high ratio of the drain current occupied by the leakage current. In comparison, SOI devices are especially suitable for operation in the high-temperature regime.

The operation temperature may affect the behavior of the SOI devices directly. In addition, it may affect self-heating, the parasitic bipolar device, impact ionization, and other parameters related to the SOI device. Figure 3.82 shows the thermal resistance versus the temperature of the SOI NMOS device with a thin film of 800 Å and a buried oxide of 3800 Å [55]. As shown in this figure, when the temperature drops, phonon-interface scattering becomes important, which raises the thermal resistance between the device and the outer environment. At a lower temperature, the drain current increases, which means a higher power dissipation. The increase in the generated heat and the higher thermal resistance results in an easier accumulation of the heat. Consequently, at a low temperature, self-heating of SOI MOS devices is especially serious.

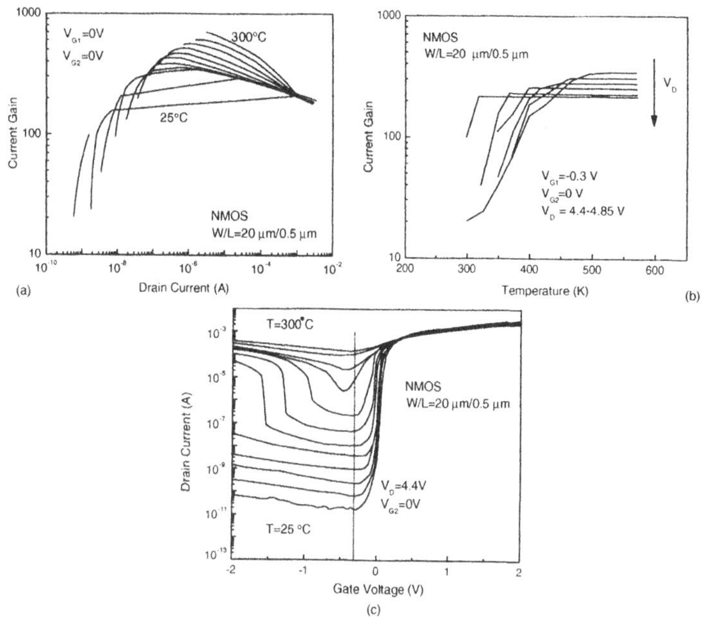

Fig. 3.83 (a) Current gain versus drain current of the parasitic bipolar device in the SOI NMOS device with a channel width of 20 μm and a channel length of 0.5 μm, biased at the front gate voltage of 0V and the back gate voltage of 0 V and (b) current gain versus temperature of the parasitic bipolar device in the SOI device biased at the front gate voltage of -0.3 V and the back gate voltage of 0 V. (c) Drain current versus gate voltage of the SOI NMOS device, biased at the drain voltage of 4.4 V and the back gate voltage of 0 V, for various temperatures. (Adapted from Reichert et al. [56].)

Figure 3.83 shows (a) the current gain versus the drain current of the parasitic bipolar device in the SOI NMOS device with a channel width of 20 μm and a channel length of 0.5 μm, biased at the front gate voltage of 0V and the back gate voltage of 0 V, (b) the current gain versus the temperature of the parasitic bipolar device in the SOI device biased at the front gate voltage of -0.3 V and the back gate voltage of 0 V, and (c) the drain current versus the gate voltage of the SOI NMOS device, biased at the drain voltage of 4.4 V and the back gate voltage of 0 V, for various temperatures [56]. As shown in this figure, at a high drain current high-level injection occurs. At a very low current, leakage current dominates. At either high or low drain current, current gain drops. Only at a medium drain current level is the current gain at its maximum value. When the temperature rises, the maximum current gain increases and the performance of the parasitic bipolar device improves. On the other hand, due to the leakage current and high-level injection, the region with the maximum current gain is limited. When the temperature rises, the current gain of the parasitic bipolar device increases. When the temperature rises to reach a certain value, the increasing trend in the current gain is saturated. At a larger drain voltage, the temperature at which the current gain becomes saturated comes earlier. At a higher temperature

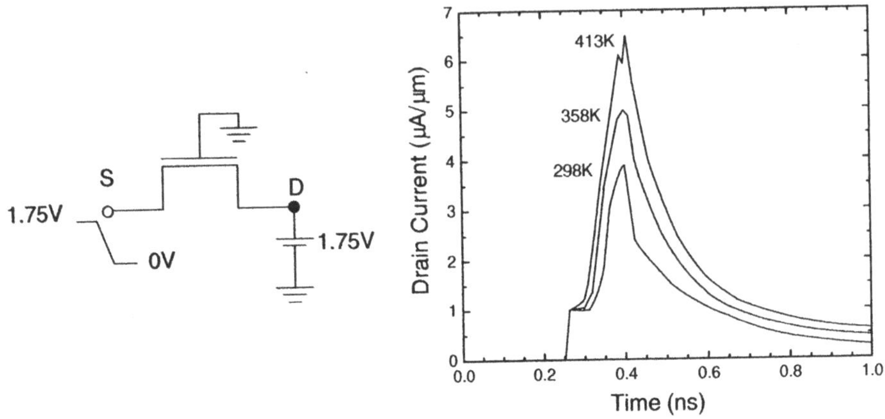

Fig. 3.84 Drain leakage current versus time during the transient of the PD SOI NMOS device serving as a pass transistor biased at the gate voltage of 0 V and the drain voltage of 1.75 V, with a negative voltage step from 1.75 to 0 V imposed at the source for various temperatures. (Adapted from Gautier et al. [57].)

and a higher drain voltage, it is easier for the device to enter the single-transistor latch regime. The high-level injection effect due to a large drain current prevents the current gain from increasing further. As shown in Fig. 3.83(c), when the temperature becomes higher, it is easier to have single-transistor latch phenomenon due to a better performance of the parasitic bipolar device at a higher temperature–the current gain is better. At the same time, the leakage current also increases such that the parasitic bipolar is easier to be triggered to turn on. Thus, it is easier for the single-transistor latch behavior to occur.

Temperature is important in determining the leakage current during the transient of a PD SOI NMOS device. Figure 3.84 shows the drain leakage current versus time during the transient of the PD SOI NMOS device serving as a pass transistor biased at the gate voltage of 0 V and the drain voltage of 1.75 V, with a negative voltage step from 1.75 to 0 V imposed at the source for various temperatures [57]. As shown in this figure, when the operating temperature rises, the drain leakage current during the transient also rises. At a higher temperature, the current gain of the parasitic bipolar device in the device becomes better. Due to the increase in the intrinsic carrier density at a higher temperature, the neutral region at the bottom of the device widens, which can accommodate more holes to trigger the turn-on of the bipolar device. With these two factors, the drain leakage current caused by the parasitic bipolar device in the PD SOI device becomes more serious at a higher temperature.

Temperature is a key factor in determining impact ionization and kink effects of SOI. Figure 3.85 shows (a) the drain current and (b) the body-emitter voltage versus the drain voltage of the PD SOI NMOS device with a front gate oxide of 250 Å, a thin film of 2500 Å doped with a p-type density of $10^{17}cm^{-3}$, a buried oxide of 4500 Å, a channel width of 50 μm, and a channel length of 25 μm, for various temperatures [58]. As shown in this figure, when the temperature rises, the drain current decreases

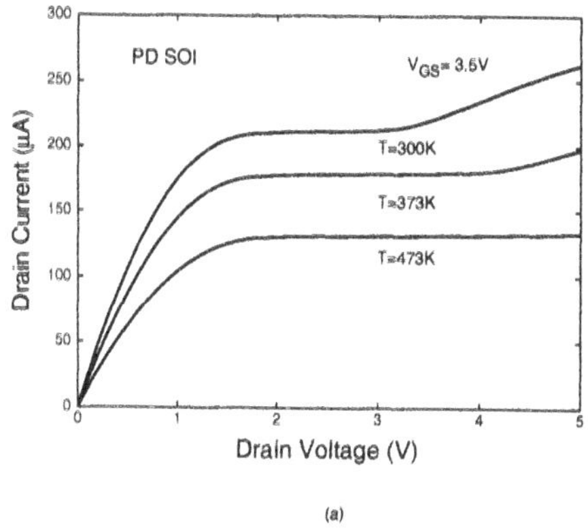

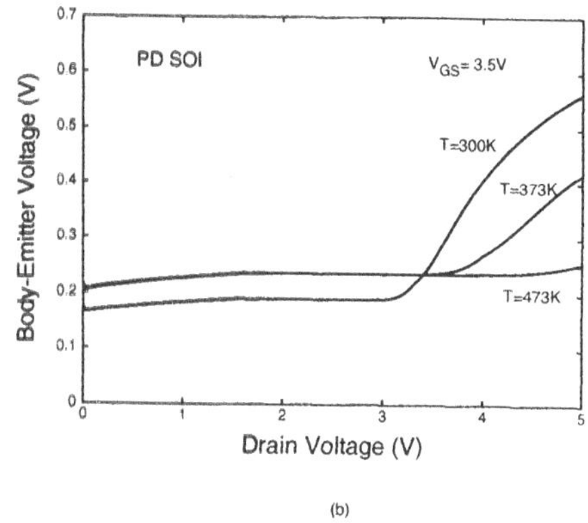

Fig. 3.85 (a) Drain current (b) body-emitter voltage versus drain voltage of the PD SOI NMOS device with a front gate oxide of 250 Å, a thin film of 2500 Å doped with a p-type density of $10^{17}cm^{-3}$, a buried oxide of 4500 Å, a channel width of 50 μm, and a channel length of 25 μm, for various temperatures. (Adapted from Lin & Kuo [58].)

at a low drain voltage due to the decrease in the mobility. At a high drain voltage, kink effects become smaller when the temperature is raised, where impact ionization is reduced due to more thermal scattering of the electrons. Therefore, the generated holes are fewer. At a higher temperature, a higher leakage current exists, thus the accumulated extra holes can be expelled from the thin film more easily. As shown in Fig. 3.85(a), at a higher temperature, it is more difficult for extra holes to be stored in the thin film. Therefore, the body-emitter voltage rises more slowly at a higher temperature. Thus less kink effects can be seen.

Figure 3.86 shows the degradation of the maximum transconductance and the drain current, and the shift in the threshold voltage versus the temperature of the SOI NMOS device with a front gate oxide of 45 Å, a thin film of 1000 Å, and a buried oxide of 800 Å, after a 50,000 s stress with the gate voltage of 0 V and the drain voltage of 2.75 V [59]. As shown in this figure, a higher temperature leads to more degradation in the transconductance. When the temperature is > 200 K, the degradation of the device after stress becomes smaller due to the decrease in the impact ionization at a higher temperature. On the other hand, when the temperature is < 200 K, due to the reduction in the leakage current and the degradation of the parasitic bipolar device, the electrons with a high energy are fewer. Thus the degradation of the device performance in terms of the maximum transconductance and the drain current is reduced when the temperature is lowered. At a temperature ~ 200 K, the degradation of the device performance is maximum, which means the reliability is worst at 200 K.

3.8.1 Noise

Noise performance of the SOI devices is also affected by the temperature. Figure 3.87 shows the gate noise power versus the frequency of the PD SOI NMOS device with a

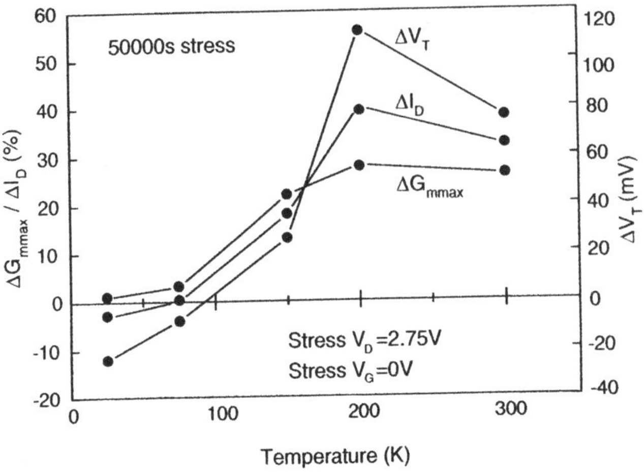

Fig. 3.86 Degradation of the maximum transconductance and the drain current, and the shift in the threshold voltage versus temperature of the SOI NMOS device with a front gate oxide of 45 Å, a thin film of 1000 Å, and a buried oxide of 800 Å, after a 50,000 s stress with the gate voltage of 0 V and the drain voltage of 2.75 V. (Adapted from Renn et al. [59].)

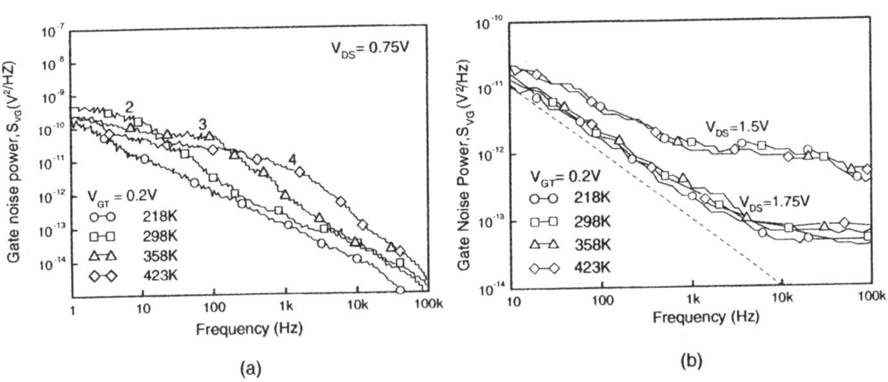

Fig. 3.87 Gate noise power versus frequency of the PD SOI NMOS device with a front gate oxide of 105 Å, a thin film of 1000 Å, and a buried oxide of 3600 Å, with its body floating, biased at the drain voltages of (a) 0.75 V and (b) 1.5 and 1.75 V, and the gate overdrive voltage of 0.2 V, for various temperatures. (From Tseng et al. [60]. ©2000 IEEE.)

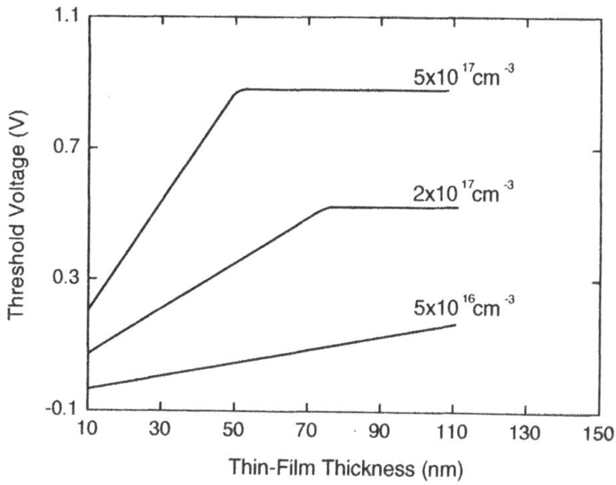

Fig. 3.88 Threshold voltage versus thin-film thickness of the FD SOI NMOS device with a front gate oxide of 80 Å and a buried oxide of 3800 Å for various thin-film doping densities for the long-channel case. (Adapted from Sherony et al. [61].)

front gate oxide of 105 Å, a thin film of 1000 Å, and a buried oxide of 3600 Å, with its body floating, biased at the drain voltages of (a) 0.75 V and (b) 1.5 and 1.75 V, and the gate overdrive voltage of 0.2 V, for various temperatures [60]. As shown in this figure, the noise spectrum is with low-frequency noise overshoot behavior. The corner frequency of the noise overshoot shifts toward a higher frequency at a higher temperature, which is not existent in the device biased at a high drain voltage. When the drain voltage is low, the reverse biased junction current, which is mainly generated by thermal generation, is the charging current for the neutral body. Thus, the noise spectrum easily varies by the change in the temperature. When the drain voltage is high (with kink effects), impact ionization current has replaced the reverse junction current to become the dominant charging current for the neutral body, which is less dependent on the temperature change.

3.9 SENSITIVITY

For FD SOI devices, the sensitivity of the threshold voltage with respect to the thin-film thickness has been a serious problem. The thin-film thickness has been scaled down especially for deep-submicron FD SOI devices. With a very thin thin film, variation of the thin-film thickness incurred from the fabrication process may be substantial. Consequently, the threshold voltage of the FD SOI device may fluctuate, which may result in difficulties for designing VLSI circuits using a low power supply voltage. In this section, sensitivity of the parameters of FD SOI devices are described.

Figure 3.88 shows the threshold voltage versus the thin-film thickness of the FD SOI NMOS device with a front gate oxide of 80 Å and a buried oxide of 3800 Å

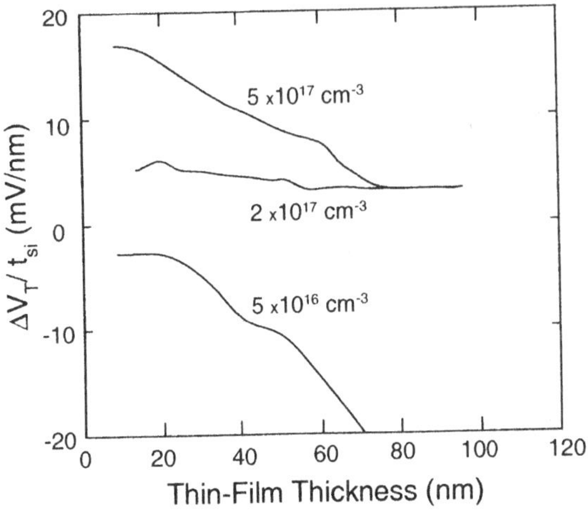

Fig. 3.89 Sensitivity of the threshold voltage with respect to the thin-film thickness versus the thin-film thickness of an FD SOI NMOS device with a front gate oxide of 80 Å, a buried oxide of 3800 Å and a channel length of 0.2 μm, for various thin-film doping densities. (Adapted from Sherony et al. [61].)

for various thin-film doping densities for the long-channel case [61]. As shown in Fig. 3.88, with a higher thin-film density, the sensitivity of the threshold voltage with respect to the thin-film thickness is higher (a larger slope). When the thin-film thickness exceeds a certain value, the device becomes PD, which is not affected by the variation of the thin-film thickness any more.

As for the short channel case, Fig. 3.89 shows the sensitivity of the threshold voltage with respect to the thin-film thickness versus the thin-film thickness of an FD SOI NMOS device with a front gate oxide of 80 Å, a buried oxide of 3800 Å and a channel length of 0.2 μm, for various thin-film doping densities [61]. As shown in this figure, with a thin-film density of $5 \times 10^{17}\text{cm}^{-3}$, the threshold voltage sensitivity is positive. With a thin-film density of $5 \times 10^{16}\text{cm}^{-3}$, due to short-channel effect and drain induced barrier lowering (DIBL), its threshold sensitivity becomes negative. In addition, with a larger thin-film thickness, its sensitivity is higher. For a thin-film density of $2 \times 10^{17}\text{cm}^{-3}$, its threshold voltage sensitivity can maintain at a relatively low value.

The sensitivity of the threshold voltage with respect to the thin-film thickness of a FD SOI NMOS device can be reduced by imposing an appropriate voltage to the back gate. Figure 3.90 shows the threshold voltage sensitivity to the thin-film thickness versus the back gate bias of the FD SOI NMOS device with a front gate oxide of 95 Å, a thin film of 500 Å doped with a p-type density of $3 \times 10^{15}\text{cm}^{-3}$, a buried oxide of 3750 Å, and a channel length of 0.1 μm [62]. As shown in this figure, since the thin-film doping density is light ($3 \times 10^{15}\text{cm}^{-3}$), before adding a back gate bias, the sensitivity of the threshold voltage to the thin-film thickness is serious because of DIBL. When the back gate bias becomes negative, the vertical electric field gets

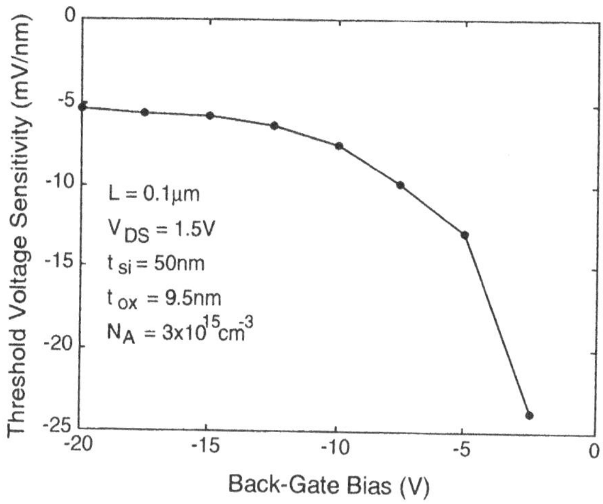

Fig. 3.90 Threshold voltage sensitivity to thin-film thickness versus back gate bias of the FD SOI NMOS device with a front gate oxide of 95 Å, a thin film of 500 Å doped with a p-type density of $3 \times 10^{15}\text{cm}^{-3}$, a buried oxide of 3750 Å, and a channel length of 0.1 μm. (Adapted from Leobandung & Chou [62].)

stronger. Thus DIBL is reduced and the threshold voltage sensitivity to the thin-film thickness has been improved to a large extent. If the back gate bias is too negative, holes are accumulated on the back surface of the thin film. In this situation, with an even more negative back gate bias, the vertical electric field in the thin film does not increase further and the improvement in the threshold sensitivity tends to be saturated.

Figure 3.91 shows the threshold voltage sensitivity to the thin-film thickness versus the thin-film thickness of several FD SOI NMOS devices with a buried oxide of 3750 Å: (a) a front gate oxide of 95 Å and a thin-film density of $3 \times 10^{15}\text{cm}^{3}$, (b) a front gate oxide of 95 Å and a thin-film density of $5 \times 10^{16}\text{cm}^{-3}$, (c) a front gate oxide of 50 Å and a thin-film density of $3 \times 10^{15}\text{cm}^{-3}$, biased at a back gate bias of -15 V [62]. As shown in this figure, since the back gate bias of -15 V has been applied, despite the doping density of the thin film [$3 \times 10^{15}\text{cm}^{-3}$ for (a) or $5 \times 10^{16}\text{cm}^{-3}$ for (b)], their sensitivities are low and almost identical. With a thinner front gate oxide (c) as compared to (a) and (b), its sensitivity is even lower than (a) and (b). Via decreasing the front gate oxide thickness, the threshold voltage sensitivity to the thin-film thickness can be effectively reduced. Although the FD device may have been regarded not suitable for further development because of the serious threshold voltage sensitivity problem, with the above techniques to reduce threshold voltage sensitivities, the future of the FD devices for VLSI is bright.

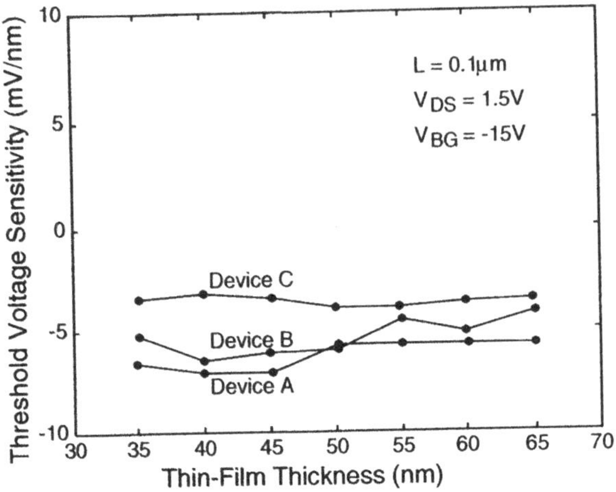

Fig. 3.91 Threshold voltage sensitivity to the thin-film thickness versus the thin-film thickness of several FD SOI NMOS devices with a buried oxide of 3750 Å: (a) a front gate oxide of 95 Å and a thin-film density of $3 \times 10^{15}\text{cm}^{-3}$, (b) a front gate oxide of 95 Å and a thin-film density of $5 \times 10^{16}\text{cm}^{-3}$, (c) a front gate oxide of 50 Å and a thin-film density of $3 \times 10^{15}\text{cm}^{-3}$, biased at the back gate bias of -15 V. (Adapted from Leobandung & Chou [62].)

3.10 RADIATION EFFECTS

Owing to the buried oxide structure, SOI devices have been regarded as having superior radiation hardness capabilities. As shown in Fig. 3.92, when high-energy particles pass through a silicon wafer, a large quantity of electron/hole pairs are generated in the substrate. For the bulk CMOS devices, these electron/hole pairs may be absorbed by the source/drain to produce a large leakage current, which may affect the operation of the related circuits [63]. It may trigger latchup to cause breakdown of the circuit. For SOI devices, due to the separation of the device from the substrate provided by the buried oxide, the above drawbacks of the bulk devices do not appear. Thus, the soft error immunity of the SOI devices is better.

Due to the buried oxide, high-energy particle-generated current can be isolated from the active device to avoid latch up. However, the high-energy particles may cause a lot of positive trapped charge in the buried oxide (or sapphire substrate for the SOS). Figure 3.93 shows (a) the cross section of the SOS NMOS device showing the effect of the trapped charge in the sapphire substrate and (b) the square root of the drain current versus the gate voltage of an SOS NMOS device for various total ionizing dose levels [64]. As shown in this figure, for an NMOS device the positive charge may make the threshold voltage of the device shift toward the negative direction. In addition, it may also attract electrons at the back surface to form a channel, which cannot be turned off by the front gate. As shown in this figure, with a larger dose of radiation, the threshold voltage shift of the device is larger, and hence the leakage current rises.

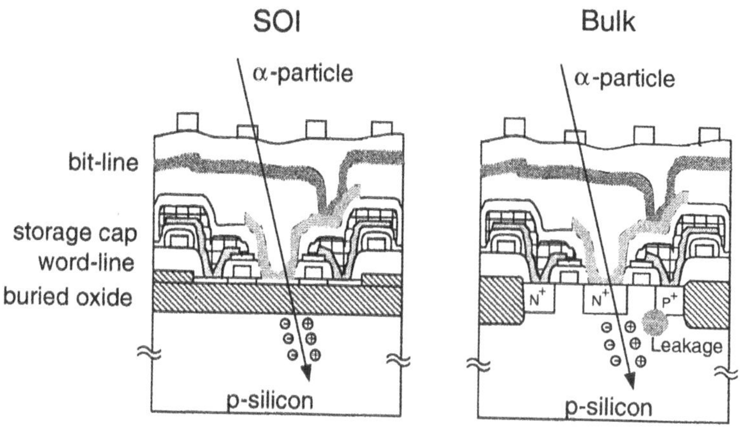

Fig. 3.92 Soft error immunity in an SOI device. (Adapted from Yamaguchi & Inoue [63].)

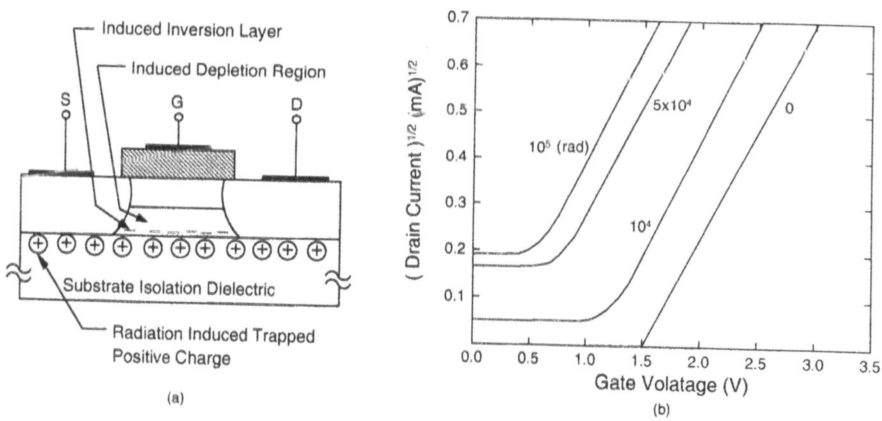

Fig. 3.93 (a) Cross-section of the SOS NMOS device showing the effect of trapped charge in the sapphire substrate. (b) Square root of the drain current versus gate voltage of an SOS NMOS device for various total ionizing dose levels. (Adapted from Buchanan et al. [64].)

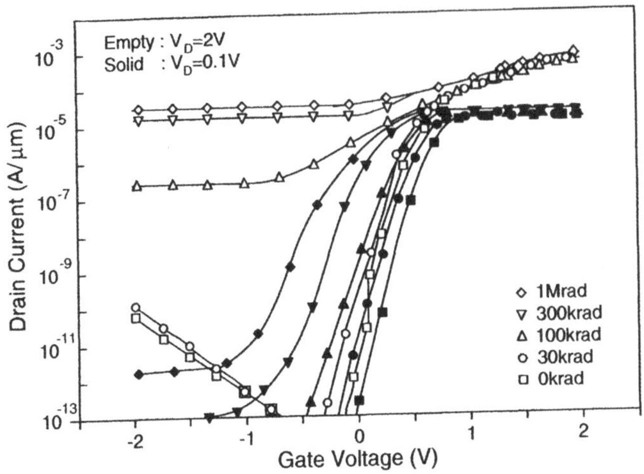

Fig. 3.94 Drain current versus gate voltage of the FD SOI NMOS device with a front gate oxide of 45 Å and a thin film of 400 Å, and a channel length of 0.2 μm, biased with the substrate grounded and the drain voltages of 0.1 and 2 V, for various irradiation doses. (Adapted from Ferlet-Cavrois et al. [65].)

Figure 3.94 shows the drain current versus the gate voltage of the FD SOI NMOS devices with a front gate oxide of 45 Å and a thin film of 400 Å, and a channel length of 0.2 μm, biased with the substrate grounded and the drain voltages of 0.1 and 2 V, for various irradiation doses [65]. As shown in this figure, when the drain voltage is small, the positive charge in the buried oxide caused by radiation may make the threshold voltage of the front channel shift toward the negative direction via a coupling effect. If the radiation dose is sufficiently large, its corresponding subthreshold slope may be worsened and the leakage current is higher. With a larger drain voltage of 2 V, the radiation-induced positive charge make the DIBL of the device more serious. Under this situation, at the high drain current, the device cannot be turned off.

The radiation effects on the PD devices are quite different from those on the FD ones. Figure 3.95 shows the drain current versus the gate voltage of the PD SOI NMOS with a front gate oxide of 45 Å and a thin film of 1000 Å and (a) with a channel length of 0.2 μm and with both body and substrate grounded, (b) with a channel length of 0.1 μm and with substrate grounded, under various irradiation doses. As shown in this figure, with the body tied to ground, the radiation-induced trapped charge in the buried oxide does not affect the front channel. Thus the PD device with body grounded has almost no threshold voltage shift characteristics for the devices with body floating as shown in the figure. As shown in Fig. 3.95(b), due to the trapped charge in the buried oxide caused by irradiation, the single-transistor latch phenomenon for the PD device with body floating becomes more serious.

Figure 3.96 shows the critical dose of the radiation to cause non-turn-off versus channel length of the FD and the PD SOI NMOS devices [65]. A lower critical dose means that the device is easily disturbed by the irradiation. As shown in this figure,

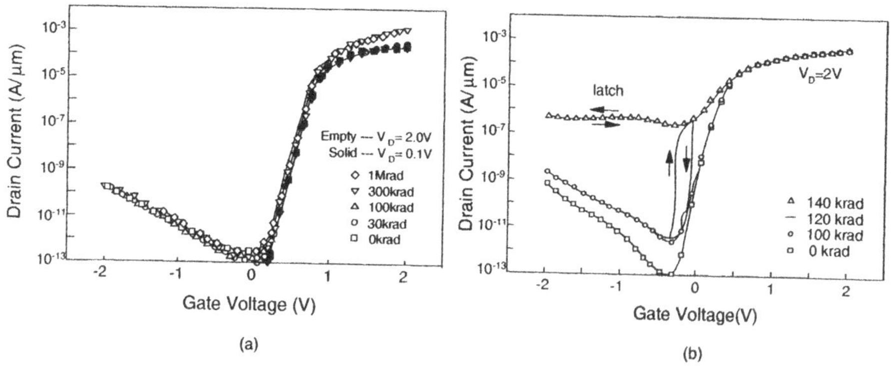

Fig. 3.95 Drain current versus gate voltage of the PD SOI NMOS device with a front gate oxide of 45 Å and a thin film of 1000 Å, under various irradiation doses. (a) with a channel length of 0.2 μm and with both body and substrate grounded. (b) with a channel length of 0.1 μm and with substrate grounded under various irradiation doses. (From Ferlet-Cavrois et al. [65]. ©1998 IEEE.)

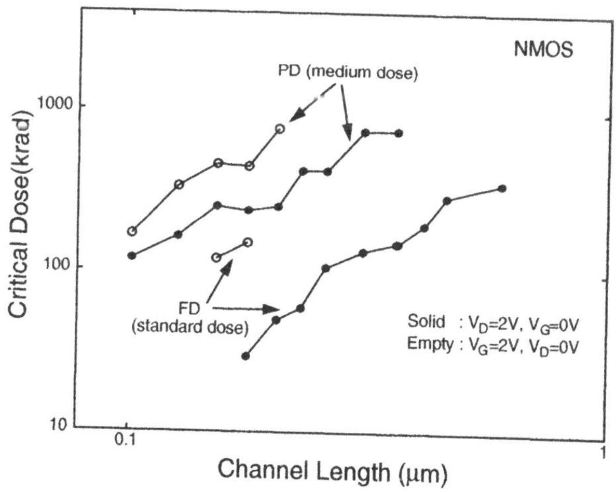

Fig. 3.96 Critical dose of the radiation to cause non-turn-off versus channel length of the FD and the PD SOI NMOS devices. (Adapted from Ferlet-Carvois et al. [65].)

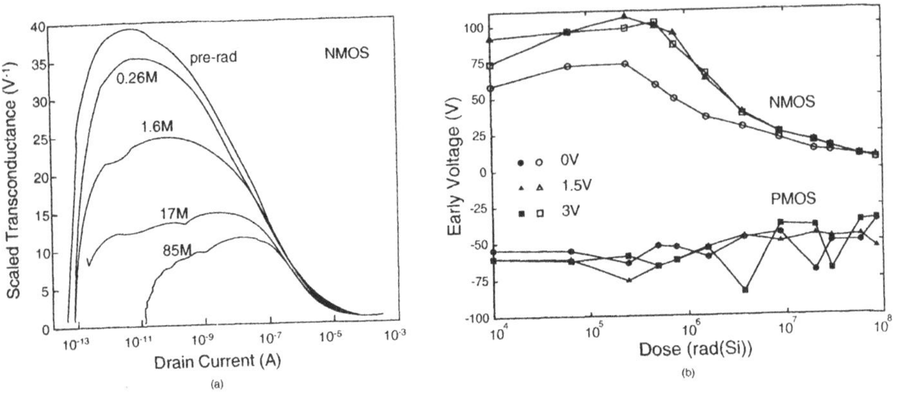

Fig. 3.97 (a) g_m/I_D versus drain current of the double-gate SOI NMOS device for various irradiation doses. (b) Early voltage versus irradiation dose received by the double-gate SOI NMOS device. (Adapted from Vandooren et al. [66].)

with a smaller channel length, the critical dose is smaller. The critical dose of the FD device is usually smaller than that of the PD one.

In addition to single-gate devices, γ-irradiation also affects double-gate devices. Fig. 3.97 shows (a) g_m/I_D versus the drain voltage of the double-gate SOI NMOS device for various irradiation doses, and (b) Early voltage versus irradiation dose received by the double-gate SOI NMOS device [66]. As shown in this figure, with γ-irradiation, the interface charge of the double-gate device increases. Therefore, the leakage current of the device increases and thus the drain current at which g_m/I_D reaches its peaks also increases. As shown in Fig. 3.97(b), the Early voltage of the double-gate SOI PMOS device is relatively independent of the radiation effects. For the double-gate SOI NMOS device, its Early voltage decreases with the increase in the irradiation dose.

The trapped oxide charge of the SOI devices caused by irradiation can be annealed in an environment with an elevated temperature. Fig. 3.98 shows the drain current versus the gate voltage of an SOS NMOS device with various annealing times at 140°C after Co^{60} irradiation of 100 krad [67]. As shown in the figure, after Co^{60} irradiation, a substantially large leakage current can be seen. After 140°C isothermal annealing, the device behavior had been recovered.

Irradiation of high-energy particles may affect the change in the DC performance of an SOI device. Irradiation may also change the transient behavior of an SOI device. When high-energy particles travel through the thin film, the quantity of the generated electron/hole pairs may not be big, but they may trigger the turn-on of the parasitic bipolar device in the PD SOI device. As shown in Fig. 3.99, the weak ion current can be amplified by the parasitic bipolar device to generate a large transient leakage current with a long duration, which may affect the operation of the related circuits [68].

Figure 3.100 shows (a) the drain and the source currents and (b) the drain collected charge and the source injected charge during the transient after the α-particle incidence

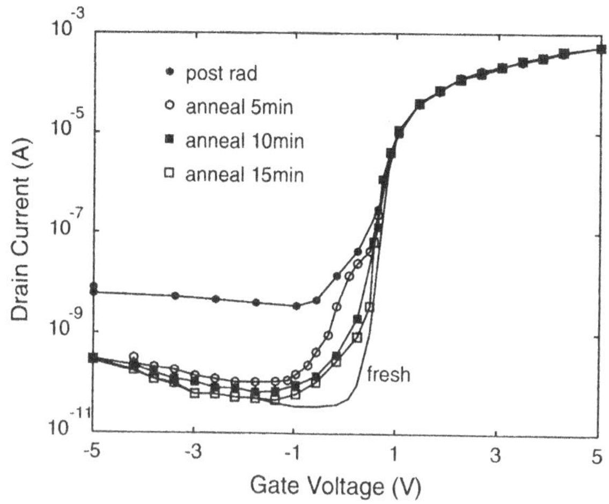

Fig. 3.98 Drain current versus gate voltage of an SOS NMOS device with various annealing times at 140°C after Co^{60} irradiation of 100 krad. (Adapted from Chao et al. [67].)

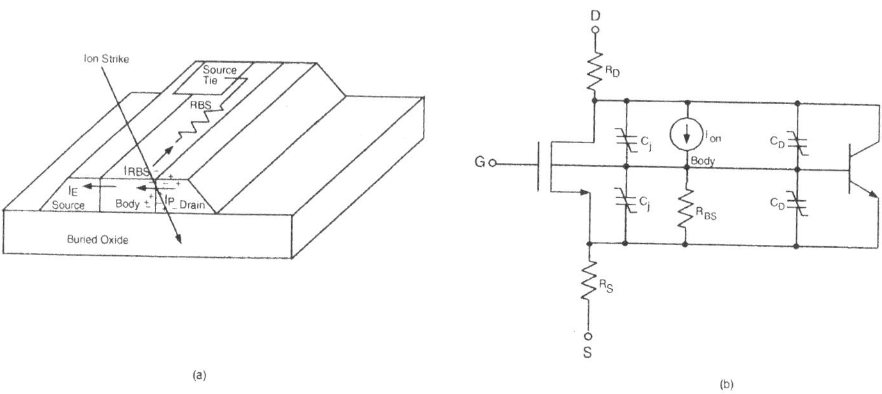

Fig. 3.99 (a) Cross-section of the PD SOI NMOS device. (b) Transient model. (Adapted from Massengill et al. [68].)

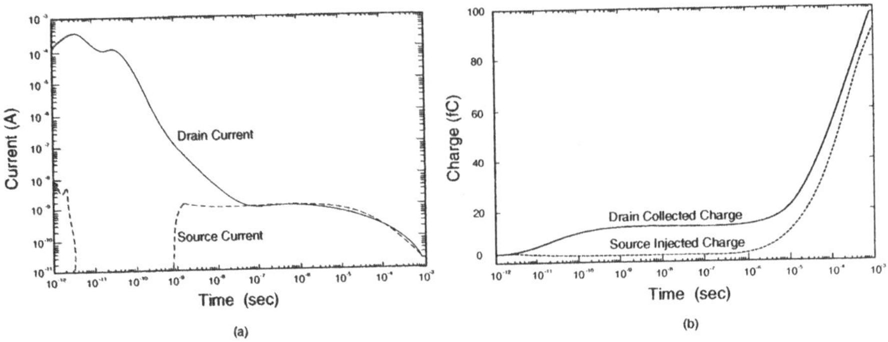

Fig. 3.100 (a) Drain and source currents and (b) drain collected charge and source injected charge during the transient after the α-particle incidence. (Adapted from Iwata et al. [69].)

[69]. After the high-energy particles penetrate the thin film, electron/hole pairs are generated. The generated electrons are collected by the drain to cause the loss of the charge near the drain during the initial 100 ps period. The remaining holes are accumulated at the bottom of the thin film to raise the body voltage. Therefore, the parasitic bipolar device is turned on. Consequently, a stable leakage current of $\sim$ 1 nA between the source and the drain is generated between 0.1 and 10 μs. Along with the recombination of the holes, the leakage current gradually drops (t $>$ 10 μs). During this stage, the loss of the charge at the drain end is the biggest. In order to reduce the soft errors of the SOI device, adopting a body contact is the most effective way.

3.11 SUMMARY

In this chapter, the hot carriers of the SOI CMOS devices, the accumulation-mode devices and the double-gate SOI devices have been analyzed. Then DTMOS devices have been introduced, followed by the scaling trends of SOI CMOS devices and the SOI SET. Then, the temperature dependence of the SOI devices have been analyzed. Finally, sensitivities and radiation effects of SOI CMOS devices have been described.

REFERENCES

1. D. E. Ioannou, F. L. Duan, S. P. Sinha, and A. Zaleski, "Opposite-Channel-Based Injection of Hot-Carriers in SOI MOSFET's: Physics and Applications," *IEEE Trans. Elec. Dev.*, **45**(5), 1147–1154 (1998).

2. B. Zhang and T. P. Ma, "Front-Channel Hot-Carrier Effect on the Drain-Source Breakdown Voltage in Thin-Film SOI/NMOSFET's," *SOI Conf. Dig.*, 126–127 (1992).

3. T. Tsuchiya and T. Ohno, "New Hot-Carrier-Degradation Mode in Thin-Film SOI nMOSFET's," *IEEE Elec. Dev. Let.*, **16**(10), 427–429 (1995).

4. J. Chen, K. Quader, R. Solomon, T. Chan, P. Ko, and C. Hu, "Hot Electron Gate Current and Degradation in P-Channel SOI MOSFET's," *SOI Conf. Dig.*, 8–9 (1991).

5. M. J. Sherony, D. A. Antoniadis, J. W. Sleight, and K. R. Mistry, "Extrapolation of DC Device Lifetime in Body-Floating and Body-Grounded SOI MOSFETs," *SOI Conf. Dig.*, 178–179 (1997).

6. F. L. Duan, D. E. Ioannou, W. C. Jenkins, H. L. Hughes, and M. S. T. Liu, "Channel Coupling Imposed Tradeoffs on Hot Carrier Degradation and Single Transistor Latch-up in SOI MOSFET's," *Symp. Reliability Conf. Dig.*, 194–202 (1998).

7. H.-J. Wann, J. King, J. Chen, P. K. Ko, and C. Hu, "Hot-Carrier Currents of SOI MOSFETs," *SOI Conf. Dig.*, 118–119 (1993).

8. R. B. Hulfachor, K. W. Kim, M. A. Littlejohn, and C. M. Osburn, "Effects of Silicon Layer Properties on Device Reliability for 0.1μm SOI n-MOSFET Design Strategies," *IEEE Trans. Elec. Dev.*, **44**(5), 815–821 (1997).

9. S.-H. Renn, C. Raynaud, J.-L. Pelloie, and F. Balestra, "A Thorough Investigation of the Degradation Induced by Hot-Carrier Injection in Deep Submicron N- and P- Channel Partially and Fully Depleted Unibond and SIMOX MOSFET's," *IEEE Trans. Elec. Dev.*, **45**(10), 2146–2152 (1998).

10. L. T. Su, H. Fang, J. E. Chung, and D. A. Antoniadis, "Hot-Carrier Effects in Fully-Depleted SOI NMOSFETs," *IEDM Dig.*, 349–352 (1992).

11. J.-W. Lee, H.-K. Kim, W.-H. Lee, M.-R. Oh, and Y.-H. Koh, "Hot-Carrier Degradation Behavior of Thin-Film SOI nMOSFET with Isolation Scheme and Buried Oxide Thickness," *IEEE Trans. Elec. Dev.*, **47**(5), 1013–1017 (2000).

12. S. H. Renn, J. Jomaah, C. Raynaud, and F. Balestra, "Analysis of Hot-Carrier-Induced Degradation in Deep Submicron Unibond and SIMOX N-MOSFETs Using Charge Pumping and Noise Techniques," *SOI Conf. Dig.*, 81–82 (1998).

13. S.-H. Renn, J.-L. Pelloie, and F. Balestra, "Hot-Carrier Effects and Reliable Lifetime Prediction in Deep Submicron N- and P-Channel SOI MOSFET's," *IEEE Trans. Elec. Dev.*, **45**(11), 2335–2342 (1998).

14. J. Wang-Ratkovic, W. M. Huang, B. Y. Hwang, M. Racanelli, J. Forestner, and J. Woo, "Novel Device Lifetime Behavior and Hot-Carrier Degradation Mechanisms Under $V_{GS} \approx V_{TH}$ Stress for Thin-Film SOI nMOSFETs," *IEDM Dig.*, 639–642 (1995).

15. S. Maeda, Y. Yamaguchi, I.-J. Kim, H.-O. Joachim, Y. Inoue, H. Miyoshi, and A. Yasuoka, "A Method for the Prediction of Hot-Carrier Lifetime in Floating SOI NMOSFET's," *IEEE Trans. Elec. Dev.*, **44**(12), 2200–2205 (1997).

16. C. Raynaud, J. L. Pelloie, O. Faynot, B. Dunne, and J. Hartmann, "Fully-Depleted Accumulation-Mode PMOSFET for 0.2μm SOI Technology," *SOI Conf. Dig.*, 12–13 (1995).

17. K.-W. Su and J. B. Kuo, "Analysis of Current Conduction in Short-Channel Accumulation-Mode SOI PMOS Devices," *IEEE Trans. Elec. Dev.*, **44**(5), 832–840 (1997).

18. K. W. Su and J. B. Kuo, "Analytical Capacitance Model for Submicron Accumulation Mode SOI MOS Devices," *Sol. St. Elec.*, **42**(4), 513–522 (1998).

19. O. Faynot, S. Cristoloveanu, A. J. Auberton-Herve, and C. Raynaud, "Performance and Potential of Ultrathin Accumulation-Mode SIMOX MOSFET's," *IEEE Trans. Elec. Dev.*, **42**(4), 713–719 (1995).

20. F. L. Duan and D. E. Ioannou, "Design and Analysis of a Novel Mixed Accumulation/Inversion Mode FD SOI MOSFET," *SOI Conf. Dig.*, 100–101 (1997).

21. A. Acovic, L. K. Wang, F. Brady, and N. Haddad, "Hot-Carrier Reliability of Fully Depleted Accumulation Mode SOI MOSFETs," *SOI Conf. Dig.*, 134–135 (1992).

22. J.-P. Colinge, M.-H. Gao, A. Romano, H. Maes, and C. Claeys, "Silicon-on-Insulator "Gate-All-Around" MOS Device," *SOI Conf. Dig.*, 137–138 (1990).

23. H.-S. P. Wong, D. J. Frank, and P. M. Solomon, "Device Design Considerations for Double-Gate, Ground-Plane, and Single-Gated Ultra-Thin SOI MOSFET's at the 25nm Channel Length Generation," *IEDM Dig.*, 407–410 (1998).

24. E. Rauly, O. Potavin, F. Balestra, and C. Raynaud, "On the Subthreshold Swing and Short Channel Effects in Single and Double Gate Deep Submicron SOI-MOSFETs," *Sol. St. Elec.*, **43**(11), 2033–2037 (1999).

25. K.-W. Su and J. B. Kuo, "A Non-Local Impact Ionization/Lattice Temperature Model for VLSI Double-Gate Ultrathin SOI NMOS Devices," *IEEE Trans. Elec. Dev.*, **44**(2), 324–330 (1997).

26. S. C. Williams, K. W. Kim, M. A. Littlejohn, and W. C. Holton, "Analysis of Hot-Electron Reliability and Device Performance in 80-nm Double-Gate SOI n-MOSFET's," *IEEE Trans. Elec. Dev.*, **46**(8), 1760–1767 (1999).

27. F. Assaderaghi, D. Sinitsky, S. A. Parke, J. Bokor, P. K. Ko, and C. Hu, "Dynamic Threshold-Voltage MOSFET (DTMOS) for Ultra-Low Voltage VLSI," *IEEE Trans. Elec. Dev.*, **44**(3), 414–422 (1997).

28. T. Ernst, D. Munteanu, S. Cristoloveanu, J. L. Pelloie, O. Faynot, and C. Raynaud, "Detailed Analysis of Short-Channel SOI DT-MOSFET," *ESSDERC Dig.*, 380–383 (1999).

29. T. Tanaka, Y. Momiyama, and T. Sugii, "Fmax Enhancement of Dynamic Threshold Voltage MOSFET (DTMOS) Under Ultra-Low Supply Voltage," *IEDM Dig.*, 423–426 (1997).

30. G. G. Shahidi, C. Blair, K. Beyer, T. Bucelot, T. Buti, P. N. Chang, S. Chu, P. Coane, J. Comfort, B. Davari, R. Dennard, S. Furkay, H. Hovel, J. Johnson, D. Klaus, K. Kiewtniack, R. Logan, T. Lii, P. A. McFarland, N. Mazzeo, D. Moy, S. Neely, T. Ning, M. Rodriguez, D. Sadana, S. Stiffler, J. Sun, F. Swell, and J. Warnock, "A Room Temperature $0.1\mu m$ CMOS On SOI," *Symp. VLSI Tech. Dig.*, 27–28 (1993).

31. F. Assaderaghi, W. Rausch, A. Ajmera, E. Leobandung, D. Schepis, L. Wagner, H.-J. Wann, R. Bolam, D. Yee, B. Davari, and G. Shahidi, "A 7.9/5.5 psec Room/Low Temperature SOI CMOS," *IEDM Dig.*, 415–418 (1997).

32. B. Davari, "CMOS Technology: Present and Future," *Symp. VLSI Ckts. Dig.*, 5–10 (1999).

33. E. Leobandung, M. Sherony, J. Sleight, R. Bolam, F. Assaderaghi, S. Wu, D. Schepis, A. Ajmera, W. Rausch, B. Davari, and G. Shahidi, "Scalability of SOI Technology into $0.13\mu m$ 1.2V CMOS Generation," *IEDM Dig.*, 403–406 (1998).

34. C. Hu, "SOI and Device Scaling," *SOI Conf. Dig.*, 1–4 (1998).

35. J. Kedzierski, P. Xuan, E. H. Anderson, J. Bokor, T.-J. King, and C. Hu, "Complementary Silicide Source/Drain Thin-Body MOSFETs for the 20nm Gate Length Regime," *IEDM Dig.*, 57–60 (2000).

36. A. Dancy and A. Chandrakasan, "Techniques for Aggressive Supply Voltage Scaling and Efficient Regulation," *CICC Dig.*, 579–586 (1997).

37. Y. Ono, Y. Takahashi, K. Yamazaki, M. Nagase, H. Namatsu, K. Kurihara, and K. Murase, "Fabrication Method for IC-Oriented Si Single-Electron Transistors," *IEEE Trans. Elec. Dev.*, **47**(1), 147–153 (2000).

38. N. Takahashi, H. Ishikuro, and T. Hiramoto, "A Directional Current Switch Using Silicon Single Electron Transistors Controlled by Charge Injection into Silicon Nano-Crystal Floating Dots," *IEDM Dig.*, 371–374 (1999).

39. K. Uchida, J. Koga, R. Ohba, and A. Toriumi, "Room-Temperature Operation of Multifunctional Single-Electron Transistor Logic," *IEDM Dig.*, 863–865 (2000).

40. A. Fujiwara, Y. Takahashi, K. Yamazaki, H. Namatsu, M. Nagase, K. Kurihara, and K. Murase," Double-Island Single-Electron Devices– a Useful Unit Device for Single-Electron Login LSI's," *IEEE Trans. Elec. Dev.*, **46**(5), 954–959 (1999).

41. P. Raha, S. Ramaswamy, and E. Rosenbaum, "Heat Flow Analysis for EOS/ESD Protection Device Design in SOI Technology," *IEEE Trans. Elec. Dev.*, **44**(3), 464–471 (1997).

42. S. H. Voldman, "The State of the Art of Electrostatic Discharge Protection: Physics, Technology, Circuits, Design, Simulation and Scaling," *BCTM Dig.*, 19–30 (1998).

43. K. Verhaege, G. Groeseneken, J.-P. Colinge, and H. E. Maes, "Double Snapback in SOI nMOSFET's and its Application for SOI ESD Protection," *IEEE Elec. Dev. Let.*, **14**(7), 326–328 (1993).

44. S. Ramaswamy, P. Raha, E. Rosenbaum, and S.-M. Kang, "EOS/ESD Protection Circuit Design for Deep Submicron SOI Technology," *Symp. EOS/ESD Conf. Dig.*, 212–217 (1995).

45. M. Chan, S. S. Yuen, Z.-J. Ma, K. Y. Hui, P. K. Ko, and C. Hu, "ESD Reliability and Protection Schemes in SOI CMOS Output Buffers," *IEEE Trans. Elec. Dev.*, **42**(10), 1816–1821 (1995).

46. C. Duvvury, A. Amerasekera, K. Joyner, S. Ramaswamy, and S. Young, "ESD Design for Deep Submicron SOI Technology," *Symp. VLSI Tech. Dig.*, 194–195 (1996).

47. P. Raha, C. Diaz, E. Rosenbaum, M. Cao, P. VandeVoorde, and W. Greene, "EOS/ESD Reliability of Partially Depleted SOI Technology," *IEEE Trans. Elec. Dev.*, **46**(2), 429–431 (1999).

48. S. Voldman, D. Hui, L. Warriner, D. Young, R. Williams, J. Howard, V. Gross, W. Rausch, E. Leobangdung, M. Sherony, N. Rohrer, C. Akrout, F. Assaderaghi, and G. Shahidi, "Electrostatic Discharge Protection in Silicon-on-Insulator Technology," *SOI Conf. Dig.*, 68–71 (1999).

49. D.-S. Jeon and D. E. Burk, "A Temperature-Dependent SOI MOSFET Model for High-Temperature Application (27°C – 300°C)," *IEEE Trans. Elec. Dev.*, **38**(9), 2101–2111 (1991).

50. G. Groeseneken, J.-P. Colinge, H. E. Maes, J. C. Alderman, and S. Holt, "Temperature Dependence of Threshold Voltage in Thin-Film SOI MOSFET's," *IEEE Elec. Dev. Let.*, **11**(8), 329–331 (1990).

51. D. Flandre, A. Terao, P. Francis, B. Gentinne, and J.-P. Colinge, "Demonstration of the Potential of Accumulation-Mode MOS Transistors on SOI Substrates for High-Temperature Operation (150 – 300°C)," *IEEE Elec. Dev. Let.*, **14**(1), 10–12 (1993).

52. G. Reichert, T. Ouisse, J. L. Pelloie, and S. Cristoloveanu, "Surface Mobility of SOI MOSFET's in the High Temperature Range: Modelling and Experiment," *SOI Conf. Dig.*, 42–43 (1995).

53. W. Jin, S. K. H. Fung, W. Liu, P. C. H. Chan, and C. Hu, "Self-Heating Characterization for SOI MOSFET Based on AC Output Conductance," *IEDM Dig.*, 175–178 (1999).

54. J.-P. Eggermont, D. De Ceuster, D. Flandre, B. Gentinne, P. G. A. Jespers, and J.-P. Colinge, "Design of SOI CMOS Operational Amplifiers for Applications up to 300°C," *IEEE J. Sol. St. Ckts.*, **31**(2), 179–186 (1996).

55. M. Asheghi, P. Sverdrup, and K. E. Goodson, "Thermal Modeling of Thin-Film SOI Transistors," *SOI Conf. Dig.*, 28–29 (1999).

56. G. Reichert, C. Raynaud, O. Faynot, F. Balestra, and S. Cristoloveanu, "Temperature Dependence (300-600K) of Parasitic Bipolar Effects in SOI-MOSFETs," *ESSDERC Dig.*, 520–523 (1997).

57. J. Gautier, M. M. Pelella, and J. G. Fossum, "SOI Floating-Body, Device and Circuit Issues," *IEDM Dig.*, 407–410 (1997).

58. S. C. Lin and J. B. Kuo, "Temperature-Dependent Kink Effect Model for Partially-Depleted SOI NMOS Devices," *IEEE Trans. Elec. Dev.*, **46**(1), 254–258 (1999).

59. S.-H. Renn, E. Rauly, J.-L. Pelloie, and F. Balestra, "Hot-Carrier Effects and Lifetime Prediction in Off-State Operation of Deep Submicron SOI N-MOSFET's," *IEEE Trans. Elec. Dev.*, **45**(5), 1140–1146 (1998).

60. Y.-C. Tseng, W. M. Huang, C. Hwang, and J. C. S. Woo, "AC Floating Body Effects in Partially Depleted Floating Body SOI nMOS Operated at Elevated Temperature: An Analog Circuit Prospective," *IEEE Elec. Dev. Let.*, **21**(10), 494–496 (2000).

61. M. J. Sherony, L. T. Su, J. E. Chung, and D. A. Antoniadis, "Reduction of Threshold Voltage Sensitivity in SOI MOSFET's," *IEEE Elec. Dev. Let.*, **16**(3), 100–102 (1995).

62. E. Leobandung and S. Y. Chou, "Threshold Voltage Sensitivity of 0.1μm Channel Length Fully-Depleted SOI NMOSFET's with Back-Gate Bias," *IEEE Trans. Elec. Dev.*, **42**(9), 1707–1708 (1995).

63. Y. Yamaguchi and Y. Inoue, "SOI DRAM: Its Features and Possibility," *SOI Conf. Dig.*, 122–124 (1995).

64. B. L. Buchanan, D. A. Neamen, and W. M. Shedd, "SOS Device Radiation Effects and Hardening," *IEEE Trans. Elec. Dev.*, **25**(8), 959–970 (1978).

65. V. Ferlet-Cavrois, S. Quoizola, O. Musseau, O. Flament, J. L. Leray, J. L. Pelloie, C. Raynaud, and O. Faynot, "Total Dose Induced Latch in Short Channel NMOS/SOI Transistors," *IEEE Trans. Nuclear Science*, **45**(6), 2458–2466 (1998).

66. A. Vandooren, P. Francis, D. Flandre, and J.-P. Colinge, "Performance of γ-Irradiated Gate-All-Around SOI MOS OTA Amplifiers," *SOI Conf. Dig.*, 62–63 (1997).

67. E. Y. Chao, C. Hu, S. Wu, and G. P. Li, "Annealing Characteristics of Radiation Induced Leakage in SOI MOSFETS," *SOI Conf. Dig.*, 84–85 (1993).

68. L. W. Massengill, D. V. Kerns, Jr., S. E. Kerns, and M. L. Alles, "Single-Event Charge Enhancement in SOI Devices," *IEEE Elec. Dev. Let.*, **11**(2), 98–99 (1990).

69. H. Iwata and T. Ohzone, "Numerical Analysis of Alpha-Particle-Induced Soft Errors in SOI MOS Devices," *IEEE Trans. Elec. Dev.*, **39**(5), 1184–1190 (1992).

Problems

1. Derive the gate-source and source-gate capacitances of the accumulation-mode SOI PMOS device in Fig.3.23. Hint: See Ref. 3.18.

2. Why does the hot carrier degradation reache its peak at $V_G = V_T$ for the SOI NMOS devices, instead of at $V_G = V_{DD}/2$ as it does for the bulk NMOS devices?

3. What is the short channel effect on the source-gate capacitance (C_{SG}) of a PD SOI NMOS device?

4. What is the narrow channel effect on the source-gate capacitance (C_{SG}) of a PD SOI NMOS device?

5. What are the advantages of the SOI CMOS devices for ESD protection?

6. What are the techniques to resolve the sensitivity problems of the FD CMOS devices, which may refrain them from sub-0.1 μm VLSI technology applications as compared to PD?

4

Fundamentals of SOI CMOS Circuits

Until now, behaviors of the SOI CMOS devices were described in Chapters 2 and 3. In the next three chapters, digital and analog circuits implemented by SOI CMOS devices are depicted. In this chapter, basic knowledge of SOI CMOS circuits is described. Starting from the basic circuit issues, the floating body effects on the performance of the SOI CMOS circuits are explained, followed by the low-voltage SOI CMOS circuits, SOI dynamic-threshold MOS (DTMOS) circuits, and SOI multithreshold MOS (MTMOS) circuits. Then, noise and self-heating problems of SOI CMOS circuits are analyzed. Finally, the SOI ESD circuits and the SOI system-on-a chip (SOC) technology are presented.

4.1 BASIC CIRCUIT ISSUES

Compared to conventional bulk circuits, SOI CMOS devices offer more advantages for designing low-voltage VLSI circuits. Owing to the buried oxide structure to isolate the thin film from the substrate, the CMOS latchup phenomena frequently encountered in bulk CMOS structures can be avoided. Owing to the buried oxide, the parasitic capacitances in the source/drain regions are reduced. Due to the thin-film structure, SOI CMOS devices have a better performance in subthreshold slope and transconductance. The leakage current is smaller for the SOI CMOS devices owing to the oxide isolation structure. In addition, the buried oxide structure can also be used to improve the radiation hardness of the SOI devices.

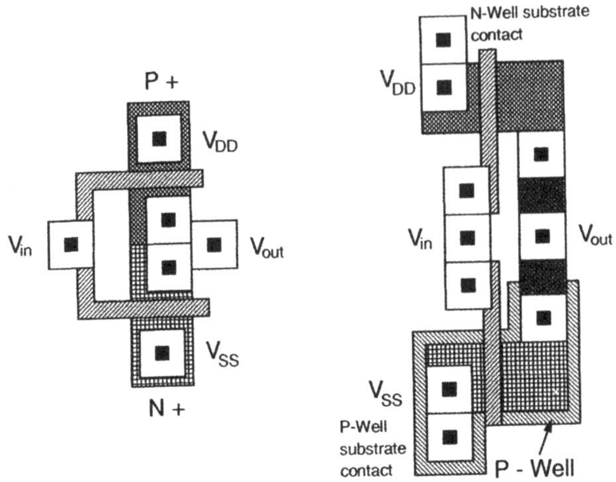

Fig. 4.1 Layout of a CMOS inverter circuit using SOI and bulk technologies. (Adapted from Stern et al. [1].)

4.1.1 Layout

Owing to the buried oxide structure, the N^+/P^+ source/drain regions of the SOI NMOS/PMOS devices can be placed against each other without worrying about the possibility of latchup. Therefore, SOI CMOS devices may have a much higher device density. Figure 4.1 shows the layout of a CMOS inverter circuit using SOI and bulk technologies [1]. As shown in this figure, based on the conventional bulk CMOS technology, since wells are needed to separate the N^+ region from the P^+ region, the device density of the circuit cannot be high. In contrast, due to the higher device density, the smaller layout area of the SOI CMOS circuits leads to a smaller leakage current and smaller parasitic capacitances. As a result, a higher speed at a smaller power consumption can be obtained from the SOI CMOS circuits. Consequently, for the next-generation VLSI designs, SOI CMOS devices are more appropriate to integrate low-power circuits.

Generally speaking, the design principles of SOI CMOS circuits are similar to those of the bulk circuits. Owing to the oxide isolation structure, CMOS circuits designed using SOI devices have a better performance. Figure 4.2 shows the layout area and the stand-by current dissipation versus the DC open loop gain of a Miller op amp implemented by SOI and bulk CMOS devices, with an output load of 10pF for a transition frequency of 10 MHz [2]. As shown in this figure, at an identical DC open loop gain, the SOI op amp has a smaller layout area and a lower stand-by current dissipation. Since SOI devices do not need junction and well isolations, thus their device density can be higher. In addition, a steeper subthreshold slope also results in a smaller leakage current, which is helpful for getting a lower stand-by current dissipation.

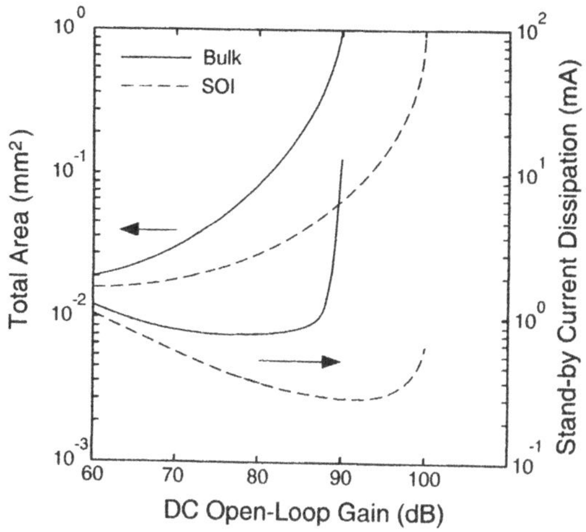

Fig. 4.2 Layout area and stand-by current versus DC open loop gain of a Miller op amp implemented by SOI and bulk CMOS devices, with an output load of 10 pF for a transition frequency of 10 MHz. (Adapted from Colinge et al.[2].)

4.1.2 High Speed and Low Power

Compared to bulk devices, the circuits built by SOI devices have a higher speed performance. Figure 4.3 shows the access time versus the power supply voltage of a 1K x 8 SRAM implemented by 0.5 μm SOI and bulk CMOS technologies [3]. As shown in this figure, when the power supply voltage is scaled down, the advantage in the speed of the SOI SRAM over the bulk device increases– SOI devices are more suitable for circuits operating at a low power supply voltage. At a lower power supply voltage, the current driving capability of the MOS device is reduced. Since the threshold voltage of the SOI device can be lowered by an appropriate design of the thin film without causing extra leakage currents, thus the reduced current driving capability due to the reduced power supply voltage can be compensated. Owing to the existence of the buried oxide, the depletion junction capacitances under the source/drain regions disappear, hence the RC delay due to the parasitic capacitance can be reduced, which can be used to further compensate for the loss resulting from the reduced power supply voltage.

In addition to the high-speed advantage, SOI devices also provide advantages for integrating low-power VLSI circuits. Since the positive body-source voltage of the SOI devices results in a decrease in the threshold voltage, at an identical low supply voltage, the drain current driving capability of the SOI devices is stronger than that of the bulk devices at the same CMOS technology level. Due to the lower leakage current and the smaller parasitic capacitances of the SOI devices, the power consumption of the circuits realized by SOI devices is less as compared to the bulk one.

Figure 4.4 shows the power consumption versus the access time of a 4M-bit SRAM using bulk and SOI CMOS technologies [4]. As shown in this figure, the SOI SRAM is suitable for operation at a low supply voltage of 1.2 V. In addition, at an identical

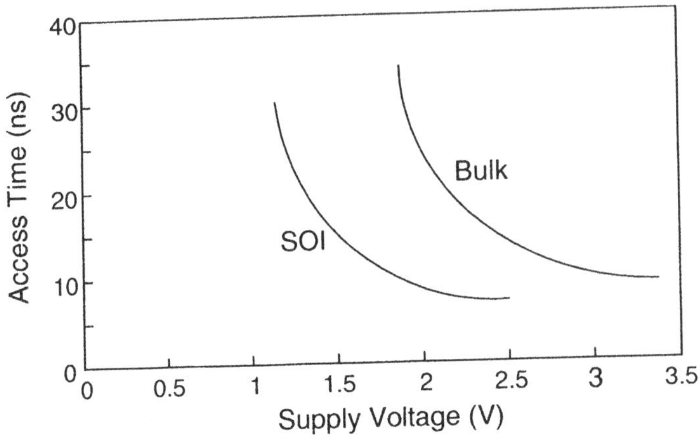

Fig. 4.3 Access time versus power supply voltage of a 1K x 8 SRAM implemented by 0.5 μm SOI and bulk CMOS technologies. (Adapted from Alles [3].)

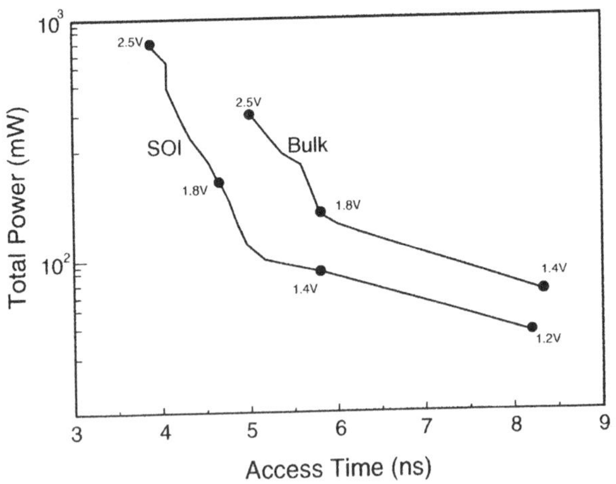

Fig. 4.4 Power consumption versus access time of a 4M-bit SRAM using bulk and SOI CMOS technologies. (Adapted from Shahidi et al. [4].)

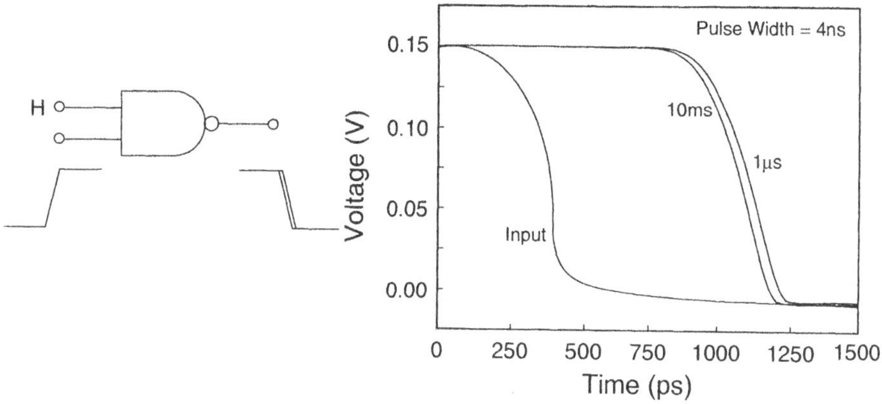

Fig. 4.5 History-dependent propagation delay of the SOI CMOS NAND gate. (Adapted from Shahidi et al. [4].)

delay time, the power consumption of the SOI SRAM is 2.3 times smaller than that of the bulk device. Although the positive body-source bias effect of the SOI device is better than that of the bulk, the floating thin film may cause accumulation of the charge, which may complicate the design of the circuits using SOI devices.

4.1.3 Floating Body

Due to the buried oxide structure, the accumulation of the charge in the floating thin film may affect the circuits made of SOI devices differently as compared to the bulk device. The floating body of the SOI devices causes (1) the history-dependent propagation delay of the related CMOS inverter circuit, (2) the leakage current of the related pass gate, and (3) self-heating effects. These effects are described as follows. The propagation delay of an SOI digital circuit can be seriously affected by the switching cycle of the clock imposed at the gates of the devices. During the switching of an SOI digital circuit, the body potential is influenced by the gate, the drain, and the source voltages. In addition, the capacitive coupling, the impact ionization, and the diode leakage current also determine the progress of the body potential reaching the steady state. By applying a very fast clock at its input gate after switching, the body potential may not reach its steady state in time when the next switching arrives. Therefore, the body charge cannot recover in time, which brings in a smaller body-source voltage, and hence an increased threshold voltage. Thus the drain current driving capability of the device is reduced, and the propagation delay time increases– history-dependent propagation delay [4][5]. The history dependence of the propagation delay affects the performance of an SOI CMOS digital circuit such as the NAND gate as shown in Fig. 4.5. Due to the history effect, the propagation delay of the NAND gate becomes larger when the switching frequency increases substantially.

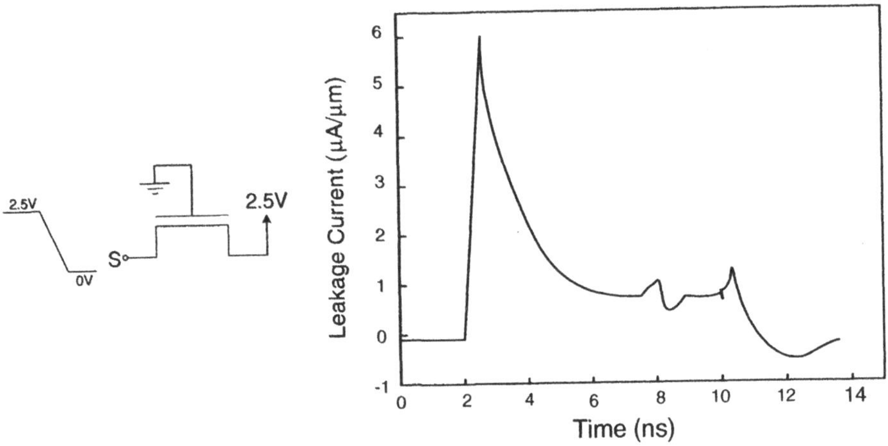

Fig. 4.6 The pass-gate leakage current during the transient of an SOI NMOS device biased at the gate voltage of 0 V, the drain voltage of 2.5 V, and with a step from 2.5 to 0 V imposed at the source. (Adapted from Shahidi et al. [4].)

For an SOI device with both the drain and the source biased at the high potential and the gate at the low potential, the body may be charged to the high potential due to the high potential of the source and the drain ends. If a voltage step from high to low potential is imposed at the source, the positive body-source voltage may lead to the holes flowing from the body to the source and the electrons injecting from the source to the drain via the body as described in Chapter 3. As a result, a drain current pulse can be seen as shown in Fig. 4.6, which is the pass-gate leakage [4]. The pass-gate leakage may cause discharge of the storage node in the DRAM memory cell and the dynamic circuit, which damages the stored data.

4.1.4 Self-Heating

Due to the buried oxide layer, the thermal conductance of the SOI device is small. During operation of its related circuit, the generated heat cannot be dissipated out of the device easily. The lattice temperature of the SOI devices in the circuit rises and the mobility decreases. Hence, the drain current becomes smaller, which is self-heating. Figure 4.7 shows the self-heating effects of the NMOS device in the SOI CMOS inverter during operation [4]. As shown in this figure, only during switching does the transient drain current exist. During switching the drain voltage is not at the power supply voltage (V_{DD}), thus the power generated is small and the rise in the lattice temperature of the device is not large. Consequently, self-heating does not cause a big impact on the switching performance of digital circuits. On the contrary, for the analog circuits, during standard operation, the devices are always biased under the ON condition, therefore the self-heating effect on lowering the drain current is more serious for analog circuits.

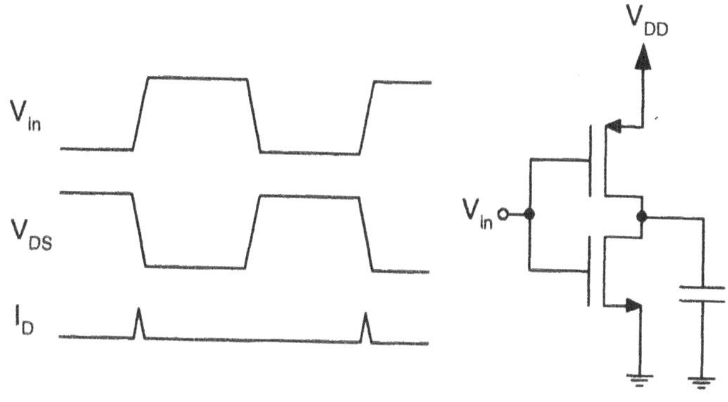

Fig. 4.7 Self-heating effects of the NMOS device in the SOI CMOS inverter during operation. (Adapted from Shahidi et al. [4].)

4.2 FLOATING BODY EFFECTS

Floating body affects the performance of analog and digital circuits using PD SOI CMOS devices. In this section, the floating body effects on the performance of digital and analog circuits are described.

4.2.1 Static Logic Circuits

Floating body effects may influence the operation of a CMOS static logic circuit made of SOI CMOS devices. Figure 4.8 shows (a) the three-way OR-AND circuit using SOI CMOS devices with a channel length of 0.25 μm, a front gate oxide of 50 Å, a buried oxide of 3500 Å, and a thin film of 1400 Å and (b) the waveforms at the output node, the internal node V_1, and the body nodes of the devices N1 and N2 during the transient after the input waveforms are imposed at $V_{in,1}$ and $V_{in,4}$ [7]. Also shown in this figure is the transient current of the parasitic bipolar device in device N2. As shown in this figure, initially the input to device N1 is high ($V_{in,1} = V_{DD}$) and the inputs to devices N2, N3, and N4 are low ($V_{in,4} = 0V$). Thus, initially the output is charged to high ($V_{out} = V_{DD}$) by device P4. Since device N1 is on ($V_{in,1} = V_{DD}$), the source end of device N1 (V_1) is pulled to $V_{DD} - V_T$. Since the drain voltage of devices N1, N2, and N3 is high ($V_{out} = 2.5$ V) and the source voltage is $V_1 = V_{DD} - V_T = 2.31$ V, thus the floating body of devices N1, N2, and N3 is pulled to a high potential of 2.33 V under the equilibrium state. Due to the positive body-source voltage of device N1, its threshold voltage is smaller than the value for the case with zero body-source voltage ($V_{BS} = 0$ V). At time $t = 0.6$ ns, the input voltage $V_{in,1}$ switches from V_{DD} to ground. Because of gate-source/gate-body capacitive coupling, the source and the body potentials of device N1 falls such that the source voltage becomes $V_1 = 1.79$ V and the body potential becomes $V_{B,N1} = 0.98$ V. Due to the body-

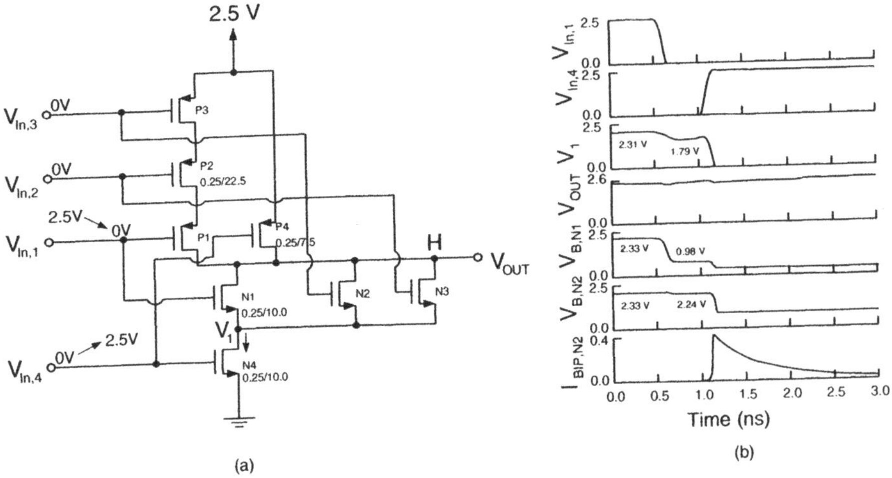

Fig. 4.8 (a) The three-way OR-AND circuit using SOI CMOS devices with a channel length of 0.25 μm, a front gate oxide of 50 Å, a buried oxide of 3500 Å, and a thin film of 1400 Å. (b) Waveforms at the output node, the internal node V_1, and the body nodes of devices N1 and N2 during the transient after the input waveforms are imposed at $V_{in,1}$ and $V_{in,4}$. Also shown in this figure is the transient current of the parasitic bipolar device in device N2. (Adapted from Lu et al. [7].)

source capacitive coupling, the body potential of devices N2 and N3 drops slightly to 2.24 V. At t = 1.1 ns, the input voltage ($V_{in,4}$) switches from ground to V_{DD} and device N4 turns on. The internal node potential V_1 is discharged to 0 V by device N4. Thus, via the body-source capacitive coupling, the body potential of devices N1, N2, and N3 falls slightly. But the body-source voltage of devices N2 and N3 is sufficiently large so that the parasitic bipolar devices in devices N2 and N3 are turned on to cause an off-state (N2 and N3 are not turned on) parasitic bipolar current of 0.38 mA as shown in the figure. Consequently, the output node is discharged. In this circuit, although devices N2 and N3 have noticeable parasitic bipolar leakage currents, due to the existing DC path, the extra bipolar leakage current does not cause an error in the output voltage but the power consumption of the circuit is increased.

The degradation in the performance of a CMOS dynamic logic circuit caused by the floating body effect of its SOI devices is more serious as compared to a static logic circuit. Figure 4.9 shows (a) Manchester carry chain circuit using a CMOS dynamic logic circuit made of SOI devices and (b) the transient waveforms at the internal carry signal nodes and the transient bipolar leakage currents of devices NP1 and NP2, after input waveforms are imposed at V_{C0} and C_{CLK} [7]. The input generate signals are low $((G_4, G_3, G_2, G_1) = (0, 0, 0, 0))$ and the input propagate signals are low $((P_4, P_3, P_2, P_1) = (0, 0, 0, 0))$. The input carry signal is high ($C_0 = 1$). As shown in this figure, the Manchester carry chain circuit is used to process the input propagate and generate signals and the input carry signal is used to produce the output carry signals. At t = 0.6 ns, V_{CLK} is low at ground and it is the precharge phase, where

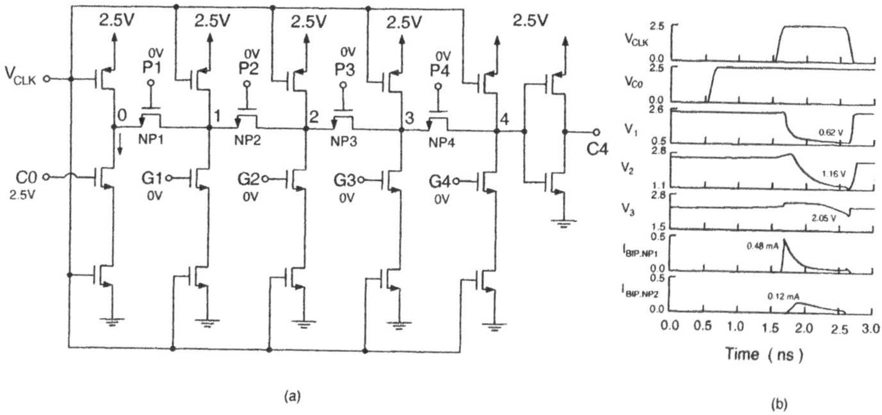

Fig. 4.9 (a) Manchester carry chain circuit using a CMOS dynamic logic circuit made of SOI devices and (b) Transient waveforms at the internal carry signal nodes and the transient bipolar leakage currents of devices NP1 and NP2, after input waveforms are imposed at V_{CO} and C_{CLK}. The input generate signals are low $(G_4, G_3, G_2, G_1) = (0, 0, 0, 0)$ and the input propagate signals are low $(P_4, P_3, P_2, P_1) = (0, 0, 0, 0)$. The input carry signal is high $(C_0 = 1)$. (From Lu et al. [7]. ©1997 IEEE.)

devices NP1, NP2, NP3, and NP4 are off. The carry signals at the internal $\overline{C_0}$, $\overline{C_1}$, $\overline{C_2}$, $\overline{C_3}$, and $\overline{C_4}$ are charged to high at $V_{DD} = 2.5$ V by the PMOS devices. Although devices NP1~NP4 are off, the source and the drain ends of these devices ($\overline{C_i}$, i = 0–4) are high. Thus, the floating body of devices NP1~NP4 is charged to high to reach the equilibrium state. At t = 1.6 ns, V_{CLK} switches from low to high (V_{DD}), the circuit enters the evaluation phase, where the carry signal at the internal node ($\overline{C_0}$) falls to low due to the discharge via the two NMOS devices below. Therefore, owing to the drop of the carry signal at the internal node ($\overline{C_0}$), the body-source voltage of device NP1 becomes positive, which triggers the turn-on of the parasitic bipolar device with a parasitic bipolar leakage current ~ 0.48 mA. Consequently, the carry signal at the internal node ($\overline{C_1}$) gradually drops. A similar situation also exists for device NP2, which causes the drop in the carry signal at the internal node $\overline{C_2}$. The discharge of the internal nodes via the parasitic bipolar leakage current of the pass-gate transistors becomes less noticeable when the chain of the pass-gate transistors becomes longer. As shown in the figure, the carry signal at the internal node ($\overline{C_3}$) falls to 2.05 V due to the parasitic bipolar leakage effects. Due to the parasitic bipolar leakage currents caused by the floating body effect of the pass-gate transistors, the carry signals at the internal nodes ($\overline{C_1}$, $\overline{C_2}$) drop from 2.5 to 0.62 V and 1.16 V, respectively, which already generates errors in the output signals.

4.2.2 Pass Gate Transistors

Floating body effects may also cause the parasitic bipolar leakage current in an SOI CMOS pass-gate transistor. Figure 4.10 shows the waveforms of the body potential

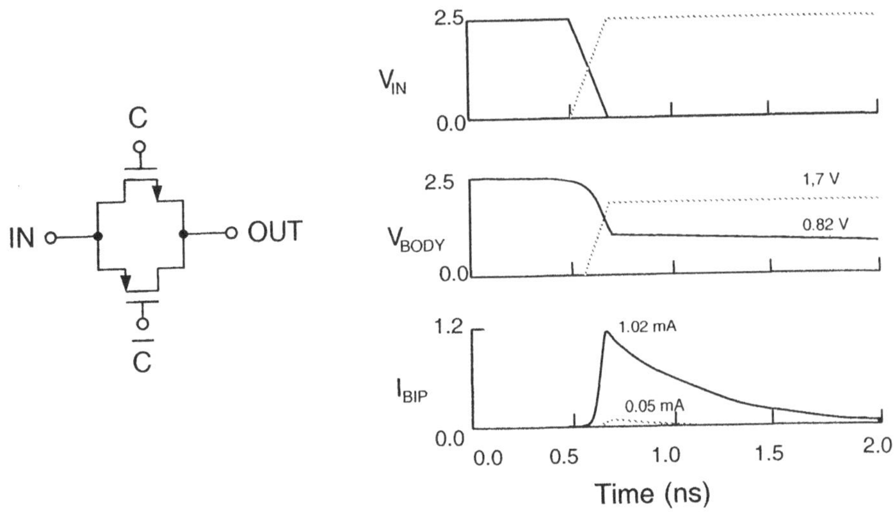

Fig. 4.10 Waveforms of the body potential and the parasitic bipolar leakage current of the NMOS device (solid lines) in the SOI CMOS pass-gate transistor when the control signal (C) turns low and the input signal (V_{IN}) from high to low is imposed. Similarly, the body potential and the parasitic bipolar leakage current of the PMOS device (dashed lines) in the device when the input signal (V_{IN}) from low to high is imposed. (Adapted from Lu et al. [7].)

and the parasitic bipolar leakage current of the NMOS device (solid lines) in the SOI CMOS pass-gate transistor when it is turned off and the input signal (V_{IN}) from high to low is imposed [7]. When the control signal (C) is high, the output of the SOI CMOS pass-gate transistor is high since the input is high. Under this situation, the body of the NMOS device is also charged to high. Once the control signal (C) switches from high to low, the SOI CMOS pass-gate transistor turns off. If the input signal switches from high to low, since the body voltage of the NMOS device is high, the parasitic NPN bipolar device turns on to cause the leakage current. Consequently, the body potential drops from 2.5 to 0.82 V since the body is discharged by the parasitic bipolar leakage current. A similar situation exists for the PMOS device (dashed lines) in Fig. 4.10. When the gate voltage (C) is low, the CMOS pass-gate transistor turns off. When the input signal (V_{IN}) switches from low to high, the parasitic PNP bipolar device of the PMOS device also turns on to cause the parasitic bipolar leakage current, which cannot be overlooked when designing circuits using SOI CMOS pass-gate transistors. Generally speaking, the leakage current caused by the parasitic NPN bipolar device in the SOI NMOS device (1.02 mA as shown in Fig. 4.10) is much larger than that caused by the parasitic PNP bipolar device in the SOI PMOS device (0.05 mA as shown in Fig. 4.10).

As described before, for the static and dynamic pass-gate transistors, during the switching transient, the potential of the floating body is determined by the capacitive coupling from the external input. In addition, the floating body effect is strongly dependent on the switching pattern and the operating states. For a logic circuit using

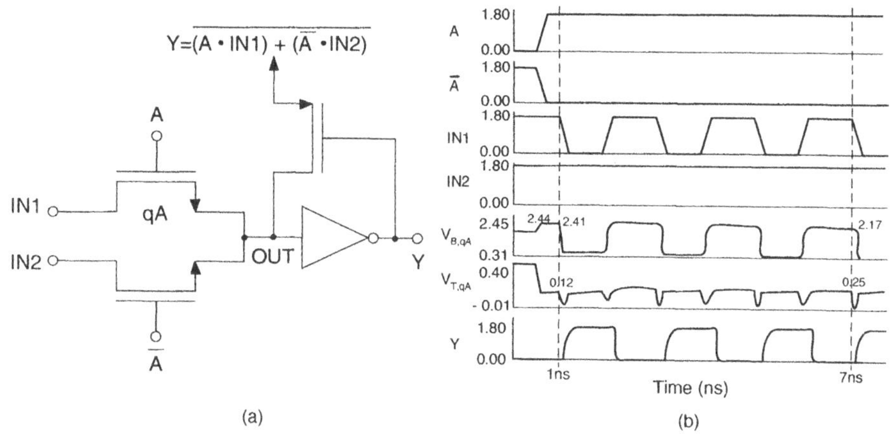

Fig. 4.11 (a) A single-ended LEAP circuit using SOI NMOS devices. (b) Waveforms of the body potential and the threshold voltage of the device qA and the output signal (Y) of the LEAP circuit with the gate signals (A, $\bar{A}$) and the input signals (IN1, IN2) imposed to the SOI NMOS devices. (Adapted from Puri & Chuang [8].)

pass-gate transistors, the capacitive coupling effect from the gate to the body is more noticeable as compared to the inverter, NAND, and NOR logic gates. Figure 4.11 shows (a) a single-ended LEAP circuit using SOI NMOS devices and (b) waveforms of the body potential and the threshold voltage of the device. Initially at $t = 0$, the select input $\bar{A}$ is high at $V_{DD} = 1.8$ V. Both data inputs IN1 and IN2 are high. Since the select input $\bar{A}$ is high, the output node (OUT) is pulled up to high. For device qA, since the select input A is low, it is off. Since both the source and the drain of qA are high, its body is charged to high. At $t = 0.5$ ns, input A switches from low to high after a rise time of 100 ps. Due to the gate-body capacitive coupling, the body of qA ($V_{B,qA}$) is elevated to $> V_{DD}$ to 2.44 V. At this time, the body-source potential of qA is $V_{BS,qA} = 0.64$ V. Due to the body effect, its threshold voltage drops from 0.4 to 0.1 V. At $t = 1$ ns, data input IN1 switches from high to low. Owing to the substantial decrease in the threshold voltage, the drain current of device qA increases substantially. Consequently, the propagation delay has been shortened. Along with the change in the switching cycle, when data input IN1 switches from low to high in the next cycle, the body potential of device qA ($V_{B,qA}$) is elevated due to the body-source capacitive coupling. During the switching cycle, the body of device qA has been charged due to impact ionization and reverse-biased PN junction. After several switching cycle, the body of device qA gradually enters the steady state as shown in the figure–$V_{B,qA}$ changes from 2.41 V at $t = 1$ ns to 2.17 V at $t = 7$ ns. Due to the change in the body potential ($V_{B,qA}$), the threshold voltage ($V_{T,qA}$) changes accordingly. The shortening of the propagation delay is most noticeable during the first switching of the data input IN1. While entering the steady state of the body potential, the propagation delay also becomes steady.

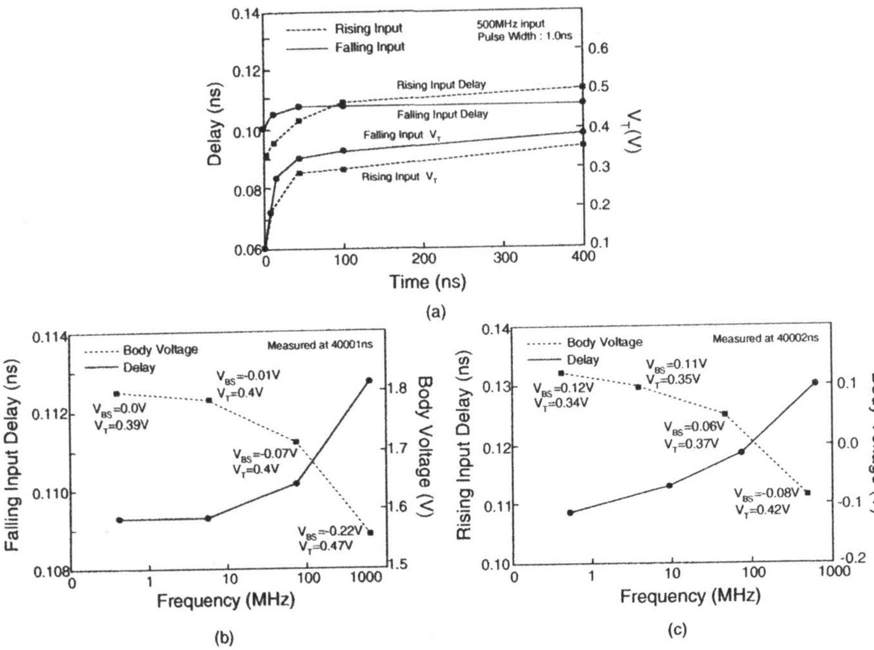

Fig. 4.12 (a) Rising/falling input delay and device qA's threshold voltage versus time of the single-ended LEAP circuit with a 500 MHz clock imposed at the input. (b) Falling input delay and (c) rising input delay, and device qA's body voltage versus input clock frequency of the single-ended LEAP circuit. (From Puri & Chuang [8]. ©2000 IEEE.)

Figure 4.12 shows (a) the rising/falling input delay and the threshold voltage versus time of the single-ended LEAP circuit with a 500 MHz clock imposed at the input, (b) the falling input delay and (c) the rising input delay, and the body voltage versus the input clock frequency of the single-ended LEAP circuit [8]. As shown Fig. 4.12(a), for the transient operation as described in Fig. 4.11, As time progresses, the body voltage of device qA falls gradually and hence its threshold voltage rises accordingly. The rising and the falling delays of the circuit become longer with the progress of time. As shown in Fig. 4.12(b) and (c), at a higher operating frequency, the change in the delay becomes larger. From the above analysis, the floating body effects may cause the instability in the performance of a circuit made of SOI devices.

4.2.3 Switch Network Logic (SNL)

The floating body effect of the PD SOI CMOS devices can be overcome by connecting the body to the source. Although this approach can be used to eliminate the floating body effect, during the circuit operation, when the source voltage is larger than the drain voltage ($V_S > V_D$), the anomalous reverse conduction (ARC) from the source to the drain may occur. Figure 4.13 shows a switch network logic (SNL) using PD SOI NMOS devices with body connected to source to eliminate floating body effects

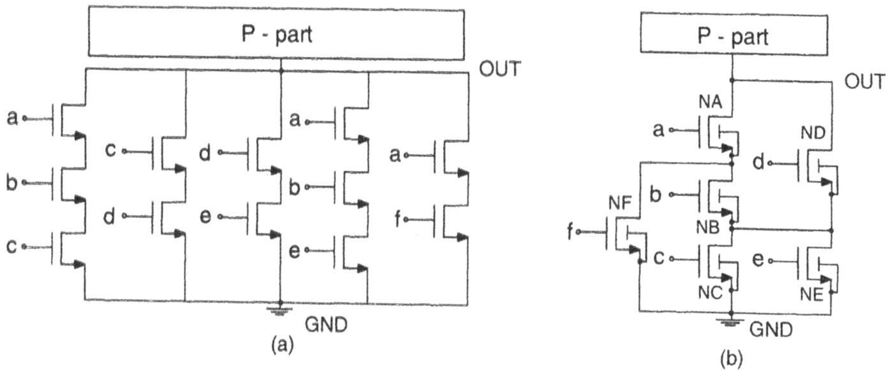

Fig. 4.13 SNL using PD SOI NMOS devices with body connected to source to eliminate floating body effects and with inputs (abcdef) = (000101), showing the ARC. (Adapted from Rajsuman et al. [9].)

[9]. With inputs (abcdef) = (000101), NA, NB, NC, and NE are off; ND and NF are on. Although NB is supposed to be off, it may be turned on due to ARC. The source of ND is pulled high and the drain of NF is pulled low. Therefore, the source-drain voltage of NB is equal to the output voltage $V_{SD} = V_{out}$ to accidentally turn on NB. As a result, due to the current conduction via ND-NB-NF, the output voltage may come down to cause an output error. From a circuit design point of view, the body tied to the source structure should be avoided for the SOI devices.

4.2.4 Hysteresis

In addition to the pass-gate based circuits, hysteresis effects may also cause hysteresis propagation delay in an inverter-based circuit made of PD SOI CMOS devices. Figure 4.14 shows four fundamental delays of inverter-based circuits made of PD SOI CMOS devices due to hysteresis effects [10]. As shown in this figure, depending on the initial (DC) state, there are two kinds of transients–(1) DC low-to-high (LH) transient and (2) DC high-to-low (HL) transient. For a basic inverter circuit, during the LH transient with the initial DC input low and with the output high, the body charge of the NMOS device reaches its peak since its drain-source (V_{DS}) is the largest. Therefore, during the LH transient, the large drain current provided by the NMOS device leads to a small propagation delay–pull-down fast (τ_{pd-f}). Similarly, during the HL transient, the pull-up fast (τ_{pu-f}) propagation delay based on the PMOS device is generated. For the subsequent transient, the accumulated body charge in the NMOS and the PMOS devices is smaller as compared to the initial DC case. Specifically, after the initial DC LH transient, during the subsequent HL transient, the output low may result in the accumulation of the body charge in the PMOS device, which is smaller than that generated by the initial DC HL pull-up slow τ_{pu-s}. Similarly, after the initial DC HL transient, the subsequent DC LH transient may lead to the pull-down slow (τ_{pd-s}). Since the supply and the removal of the

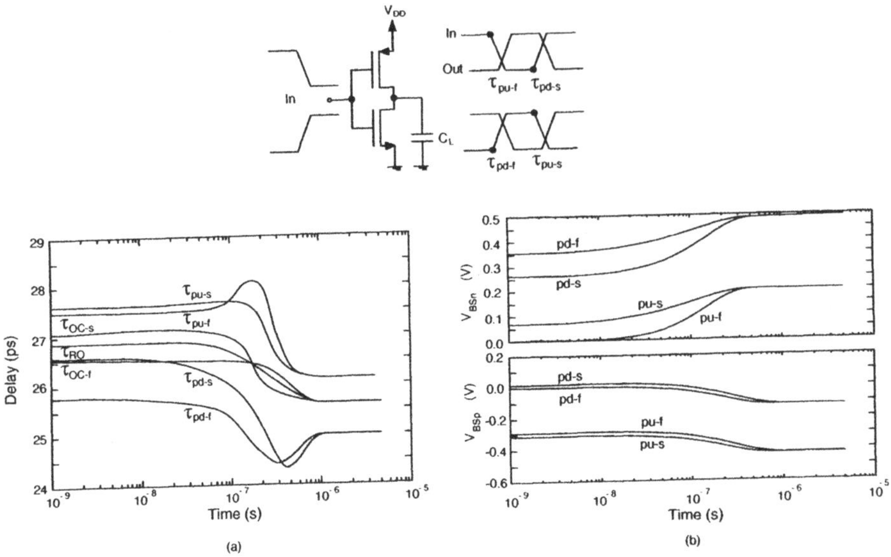

Fig. 4.14 Four fundamental delays of inverter-based circuits made of PD SOI CMOS devices due to hysteresis effects. (From Pelella et al. [10]. ©1999 IEEE.)

internal charge in the SOI device are done via thermal generation/recombination and impact ionization, to reach the dynamic steady state it takes several μs. Therefore, along with the body-source voltage (V_{BS}) reaching its steady state, the propagation delay may be shortened gradually to cause hysteresis.

Figure 4.15 shows (a) the data path of a microprocessor-related logic circuit with latches made of PD SOI CMOS devices and (b) the transients showing the pass or the fail of the D0 signal depending on the hysteresis [10]. As shown in this figure, when the data D0 switches from low to high and the clock is high, within a clock cycle, data is transferred to D2. Due to the hysteresis phenomenon of the propagation delay in the inverter-based circuit, after several μs, the propagation delay gradually shortens. As shown in this figure, at t = 1 μs, at the end of the clock cycle, the clock becomes low to turn off the transmission gate to prevent D2 from reaching D3. Due to the shortening of propagation delay in the D2 signal, before the transmission gate turns off, the D2 signal may already be transferred to D3 to cause the pull-low of D3 to generate an error. To eliminate the error, the clock timing should be properly designed.

4.2.5 Analog Circuits

Floating body effects are not just for digital circuits, they also affect the performance of analog circuits. For an amplifier, its gain is determined by the output conductance of the input device biased in the saturation region. As shown in Fig. 4.16(a), due to the kink effects caused by the floating body effect, the output conductance is worsened

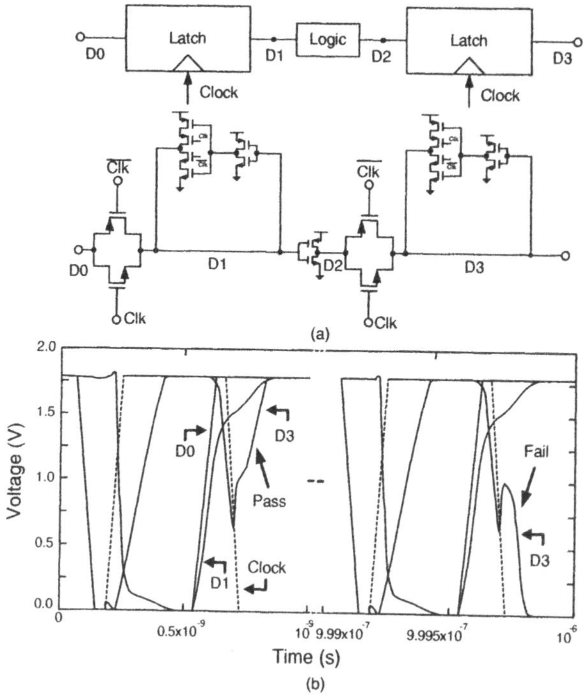

Fig. 4.15 (a) Data path of a microprocessor-related logic circuit with latches made of PD SOI CMOS devices. (b) Transients showing the pass or the fail of the D0 signal depending on the hysteresis. (From Pelella et al. [10]. ©1999 IEEE.)

in some regions, which should be avoided for the amplifier design. In addition, the floating-body dependent output conductance is also sensitive to the operating frequency. Due to the low-pass filtering for the impact ionization current caused by the body-source capacitance and the junction conductance, the variation of the kink-effect-related conductance at low frequency is large. As shown in Fig. 4.16, for an amplifier stage using an SOI PD NMOS device as the input device, at low frequency, the amplifier gain is degraded substantially due to the frequency-dependent kink effects. At high frequency, the amplifier gain reaches its ideal gain due to a close to ideal output conductance at high frequency.

4.3 LOW-VOLTAGE CIRCUIT TECHNIQUES

Nowadays, in order to meet the requirements for implementing portable computer and communication equipment, circuit designs using low-voltage and low-power techniques become a necessity. SOI CMOS devices are especially suitable for low-voltage and low-power circuit designs owing to their intrinsic thin-film structure above the buried oxide. Both FD and PD SOI technologies provide CMOS devices

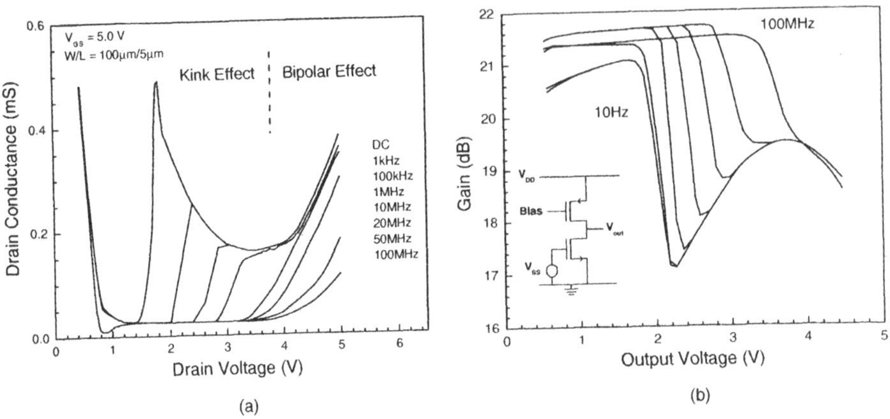

Fig. 4.16 (a) Drain conductance versus drain voltage of an SOI NMOS device at various operating frequencies. (b) Gain versus output voltage of an amplifier stage using an SOI NMOS device as the input device for various operating frequencies. (Adapted from Assaderaghi et al. [11].)

with small threshold voltages and small junction capacitances. In addition, SOI CMOS devices offer good subthreshold characteristics, small leakage currents, and RC delays. Hence, SOI CMOS devices are especially suitable for low-voltage, low-power, high-speed VLSI circuits. The realization of SOI CMOS low-voltage VLSI circuits has been achieved by device technology and circuit techniques. In Section 4.3.1, the SOI CMOS device technology for low-voltage circuit techniques and circuit techniques are described.

4.3.1 Low-Voltage Technology

Figure 4.17(a) shows the cross section of $0.35\ \mu m$ FD SOI CMOS devices with a front gate oxide of 70 Å, a thin film of 500 Å, and a buried oxide of 1000 Å for implementing low-voltage analog-digital mixed-mode VLSI circuits. As shown in this figure, salicided dual polysilicon gate and two-level metal interconnect layers have been adopted to facilitate circuit designs. In order to realize analog circuits, resistances realized by n-type polysilicon and capacitances implemented by the gate oxide sandwiched by the polysilicon gate and the n^+ doped thin film have been achieved with the capability of realizing a 10-bit analog-digital converter (ADC). The $0.35\ \mu m$ FD SOI CMOS devices are especially useful for low-voltage, low-power circuit applications. Fig. 4.17(b) shows the propagation delay versus the normalized power dissipation of the inverter realized by the $0.35\ \mu m$ FD SOI CMOS devices for various power supply voltages [12]. Also shown in this figure are the results for the data using bulk devices. Owing to smaller threshold voltages, smaller subthreshold leakage currents, and smaller parasitic capacitances, the SOI inverter demonstrates a much better propagation delay at a smaller power dissipation. This figure shows that the SOI circuits are especially suitable for low-voltage (as low as 0.6 V) applications. For

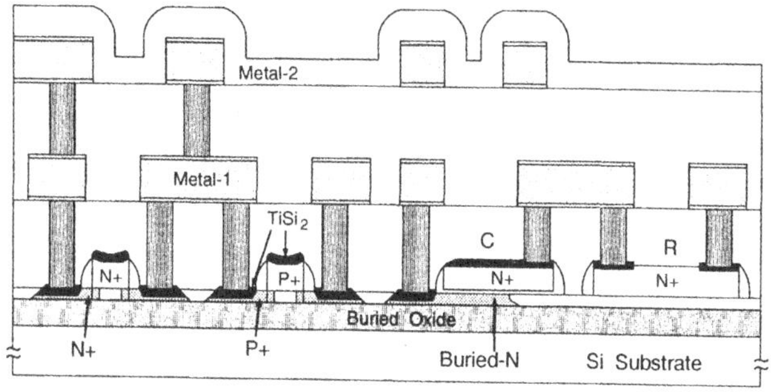

(a)

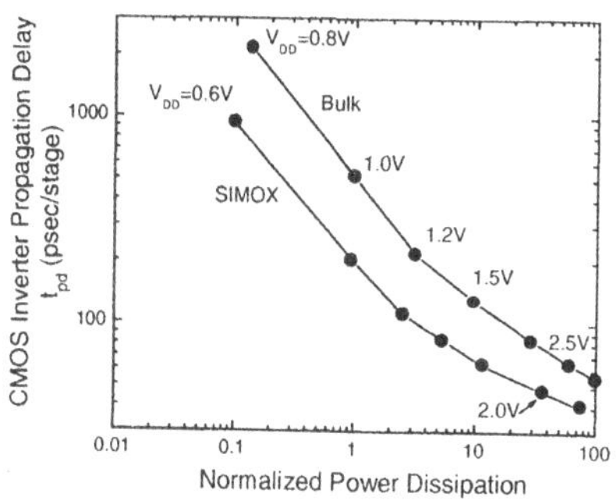

(b)

Fig. 4.17 (a) Cross-section of 0.35 μm FD SOI CMOS devices with a front gate oxide of 70 Å, a thin film of 500 Å, and a buried oxide of 1000 Å. (b) Propagation delay versus normalized power dissipation of the inverter realized by 0.35 μm bulk and FD SOI CMOS devices for various power supply voltages. (Adapted from Adan et al. [12].)

analog application, the 0.35 μm FD SOI CMOS devices show a unity gain bandwidth of 1.6 GHz at the power supply voltage of 1.5 V and 0.7 GHz at 1 V.

4.3.2 Dynamic Body Control

For VLSI circuits using a low power supply voltage, the magnitude of the threshold voltage of the CMOS devices cannot be large. For standard CMOS devices the magnitude of the threshold voltage cannot be lowered at will due to the subthreshold slope and leakage current considerations. If the magnitude of the threshold voltage of standard CMOS devices is lowered without considering the subthreshold slopes, the increased leakage current in the subthreshold region may disqualify the devices for dynamic logic circuits and low-power applications. The floating body effect described in Section 4.3.1 seems to have negative impacts on the performance of an SOI CMOS circuits. Here, the floating body is taken advantage to facilitate low-voltage CMOS digital circuits using a power supply voltage < 1 V.

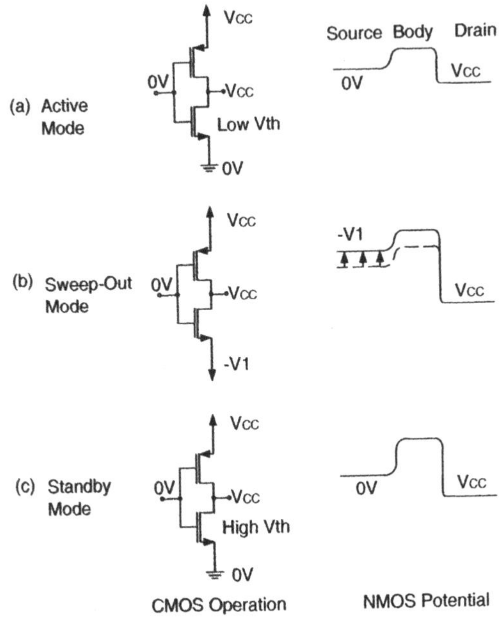

Fig. 4.18 Operation principles of the dynamic floating body-control SOI CMOS circuits. (Adapted from Morishita et al. [13].)

Figure 4.18 shows the operation principles of the dynamic floating body-control SOI circuits [13]. For the dynamic floating body-control SOI CMOS circuits, their operation is composed of three modes– active, sweep-out, and standby. During the active mode operation, the accumulation of the floating body charge has been taken advantage to obtain a small threshold voltage based on body effect. After accomplishing the active mode operation, prior to entering the standby mode, the source voltage of the device has been controlled to achieve the sweep-out of the floating body charge. The operation of the sweep-out is carried out after the active mode to lower the source voltage to a negative voltage ($-V_1$) such that the source-body junction of the device is forward biased. Consequently, the charge accumulated in the floating body can be discharged to lower the body voltage quickly. After the sweep-out mode operation, in the standby mode, the source voltage is back to ground. At this time, since the body potential still sustains at a relatively lower potential, thus a higher threshold voltage results in a reduced leakage current and a smaller power consumption. These are the operation principles of the dynamic floating-body control SOI CMOS circuits.

Figure 4.19(a) shows a 100-stage inverter chain using the SOI dynamic floating body control techniques [13]. As shown in the figure, in the inverter chain circuit, both V_{BCC} and V_{BSS} are the additional power supply voltages, which are used to control the source voltages of the PMOS and the NMOS devices, respectively, for the sweep-out operation. During the active mode operation, both V_{BCC} and V_{CC} are at 1

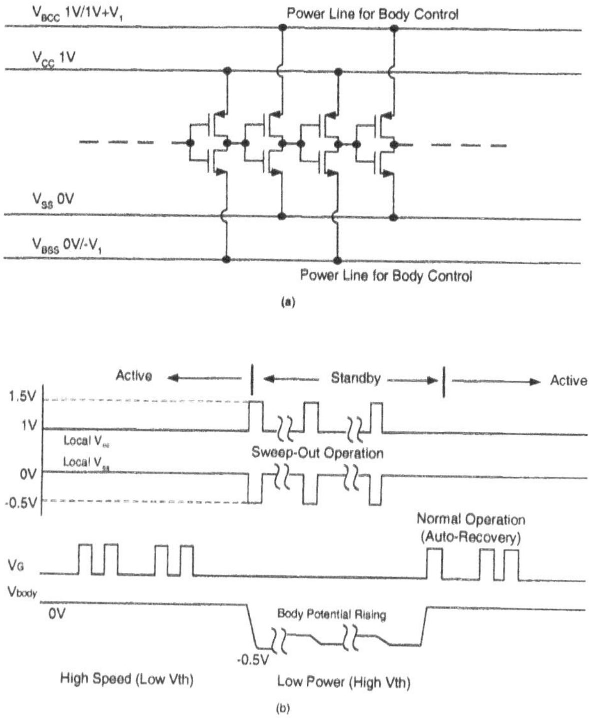

Fig. 4.19 (a) 100-stage inverter chain circuit using the SOI dynamic floating body control techniques. (b) Timing diagram of the SOI dynamic floating body control techniques. (From Morishita et al. [13]. ©1997 IEEE.)

V and both V_{BSS} and V_{SS} are at 0 V. $-V_1$ and $V_1 + 1$ V are used as the source voltage for NMOS and PMOS devices, respectively, during the sweep-out mode operation. Figure 4.19(b) shows the timing diagram of the SOI dynamic floating body control techniques [13]. During the sweep-out mode operation, the additional power supply voltages V_{BCC} and V_{BSS} are at 1.5 and -0.5 V, respectively, for a period of 10 ns to discharge the accumulated charge in the floating body. For the NMOS device, during the sweep-out mode operation, the additional power supply voltage V_{BSS} is at -0.5 V such that the source voltage of the NMOS device is lowered and the source/body junction is forward biased. For the PMOS device, during the sweep-out mode operation, the additional power supply voltage V_{BCC} is at 1.5 V such that the source voltage of the PMOS device is raised and the source/body junction is forward biased. In the standby mode operation, the additional power supply voltages are back to their original values. The sustained lower/higher voltage in the body of the NMOS/PMOS device increases the magnitude of the threshold voltage of the NMOS/PMOS device. Consequently, the leakage current in the NMOS/ PMOS device is reduced, and hence a smaller power consumption. During the standby mode, due to thermal generation and recombination, the body potential of the NMOS/PMOS device gradually recovers to its original value 0 V/1 V, which results in a decrease

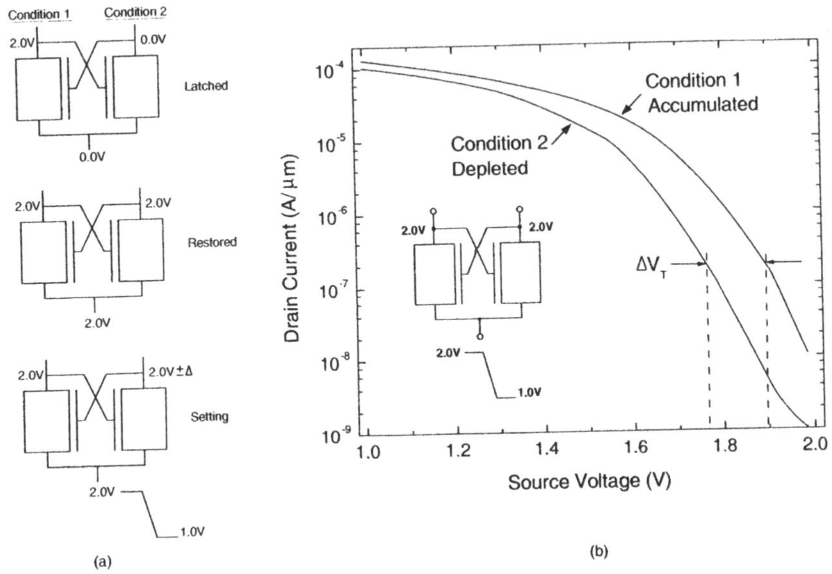

Fig. 4.20 (a) Sense amp made of cross-coupled NMOS devices during latched, restored, and setting phases. (b) Drain current versus source voltage of the NMOS device in the sense amp during the setting phase. (From Mandelman et al. [14]. ©1997 IEEE.)

in the magnitude of its threshold voltage–the leakage current may increase in the standby mode. Therefore a periodic sweep-out operation during the standby mode is adopted to maintain a high magnitude of the threshold voltage for the NMOS/PMOS devices. After the conclusion of the standby mode, in the active mode, via gate-to-body capacitive coupling, the body potential of the NMOS/PMOS device has been raised/lowered to obtain a low threshold voltage for NMOS and PMOS devices. At the boundary of the standby and the active mode, a positive input pulse imposed at the gate of the NMOS device brings an increase of the body potential from -0.5 to 0 V, which is the so-called auto-recovery. After auto-recovery, the gate pulse is over and the body potential may fall slightly with a slight effect on the drain current. Therefore, using the SOI dynamic floating body control techniques, sub-1V operation of an SOI circuit with a low power consumption can be obtained via controlling the body potential dynamically without extra body contacts. Thus, a small layout area of the circuits can be achieved.

4.3.3 Sense Amp

Floating body effect may also influence the operation of a sense amp in a digital circuit such as DRAM. Figure 4.20(a) shows a sense amp made of cross-coupled NMOS devices during latched, restored, and setting phases [14]. During the latched phase, for the NMOS device at the left (condition 1) biased at the gate-source voltage of 0 V and the drain-source voltage of 2 V, due to impact ionization and thermal generation,

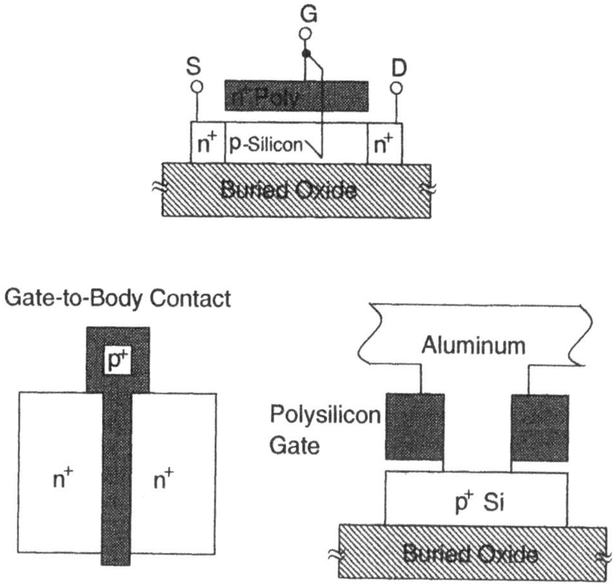

Fig. 4.21 Cross-section of the DTMOS device with its gate tied to the body of a thin film. (Adapted from Assaderaghi et al. [15].)

charge accumulates in the body. For the NMOS device at the right (condition 2) biased at the gate-source voltage of 2 V and the drain-source voltage of 0 V, a depletion region exists in the body of the device, thus body charge decreases. During the restored phase, the biasing conditions of the devices become identical, but the internal body charge in each device is different since it takes several ms for the internal body charge in the two devices to reach equilibrium. Before equilibrium is reached, if the circuit enters the setting phase, the driving drain currents of the two devices are different because of the different body charge. As a result, as shown in Fig. 4.20(b), a mismatch in the threshold voltage $\sim$ 0.135 V can be seen.

4.4 DTMOS CIRCUITS

In order to prepare the SOI devices for the VLSI circuits using an ultra-low supply voltage, as described in the Chapter 3, DTMOS devices with the neutral body of the thin film tied to the gate has been used to control the threshold voltage of the front channel for a better performance.

4.4.1 DTMOS Device

Figure 4.21 shows the cross-section of the DTMOS device with its gate tied to the body of a thin film [15]. By tying the body to the gate, both the body and the gate are biased

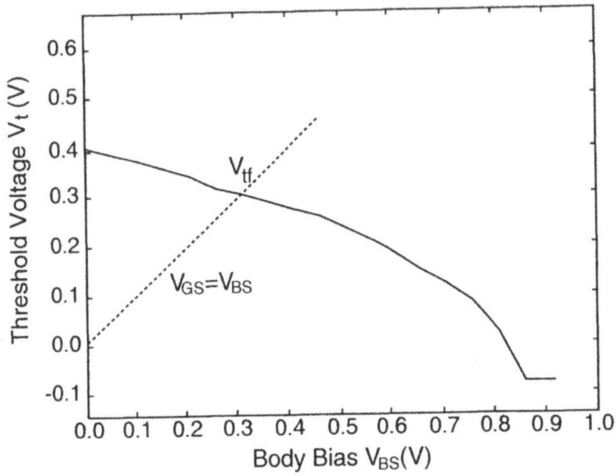

Fig. 4.22 Threshold voltage versus body-source voltage of an SOI NMOS device with a front gate oxide of 64 Å and a p-type thin-film doping density of $3 \times 10^{17}\mathrm{cm}^{-3}$. (Adapted from Assaderaghi et al. [15].)

at the same potential. When the gate voltage is high, the front channel of the device is turned on. Owing to the body-tied-to-gate structure, the body potential becomes high too. Due to body effect, the threshold voltage of the front channel becomes smaller and thus the current driving capability of the front channel increases. When the gate voltage becomes low, the front channel turns off. Since the body voltage also becomes low, via body effect, the threshold voltage of the front channel is recovered, thus the leakage current is small. From the above description, the DTMOS device provides a good performance for both turn-on and turn-off operations. As shown in this figure, the connection of the body and the gate is via the metal through the hole in the polysilicon layer to the P^+ thin-film region for the NMOS device and to the N^+ thin-film region for the PMOS device. From the layout point of view, no extra area is needed for the DTMOS device.

Figure 4.22 shows the threshold voltage versus the body-source bias voltage of the SOI NMOS device with a front gate oxide of 64 Å and a p-type thin-film doping density of $3 \times 10^{17}\mathrm{cm}^{-3}$ [15]. The threshold voltage of the SOI NMOS device with its body-source junction forward biased ($V_{BS} > 0$), as shown in this figure is used to see the case with the body tied to the gate. When the body-source junction is zero biased, the DTMOS device has the largest threshold voltage as for the SOI NMOS device with its body tied to the ground. When the gate voltage increases, for the DTMOS device, its body voltage also increases. As a result, its threshold voltage decreases accordingly. When the body-source voltage reaches $2\phi_b$, where ϕ_b is the electrostatic potential of the thin film ($\phi_b = \frac{kT}{q}\ln\frac{N_A}{n_i}$), where n_i is the intrinsic doping density, and N_A is the doping density of the thin film, the threshold voltage reaches its minimum value $V_{th} = 2\phi_b - V_{fb}$, where V_{fb} is the flat-band voltage. When the DTMOS device turns on, due to the decreased threshold voltage, the drain current

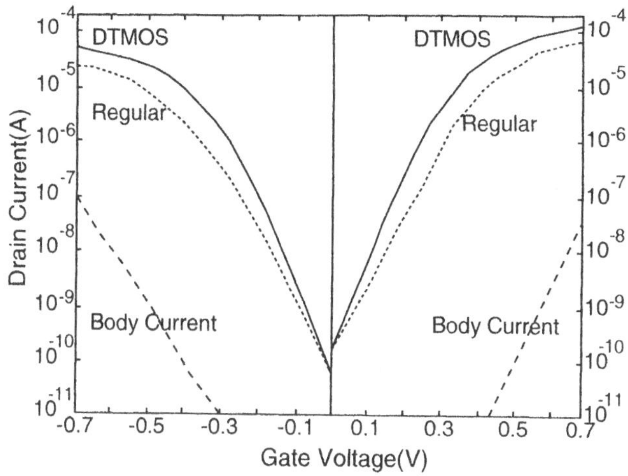

Fig. 4.23 Subthreshold characteristics of SOI NMOS and PMOS devices with body grounded and body tied to the gate (DTMOS). (Adapted from Assaderaghi et al. [16].)

becomes larger. When the device is turned off, its threshold voltage stays at the peak value, thus the leakage current and the subthreshold slope are optimized.

As described in the Chapter 3, DTMOS devices offer a better subthreshold slope. Figure 4.23 shows the subthreshold characteristics of SOI NMOS and PMOS devices with body grounded and body tied to the gate (DTMOS) [16]. As shown in this figure, at the zero gate voltage, both the regular SOI device (with body grounded) and the DTMOS device have an identical leakage current. When the gate voltage increases, the DTMOS device has a larger drain current. At the gate voltage of 0.3 V, the drain current of the DTMOS device is 5.5 times as compared to the regular SOI device. At the gate voltage of 0.6 V, it is 2.5 times. The increase in the drain current of the DTMOS device is due to the decrease in the threshold voltage from body effect and the enhanced mobility caused by the decreased electric field from the shrunk depletion region in the thin film with a positive back gate bias. In addition, the subthreshold swing becomes better (60 mV/dec), which implies a stronger gate control over the channel for the body-tied-to-gate structure. The drawback of the DTMOS device is that the gate cannot be biased at a voltage > 0.7 V. If the gate voltage is > 0.7 V, the body-source junction is forward biased. As a result, a large amount of extra leakage current may deteriorate the subthreshold performance.

4.4.2 DTMOS Inverter

The DTMOS device described before is applicable for circuits using a power supply voltage < 0.7 V. To apply the DTMOS techniques to circuits using a power supply voltage > 0.7 V, by adding a diode between the gate and the body, the applicable power supply voltage can be increased. Figure 4.24 shows the DTMOS inverter cir-

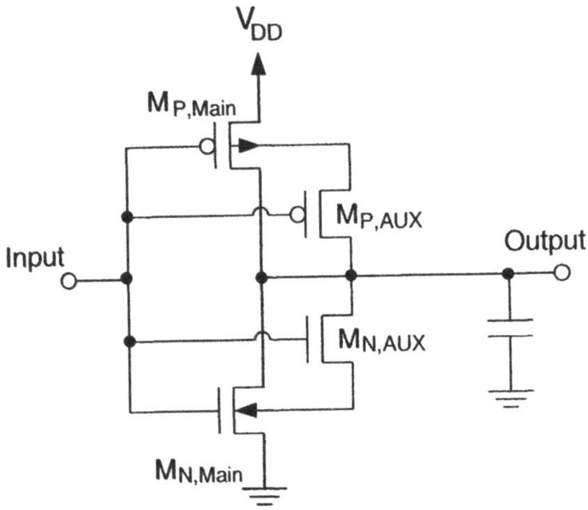

Fig. 4.24 DTMOS inverter circuit using auxiliary transistors to increase the applicable power supply voltage. (Adapted from Chung et al. [17].)

cuit using auxiliary transistors to increase the applicable power supply voltage [17]. As shown in this figure, by adding auxiliary transistors to connect the drain to the body controlled by the gate of the main transistor, this DTMOS inverter can work at a higher power supply voltage. The advantage of this circuit can be understood by studying the pull-down operation. When the input signal switches from low to high, both the main and the auxiliary transistors $M_{N,main}/M_{N,aux}$ turn on. At this time, the output voltage is high. Via the auxiliary transistor $M_{N,aux}$, the body of the main transistor ($M_{N,main}$) is charged by the auxiliary transistor ($M_{N,aux}$). Therefore, the body potential of the main transistor ($M_{N,main}$) increases, which decreases its threshold voltage. Consequently, the driving capability of the main transistor ($M_{N,main}$) is enhanced and a faster pulldown can be obtained. During the pulldown process, the source of the auxiliary transistor ($M_{N,aux}$), which is also the body of the main transistor ($M_{N,main}$), is larger than its drain voltage, which is the output voltage of the inverter, therefore the charge accumulated in the body of the main transistor is injected back to the drain. As a result, the body potential of the main transistor is recovered to the ground level. Hence its threshold voltage is back to its original value. A small leakage current in the NMOS transistor expedites the pullup cycle. Since the body of the main transistor is connected to the gate via the auxiliary transistor– the body is not connected to the gate directly any longer, thus the operating voltage can exceed a diode voltage without worrying about the extra leakage current in increasing its power consumption. The principle of pullup operation is similar. Only the NMOS devices are replaced by the PMOS.

4.4.3 DTMOS Buffer

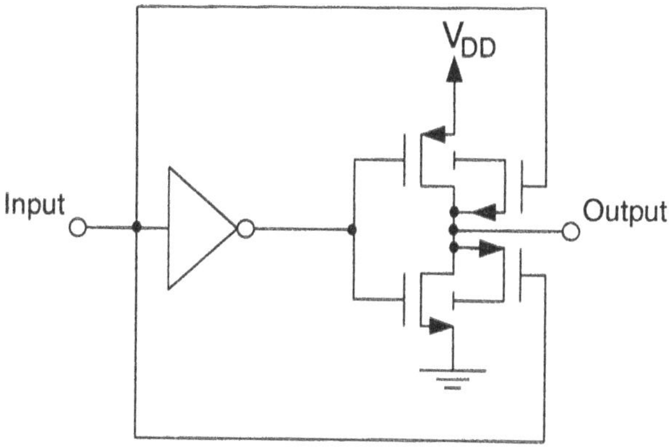

Fig. 4.25 DTMOS buffer circuit. (Adapted from Houston [18].)

For the DTMOS inverter circuit described in the Fig. 4.24, during the pull down operation, both the main and the auxiliary NMOS devices turn on at the same time. Before the body potential of the main transistor is pulled to high by the auxiliary transistor, the main transistor already turns on. Under this condition, the advantage of the low-threshold voltage for the main transistor shows up only in the latter portion of the pull down process. Hence, the advantage of the dynamic threshold techniques has not been expedited fully.

Fig. 4.25 shows a DTMOS buffer circuit, which is improved from the DTMOS inverter presented in Fig. 4.24 [18]. As shown in this figure, the DTMOS buffer circuit is applicable to a chain of inverters with a gradually increasing size used as a driver. In the present-stage inverter, considering the pull down portion, its auxiliary transistor, which is used to connect the body of the main transistor to the output node, is with its gate controlled by the input to the previous stage instead of the current stage. The auxiliary transistor has been changed to PMOS instead of NMOS. The previous stage input has been used to control the turn-on of the auxiliary transistor to provide advantages. Before the main transistor of the present-stage inverter turns on, the body of the main transistor has been raised to a high level by the auxiliary transistor controlled by the previous-stage input. Once the main transistor of the current stage starts to operate, the SOI dynamic threshold technique functions fully at the very beginning. In addition, the PMOS auxiliary transistor is more effective in raising the body potential of the NMOS as compared to the NMOS auxiliary transistor. Similarly, in the pull up portion, the NMOS auxiliary transistor has been used to replace the PMOS auxiliary transistor to raise the body potential of the PMOS main transistor.

The advantages of the NMOS main transistor with the PMOS auxiliary transistor (N-p) for the pull down operation described in Fig. 4.25 can be perceived from Fig. 4.26. In the N-p configuration, its gate voltage falls from V_{DD} to $V_{DD} - |V_{tp}|$ to charge the body of the NMOS main transistor, where V_{tp} is the threshold voltage of the auxiliary PMOS device [19]. Since the gate of auxiliary PMOS transistor in

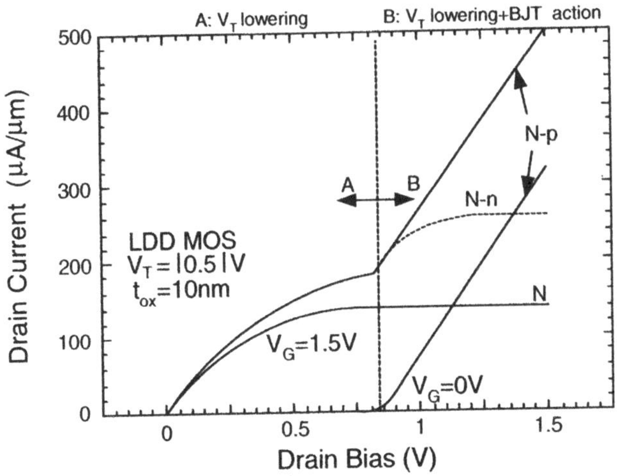

Fig. 4.26 Drain current versus drain bias of the NMOS main transistor with the PMOS auxiliary transistor (N-p) for the pull down operation described in Fig. 4.25, the NMOS main transistor with the NMOS auxiliary transistor (N-n) described in Fig. 4.24, and the NMOS main transistor only. (Adapted from Lee et al. [19].)

the N-p configuration is connected to the previous-stage input. When the input of the previous stage reaches $V_{DD} - |V_{tp}|$, the PMOS auxiliary transistor starts to charge the body of the NMOS maintain transistor. In the N-n configuration (NMOS main device with the NMOS auxiliary transistor), its gate voltage needs to rise from 0 V to $V_{tn} + V_{BE}$ for the auxiliary device to be active, where V_{tn} is the threshold voltage of the NMOS auxiliary transistor and V_{BE} is the voltage of the body-source junction of the NMOS main transistor. Therefore, the current driving capability of the N-p configuration is stronger than that of the N-n configuration as seen in the figure.

4.4.4 Advanced DTMOS Devices

Until now, SOI devices with various body control techniques have been presented. As shown in Fig. 4.27, in addition to (a) the conventional SOI NMOS devices with body grounded and (b) the conventional SOI DTMOS device with body tied to gate, there are (c) the SOI DTMOS device with an NMOS auxiliary device sharing gate and drain to control its body and (d) the DTMOS device with an NMOS auxiliary device for its drain and gate tied to the gate of the main device to control its gate[20]. As shown in Fig. 4.27(d), in the SOI DTMOS device with an NMOS auxiliary device with drain and gate tied to gate of the main device to control its gate, the source of the NMOS auxiliary transistor is connected to the body of the main transistor. Different from Fig. 4.27(c), the drain of the NMOS auxiliary transistor is not connected to the drain of the main transistor any longer. Instead, along with the gate of the NMOS auxiliary transistor, its drain is also connected to the gate of the main transistor to

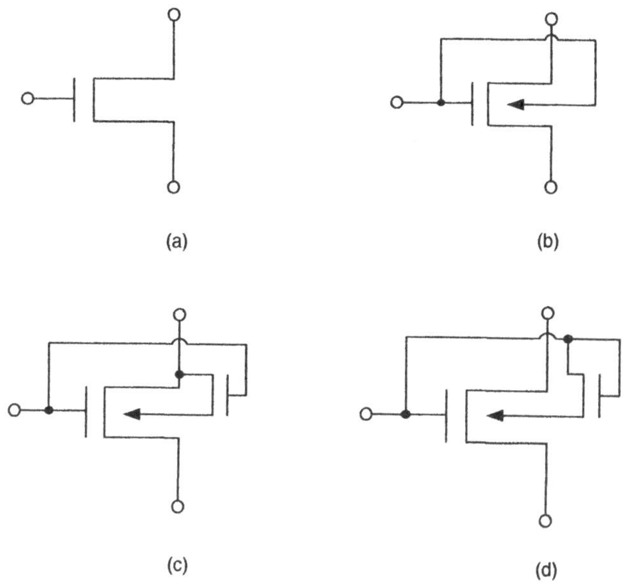

Fig. 4.27 Four types of SOI DTMOS devices (a) Conventional SOI NMOS device, (b) conventional DTMOS, (c) DTMOS device with an NMOS auxiliary sharing gate and drain to control body, and (d) DTMOS device with an NMOS auxiliary with drain and gate tied to gate to control body. (Adapted from Lee & Park [20].)

achieve the dynamic threshold voltage control. In this new DTMOS configuration, the body of the main transistor is controlled directly by the main transistor via the auxiliary transistor, which is similar to the DTMOS device with its body tied to gate. Owing to the addition of the NMOS auxiliary transistor between the gate and the body of the main transistor, the applicable power supply voltage can be high and the gate leakage current can be improved. The maximum supply voltage applicable to this new device is ~ 2 V. A large power supply voltage may bring in the saturation of the auxiliary transistor to cause a large amount of the drain current. Consequently, the gate leakage current of the main transistor increases substantially. Compared to the DTMOS with an auxiliary transistor sharing gate and drain to control its body (c), the threshold voltage of the new DTMOS device (d) with its body directly controlled by its gate via the auxiliary transistor is smaller and the drain current is larger. In addition, its subthreshold behavior is improved and the leakage current is reduced. Furthermore, a larger effective mobility of the new DTMOS device (d) is obtained. The new DTMOS device (d) still has a power supply voltage limitation. Since the drain of the auxiliary transistor is connected to the gate of the main transistor instead of the drain. For an inverter realized by the new DTMOS devices, when its output is pulled down to low, since its input still stays high if the power supply voltage is $> V_{th} + V_{BE}$, the leakage current may increase to a large extent.

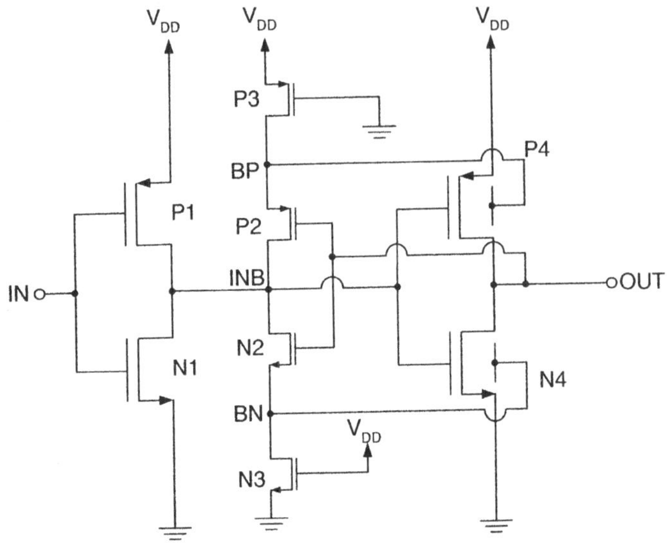

Fig. 4.28 Active body bias CMOS driver circuit using another DTMOS technique. (Adapted from Wada et al. [21].)

4.4.5 Active Body Control

Figure 4.28 shows the active body bias CMOS driver circuit using another DTMOS technique [21]. As shown in this figure, the driver circuit is composed of six SOI CMOS devices connected in feedback configuration. The body node of transistor P4 is connected to the drain node of P3 (BP) and the body node of transistor N4 is connected to the drain node of transistor N3 (BN). When the input signal IN is low, the output of the inverter (P_1, N_1) INB is high and its final output OUT is low. Via the feedback of the output signal OUT, transistor P2 is on and transistor N2 is off. When the input signal switches from low to high. At this time, transistor N1 turns on to form a DC current path by transistors P3, P2, and N1. Via this DC current path, node INB is gradually pulled to low. Node BP is charged by transistor P3 and discharged by transistors P2 and N1. If transistor P3 is designed to have a small current driving capability, node BP is pulled down to a low voltage. Thus the body voltage of transistor P4 drops substantially to lower the magnitude of its threshold voltage. As a result, the output node OUT is charged quickly to high by transistor P4. After the output node OUT reaches high, transistor P2 is turned off and the DC current path made of transistors P3, P2, and N1 is broken. Node BP is charged to high quickly, and thus the magnitude of the threshold voltage of transistor P4 increases to reduce extra power consumption. During the pulldown operation, transistor N4 has a similar operation. Figure 4.29 shows the waveforms of the internal and the output nodes of the active body bias CMOS driver circuit as shown in Fig. 4.28 during the transient [21]. As shown in this figure, the speed performance of the CMOS driver circuit using the active body bias is faster as compared to the bulk one. The body

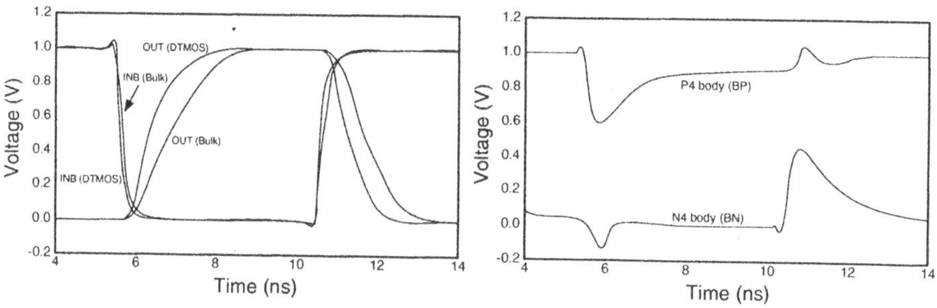

Fig. 4.29 Waveforms of the internal and the output nodes of the active body bias CMOS driver circuit with an output load of 60 pF and at a power supply voltage of 1V as shown in Fig. 4.28 during the transient. (Adapted from Wada et al. [21].)

potential (BP/BN) of the output transistor (P4/N4) changes almost at the same pace with node INB due to the discharge/charge via transistor P2/N2 and the gate-to-body capacitive coupling. Based on this behavior, the output transistors can drive the output load quickly. Note that the body voltage (BP/BN) of the output transistor P4/N4 is pulled low/high only during the switching. During another period, BP/BN stays at V_{DD}/0V such that the leakage current during idle can be reduced, which is similar to the operation of a 1.5 V BiCMOS dynamic logic circuit with a BiPMOS pull down scheme [22].

4.5 MTCMOS CIRCUITS

In addition to DTMOS techniques, SOI multithreshold CMOS (MTCMOS) logic circuit techniques have also been used for low-power high-speed VLSI circuits. In an MTCMOS circuit, there are two kinds of devices– (1) low-threshold devices for CMOS logic gates and (2) high-threshold devices for power switches. Figure 4.30 shows the SOI multithreshold CMOS (MTCMOS) logic circuit [23]. As shown in this figure, in the MTCMOS circuit, the logic gates are made of FD SOI CMOS devices with a low threshold voltage. The power switches are made of PD SOI PMOS devices with a high magnitude in the threshold voltage and with its body connected to its gate via a reverse-biased diode-connected FD device with a low threshold voltage. When SL is high, the power-switch transistor turns off and the circuit is in the sleep mode. The logic gates are disconnected from the power supply to save power. When SL is low, power-switch transistor turns on and the CMOS logic gates are connected to the power supply. Under this situation, the FD SOI devices with a small threshold voltage and a steep subthreshold slope offer a high-speed logic operation. At this time, the circuit is in the active mode. The design of the power-transistor switch is used to reduce the leakage current during the sleep mode. Thus, the high-threshold PD SOI PMOS device is adopted in the power-transistor switch. In order to increase the current conduction capability of the power-transistor switch during the active mode, DTMOS techniques described in the previous section have been used.

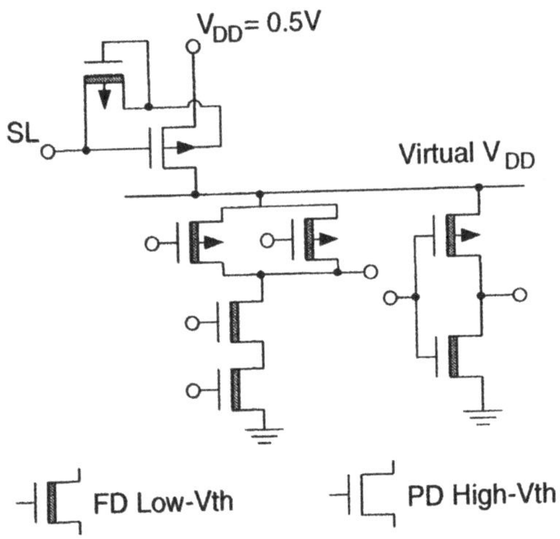

Fig. 4.30 SOI multithreshold CMOS (MTCMOS) logic circuit. (Adapted from Douseki et al. [23].)

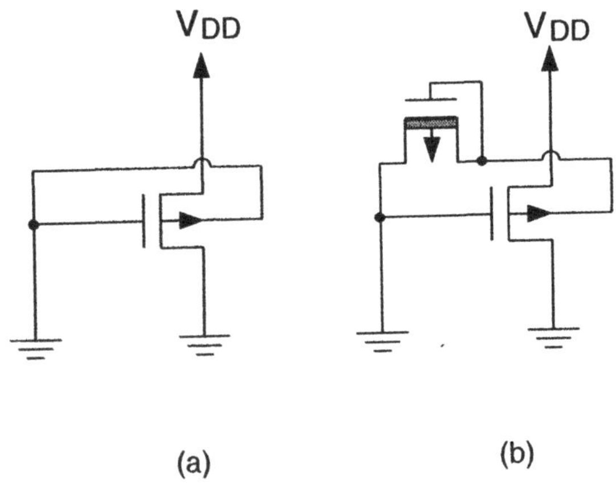

Fig. 4.31 (a) Conventional DTMOS device. (b) power switch transistor using the variable-threshold MOSFET technique used in the MTCMOS circuit when SL is low. (Adapted from Douseki et al. [23].)

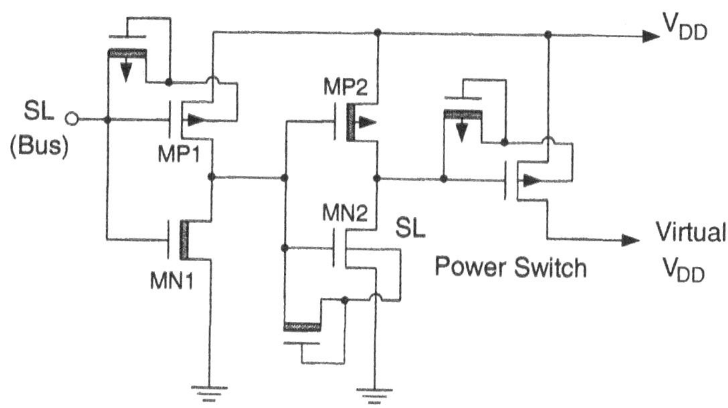

Fig. 4.32 Sleep control circuit. (Adapted from Douseki et al. [23].)

Figure 4.31 shows (a) the conventional DTMOS device and (b) the power switch transistor using variable-threshold MOSFET technique used in the MTCMOS circuit when SL is low [23]. As shown in Fig.4.31 via a reverse-biased diode-connected FD device, which is called the variable-threshold MOSFET, the body of the PD SOI PMOS device is connected to its gate. For the conventional DTMOS device as shown in Fig. 4.31(a), when V_{DD} exceeds 0.8 V, the body-source pn junction is forward biased, thus a large gate leakage current may degrade the performance. For the variable threshold device as shown in Fig. 4.31(b), if V_{DD} exceeds 0.8 V, the body-source voltage of the PD SOI PMOS device is clamped at 0.8 V by the reverse-biased diode made of the FD device to reduce the gate leakage current. When SL is high, the FD diode turns on and the body of the PD SOI PMOS is raised to a high value. Consequently, the magnitude of the threshold voltage of PD SOI PMOS device is raised and the leakage current is reduced during the sleep mode.

In general, power-switch transistors are large-size devices. When the SL signal is passed to the power-switch transistor, a set of sleep-control circuit as shown in Fig. 4.32 is used. The sleep control circuit is composed of two sets of inverters served as a buffer, which is used to enhance the switching speed. When driving power-switch transistors, low leakage current can be kept. In the sleep control circuit, variable-threshold MOSFETs MP1 and MN2 are used to reduce leakage currents when the circuit enters the sleep mode resulting from a higher threshold voltage of MP1 and MN2. When SL is high, both MP1 and MN2 are off, hence a low leakage current situation is achieved. When SL is low, both MP1 and MN2 are on, the PD SOI CMOS devices have large driving currents.

Figure 4.33 shows (a) the propagation delay versus V_{DD} of a two-input NAND logic circuit with a fan-out of 3, using SOI and bulk MTCMOS devices with channel lengths of 0.25 and 0.5 μm and (b) the power consumption versus the propagation delay of the SOI MTCMOS VLSI circuits operating at V_{DD} of 1 and 0.5 V [23]. As shown in Fig. 4.33(a), the speed performance of the SOI MTCMOS is faster as

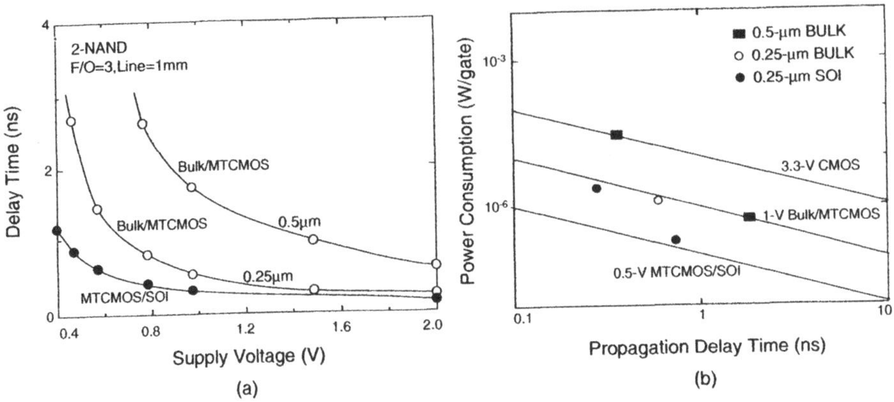

Fig. 4.33 (a) Propagation delay versus V_{DD} of a two-input NAND logic circuit with a fan-out of 3, using SOI and bulk MTCMOS devices with channel lengths of 0.25 and 0.5 μm. (b) Power consumption versus propagation delay of the SOI MTCMOS VLSI circuits operating at V_{DD} of 1 and 0.5 V. (Adapted from Douseki et al. [23].)

compared to the bulk MTCMOS. When the power supply voltage (V_{DD}) is < 0.5 V, only the SOI MTCMOS circuit can work. As shown in Fig. 4.33(b), the power-delay product of the 0.5 V SOI MTCMOS circuits is smallest among all technologies listed in this figure.

In the SOI MTCMOS circuit described above, logic gates are realized by low-threshold FD devices. Although high-speed operation can be achieved using low-threshold FD devices, their leakage currents cannot be lowered during the active mode, which is a major drawback. Figure 4.34(a) shows the triple-threshold SOI CMOS circuit scheme. As shown in this figure, in the triple-threshold SOI CMOS circuit scheme, there is still a power-switch transistor made of a high-threshold PD SOI device, which is used to connect the power supply line to the virtual power supply line. In the logic gate portion, low- and medium-threshold FD SOI CMOS devices have been used. Low-threshold FD devices are used for implementing the logic gates along the critical path and the medium-threshold FD devices are used in the noncritical path. The design of the threshold voltages for the medium- and low-threshold FD devices is based on the subthreshold slope consideration. For an ideal FD device, its subthreshold swing is $\sim$ 60 mV/decade. By increasing the threshold voltage by 0.12 V, its leakage current decreases by two orders of magnitude based on the subthreshold slope consideration. Based on this reasoning, the threshold voltages of the medium- and the low-threshold FD devices used in the logic gates have been designed such that during the active mode, leakage current and power consumption can be reduced without affecting its operation speed. Fig. 4.34(b) shows the the power consumption versus frequency of a 16-bit adder realized by the triple-threshold scheme [24]. As shown in this figure, operating at an identical speed, the circuit realized by the triple-threshold scheme consumes less power.

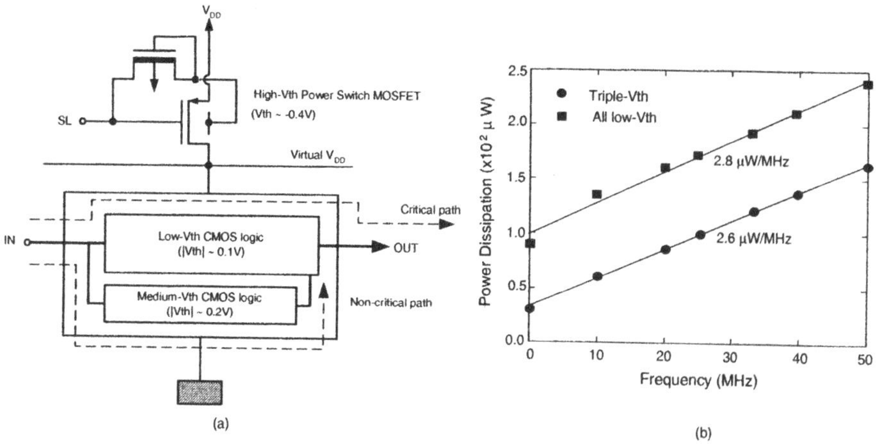

Fig. 4.34 (a) Triple-threshold SOI CMOS circuit scheme. (b) Power dissipation versus frequency of a 16-bit adder realized by the triple-threshold SOI CMOS circuit scheme. (Adapted from Fujii et al. [24].)

If we use the SOI triple-threshold circuit scheme, the leakage current of the VLSI circuits during the active mode can be substantially reduced. While designing the SOI triple-threshold circuits based on the conventional SOI MTCMOS circuits, in the logic gates many low-threshold FD devices are replaced by medium-threshold FD devices. Thus, the operation speed may be sacrificed. In order to compensate for this drawback, a boost circuit using single-channel pass transistor can be adopted. Since the output load capacitance of the single-channel pass transistor is smaller than that of the CMOS transmission gate, the boost circuit using the single-channel pass transistor has advantages in low power and high speed. Figure 4.35 shows a full adder and an exclusive NOR using the triple-threshold SOI CMOS circuit scheme with the boost circuit made of the single-channel pass transistor [24]. As shown in this figure, the boost circuit is composed of the single-channel pass transistor and the low-threshold FD SOI device M1, which have a source-drain overlap capacitance as marked. After data A1 and A2 are available at the input, when the carry input signal C1 changes from low to high due to the coupling of the source-drain overlap capacitance, the gate voltage of the single-channel pass transistor is boosted to exceed V_{DD}. Therefore, the driving capability of the pass transistor has been substantially enhanced and the propagation of the input signal over the critical path has been increased. As shown in the figure, the exclusive NOR circuit realized by six low- or medium-threshold FD devices has been used in the adder. By using the triple-threshold SOI CMOS circuit scheme and the boost circuit, the speed performance of the full adder has been improved by 15% and the power consumption has been reduced by 40%.

4.6 NOISE

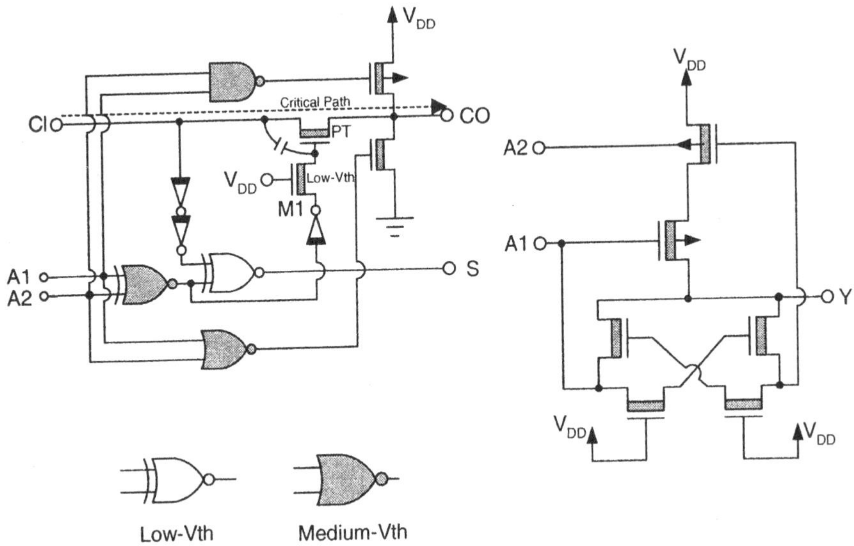

Fig. 4.35 Full adder and exclusive NOR using the triple-threshold SOI CMOS circuit scheme with the boost circuit made of the single-transistor pass transistor. (Adapted from Fujii et al. [24].)

For SOI devices, noise analysis is important especially for analog circuit applications. An accurate noise model can be used to predict the performance of an analog circuit precisely. Due to floating body effects, noise of the PD SOI devices is unique. Fig. 4.36 shows (a) the drain current versus the drain voltage and (b) the normalized drain current noise spectral density at 100Hz versus the drain voltage, of the PD SOI NMOS device with a front gate oxide of 115Å [25]. As shown in Fig. 4.36(b), the drain current noise spectrum also shows a behavior similar to the DC kink effects. Holes generated by the impact ionization in the PD SOI NMOS device, which accumulate in the body, cause trapping and detrapping at the buried oxide/thin-film interface. As a result, the body potential fluctuates to generate the drain current noise. Noise overshoot occurs at the drain voltage, where the DC kink effects exist. The noise overshoot phenomenon can be understood from the following drain current noise in terms of the perturbed drain current (dI_D) as follows:

$$dI_D = \frac{dI_D}{dV_B}\frac{dV_B}{dI_B}dI_B$$

where I_D, V_B, and I_B are the drain current, the body voltage, and the body current, respectively. Due to trapping and detrapping at the thin-film/buried oxide interface, a peak value in $\frac{dV_B}{dI_B}$ occurs. Susceptible to the $\frac{dV_B}{dI_B}$ influence, the drain current noise, which is the magnitude of the perturbation of the drain current– dI_D also has the noise overshoot and the noise peak. Compared to the average noise level, the noise peak is about two orders higher.

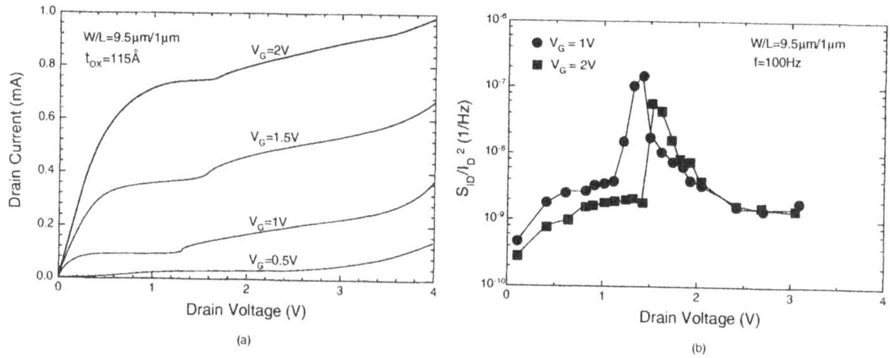

Fig. 4.36 (a) Drain current versus drain voltage and (b) normalized drain current noise spectral density versus drain voltage at 100 Hz of the PD SOI NMOS device with a front gate oxide of 115 Å. (Adapted from Chen et al. [25].)

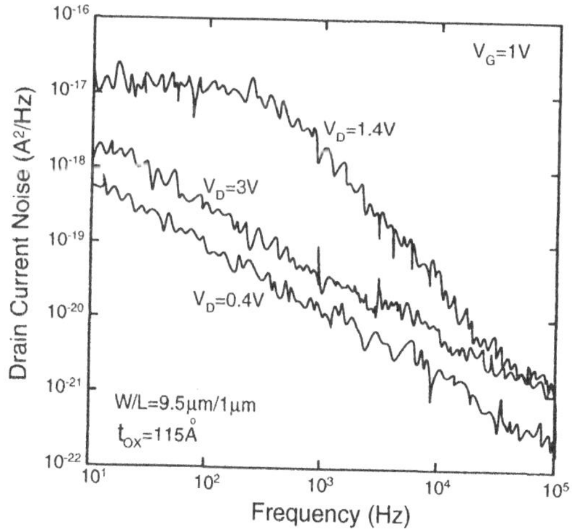

Fig. 4.37 Drain current noise power spectral density versus frequency of the PD SOI NMOS device with a front gate oxide of 115 Å, biased at the gate voltage of 1 V and the drain voltage of 0.4 V (before the kink), 1.4 V (at the kink), and 3 V (beyond the kink). (Adapted from Chen et al. [25].)

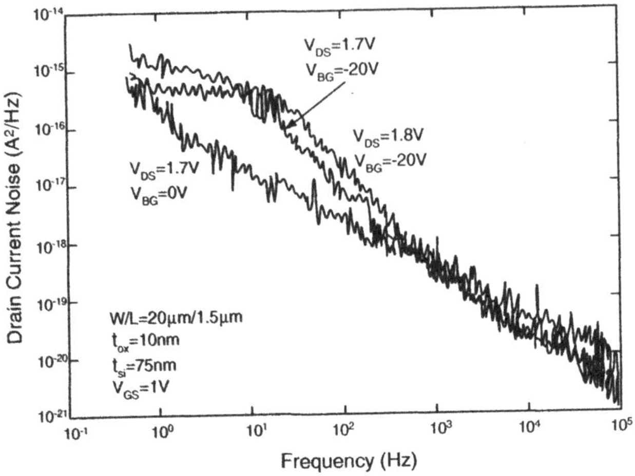

Fig. 4.38 Drain current noise spectral density versus frequency of the FD SOI NMOS device with a front gate oxide of 100 Å, a thin film of 750 Å, a channel width of 20 μm, and a channel length of 1.5 μm, biased at the gate-source voltage of 1 V and (1) the drain-source voltage of $V_{DS} = 1.7$ V and the back gate voltage of $V_{BG} = -20$ V, (2) the drain-source of voltage $V_{DS} = 1.8$ V and the back gate voltage of $V_{BG} = -20$ V, and (3) the drain-source voltage of $V_{DS} = 1.7$ V and the back gate voltage of $V_{BG} = 0$ V. (Adapted from Jin et al. [26].)

Figure 4.37 shows the drain current noise power spectral density versus the frequency of the PD SOI NMOS device with a front gate oxide of 115 Å, biased at the gate voltage of 1 V and the drain voltage of 0.4 V (before the kink), 1.4 V (at the kink), and 3 V (beyond the kink) [25]. As shown in this figure, biased before the kink at the drain voltage of 0.4 V and beyond the kink at 3 V (not in the noise overshoot region), its noise spectrum is similar to the bulk one, which has the 1/f frequency dependency. At the kink with the drain voltage equal to 1.4 V, the noise spectrum shows the Lorentzian-like profile.

The Lorentzian-like profile of the drain current noise spectrum at the kink for the PD SOI NMOS devices also exists for the FD device. When the back gate bias of an FD SOI device is sufficiently negative to cause the carrier accumulation in the thin film, a Lorentzian-like profile can be seen in the noise spectrum. Figure 4.38 shows the drain current noise spectral density versus frequency of the FD SOI NMOS device with a front gate oxide of 100 Å, a thin film of 750 Å, biased at the gate-source voltage of 1 V and (1) the drain-source voltage of $V_{DS} = 1.7$ V and the back gate voltage of $V_{BG} = -20$ V, (2) the drain-source voltage of $V_{DS} = 1.8$ V and the back gate voltage of $V_{BG} = -20$ V, and (3) the drain-source voltage of $V_{DS} = 1.7$ V and the back gate voltage of $V_{BG} = 0$ V [26]. As shown in Fig. 4.38, when the body-gate voltage is equal to $V_{BG} = 0$ V, the thin film of the PD SOI NMOS device is fully depleted. Therefore the noise spectrum is conventional 1/f frequency dependent. When the body-gate voltage is equal to $V_{BG} = -20$ V, at the back surface in the thin film of the PD SOI NMOS device, accumulation of holes appears. As a result, noticeable kink effects due to floating body effect appear in the

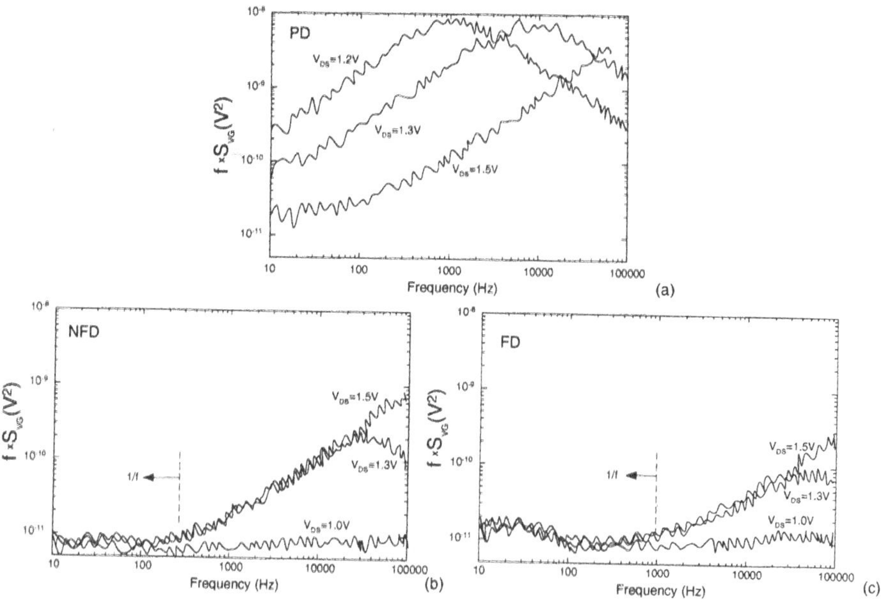

Fig. 4.39 Equivalent gate noise power spectral density multiplied by the frequency versus frequency of (a) PD, (b) nearly FD (NFD), and (c) FD SOI NMOS devices with a channel length of 0.45 μm, biased at $V_G - V_T = 0.2$ V and with its body floating. (From Tseng et al. [27]. ©1998 IEEE.)

DC current. As shown in Fig. 4.38, when the drain voltage is around the kink onset voltage, the Lorentzian-type noise spectrum shows up.

Low-frequency noise is important in RF analog circuits. Due to the floating body, noise overshoot exists for the PD SOI devices. Noise overshoot does not exist just for the PD devices. It also occurs to the FD devices or the SOI devices with the body-tied-to-source structure at high frequency. At high frequency, due to the depletion capacitance of the source/body junction in the SOI device, the generated AC hole current becomes larger than the leakage current of the source/body diode. Therefore, a net charge current due to the AC hole current results in the frequency dependent AC kink effects, which trigger the high-frequency noise overshoot behavior. Figure 4.39 shows the equivalent gate noise power spectral density multiplied by the frequency versus frequency of (a) PD, (b) nearly FD (NFD), and (c) FD SOI NMOS devices with a channel length of 0.45 μm, biased at $V_G - V_T = 0.2$ V and with its body floating [27]. As shown in this figure, the low-frequency noise overshoot is noticeable for the PD device. In contrast, for NFD and FD devices, the low-frequency noise spectrum shows the 1/f frequency dependence shape. At high frequency, due to AC kink effects, the noise overshoot behavior appears.

SOI devices may have Lorentzian-like noise overshoot in the low-frequency noise power spectrum. For PD SOI devices, due to kink effects, the noise overshoot is more noticeable as compared to NFD and FD devices. Figure 4.40 shows the equivalent gate

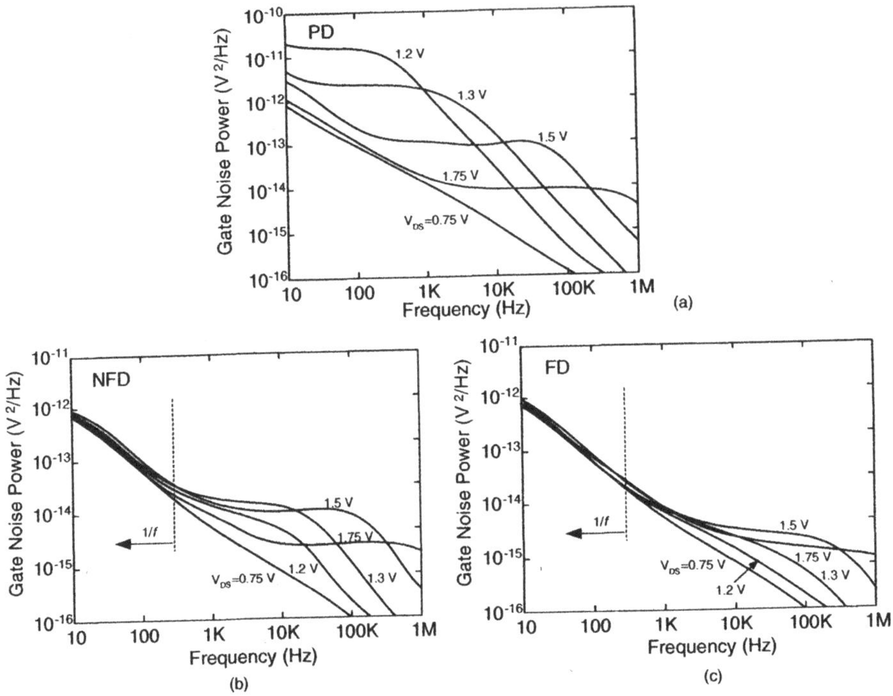

Fig. 4.40 Equivalent gate noise power spectral density versus frequency of (a) PD, (b) nearly FD (NFD), and (c) FD SOI NMOS devices with a channel length of 0.45 μm, biased at $V_G - V_T = 0.2$ V and with its body floating, showing the corner frequencies. (From Tseng et al. [28]. ©1999 IEEE.)

noise power spectral density versus frequency of (a) PD, (b) nearly FD (NFD), and (c) FD SOI NMOS devices with a channel length of 0.45 μm, biased at $V_G - V_T = 0.2$ V and with its body floating, showing the corner frequencies [28]. As shown in Fig. 4.40(a), Lorentzian-like noise spectrum can be seen from the noise spectrum with the $1/f$ frequency dependence at very low frequency. Over the flat noise overshoot region, via the corner frequency the spectrum extends to another $1/f^2$ frequency-dependent region. With the device biased at a higher drain-source voltage, its corner frequency becomes higher. In NFD and FD devices, the neutral body region shrinks or disappears, thus the disabled kink effects eliminate the low-frequency noise overshoot region as shown in Fig. 4.40(b) and (c). As a result, from noise consideration, NFD and FD devices are more suitable for RF circuits. However, at a high drain-source voltage (V_{DS}), drain-bias dependent noise still occurs at high frequency. At the drain-source voltage of 1.5 V, the FD device has a corner frequency of $\sim$ 100 kHz with the noise overshoot.

In addition to the 1/f noise, which is not dependent on bias, generation/ recombination (G/R) noise is also important in SOI devices. Figure 4.41 shows the equivalent gate noise power spectral density versus the gate-source overdrive volt-

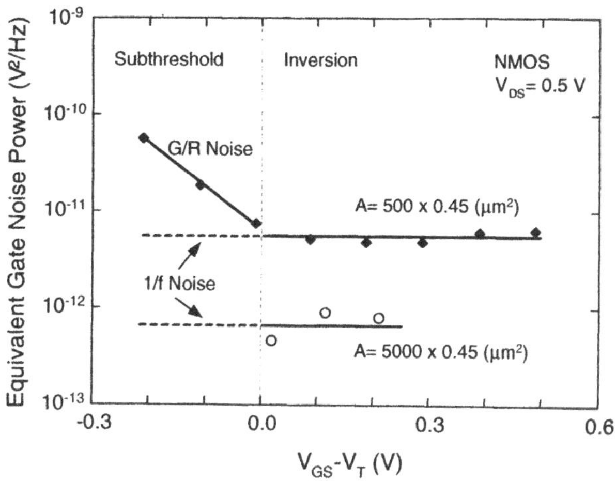

Fig. 4.41 Equivalent gate noise power spectral density multiplied by frequency versus gate-source overdrive voltage ($V_{GS} - V_T$) of the SOI NMOS device with gate areas of $500 \times 0.45\ \mu m^2$ and $5000 \times 0.45\ \mu m^2$, biased at the drain-source voltage of $V_{DS} = 0.5$ V, showing the generation/recombination noise in addition to the 1/f noise in the subthreshold and the inversion regions. (Adapted from Babcock et al. [29].)

age ($V_{GS} - V_T$) of the SOI NMOS device with gate areas of $500 \times 0.45\ \mu m^2$ and $5000 \times 0.45\ \mu m^2$, biased at the drain-source voltage of $V_{DS} = 0.5$ V, showing the generation/recombination noise in addition to the 1/f noise, in the subthreshold and the inversion regions [29]. For the SOI NMOS device, after inversion of channel occurs, the noise of the device is mainly 1/f noise, which is not dependent on the bias. The 1/f noise is a function of the gate area. A larger gate area leads to a smaller 1/f noise. When the device biased in the subthreshold region, along with the decrease in the gate-source overdrive voltage ($V_{GS} - V_T$), the generation/recombination (G/R) noise dominates.

Figure 4.42 shows the equivalent power spectral density multiplied by frequency versus frequency of SOI (a) NMOS and (b) PMOS devices, with a gate area of $500 \times 0.45\ \mu m^2$, biased at the drain-source voltage of $V_{DS} = 0.5$ V and the gate-source voltages of $V_{GS} = 0.3$ V (subthreshold region) and $V_{GS} = 1$ V (saturation region) [29]. As shown in this figure, for the SOI devices biased in the saturation region, 1/f noise dominates. In addition, the 1/f noise is not dependent on the frequency. For the SOI devices biased in the subthreshold region, noise with a Lorentzian-shaped hump at 3.5 Hz and 6 kHz (for the NMOS device), which is due to the existence of the generation/recombination noise centers, can be seen. Therefore, in the subthreshold region, the generation/recombination noise is more important than the 1/f noise. In addition, in the subthreshold region, for the SOI device operating at low frequency, noise overshoot may appear.

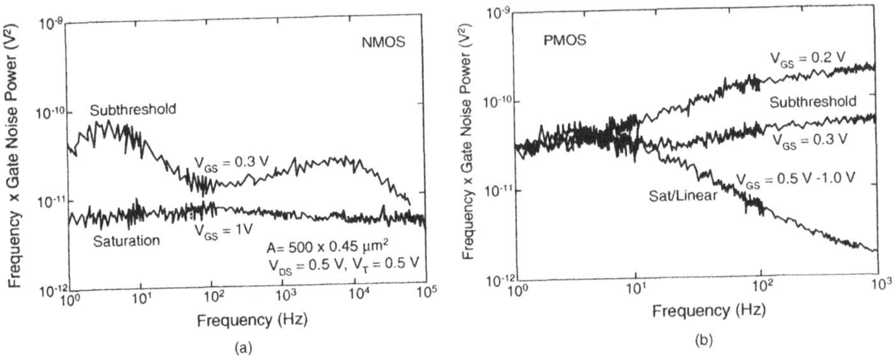

Fig. 4.42 Equivalent power spectral density multiplied by frequency versus frequency of SOI (a) NMOS and (b) PMOS devices, with a gate area of $500 \times 0.45\ \mu m^2$, biased at the drain-source voltage of $V_{DS} = 0.5$ V and the gate-source voltages of $V_{GS} = 0.3$ V (subthreshold region) and $V_{GS} = 1$ V (saturation region). (From Babcock et al. [29]. ©1998 IEEE.)

4.7 SELF-HEATING

Due to the existence of the buried oxide beneath the thin film, the thermal conductance of the SOI devices is small, which brings in a poor heat dissipation capability. The rise in the lattice temperature of the SOI devices, which is called self-heating, degrades the drain current driving capability and causes reliability problems. From the circuit design point of view, the influence of self-heating in digital circuits is small [4]. For digital circuits, only during rise and fall transients do the devices conduct currents. During these very short transients, the devices are conducting current not in the full scale of the power supply voltage. Therefore, the power consumption in these devices is small. The rise in the internal lattice temperature of these devices is not significant.

Figure 4.43 shows the lattice temperature of a nearly FD (NFD) SOI NMOS device with a channel length of $0.2\ \mu m$ and with its body floating, used in a CMOS NAND ring oscillator, oscillating at a frequency of 2.35 GHz, at $V_{DD} = 1.8$ V [30]. When the NMOS device turns on, its output switches from high to low. The lattice temperature of this NMOS device rises quickly due to thermal resistance and thermal capacitance. When the NMOS device turns off, its lattice temperature falls slowly, since in the NAND gate, the actual time for the switching transient is very short. Thus, the consumed power is small. In each cycle, the NMOS device in the NAND gate consumes ~ 0.13 mW. After a steady-state condition has been reached for the device after switching for several times, the variation in the lattice temperature is only 0.7 K. Therefore, the dynamic self-heating has a small influence in the performance of a digital circuit. Even when the SOI devices are scaled down toward the deep-submicron regime with the device size and the supply voltage is scaled down by a factor of s, the power consumption of the down-scaled devices becomes $1/s^2$ times smaller. Since the area of the device available for heat dissipation also becomes $1/s^2$ times smaller, the thermal resistance of the device becomes s^2 times larger. Therefore, the increase in the thermal resistance of the device has been compensated

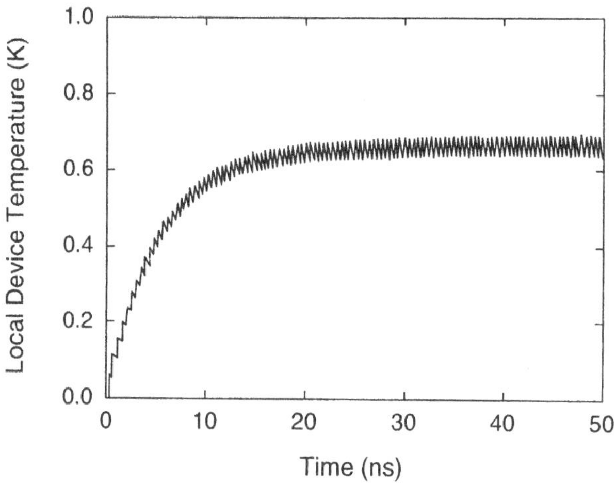

Fig. 4.43 Lattice temperature of a nearly FD (NFD) SOI NMOS device with a channel length of 0.2 μm and with its body floating, used in a CMOS NAND ring oscillator, oscillating at a frequency of 2.35 GHz, at $V_{DD} = 1.8$ V. (Adapted from Workman et al. [30].)

by the decrease in the dissipated power. For deep-submicron SOI CMOS devices, the effect of self-heating on the digital circuits is still not significant.

Figure 4.44 shows the lattice temperature of the PD NMOS device during switching of a 5-stage and a 51-stage ring oscillator using PD SOI NMOS devices with a channel length of 0.7 μm [31]. As shown in this figure, for the ring oscillator with more stages, each cycle becomes smaller. The conducting cycle to cause self-heating of the PD SOI NMOS becomes shorter, hence self-heating is less. No matter how many stages are in the ring oscillator due to self-heating, the variation in the lattice temperature is only several kelvins, which can be neglected for digital circuits.

The influence of the self-heating of SOI devices in the performance of digital circuits has been analyzed. The influence of self-heating in the performance of the analog circuits is described here. Figure 4.45 shows (a) the output voltage and the lattice temperature of the driver during the operation of the class-A amplifier made of PD SOI NMOS devices with a channel length of 0.7 μm and with a 1 kHz sine wave imposed at its input and (b) the frequency response of the amplifier gain considering self-heating [31]. As shown in this figure, during the operation, since the output of the amplifier is continuously changing due to the corresponding change in the power consumption, the lattice temperature of the PD SOI driver device also changes with variation in the output voltage. Due to self-heating, the gain of the amplifier becomes smaller as compared to the case without self-heating.

Figure 4.46(a) shows the 7-bit digital-to-analog (DAC) circuit using a binary-weighted current source approach, made of PD SOI NMOS devices. For a binary-weighted current source not use, which is referred to the zero of that bit, its drain is connected to V_{DD}. If a binary-weighted current source is selected, which implies logic-1 of that bit, its drain is connected to the output resistive load. The analog

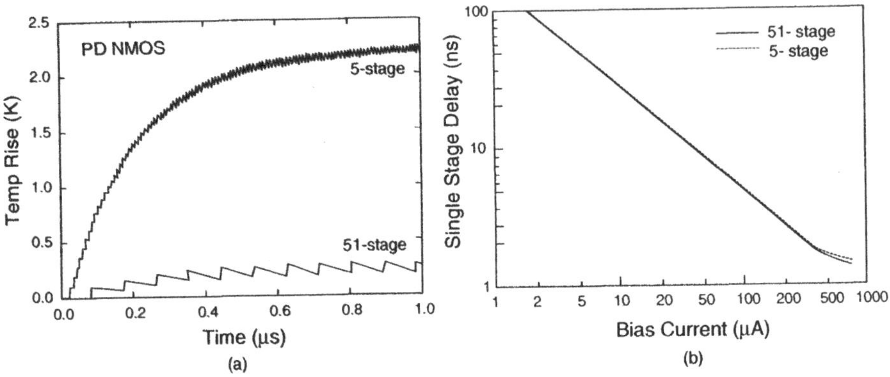

Fig. 4.44 Lattice temperature of the PD NMOS device during switching of a 5-stage and a 51-stage ring oscillator using PD SOI NMOS devices with a channel length of 0.7 μm. (Adapted from Tenbroek et al. [31].)

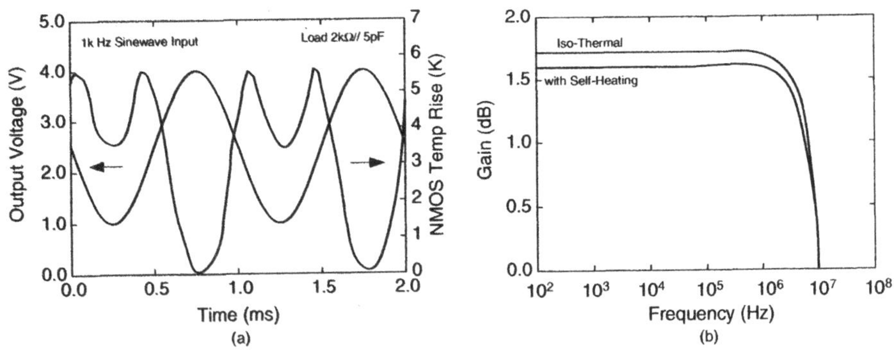

Fig. 4.45 (a) Output voltage and lattice temperature of the driver during the operation of the class-A amplifier made of PD SOI NMOS devices with a channel length of 0.7 μm and with a 1 kHz sine wave imposed at its input. (b) Frequency response of the amplifier gain considering self-heating. (Adapted from Tenbroek et al. [31].)

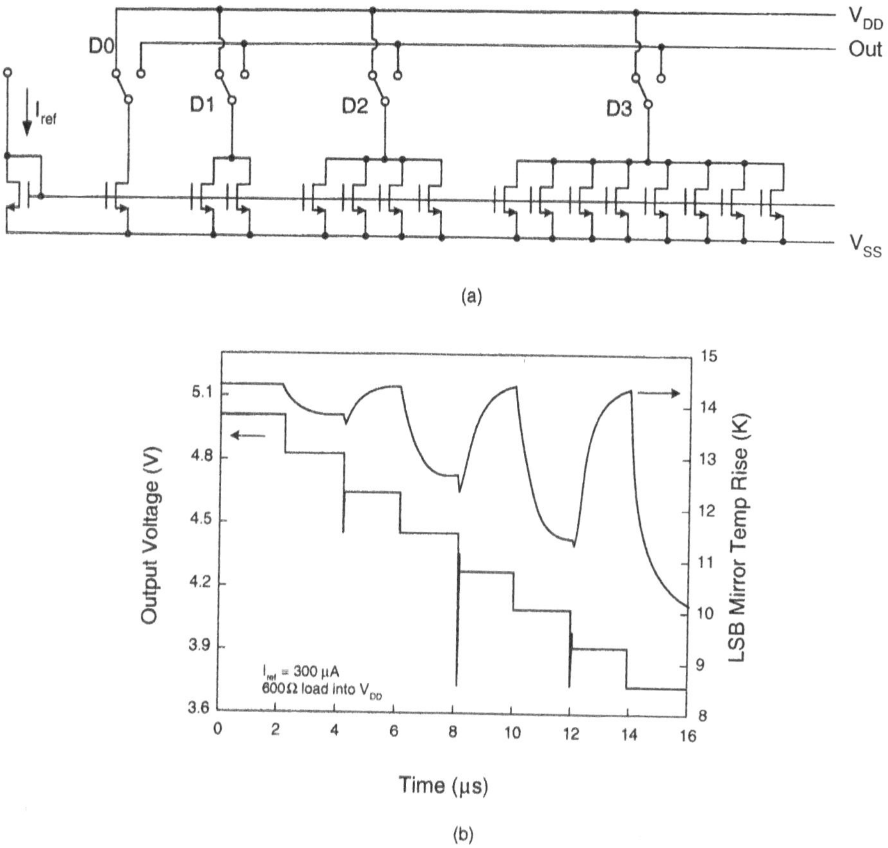

Fig. 4.46 (a) 7-bit DAC circuit made of PD SOI NMOS devices. (b) Output voltage and lattice temperature of the LSB current source device during the operation of increasing the digital input. (From Tenbroek et al. [31]. ©1997 IEEE.)

output voltage can be determined by controlling the connection of the drains of the binary weighted current sources depending on the input digital data. Figure 4.46(b) shows the output voltage and the lattice temperature of the least-significant-bit (LSB) current source during the operation of increasing the digital input [31]. As shown in this figure, initially, the drain of the LSB current source is connected to V_{DD}, hence its lattice temperature is high due to its dissipated power. When the LSB current source is selected, its drain is connected to the output resistive load. Due to the voltage drop over the output resistive load, its drain voltage drops, which brings in a decrease in its dissipated power. Therefore, its lattice temperature falls accordingly. Due to the variation in the lattice temperature of the LSB current source, the step size of the output voltage of the DAC becomes nonlinear.

The thermal effect due to self-heating of SOI devices influences the performance of the related analog circuits substantially. The increased lattice temperature due

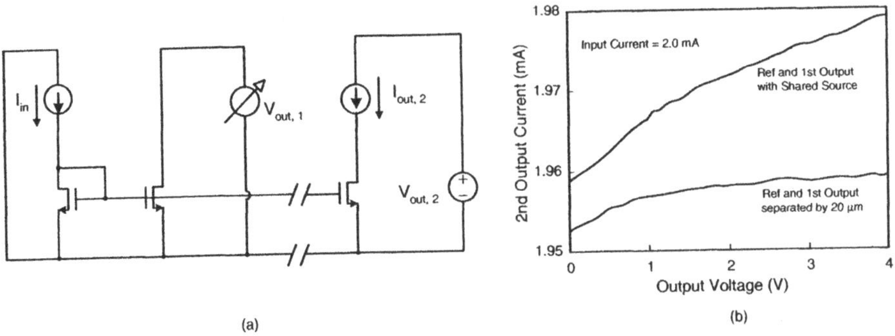

Fig. 4.47 (a) The current mirror with two outputs, using PD SOI NMOS devices. (b) Output current of the V_{out2} device versus output voltage (V_{out1}) of the DAC current mirror. (Adapted from Redman-White et al. [32].)

to the internal dissipated power may lead to mismatch problems for analog circuits. Figure 4.47(a) shows the current mirror with two outputs V_{out1}/V_{out2}, using PD SOI NMOS devices. The V_{out1} is close to the reference device and the V_{out2} is far away from the reference. The reference device offers a fixed current source. The drain of the V_{out1} device is connected to a variable voltage source and the drain of the V_{out2} device is connected to a fixed voltage source. If the drain voltage of the V_{out1} device is changed by the variable voltage source, the dissipated power of the V_{out1} device changes. Hence, the lattice temperature and the drain current of the V_{out1} device change accordingly. The lattice temperature of the reference device also changes accordingly through thermal coupling from the V_{out1} device to the reference device. The gate voltage of the reference device changes and hence the drain current of the V_{out2} device changes correspondingly. Figure 4.47(b) shows the output current of the V_{out2} device versus the output voltage V_{out1} of the current mirror [32]. Two cases have been studied. One is with its V_{out1} device adjacent to the reference with a shared source. The other one is with its V_{out1} device separated from the reference device by 20 μm. For the case with the V_{out1} device adjacent to the reference device with a shared source, the self-heating of the V_{out1} device is thermally coupled to the reference device, which is reflected in the output current. A change in the V_{out1} voltage leads to change in the V_{out2} current due to the thermal coupling from the V_{out1} device to the reference device. In contrast, for the case with the V_{out1} device separated from the reference device by 20 μm, the change in the V_{out1}, which leads to the change in lattice temperature, cannot be coupled to the reference device as well as compared to the case with the V_{out1} adjacent to the reference device.

4.8 ESD CIRCUITS

As described in the previous chapter, as for bulk CMOS circuits, electrostatic discharge (ESD) protection circuits to the input/output pads are important for SOI CMOS

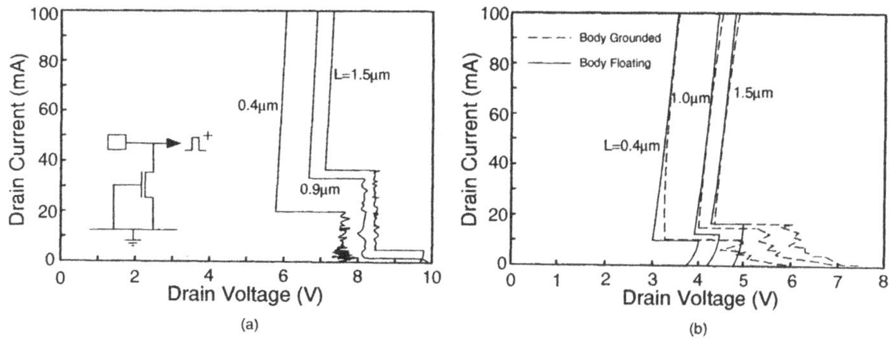

Fig. 4.48 Breakdown/snapback characteristics of (a) bulk and (b) SOI NMOS devices with its body floating and grounded, for various channel lengths. (Adapted from Chan et al. [33].)

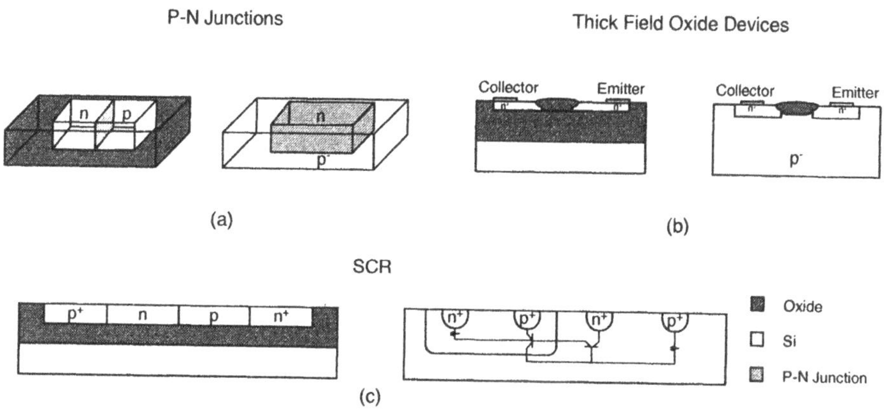

Fig. 4.49 Compared to bulk, the challenges of ESD protection circuits for the SOI technology: (a) no large-area vertical PN junction, (b) no thick field oxide (TFO), and (c) no large-area pnpn silicon controlled rectifier (SCR) structure. (Adapted from Raha et al. [35].)

circuits. From the breakdown/snapback characteristics as shown in Fig. 4.48, both bulk and SOI NMOS devices have a similar ESD performance [33]. The main difference in the ESD performance between bulk and the SOI devices is that the second snapback of the SOI device occurs at a smaller holding voltage and a smaller drain current. Due to the smaller thermal conductance from the buried oxide layer, a large heat dissipation results in a higher joule heating condition to cause the negative resistance, which brings in the second snapback behavior. The higher joule heating is also evidence of the smaller second snapback current.

The ESD condition of the SOI devices is different from that of the bulk devices. In general, the ESD protection circuits designed for the bulk CMOS technology are not applicable for the SOI devices due to the buried oxide layer used for isolation, which brings in different heat dissipation problems [34].

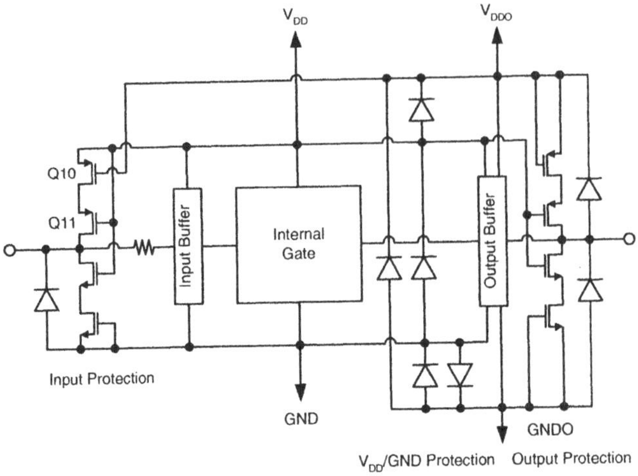

Fig. 4.50 SOI ESD protection circuit based on a 0.25 μm SOI CMOS technology using double supply voltages of not larger than $V_{DD} \leq 2$ V and $V_{DDO} = 3.3$ V. (Adapted from Ohtomo et al. [36].)

As shown in Fig. 4.49, compared to bulk, ESD protection circuits for the SOI technology are difficult to build because of (a) no large-area vertical pn junction, (b) no thick field oxide (TFO), and (c) no large-area pnpn silicon control rectifier (SCR) structure [35]. As shown in this figure, for the bulk technology, large-area vertical pn junction is easily available for ESD protection circuits. In contrast, for SOI technology, due to the restriction from the thin-film thickness, the cross-section of the pn junction is small and no vertical large-area pn junction exists. The SOI pn junction used for the ESD protection circuits needs to sustain a much higher power density. Thus the heating effect is more serious, which lowers the effectiveness. For the bulk technology, the thick field oxide (TFO) devices are available for implementing ESD protection circuits. In contrast, for SOI technology, due to the thickness of the thin film, TFO devices are not available. For bulk technology, large-area pnpn silicon control rectifier (SCR) structures are common. For SOI technology, if SCR structures are to be implemented for the ESD protection circuits, it takes a much larger-area, which may not be practical. >From the above reasoning, the ESD protection circuits created for bulk technology cannot be used for SOI technology. ESD protection circuits for the SOI technology are described below.

Figure 4.50 shows an SOI ESD protection circuit based on a 0.25 μm SOI CMOS technology using the supply voltages of not larger than $V_{DD} \leq 2$ V and $V_{DDO} = 3.3$ V [36]. As shown in this figure, by taking advantage of the low resistance path provided by the pn junction, the positive or the negative ESD current could be bypassed via the diode from the input pad to V_{DD} and ground to increase the ESD failure voltage. Double power supply lines– V_{DD}/V_{DDO} and GND/GNDO are used. V_{DD} equal to or less than 2 V is used for the internal circuits. V_{DDO} of 3.3 V is used for the external circuit. Between power lines V_{DD} and V_{DDO} and between ground

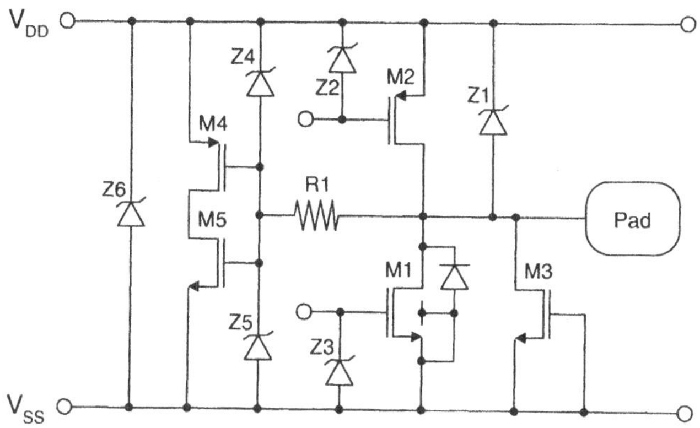

Fig. 4.51 SOI ESD protection circuit. (Adapted from Smith et al. [37].)

lines GND and GNDO diodes have been added to increase the effectiveness of the ESD protection circuit. The adoption of the diodes in the ESD protection circuits provides advantages in a concise layout area since the diodes can merge directly with other SOI devices directly.

Figure 4.51 shows the SOI ESD protection circuit [37]. As shown in this figure, a grounded-gate NMOS device M3, a Zener diode Z1, a bus Zener diode Z6, and the output driver device with a drain diode M1 constitute most of the protection path against the ESD positive and negative stresses. Zener diodes Z5, Z4, Z3, and Z2 are used to protect the gate oxide of devices M5, M4, M1, and M2, respectively. For the pad-to-V_{SS} positive stress, due to a large contact-to-gate spacing for the grounded-gate NMOS device M3 to increase heat dissipation, a good current path has been created to relieve the ESD current. For the pad-to-V_{SS} negative stress, device M1 has been used to provide a current path. Device M1 with its source/body tied together provides a high breakdown voltage. In addition, the parallel drain diode of the M1 device also offers a small resistance, which can be used to bypass the ESD current. For the pad-to-V_{DD} positive stress, the forward-bias Zener diode Z1 has been used. In contrast to the high-resistive PMOS device M2, Zener diode Z1 can provide a large diode edge width such that a large portion of the ESD current flows via Z1 instead of M2. Therefore, with Z1 the ESD protection circuit has provided protection against pad-to-V_{DD} positive and negative stresses. In addition, the extra Zener diode Z6 and device M1 constitute an auxiliary ESD current path.

In addition to the grounded-gate NMOS (GGNMOS) approach described before, SOI ESD protection circuits can also be implemented by the gated double-diode networks using T-shaped SOI CMOS, PMOS, and NMOS devices as shown in Fig. 4.52. As shown in this figure, the floating body contact of a PD SOI MOS device has been used as a cathode or an anode of a diode to realize the SOI diode-like structure for ESD protection [38]. This gated double-diode network can be implemented in CMOS, PMOS, and NMOS devices. For the CMOS gated double-diode network, the drain/source end of the NMOS/PMOS devices is connected to the pad to form the

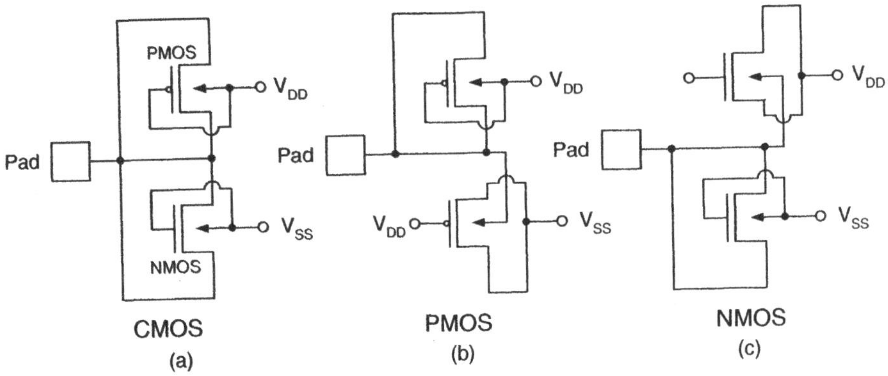

Fig. 4.52 Gated double-diode networks using T-shaped body-contacted SOI (a) CMOS, (b) PMOS, and (c) NMOS devices. (Adapted from Voldman et al. [38].)

cathode/anode of the diode. In addition, the body is connected to the gate and the power supply V_{SS}/V_{DD} to form the anode/cathode of the diode. For the PMOS gated double-diode network, the diode-to-V_{DD} path is realized by connecting the source and the drain of the upper section to the pad to form the anode of the diode, and the body and the gate are connected to the V_{DD} to form the cathode. In the lower section, the diode-to-V_{SS} path is formed by connecting the source/drain implant region to V_{SS} and the n-type body is connected to the input pad with the gate connected to V_{DD}. For the NMOS gated double-diode network, it is similar to the PMOS one except that the connection for the diode-to-V_{DD} path and the diode-to-V_{SS} path is reversed as shown in the figure. The gated double-diode network provides a low resistive path to bypass the ESD current– the effectiveness of the ESD protection is robust. When ESD occurs, the surface channels of the related devices also exist. In addition to the current path from the body to source/drain, the current also flows from the body to the surface channel. Due to the increase in the current paths, the resistance of the ESD current path in the gated double-diode networks is small. Thus a larger ESD current can be tolerated, which are similar to lubistor described in the Chapter 3.

Figure 4.53 shows (a) the cross-section of the SOI lubistor and (b) the SOI ESD protection circuit and the receiver network[39]. As shown in this figure, in this ESD protection circuit the SOI lubistor configuration has been used to replace the SOI diode. The SOI lubistor, which has better ESD properties, has been used to reduce the SOI ESD problems. Between the pad and the 3.3 V power supply line and between the pad and V_{SS}, ESD protection has been implemented. Between the 2.5 and 3.3 V power lines, a diode string composed of lubistors has been used to provide ESD protection. In the receiver network, a floating-body NMOS pass transistor has been used to avoid early failure to increase the ESD protection level. Using this approach, the ESD robustness exceeds 4 kV.

In this section, gate-coupled device, Zener diode, SOI diode, and SOI diode string configuration for ESD protection circuits have been described. Another approach using a DTMOS SOI device for ESD protection circuits is described. Figure 4.54

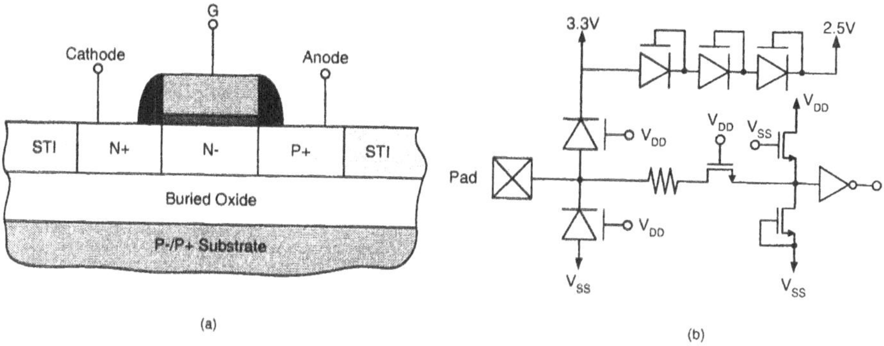

Fig. 4.53 (a) Cross-section of the SOI lubistor and (b) SOI ESD protection circuit and receiver network. (Adapted from Voldman et al. [39].)

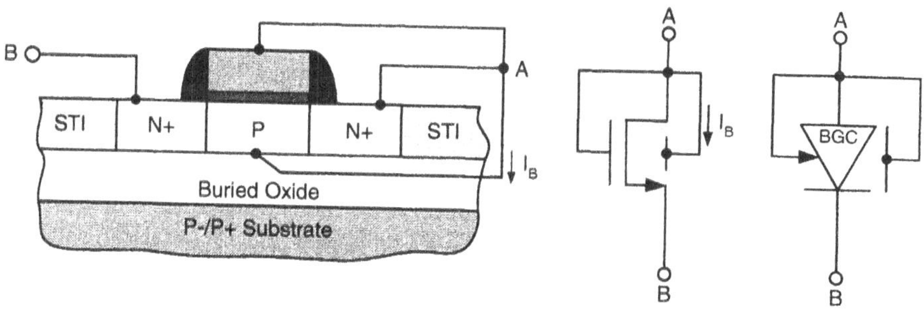

Fig. 4.54 Body Gate-coupled (BGC) DTMOS SOI diode for ESD protection circuits. (Adapted from Voldman et al. [40].)

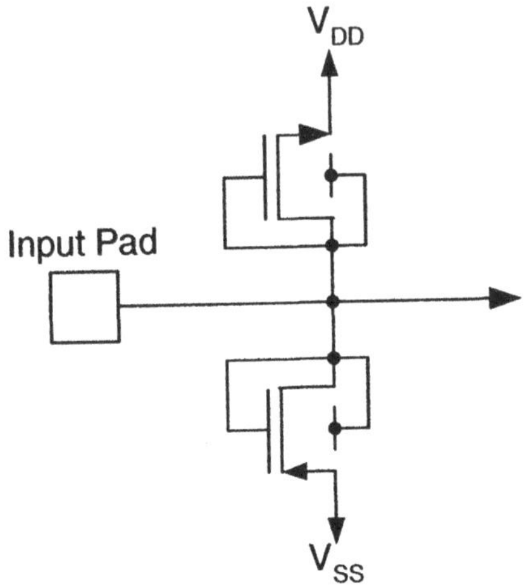

Fig. 4.55 BGC SOI diode ESD network. (Adapted from Voldman et al. [40].)

shows the body gate-coupled (BGC) DTMOS diode for ESD protection circuits [40]. The BGC DTMOS diode is formed by connecting the body and the drain to the gate to form the anode and the source served as the cathode. The advantages of the BGC DTMOS diode are that it can be regarded as a parallel combination of the MOS device and the diode having both properties. If the node with the gate/body/drain is biased positively, the threshold voltage of the device becomes smaller, thus the MOS device turns on earlier to increase its current driving capability. When the positive bias increases, the body diode between the body and the source is forward biased to generate a substantial amount of current. In the ESD environment, the body/gate/drain node of the BGC DTMOS SOI diode is connected to the input pad and its source node is grounded. When the pad voltage increases consistently toward the positive direction, the body-source diode may provide a low resistive path to bypass the current from the pad to the ground. In addition, the body-source voltage also lowers the threshold voltage of the SOI device to cause an early turn-on of the device. Therefore, a large driving current leads to an increased ESD robustness.

Figure 4.55 shows a body gate-coupled (BGC) SOI diode ESD network [40]. As shown in this figure, the BGC NMOS SOI diode is placed between the power supply (V_{DD}) line and the input pad and the BGC PMOS SOI diode is arranged between the input pad and the power supply V_{SS} line. This ESD network formed by the BGC CMOS SOI diodes provides several current conducting paths. The BGC NMOS SOI diode with its built-in DTMOS and npn bipolar properties provides the ESD current path from the pad to V_{DD}. The BGC PMOS SOI diode with its built-in DTMOS and pnp bipolar properties provides the ESD current path from V_{SS} to the pad. When the external and the internal operating voltages in the SOI CMOS circuits and the related ESD protection circuits are not identical, a diode string network formed by the BGC SOI diodes can serve as a mixed-voltage interface ESD circuit. This interface ESD circuit can be placed between the input pad and the power supply or between different

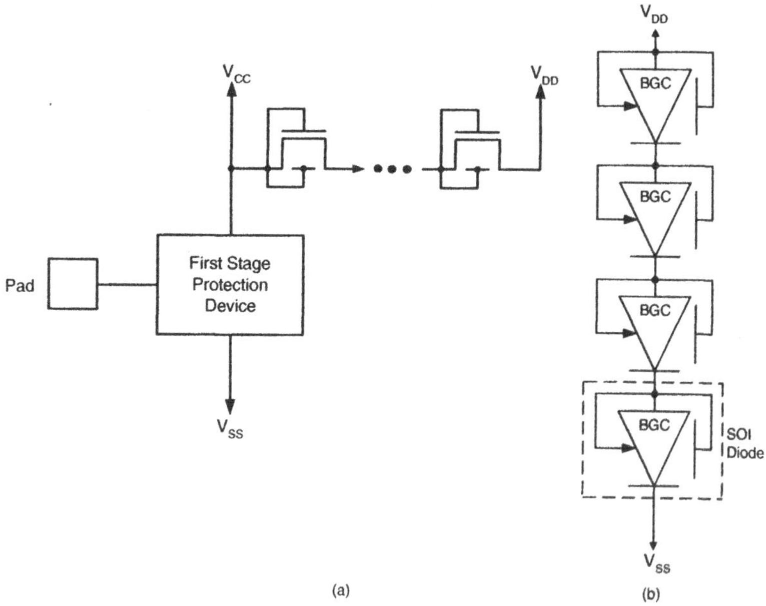

Fig. 4.56 (a) BGC SOI diode string for mixed voltage I/O interface. (b) BGC SOI diode string between V_{DD} and V_{SS} power lines. (Adapted from Voldman et al. [40].)

power supply lines. In addition, it can be placed between the power supply line and the chip substrate as shown in Fig. 4.56. In Fig. 4.56(a), in the BGC diode string circuit, the source of the previous diode is connected to the drain/gate/body of the current diode and the source of the current diode is linked to the drain/gate/body of the next diode to form a string.

4.9 SYSTEM-ON-A CHIP (SOC) TECHNOLOGY

Owing to the consistent progress of the processing technology and the needs of the integrated circuits and systems, system on a chip (SOC) is the future trend on VLSI. The SOI technology offers a feasible technology for integrating SOC. Since SOI technology could provide low-voltage, low-power, and high-speed capabilities for circuit designs, which are helpful for implementing portable systems using a power supply voltage under 1 V. For low-voltage, low-power systems using a single battery or a solar cell power supply, due to the superior isolation of the SOI devices, SOI technology can provide a high integration density capability, which is suitable for integrating mixed signal function logic and memory circuits to achieve a personal portable system on a chip.

For integrating communication systems, an advanced SOI CMOS technology is already capable of providing a 1 V amplifier with an f_T of 19 GHz and an f_{max} of 27 GHz [41]. With this kind of performance, GaAs hybrid circuits can be replaced

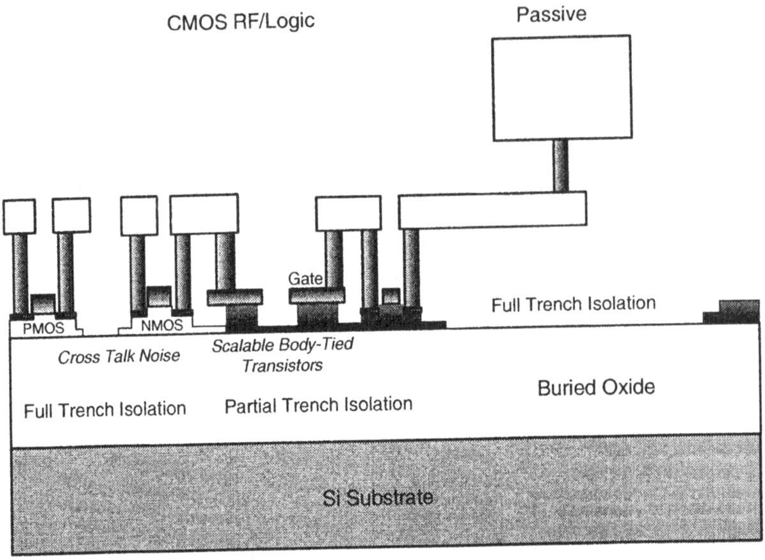

Fig. 4.57 Cross-section of the 0.18 μm SOI CMOS technology with digital, embedded RF/analog, and SOC capabilities. (Adapted from Maeda et al. [42].)

by SOI CMOS circuits to achieve 2 GHz L-Band monolithic microwave integrated circuits (MMIC) with a low-cost SOC capability. In addition to communication systems, combining the SOI technology with the smart sensor using the microelectro-mechanical system (MEMS) technology, we can produce the MEMS system on a chip.

Figure 4.57 shows the cross-section of 0.18 μm SOI CMOS technology with digital, embedded RF/analog, and SOC capabilities [42]. In this technology, hybrid trench isolation and high resistive substrate (HRS) have been adopted. As shown in the figure, the SOI layer is grown on the top of the HRS. Partial trench isolation and full trench isolation have been used. Full trench isolation has been used in the transistor area and the passive portion. This SOI technology is suitable for integrating digital circuits, embedded RF/analog circuits, and SOC. Owing to reduced junction capacitances, SOI technology is useful to integrate RF circuits with an improved power loss, gain, and noise figure. The adoption of HRS also reduces the signal loss due to the electric field penetrating the buried oxide into the substrate at high frequency. In this SOI technology, partial trench isolation has been used to form the body-tied structure to remove floating body and to reduce noise overshoot such that the stability of AC conductance can be achieved. In addition, it can be used to reduce hot carrier effect due to impact ionization. On the other hand, full trench isolation and HRS can be combined to provide a high-quality on-chip inductor. In addition to low-voltage low-power high-speed digital systems, the SOI CMOS technology could also integrate embedded RF/analog circuits with low-noise, stable, linear, high-quality on-chip inductors for wireless communication portable system designs.

4.10 SUMMARY

In this chapter, basic knowledge of SOI CMOS circuits has been described. Starting from the basic circuit issues, the floating body effects on the performance of the SOI CMOS circuits have been explained, followed by the low-voltage SOI CMOS circuits, SOI dynamic-threshold MOS (DTMOS) circuits, and SOI multithreshold MOS (MTMOS) circuits. Then, noise and self-heating problems of SOI CMOS circuits have been analyzed. Finally, the SOI ESD circuits and the SOI system-on-a chip (SOC) technology have been presented.

REFERENCES

1. J. M. Stern, P. A. Ivey, S. Davidson, and S. N. Walker, "Silicon-On-insulator (SOI): A High Performance ASIC Technology," *CICC Dig.*, 9.2.1–9.2.4 (1992).
2. J. P. Colinge, J. P. Eggermont, D. Flandre, P. Francis, and P. G. A. Jespers, "Potential of SOI for Analog and Mixed Analog-Digital Low-Power Applications," *ISSCC Dig.*, 194–195 (1995).
3. M. L. Alles, "Thin-Film SOI Emerges," *IEEE Spectrum*, **34**(6), 37–45 (1997).
4. G. G. Shahidi, A. Ajmera, F. Assaderaghi, R. J. Bolam, H. Hovel, E. Leobandung, W. Rausch, D. Sadana, D. Schepis, L. F. Wagner, L. Wissel, K. Wu, and B. Davari, "Device and Circuit Design Issues in SOI Technology," *CICC Dig.*, 339–346 (1998).
5. F. Assaderaghi, G. G. Shahidi, M. Hargrove, K. Hathorn, H. Hovel, S. Kulkarni, W. Rausch, D. Sadana, D. Schepis, R. Schulz, D. Yee, J. Sun, R. Dennard, and B. Davari, "History Dependence of Non-Fully Depleted (NFD) Digital SOI Circuits," *Symp. VLSI Tech. Dig.*, 122–123 (1996).
6. F. Assaderaghi, G. G. Shahidi, L. Wagner, M. Hsieh, M. Pelella, S. Chu, R. H. Dennard, and B. Davari, "Transient Pass-Transistor Leakage Current in SOI MOSFETs," *IEEE Elec. Dev. Let.*, **18**(6), 241–243 (1997).
7. P.-F. Lu, C.-T. Chuang, J. Ji, L. F. Wagner, C.-M. Hsieh, J. B. Kuang, L. L.-C. Hsu, M. M. Pelella Jr., S.-F. S. Chu, and C. J. Anderson, "Floating-Body Effects in Partially Depleted SOI CMOS Circuits," *IEEE J. Sol. St. Ckts.*, **32**(8), 1241–1252 (1997).
8. R. Puri and C.-T. Chuang, "Hysteresis Effect in Pass-Transistor-Based, Partially Depleted SOI CMOS Circuits," *IEEE J. Sol. St. Ckts.*, **35**(4), 625–631 (2000).
9. R. Rajsuman, "Implementation of Switch Network Logic in SOI," *IEEE J. Sol. St. Ckts.*, **25**(3), 874–877 (1990).

10. M. M. Pelella, J. G. Fossum, M.-H. Chiang, G. O. Workman, and C. R. Tretz, "Analysis and Control of Hysteresis in PD/SOI CMOS," *IEDM Dig.*, 831–834 (1999).

11. W. Redman-White, B. M. Tenbroek, M. S. L. Lee, C. F. Edwards, M. J. Uren, and R. J. T. Bunyan, "Analogue Design Issues for SOI CMOS," *SOI Conf. Dig.*, 6–8 (1996).

12. A. O. Adan, T. Naka, S. Kaneko, D. Urabe, K. Higashi, and A. Kagisawa, "Low-Voltage 0.35μm CMOS/SOI Technology for High-Performance ASIC's," *CICC Dig.*, 427–430 (1997).

13. F. Morishita, M. Tsukude, and K. Arimoto, "Dynamic Floating Body Control SOI CMOS Circuits for Power Managed Multimedia ULSIs," *CICC Dig.*, 263–266 (1997).

14. J. A. Mandelman, F. Assaderaghi, and L. L. Hsu, "SOI MOSFET Mismatch Due to Floating-Body Effects," *SOI Conf. Dig.*, 164–165 (1997).

15. F. Assaderaghi, S. Parke, D. Sinitsky, J. Bokor, P. K. Ko, and C. Hu, "A Dynamic Threshold Voltage MOSFET (DTMOS) for Very Low Voltage Operation," *IEEE Elec. Dev. Let.*, **15**(12), 510–512 (1994).

16. F. Assaderaghi, D. Sinitsky, S. A. Parke, J. Bokor, P. K. Ko, and C. Hu, "Dynamic Threshold-Voltage MOSFET (DTMOS) for Ultra-Low Voltage VLSI," *IEEE Trans. Elec. Dev.*, **44**(3), 414-422 (1997).

17. I.-Y. Chung, Y. J. Park, and H. S. Min, "A New SOI Inverter Using Dynamic Threshold for Low-Power Applications," *IEEE Elec. Dev. Let.*, **18**(6), 248–250 (1997).

18. T. W. Houston, "A Novel Dynamic Vt Circuit Configuration," *SOI Conf. Dig.*, 154–155 (1997).

19. J.-H. Lee and Y.-J. Park, "High Speed SOI Buffer Circuit with the Efficient Connection of Subsidiary MOSFET's for Dynamic Threshold Control," *SOI Conf. Dig.*, 152–153 (1997).

20. J.-W. Lee, H.-K. Kim, J.-H. Oh, J.-W. Yang, W.-C. Lee, J.-S. Kim, M.-R. Oh, and Y.-H. Koh, "A New SOI MOSFET for Low Power Applications," *SOI Conf. Dig.*, 65–66 (1998).

21. Y. Wada, K. Ueda, T. Hirota, Y. Hirano, K. Mashiko, and H. Hamano, "Active Body-Bias SOI-CMOS Driver Circuits," *Symp. VLSI Ckts. Dig.*, 29–30 (1997).

22. J. B. Kuo, K. W. Su, J. H. Lou, S. S. Chen, and C. S. Chiang, "A 1.5V Full-Swing BiCMOS Dynamic Logic Gate Circuit Suitable for VLSI Using Low-Voltage BiCMOS Technology," *IEEE J. Sol. St. Ckts.*, **30**(1), 73–75 (1995).

23. T. Douseki, S. Shigematsu, J. Yamada, M. Harada, H. Inokawa, and T. Tsuchiya, "A 0.5-V MTCMOS/SIMOX Logic Gate," *IEEE J. Sol. St. Ckts.*, **32**(10), 1604–1609 (1997).

24. K. Fujii, T. Douseki, and M. Harada, "A Sub-1V Triple-Threshold CMOS/SIMOX Circuit for Active Power Reduction," *ISSCC Dig.*, 190–191 (1998).

25. J. Chen, P. Fang, P. K. Ko, C. Hu, R. Solomon, T.-Y. Chan, and C. G. Sodini, "Noise Overshoot at Drain Current Kink in SOI MOSFET," *SOI Conf. Dig.*, 40–41 (1990).

26. W. Jin, P. C. H. Chan, S. K. H. Fung, and P. K. Ko, "Shot-Noise-Induced Excess Low-Frequency Noise in Floating-Body Partially Depleted SOI MOSFET's," *IEEE Trans. Elec. Dev.*, **46**(6), 1180–1185 (1999).

27. Y.-C. Tseng, W. M. Huang, D. Monk, D. Diaz, J. M. Ford, and J. C. S. Woo, "Comprehensive Study on AC Characteristics in SOI MOSFETs for Analog Applications, " *Symp. VLSI Tech. Dig.*, 112-113 (1998).

28. Y.-C. Tseng, W. M. Huang, D. J. Monk, P. Welch, J. M. Ford, and J. C. S. Woo, "AC Floating Body Effects and the Resultant Analog Circuit Issues in Submicron Floating Body and Body-Grounded SOI MOSFET's," *IEEE Trans. Elec. Dev.*, **46**(8), 1685–1692 (1999).

29. J. A. Babcock, D. K. Schroder, and Y.-C. Tseng, "Low-Frequency Noise in Near-Fully-Depleted TFSOI MOSFET's," *IEEE Elec. Dev. Let.*, **19**(2), 40–43 (1998).

30. G. O. Workman, J. G. Fossum, S. Krishnan, and M. M. Pelella Jr., "Physical Modeling of Temperature Dependences of SOI CMOS Devices and Circuits Including Self-Heating," *IEEE Trans. Elec. Dev.*, **45**(1), 125–132 (1998).

31. B. M. Tenbroek, M. S. L. Lee, W. Redman-White, C. F. Edwards, R. J. T. Bunyan, and M. J. Uren, "Measurement and Simulation of Self-Heating in SOI CMOS Analogue Circuits," *SOI Conf. Dig.*, 156–157 (1997).

32. W. Redman-White, B. M. Tenbroek, M. S. L. Lee, C. F. Edwards, M. J. Uren and R. J. T. Bunyan, "Analogue Design Issues for SOI CMOS," *SOI Conf. Dig.*, 6-8 (1996).

33. M. Chan, S. S. Yuen, Z.-J. Ma, K. Y. Hui, P. K. Ko, and C. Hu, "ESD Reliability and Protection Schemes in SOI CMOS Output Buffers," *IEEE Trans. Elec. Dev.*, **42**(10), 1816–1821 (1995).

34. K. Verhaege, G. Groeseneken, J.-P. Colinge, and H. E. Maes, "The ESD Protection Capability of SOI Snapback NMOSFETs: Mechanisms and Failure Modes," *Symp. EOS/ESD Conf. Dig.*, 215–219 (1993).

35. P. Raha, J. C. Smith, J. W. Miller, and E. Rosenbaum,"Prediction of ESD Protection Levels and Novel Protection Devices in Thin Film SOI Technology," *Symp. EOS/ESD Conf. Dig.*, 356–365 (1997).

36. Y. Ohtomo, T. Mizusawa, K. Nishimura, H. Sawada and M. Ino, "A Quarter-Micron SIMOX-CMOS LVTTL-Compatible Gate Array with an Over 2000V ESD-Protection Circuit," *CICC Dig.*, 57–60 (1996).

37. J. C. Smith, M. Lien, and S. Veeraraghavan, "An ESD Protection Circuit for TFSOI Technology," *SOI Conf. Dig.*, 170–171 (1996).

38. S. Voldman, R. Schulz, J. Howard, V. Gross, S. Wu, A. Yapsir, D. Sadana, H. Hovel, J. Walker, F. Assaderaghi, B. Chen, J. Y.-C. Sun, and G. Shahidi, "CMOS-On-SOI ESD Protection Networks," *Symp. EOS/ESD Conf. Dig.*, 291–301, (1996).

39. S. Voldman, D. Hui, L. Warriner, D. Young, J. Howard, F. Assaderaghi, and G. Shahidi, "Electrostatic Discharge (ESD) Protection in Silicon-on-Insulator (SOI) CMOS Technology with Aluminum and Copper Interconnects in Advanced Microprocessor Semiconductor Chips," *Symp. EOS/ESD Conf. Dig.*, 105–115 (1999).

40. S. Voldman, F. Assaderaghi, J. Mandelman, L. Hsu, and G. Shahidi, "Dynamic Threshold Body- and Gate-Coupled SOI ESD Protection Networks," *Symp. EOS/ESD Conf. Dig.*, 210–220 (1997).

41. A. J. Auberton-Herve, "SOI: Materials to Systems," *IEDM Dig.*, 3–10 (1996).

42. S. Maeda, Y. Wada, K. Yamamoto, H. Komurasaki, T. Matsumoto, Y. Hirano, T. Iwamatsu, Y. Yamaguchi, T. Ipposhi, K. Ueda, K. Mashiko, S. Maegawa, and M. Inuishi, "Impact of 0.18μm SOI CMOS Technology Using Hybrid Trench Isolation with High Resistivity Substrate on Embedded RF/Analog Applications," *Symp. VLSI Tech. Dig.*, 154-155 (2000).

Problems

1. In Fig.4.54, the body gate-coupled (BGC) DTMOS SOI diode has been used in the ESD protection circuits. How can the DTMOS techniques be used to improve the ESD protection using PD SOI CMOS devices?

2. If the switch network logic (SNL) is to be used for VLSI circuits, how do we avoid the anomalous reverse conduction (ARC) problem?

5

SOI CMOS Digital Circuits

SOI CMOS technology has been used to integrated VLSI digital circuits owing to its advantages. In this chapter, starting from fundamental SOI CMOS static and dynamic logic circuits, DRAM and SRAM circuits using SOI CMOS technology are described. Then, SOI cache memory and content addressable memory (CAM) are depicted, followed by SOI gate arrays. SOI CPU and embedded memory are introduced, and finally SOI multipliers/digital signal processing (DSP) circuits, and SOI frequency dividers.

5.1 STATIC LOGIC CIRCUITS

As shown in Fig. 5.1, a static CMOS logic circuit is composed of pullup and pull-down switches S_u/S_d, cross-coupled element (CC), an extra resistive load and an active path to the power supply voltages V_{DD}/GND. Using the concept of the generic static CMOS logic gate, Fig. 5.2 shows the various differential static CMOS gates implementing XOR logic in terms of (a) standard CMOS, (b) push~pull cascode logic (PPCL), and (c) complementary pass transistor logic (CPL). Among three differential static CMOS gates, CPL has the smallest transistor count, since the input signal is passed via the gate or the source/drain of a single NMOS device. It has small parasitic capacitances. Due to the reduced threshold voltage using the body control schemes, PD SOI CMOS devices are suitable for use to integrate CPL circuits since the output swing of the CPL is close to the power supply voltage, which is good for driving the next stage. In this section, several SOI CMOS static logic circuits including CPL, ADTPT, CVSL, and adiabatic CVSL are described.

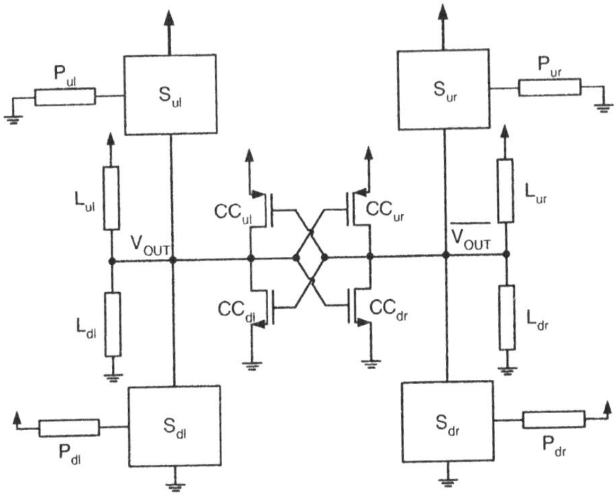

Fig. 5.1 Generic static CMOS logic gate. (Adapted from Tretz et al. [1].)

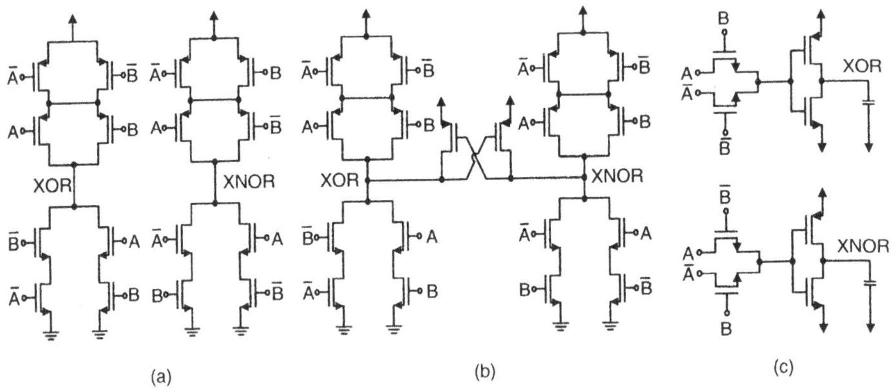

Fig. 5.2 Various differential static CMOS gates implementing XOR logic: (a) standard CMOS, (b) push~pull cascode logic (PPCL), and (c) complementary pass transistor logic (CPL). (Adapted from Tretz et al. [1].)

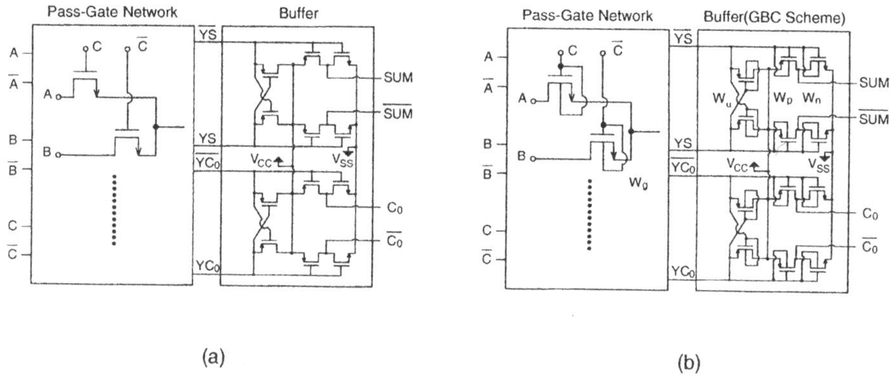

Fig. 5.3 (a) CPL. (b) Gate-body-connected (GBC) pass transistor logic. (Adapted from Fuse et al. [2].)

5.1.1 SOI CPL

CPL is an important logic circuit for realizing low-voltage SOI CMOS VLSI systems. Pass-transistor logic circuits have been known for their advantages in high density, high speed and low power. As shown in Fig. 5.3, in the SOI CPL circuit, due to the channel resistance of the SOI pass gates, the limited current driving capability may not be effective in driving its output load [2]. Hence, an output buffer has been added to the pass-gate networks such that the driving capability of the circuit can be enhanced. Due to the floating body effect of the SOI devices, the high signal at the output of the pass gate can only reach $V_{DD} - V_{TH}$. In order to improve the output signal swing due to floating body effects, the GBC scheme using SOI DTMOS techniques has been used to realize the pass transistor logic as shown in Fig. 5.3(b). By connecting the body to the gate, the threshold voltage of the pass gate can be reduced when it is turned on such that the drain current driving capability can be improved.

5.1.2 DTPT

The limited logic-1 signal of a pass-transistor may refrain itself from low-voltage applications. Based on SOI DTMOS techniques, Fig. 5.4 shows the SOI symmetrical dynamic threshold pass-transistor (DTPT) technique using two extra auxiliary transistors for each pass transistor to solve the reduced logic-1 signal problem frequently encountered in a pass-transistor logic circuit using a low supply voltage [3]. By connecting the body of the main transistor to the source/drain ends via two auxiliary transistors with their gates controlled by the gate of the main transistor, the threshold voltage of the pass transistor can be reduced to increase its signal swing. Compared to the conventional pass-transistor logic circuit, the layout area of the DTPT increases substantially. As shown in the figure, the body node is in the middle of the two extra auxiliary devices. The enhancement of the body bias provided by the structure for the DTPT circuit may not be enough.

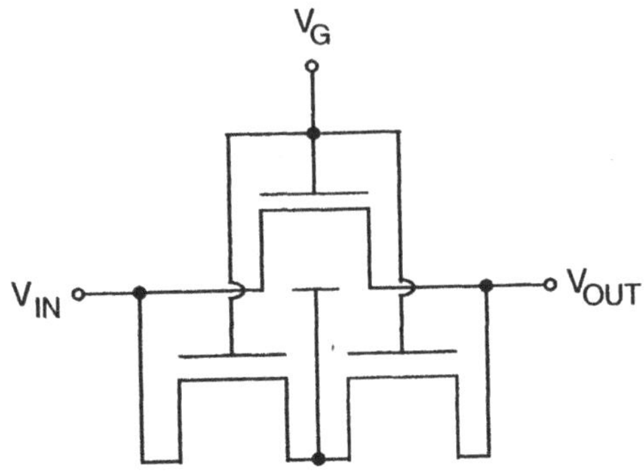

Fig. 5.4 SOI symmetrical DTPT circuit with two auxiliary transistors. (Adapted from Lindert et al. [3].)

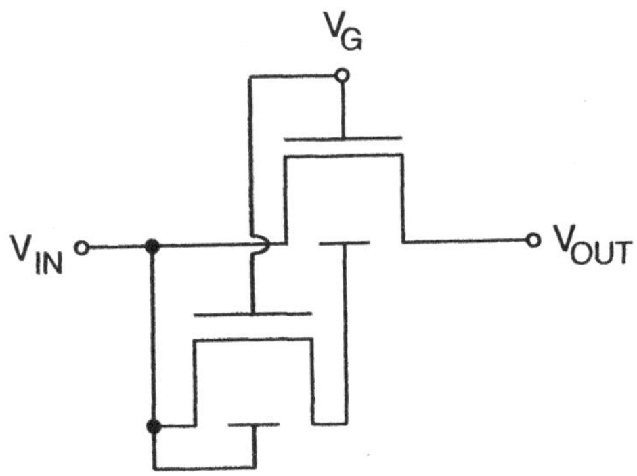

Fig. 5.5 The ADTPT circuit. (Adapted from Wang & Kuo [4].)

Figure 5.5 shows an asymmetrical dynamic threshold pass-transistor (ADTPT) technique using SOI CMOS DTMOS techniques to dynamically control the body bias of the pass-transistor via only one auxiliary transistor, for high-speed operation at a low supply voltage [4]. As shown in the figure, in the ADTPT circuit, the body of the main pass-transistor is connected to the source/drain node of the auxiliary transistor, whose gate is also tied to the gate of the main pass-transistor and its body is tied to the source/drain node of the main pass-transistor, instead of floating as in the DTPT case. The advantages of the ADTPT circuit can be understood by considering its logic operation. When V_G is high (V_{DD}), both the pass-transistor and the auxiliary transistor are on. During the pass-logic-1 operation, the logic-1 level is propagated from the input V_{IN} to the output V_{OUT}. When the input V_{IN} increases from low to high, due to the function of the auxiliary transistor, the body of the main pass-transistor (V_B) is raised to $V_{DD} - V_{TN}(V_B = V_{DD})$, where $V_{TN}(V_B = V_{DD})$ is the threshold voltage of the auxiliary transistor biased with the body bias of V_{DD}. Compared to the DTPT circuit, the ADTPT circuit has a faster speed owing to a higher body voltage provided by its auxiliary transistor– in the DTPT circuit, due to its two auxiliary

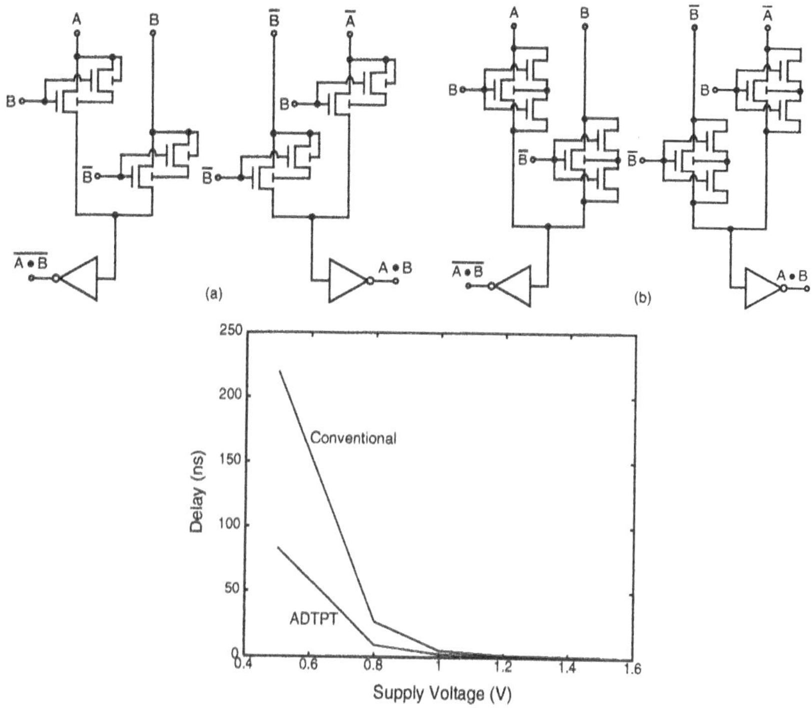

Fig. 5.6 AND/NAND CPL circuit implemented by (a) ADTPT and (b) DTPT. Fall time versus power supply voltage. (Adapted from Wang & Kuo [4].)

transistor structure, the body bias of the main pass-transistor is half-way between the input $[V_{IN}(= V_{DD})]$ and the output V_{OUT}, which rises from 0 V to $V_{DD} - V_{TN}$. In the ADTPT circuit, owing to the single-auxiliary transistor structure, the body bias of the main pass-transistor is tied to a higher level $V_{DD} - V_{TN}(V_B = V_{DD})$. Therefore, the effective threshold voltage of the main pass-transistor of the ADTPT circuit is much smaller. As a result, a higher speed in passing the logic-1 signal from the input V_{IN} to the output V_{OUT} can be obtained. Figure 5.6 shows AND/NAND CPL circuit implemented by (a) ADTPT and (b) DTPT [4]. As shown in Fig. 5.6(c), among two cases–ADTPT and conventional PT, the ADTPT one has the smaller propagation delay time, which is especially noticeable at a low supply voltage.

5.1.3 SOI CVSL

Cascode voltage switch logic (CVSL) circuit, which is also called cascoded differential voltage switch logic (DCVS) [5][6], is a major logic circuit in the CMOS digital static logic circuit. Figure 5.7 shows the SOI CMOS static cascode voltage switch logic (CVSL) circuit for realizing OR-AND function [7]. As shown in this figure, the SOI DCVS logic circuit is composed of two parts– the DCVS logic tree and the

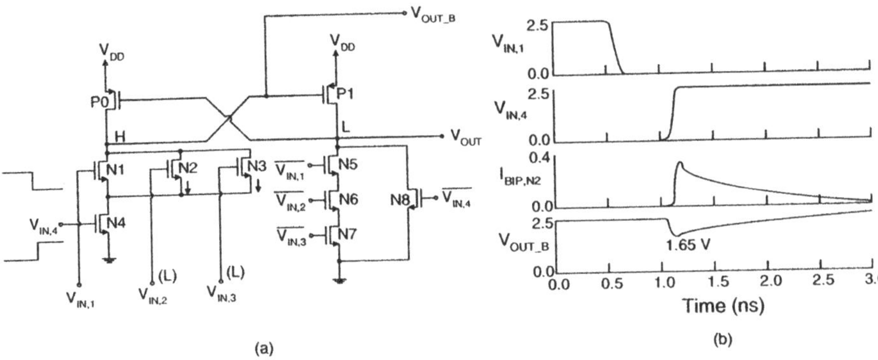

Fig. 5.7 (a) SOI CMOS static cascode voltage switch logic (CVSL) circuit for realizing OR-AND function. (b) Switching waveforms. (Adapted from Lu et al. [7].)

PMOS push~pull load. The DCVS logic tree, which is implemented in SOI NMOS devices, is in the bottom portion of the DCVS logic circuit. Since the complementary results are available at the output nodes, both 0-tree and 1-tree are required in the DCVS logic tree. The outputs of the SOI DCVS logic trees are connected to the PMOS push~pull loads, which are at the top of the DCVS logic circuit. Due to the floating body effect of the SOI devices, the output high voltage of the SOI static CVSL circuit may drop. At the beginning of the operation, the output node V_{outb} is high, the gate voltage of the NMOS device N1 is high, and the gate of the NMOS device N4 is low. Thus, the source and the floating body of the NMOS device N1 are charged to high. Although NMOS devices N2 and N3 are off at this time, their internal floating body is gradually charged to high since both the source and the drain of the devices are high. When NMOS device N1 is off and NMOS device N4 is on, the source potential of the NMOS devices N2 and N3 is pulled low. As a result, their internal parasitic bipolar devices turn on to cause the leakage current, which leads to a drop in the potential of the output node V_{outb}. These extra leakage currents cause additional power consumption. Note that the N1/N2/N3/N4 complementary tree N5/N6/N7/N8 is not affected by the floating body effect of the SOI devices.

5.1.4 SOI Adiabatic CVSL

Adiabatic CMOS logic circuits based on recycling of energy are useful for potentials in low-power VLSI applications [8][9]. CVSL has been used to realize the adiabatic CMOS logic circuits [10][11]. Figure 5.8 shows an SOI CMOS adiabatic differential cascode voltage switch logic (ADCVSL)using a two-phase nonoverlapping clock, which is based on the SOI MTCMOS techniques [12]. As shown in this figure, during the first clock cycle PWR1, via an NMOS logic tree, F and FB are elevated. Via devices M5~M8, S and SB are raised to have a sufficient voltage swing. At this time, devices M1 and M2 need to be off, hence high-threshold SOI NMOS devices are used. A conduction path in the NMOS logic tree lowers the potential of node

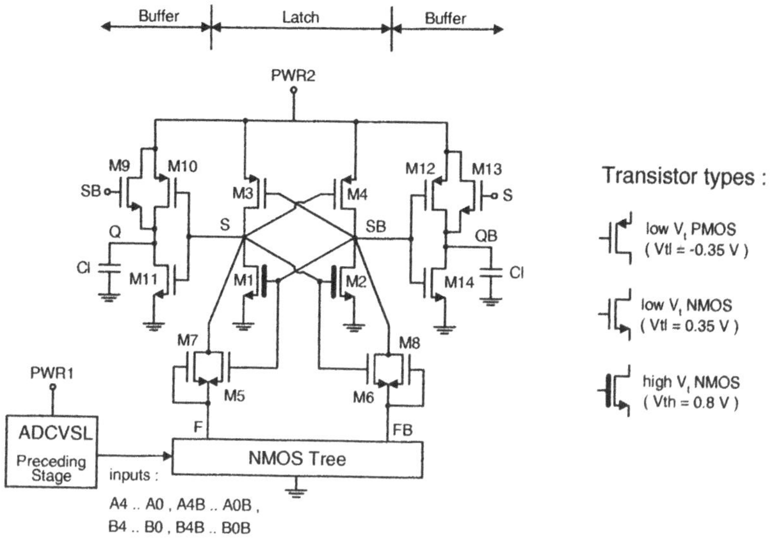

Fig. 5.8 Adiabatic differential cascode voltage switch logic (ADCVSL) using a two-phase nonoverlapping clock and based on the SOI MTCMOS techniques. (Adapted from Suvakovic & Salama [12].)

F and thus node S becomes lower. At this time FB and SB stay unchanged. Via this approach, during the clock cycle PWR2, the differential signals generated in the ADCVSL are provided to the latch circuit for further processing. SOI MTCMOS techniques are appropriate for integrating the ADCVSL circuits.

5.2 DYNAMIC LOGIC CIRCUITS

SOI CMOS technology is appropriate for realizing VLSI dynamic logic circuits. In this section, SOI DTMOS logic, instability problems of the SOI dynamic logic circuits, dynamic CVSL, pseudo-two-phase CVSL, and race problems of the SOI dynamic logic circuits are described.

5.2.1 SOI DTMOS Dynamic Logic Circuit

Manchester carry chain circuit is a well-known dynamic logic circuit for generating the carry look ahead signals in an adder [13]. SOI CMOS technology has been used to design the Manchester carry chain circuit. Figure 5.9 shows a 0.7 V two-bit Manchester carry chain circuit using SOI DTMOS and ADTPT techniques [14]. As shown in this figure, the Manchester carry chain circuit is used to process the propagate and the generate signals produced by half-adders to generate the carry signals, which are important for realizing a full adder. With the pass transistor structure, Manchester carry chain circuit is the most concise among all carry look-ahead circuits. In Manch-

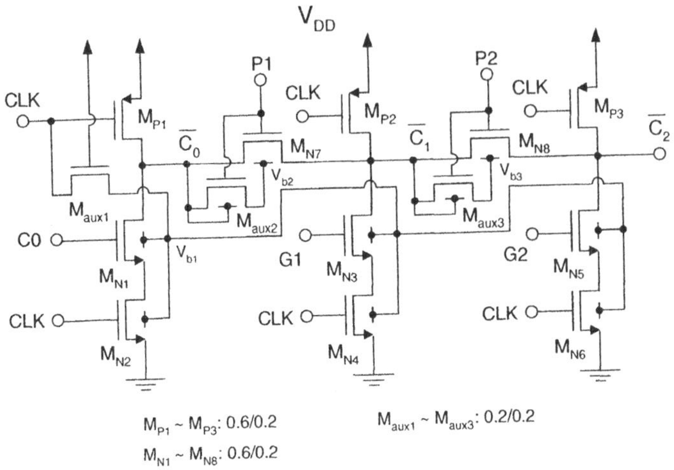

Fig. 5.9 0.7 V two-bit Manchester carry chain circuit using SOI DTMOS and ADTPT techniques. (Adapted from Kuo et al. [14].)

ester carry chain circuit, the carry signal of the present bit C_i is high if the generate signal G_i is high or if the carry signal of the previous bit C_{i-1} and the propagate signal P_i are high: $C_i = G_i + C_{i-1}P_i$, for $i = 1 - n$, where n is the bit number, G_i and P_i are the generate and the propagate signals $G_i = X_iY_i$ and $P_i = X_i \oplus Y_i$ produced from two inputs X_i and Y_i to the half-adder.

As shown in Fig. 5.9, when clock CLK is low, it is the precharge phase– the internal output nodes $\overline{C_0}$-$\overline{C_2}$ are set to high. At this time, since the auxiliary transistor M_{aux1} is on, the body voltage V_B of the pull-down devices in the dynamic logic circuit M_{n1}–M_{n6} is low. The body voltage V_{b2}/V_{b3} of the main pass-transistor in the ADTPT is controlled by the auxiliary transistor M_{aux2}/M_{aux3}. When the propagate signal (P1/P2) is high, the auxiliary transistor M_{aux2}/M_{aux3} is on– the body voltage V_{b2}/V_{b3} is charged to high. When the propagate signal (P1/P2) is low, the body of the main pass-transistor M_{n7}/M_{n8} is floating. When clock CLK is high, it is the evaluation phase. During this time, the body voltage V_{b1} of the pull-down devices M_{n1}-M_{n6} in the dynamic logic circuits is charged to $V_{DD} - V_{TH}$, where V_{TH} is the threshold voltage of the auxiliary device M_{aux1} such that the threshold voltage of the pull-down devices is lowered to enhance the current driving capability. When the propagate signal P1/P2 is high, the body voltage V_{b2}/V_{b3} of the main pass-transistor in the ADTPT is charged to high via the auxiliary transistor M_{aux2}/M_{aux3} throughout the internal node $\overline{C_0}/\overline{C_1}/\overline{C_3}$. Therefore, the threshold voltage of the main pass-transistor in the ADTPT is lowered to decrease the RC delay time associated with the pass transistors. As a result, a higher speed has been achieved.

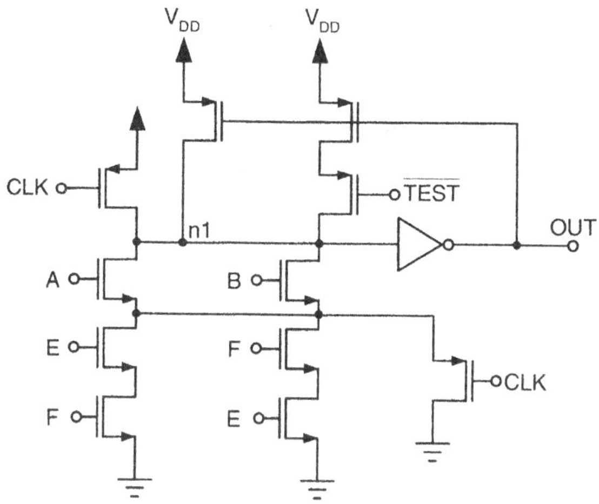

Fig. 5.10 SOI dynamic logic circuit. (Adapted from Aipperspach [15].)

5.2.2 Instability of SOI Dynamic Logic

Although PD SOI CMOS devices have advantages in speed performance, there are circuit problems associated with them. Due to the floating body effect, during dynamic operation the body-source voltage may have a substantial drop to generate the leakage current. Consequently, there may be dynamic instability problems– failures with wrong logic states generated at the dynamic circuit output. In order to solve this problem, the PD SOI dynamic circuit as shown in Fig. 5.10 has been used. As shown in this figure, the NMOS tree portion of the dynamic logic circuit has been rearranged– the NMOS devices with a larger channel width are placed in the bottom of the NMOS string [15]. In addition, a parallel multiple-fingered arrangement of these NMOS devices have been adopted to reduce the current of the parasitic bipolar device. In addition, at the intermediate node, a predischarged PMOS device with its gate controlled by a clock identical to the one for the precharge PMOS device has been added such that the charge accumulating in the body of the NMOS devices above the intermediate node during the previous precharge period, can be discharged. Thus the potential of the node n1 does not drop to an error logic state at the output for a sudden drop from high to low in the intermediate node during the evaluation phase. This NMOS tree structure is applicable for SOI, which does not cause any charge sharing problems due to the smaller junction capacitances as compared to the bulk counterpart.

Figure 5.11 shows the SOI dynamic inverter circuit with a PMOS feedback MP3. During the evaluation phase, if V_{IN} is low, device MN1 is off. Due to the leakage current from the parasitic bipolar devices in the devices, the internal node voltage V_1 may be pulled from high to low. As a result, an error state at the output is occurred. The PMOS feedback device MP3 has been used to correct this problem. The size of the PMOS device MP3 is critical. With a width of 1 μm (weak feedback), a

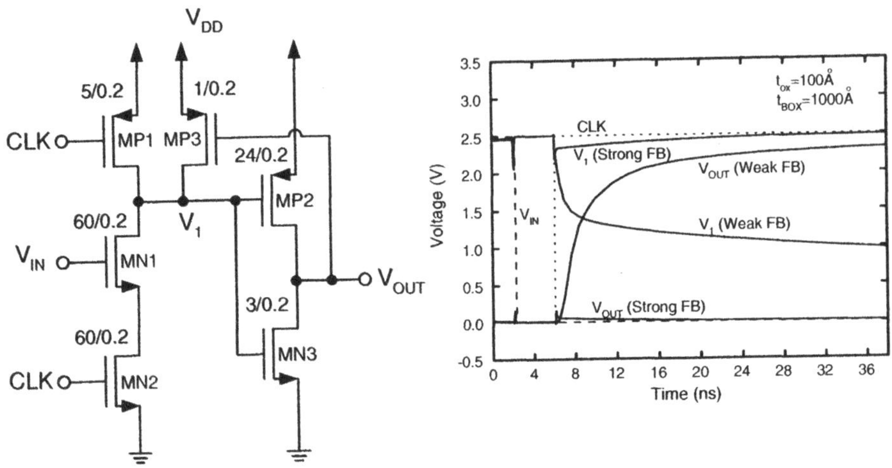

Fig. 5.11 SOI dynamic inverter circuit with a PMOS feedback MP3.

logic error at the output still occurs. With a width of > 4 μm (strong feedback) for the feedback PMOS device MP3, the output error problems due to parasitic bipolar leakage effect can be corrected. In addition, the number of input devices should not be too large. If too large, the width of the feedback PMOS device MP3 needs to be increased accordingly for dealing with the leakage current due to the parasitic bipolar device in all input NMOS devices. On the other hand, with a large width MP3, the driving capability of a single NMOS device is not sufficient to pull down the internal node V_1 and consequently another error occurs.

Figure 5.12 shows a low noise circuit, which is composed of MP, MT, and MN, to eliminate the leakage current problem of MA in the SOI dynamic logic circuit [16]. As shown in this figure, after the precharge phase, the internal node is charged to high. When the clock turns high, device MB is on and the internal node N1 is pulled low. If input A is low, MA is off and the leakage current due to the parasitic bipolar device in its floating body may pull down the dynamic node N3. Using MP, MT, and MN as the low noise circuit to replace MA, the leakage current is bypassed via MP and MN. As a result, the internal node N2 maintains high and MT is off– no logic state error occurs at the output.

5.2.3 SOI Dynamic CVSL Circuit

Figure 5.13 shows the SOI CMOS dynamic CVSL OR-AND circuit with the transient waveforms [7]. As shown in this figure, the cross-coupled PMOS load used in the static CVSL circuit has been replaced by the feedback PMOS devices at the two internal nodes to avoid the leakage current problem due to the parasitic bipolar devices in the input devices. After the precharge phase, both internal nodes V_1 and V_2 are high. When the clock signal turns high, it is the evaluation phase. If all three inputs V_{in1}, V_{in2}, and V_{in3} are low, the output voltage (V_{out}) is supposed to be high and the

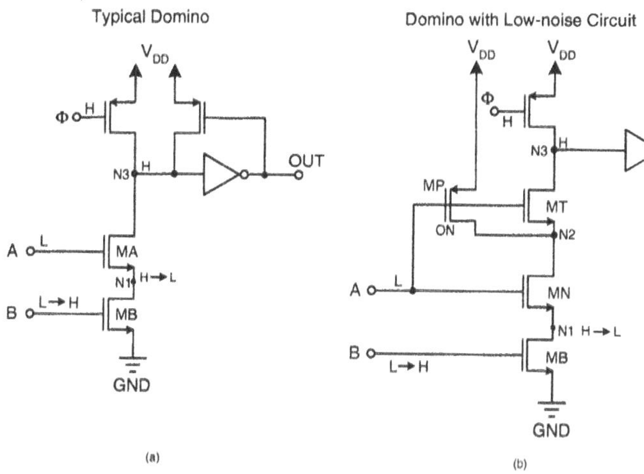

Fig. 5.12 Low noise circuit to eliminate the leakage current problem in the SOI dynamic logic circuit. (Adapted from Bobba & Hajj [16].)

complementary output voltage ($\overline{V_{out}}$) is low. Due to the leakage current caused by the parasitic bipolar device in the floating body of the three input devices as described before, the internal voltage V_1 may come down. If the internal voltage V_1 drops to below the threshold voltage of the output inverter (MP10/MN10), the complementary output $\overline{V_{out}}$ becomes high– a wrong logic state. The design of the feedback PMOS devices is very important for the stability of SOI dynamic CVSL circuit.

5.2.4 SOI Pseudo Two-Phase Dynamic Logic Circuit

SOI technology has also been used to design pseudo two-phase SOI dynamic logic circuit. Fig. 5.14 shows the pseudo two-phase SOI dynamic logic circuit with switching waveforms[7]. As shown in this figure, via a transmission gate controlled by a clock, the data of each stage can be latched at the input during a certain clock phase. This data latch capability is challenged by the leakage current caused by the parasitic bipolar device due to the floating body of the SOI devices. As shown in this figure, when the clock signal V_{c2} is low–both P2 and N2 are off, the input node INT3 is not supposed to be affected. When V_{c1} turns high, the internal node INT2 is pulled high due to precharge. If INT3 is low originally, after the turn-high of the internal node INT2, the parasitic bipolar device in the floating body of the P2 turns on to cause a leakage current such that the potential of the internal node INT3 is disturbed to rise. For more serious situations, it may rise further to cause an error state. Consider another situation. When V_{c2} is low and V_{c1} turns from high to low, it enters the evaluation phase. If the NMOS-tree is conducting current to lower the internal INT2, and if the internal node INT3 is high originally, the turn-low of the internal node INT2 may trigger the turn-on for the parasitic bipolar device in the floating body of N2. As a result, a large amount of leakage current lowers the internal node voltage V_{int3} to

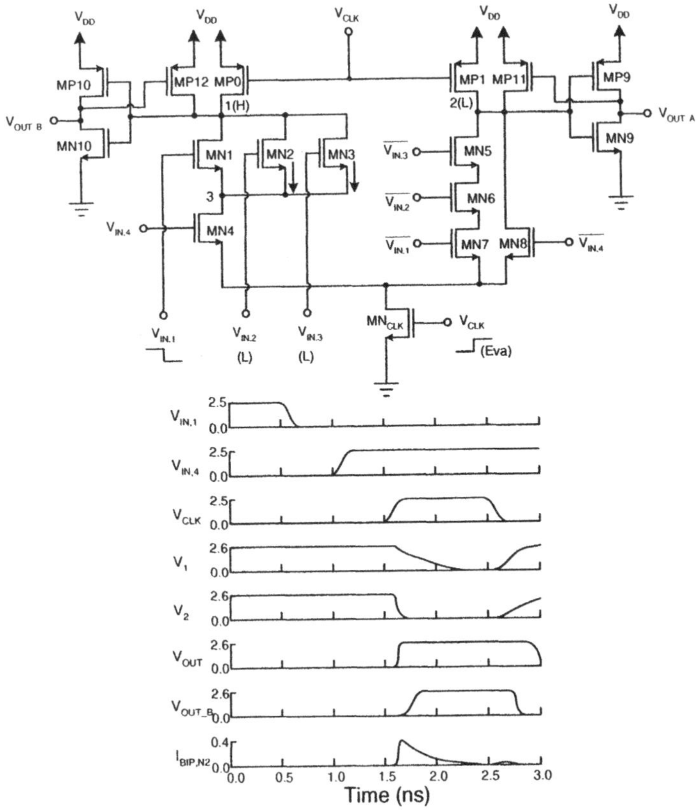

Fig. 5.13 SOI CMOS dynamic CVSL OR-AND circuit with the transient waveforms. (From Lu et al. [7]. ©1997 IEEE.)

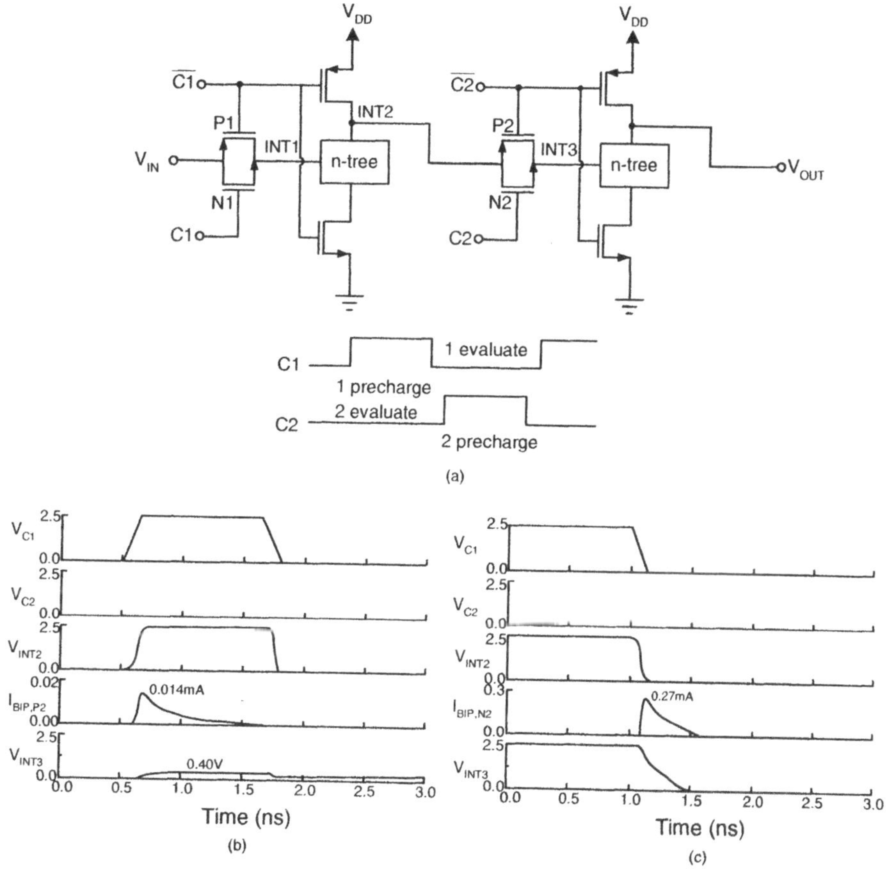

Fig. 5.14 Pseudo two-phase SOI dynamic logic circuit with switching waveforms. (From Lu et al. [7]. ©1997 IEEE.)

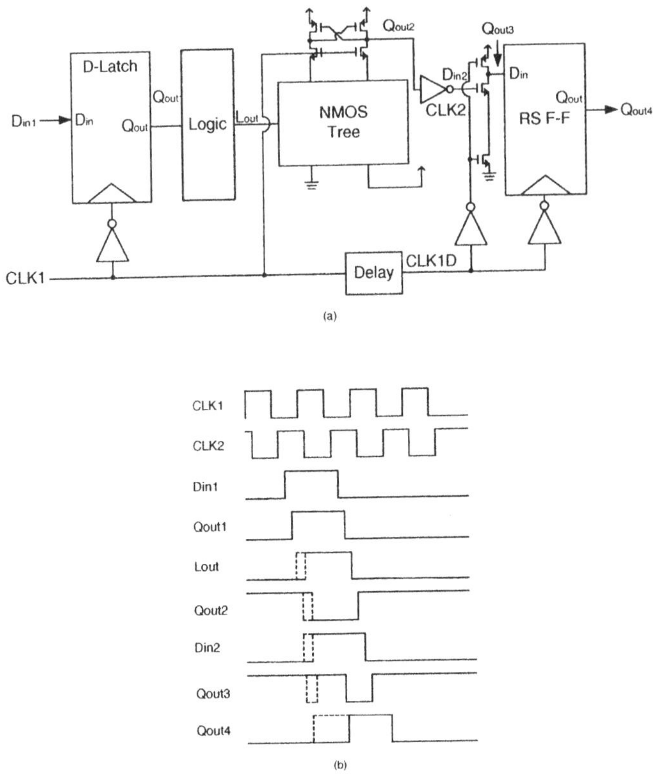

Fig. 5.15 SOI dynamic logic circuit illustrating the race problem. (From Park et al. [17]. ©1999 IEEE.)

cause an error state. From the above analysis, transmission gates serving as a latch should be avoided– latch with a feedback loop should be adopted.

5.2.5 Race Problems in SOI Dynamic Logic Circuits

As for a bulk counterpart, race problems are also important in SOI dynamic logic circuits. Figure 5.15 shows an SOI dynamic logic circuit illustrating the race problem [17]. As shown in this figure, in this SOI dynamic logic circuit, there are a CVSL latch controlled by clock CLK1, a multistage domino circuit controlled by clock CLK2, and a logic function circuit. The output of the CVSL latch circuit is passed to the multistage domino logic, whose output is provided to the output via an RS flip-flop. This circuit is designed with timing for the logic function circuit to generate an output Lout when CLK1 is high. If before CLK2 turns high, the logic function circuit already provides its output to the CVSL latch. Due to the strong current driving capability and the small junction capacitances of the NMOS-tree in the CVSL latch and the following domino circuit, during the overlapping period of CLK1 and CLK2, the

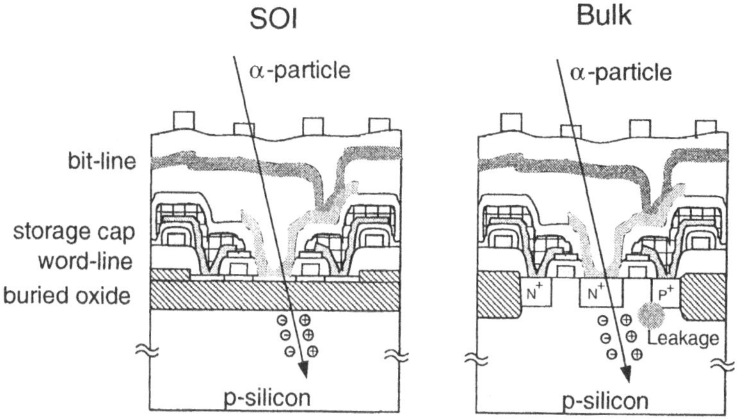

Fig. 5.16 Soft error immunity in SOI and bulk DRAMs subject to α-particles. (Adapted from Yamaguchi & Ionue [18].)

logic circuit's output is directly passed to the final output node Q_{out4} to cause a logic state error as a race problem.

5.3 DRAM

Owing to high density, low parasitic capacitances, low leakage current, and good radiation hardness, SOI CMOS devices are suitable for implementing low-power, low-voltage, large-size, high-speed DRAMs. Figure 5.16 shows the soft error immunity in SOI and bulk DRAMs [18]. As shown in this figure, for bulk devices, the electron/hole pairs generated by α-particles are received by the source/drain. As a result, the potential of the storage node changes, which causes soft error immunity problems. Owing to the isolation of the thin film from the substrate by the buried oxide, a large amount of electron/hole pairs generated by the α-particles do not affect the storage node substantially, which implies a good soft error immunity.

Figure 5.17 shows the processing steps of an SOI DRAM technology with LOCOS and local implantation post-field (LIF) oxidation based on a 0.6 μm technology [19], where SIMOX or BESOI wafers using LOCOS isolation, two-level metal interconnects, and semi-double-fin capacitors have been adopted. Using this technology, a 16 Mb SOI DRAM has been fabricated. At a power supply voltage of 3 V, the access time is 50 ns, which is 20% faster as compared to the DRAM made by the same level bulk technology.

5.3.1 Floating Body Effects

Although SOI devices provide many advantages for the DRAM design, the floating body effects of the SOI devices also bring in some drawbacks. As described in the

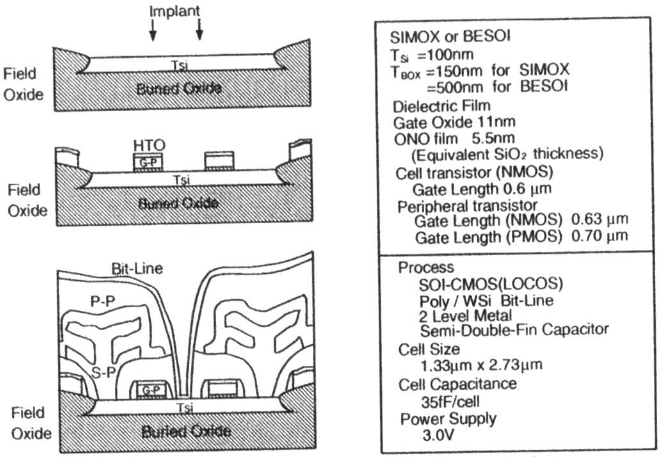

Fig. 5.17 Processing steps of an SOI DRAM with LOCOS and LIF oxidation based on a 0.6 μm technology. (From Kim et al. [19]. ©1995 IEEE.)

Chapter 4, the floating body may cause the dynamic instability problems in the DRAM sense amp. Figure 5.18(a) shows a DRAM sense amp circuit made of non-FD (NFD) SOI devices[20]. As shown in the figure, complementary bit line pairs are connected to the each side of the DRAM sense amp, which is made of a pair of cross-coupled CMOS inverters. At each bit line, a DRAM cell made of a storage capacitor and a pass transistor controlled by its world line is connected. The operation of the DRAM sense amp circuit is divided into two phases– the precharge phase and the sensing phase. During the precharge phase, the cross-coupled inverters are disconnected to the top and the bottom power supplies. Due to the turn-on of device P6, the voltages on both bitlines are equalized to $V_{DD}/2$. Then the storage node of the accessed DRAM cell is connected to the bit line when its word line turns on. During the sensing phase, P6 is turned off and the cross-coupled sense amp is turned on. Then, the voltage difference on the bit lines due to the charge in the DRAM capacitor is amplified. Because the biasing voltages of devices N1 and N2 in the cross-coupled inverter of the sense amp are not identical during operation, the generation/recombination rates in these two transistors are not identical, which leads to different floating body voltages V_{BS}, and hence different dynamic threshold voltages. Due to the different threshold voltages of devices N1 and N2, dynamic instability exists in the operation of the sense amp. Figure 5.18(b) shows the bitline voltage (V_{BL}) and the body-source voltages (V_{BS}) of devices N1 and N2 during read 0/read 1/read 0 sensing operation of the DRAM after an extended period of read/write 0 operation. As shown in this figure, initially, the potential of the bit line at right (V_{BL}) is low and the potential of the complementary bit line ($\overline{V_{BL}}$) at the left is high. Thus, device N1 turns on and device N2 turns off. Due to the gate-body/source capacitive coupling and the difference in the charge and the discharge of the floating body, the internal charge of device N2 is larger than that of device N1 [$Q_B(N2) > Q_B(N1)$]. Thus, the body-source voltage

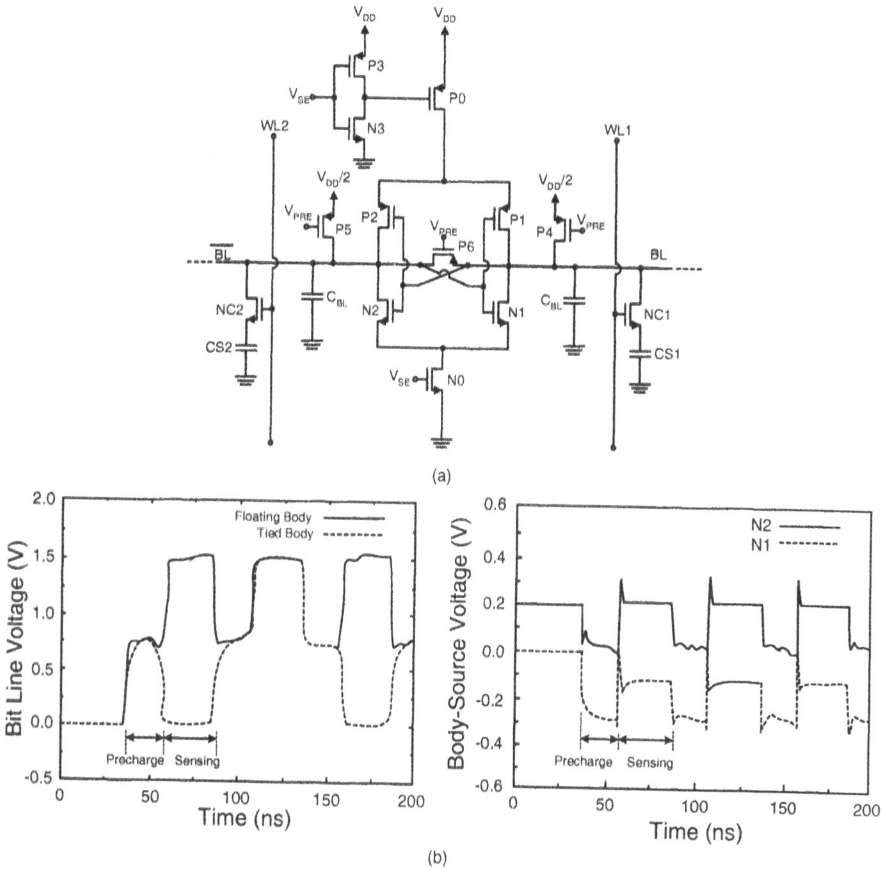

Fig. 5.18 (a) DRAM sense amp circuit formed by non-FD (NFD) SOI devices. (b) Bit line voltage (V_{BL}) and body-source voltages (V_{BS}) of devices N1 and N2 during read 0/read 1/read 0 sensing operation of the DRAM after an extend period of read/write 0 operation. (From Fossum et al. [20]. ©1998 IEEE.)

of N2 is larger than that of N1 [$V_{BS}(N2) > V_{BS}(B1)$]. As shown in this figure, in the subsequent precharge phase, both bit lines V_{BL} and $\overline{V_{BL}}$ are equalized to $\frac{V_{DD}}{2}$. Because of the floating drain of device N0, via gate-body capacitive coupling, the $V_{BS}(N2) > V_{BS}(N1)$ condition continues. When V_{SE} becomes high, the sense amp begins to operate and the drain voltage of device N0 ($V_{DS} = 0$ V) is 0 V. The imbalance of the dynamic threshold voltages caused by the difference in the body voltage (V_{BS}) between devices N1 and N2 increases noticeably–$V_T(N2) < V_T(N1)$. Consequently, it causes an error for the read 0 operation.

The floating body induced parasitic BJT leakage current affects the performance of a DRAM circuit substantially. Figure 5.19 shows the transient leakage mechanism in the SOI DRAM cell composed of the storage capacitor and the access pass transistor [21]. During the write high operation (1), the storage node of the DRAM cell is

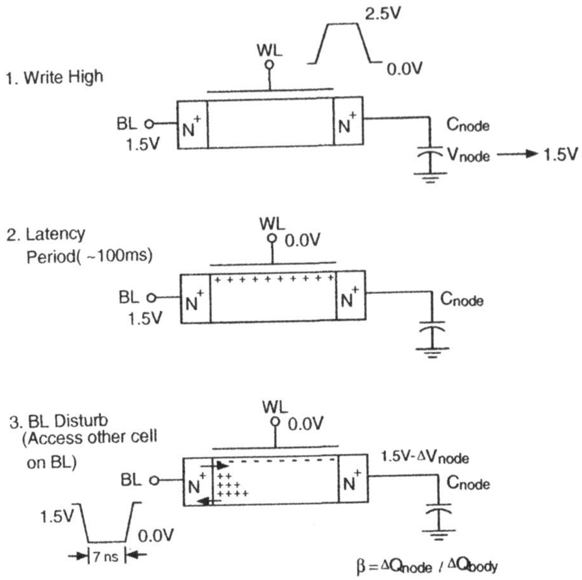

Fig. 5.19 Transient leakage mechanism in the SOI DRAM cell composed of the storage capacitor and the access pass transistor. (From Mandelman et al. [21]. ©1996 IEEE.)

charged to 1.5 V. During the latency period (2), since all cells of the same column are not accessed due to thermal generation or diode leakage, a certain amount of holes stored in the body lead to the rise of the body potential to a certain level. If one cell of the same column is accessed for the write 0 operation (3), the related bit line is pulled low suddenly. Although this DRAM cell is not selected and its pass transistor is off, its source voltage is lowered suddenly, which causes the turn-on of the body-to-source diode. As a result, the holes in the body are expelled. In addition, the parasitic bipolar device is triggered to turn on and a large amount of current is injected from the drain (the storage node). This leakage current continues until the holes in the body are totally removed and the corresponding body potential falls. As a result, the charge and the potential at the storage node are changed to cause an error in the data stored. If the latency period is sufficiently long, the rise in the body voltage during this period increases the subthreshold leakage current. Therefore, the data retention time of the DRAM may be worsened, which should be taken care of when designing the DRAM circuit.

The data retention time of a DRAM cell can be improved using the body refresh operation. Before the body potential of the access pass transistor in the DRAM cell rises to a certain value, the related bit line is pulled low such that before the holes can be accumulated in the body to a sufficiently large amount, first the holes are expelled. Then, the body potential can remain low, which is the so-called body refresh scheme. From the potential distribution in the lateral direction of the access pass transistor in the DRAM cell as shown in Fig. 5.20, with the body refresh scheme, after the body fresh procedure, the body-source junction cannot be forward biased–the parasitic

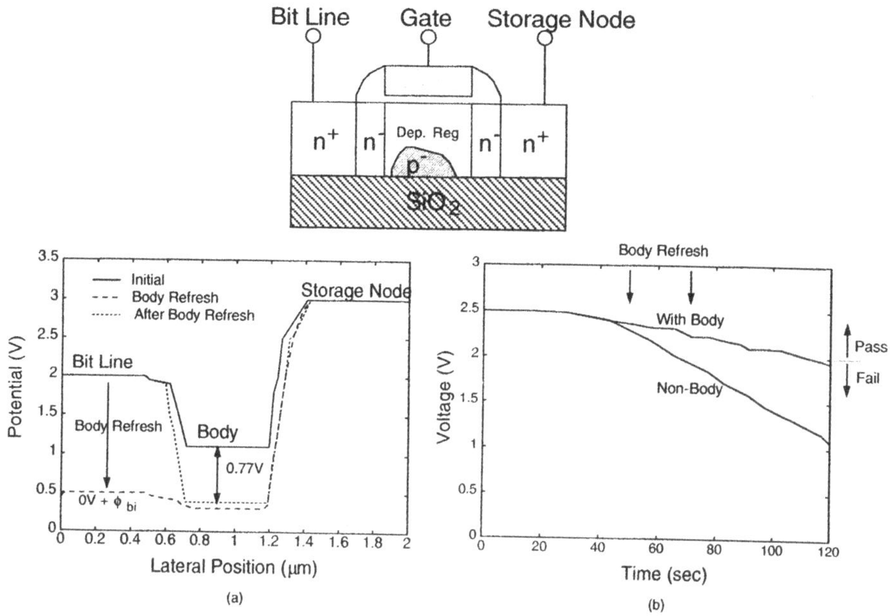

Fig. 5.20 (a) Potential distribution in the lateral direction of the access pass transistor in the DRAM cell. (b) Voltage of the storage node during the retention phase. (From Tomishima et al. [22]. ©1996 IEEE.)

bipolar device cannot turn on [22]. In addition, since the body voltage is sustained, the leakage current during the latency period is much smaller, thus the data retention time is much better as shown in Fig. 5.20(b).

The data retention time of an SOI DRAM cell can be improved by the adoption of the PD SOI device with a nonuniformly doped thin film. At the bottom of the thin film, a more heavily doped P^+ region is created to lessen the function of the parasitic bipolar device. Fig. 5.21 shows the current gain of the parasitic bipolar device versus the doping density of the P^+ thin film region of the PD SOI NMOS device [23]. As shown in this figure, with the nonuniformly doped thin film, the gain of the parasitic bipolar device has been lowered substantially. As a result, the leakage current is reduced and the data retention time is improved.

The leakage current of the access pass transistor determines the data retention time of a DRAM cell. As described above, with the body of the thin-film floating, the leakage current of the access pass transistor can be large. The most straightforward way to reduce the leakage current of the access pass transistor is by adding body contact to avoid the floating body structure. As described in the previous chapters, adding body contacts by using interconnect may lead to an increase in the layout area. Figure 5.22 shows the cross section of the body-contacted (BC) SOI CMOS device used in a 64 Mb SOI DRAM [24]. As shown in this figure, the field oxide does not isolate the thin film completely. The silicon space under the field oxide is reserved for connecting the thin-film body to the body contact. For the DRAM cell with the

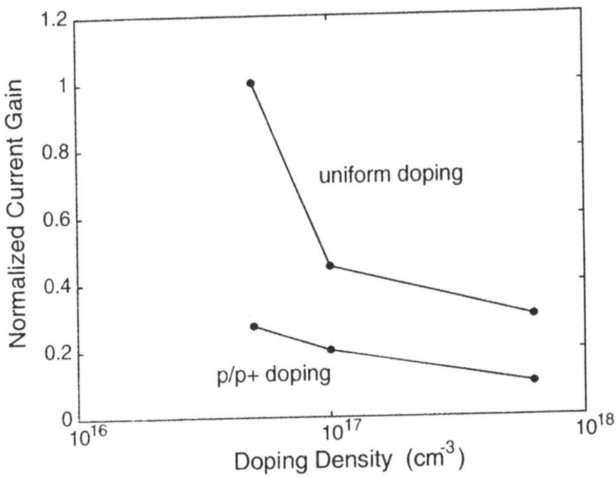

Fig. 5.21 Current gain of the parasitic bipolar device versus doping density of the P^+ thin film region of the PD SOI NMOS device. (Adapted from Kim et al. [23].)

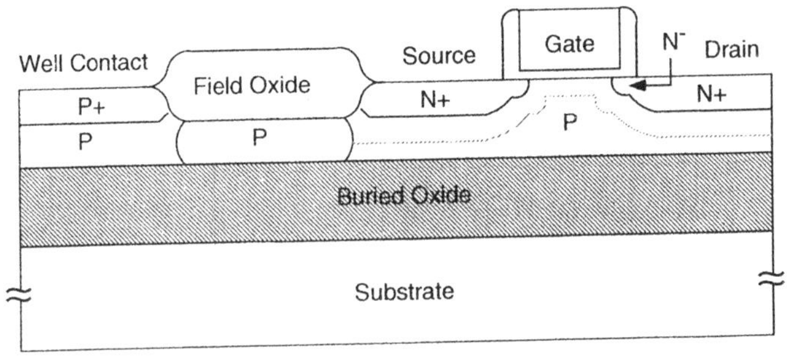

Fig. 5.22 (a) Cross-section of body-contacted (BC) SOI CMOS Devices. (Adapted from Koh et al. [24].)

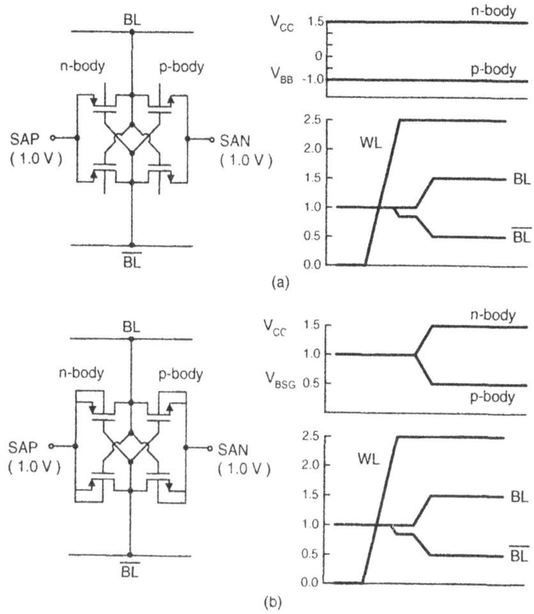

Fig. 5.23 (a) Conventional sensing scheme and (b) body-synchronous sensing scheme used in a 1.5 V multi-Giga-scale SOI DRAM using a 0.15 μm PD SOI CMOS technology. (From Kuge et al. [25]. ©1996 IEEE.)

body-contacted (BC) SOI CMOS device, the synchronous operation is 10% faster than that of the DRAM cell using the bulk device. Although with the body contact the density of the SOI devices is sacrificed a little, the DRAM circuits using BC SOI CMOS devices are compatible to those using the bulk ones. Therefore, the designs of the DRAM circuits for bulk technology can be used for BC SOI CMOS technology.

Owing to the floating body structure, body control can be used to enhance the sensing speed of a sense amp in an SOI DRAM. Figure 5.23 shows (a) the conventional sensing scheme and (b) the body-synchronous sensing scheme used in a multi-Giga-scale SOI DRAM [25]. As shown in this figure, for the conventional sensing scheme, the body bias of the NMOS/PMOS device is at V_{CC}/V_{BB} (1.5/-1 V). By considering the body effect, the threshold voltage of the sense amp is 0.71 V. If the body is tied to the source as shown in Fig. 5.23(b), the source-body voltage is 0 V without body effect. Thus the threshold voltage of the sense amp becomes smaller at 0.44 V, which leads to a faster sensing owing to an enhanced current driving capability. This sensing scheme is called body-synchronous sensing.

Figure 5.24 shows the super body-synchronous sensing scheme used in a multi-Giga-scaled SOI DRAM [25]. As shown in this figure, before the sensing operation, the body potentials of the both PMOS and NMOS devices are maintained at V_{CC} and V_{BSG} (boosted sense ground), respectively. Due to body effect, the threshold voltage of the sense amp is even smaller than that of the body-synchronous one (0.35

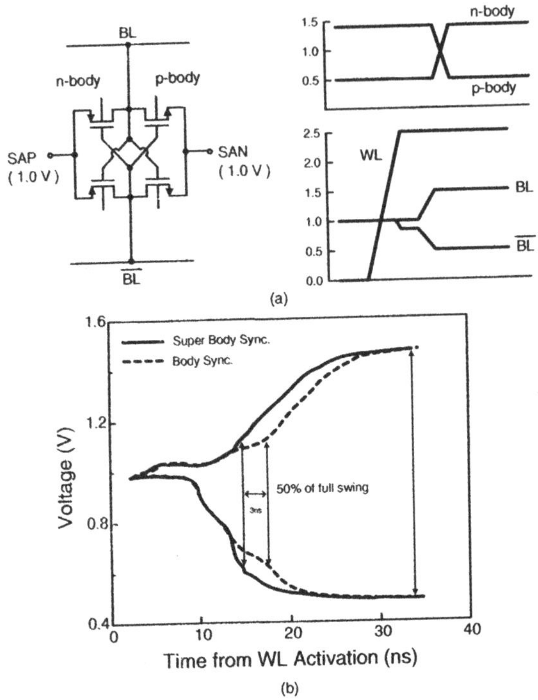

Fig. 5.24 (a) Super body-synchronous sensing scheme used in a 1.5 V multi-Giga-scaled SOI DRAM using a 0.15 μm PD SOI CMOS technology. (b) Bit line waveforms during the sensing transient. (From Kuge et al. [25]. ©1996 IEEE.)

V), which brings in an even larger drain conducting current. After sensing, the body potentials of the NMOS and the PMOS devices are reversed, which results in the threshold voltage of the sense amp of 0.44 V as for the body-synchronous sensing scheme. Compared to the body-synchronous sensing scheme, the sensing time of the super body-synchronous sensing scheme is $\sim$ 3 ns faster as shown in Fig. 5.24(b). In this multi-Giga-scaled DRAM, body and super body synchronous sensing schemes using the boosted sense ground (BSG) have been adopted. Thus, the boosted sense ground of $V_{BSG} = 0.5$ V has been provided for the body of both NMOS and PMOS devices, which is different from the body bias of $V_{BB} = -1$ V for the conventional DRAM sense amp. If we use the boosted sense ground (V_{BSG}) no additional body power supply is required.

Figure 5.25(a) shows the cross section of a 0.5 μm PD SOI NMOS device with a body contact via a polysilicon pad [26]. Fig. 5.25(b) shows the critical path during the read operation of the DRAM circuit using the PD SOI CMOS devices with body contacts. As shown in this figure, in the critical path, body contacts are applied to the devices used in the word-line driver, the sense amp, and the output buffer, which are susceptible to floating body effects easily. As for the memory cells and logic gates, floating-body devices are still used since in these portions, the floating body effects

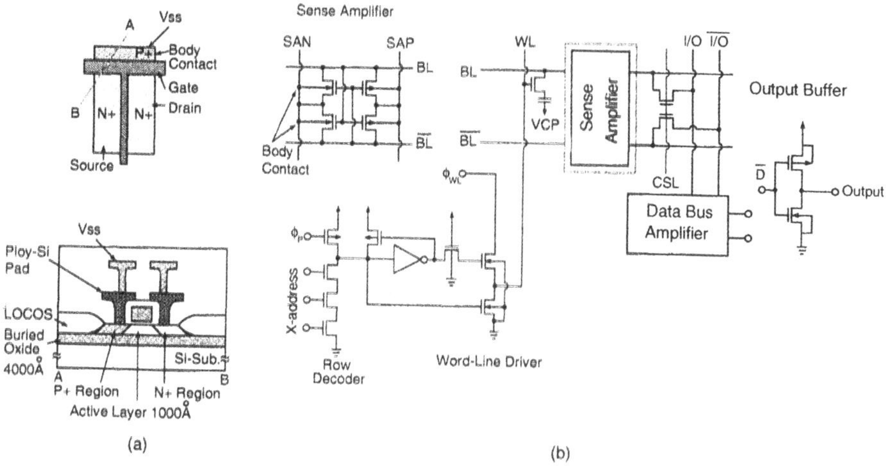

Fig. 5.25 (a) Cross-section of the 0.5 μm PD SOI NMOS device with a body contact via a polysilicon pad. (b) Critical path during the read operation of the DRAM. (From Suma et al. [26]. ©1994 IEEE.)

are smaller, which can be avoided using techniques described before. In order to facilitate high layout density for the these portions, body contacts are not used for the devices.

In addition to elimination of the floating body effects, the body contacts also provide advantages for low-voltage operation. Figure 5.26 shows the body-pulsed sense amp (BPS) used in a 16-Mb SOI DRAM [27]. As shown in this figure, the body voltage of the devices used in the BPS is adjusted dynamically to increase the current conductivity of the devices at the appropriate time. During other times, the body voltage is maintained at its original value. Thus the speed performance can be enhanced without increasing the power consumption substantially. During period (1), word line (WL) turns high and bit line (BL) begins to change. During period (2), SON turns high and the sense amp is active. At this time, the body voltage of devices M1 and M2 (SBN) is adjusted to a high value and similarly for the body voltage of devices M3 and M4 (SBP). Thus the magnitude of the threshold voltages of devices M1-M4 becomes smaller, their currents become larger, and the function of the sense amp speeds up. In period (3), which is the restoring period, the body voltages of devices M5 and M6 (SWP & SWN) are adjusted dynamically such that the current conductivity of devices M5 and M6 could be raised temporarily to speed up the restoring process.

Fig. 5.27 shows the body driven equalizer (BDEQ) circuit used in a 16-Mb SOI DRAM [27]. When the equalizer is active (BLEQ is high), the body voltage of the NMOS device is raised to decrease the threshold voltage of the NMOS device. Thus, the current driving capability of the device is enhanced to speed up the equalizing time.

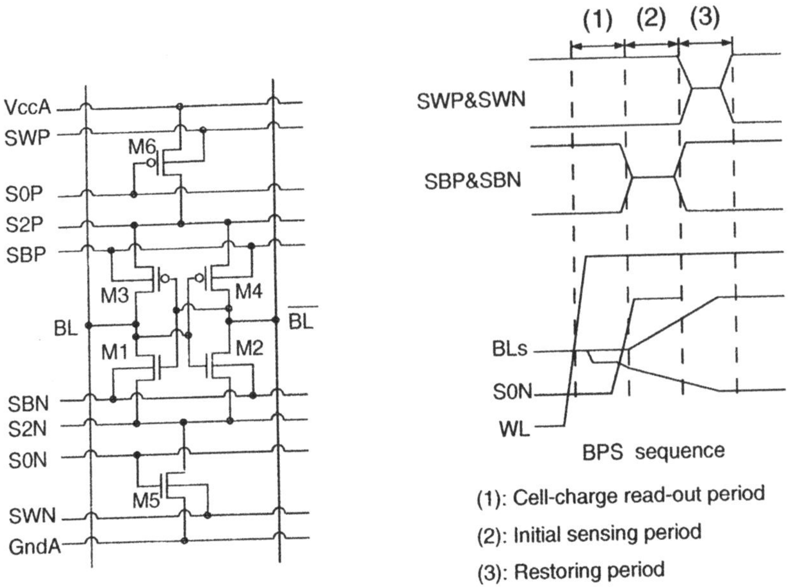

Fig. 5.26 Body-pulsed sense amp (BPS) used in a 16-Mb SOI DRAM. (From Shimomura et al. [27]. ©1997 IEEE.)

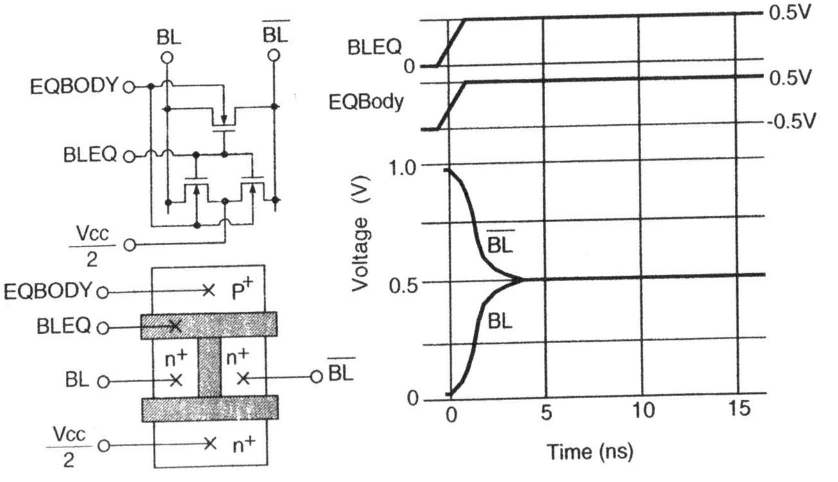

Fig. 5.27 Body driven equalizer (BDEQ) circuit used in a 16-Mb SOI DRAM. (Adapted from Shimomura et al. [27].)

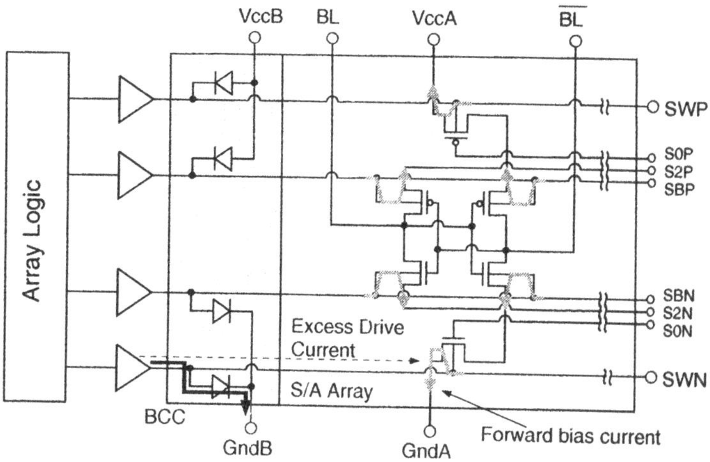

Fig. 5.28 Body current clamper (BCC) used in a 16-Mb SOI DRAM which is based on stabilizing the body voltages of devices used in the sense amp. (Adapted from Shimomura et al. [27].)

Figure 5.28 shows the body current clamper (BCC) used in a 16-Mb SOI DRAM [27], which is based on stabilizing the body voltages of devices used in the sense amp. If the body voltage is too high (for NMOS) or too low (for PMOS), the body-source diode may turn on to cause a large amount of current flowing over V_{CCA} and GNDA, which brings in a fluctuation in V_{CCA} and GNDA. As a result, the operation of the sense amp is affected. As shown in the figure, by using the diodes to connect the body-bias line to V_{CCB} and GNDB, the body voltage can be clamped to an appropriate value to reduce the fluctuation in V_{CCA} and GNDA.

Figure 5.29 shows the body pulsed transistor logic (BPTL) circuit [27], which is used in the peripheral circuits of a 16-Mb SOI DRAM. The BPTL circuit is based on the dynamic adjustment of the body potential of the active logic gate to accelerate the speed performance temporarily. Since the arrival time for the active signal (CK) to reach each block is different and the time needed for each block to operate is limited, only during the time from the beginning to the end of the operation of each block the body bias (BP) need to be adjusted. As shown in the figure, using an AND gate with an appropriate delay line as marked, the needed BP signal can be accomplished. When BP is high, it is the time for the operation of the block. Figure 5.30 shows the access time and the performance of the 16-Mb SOI DRAM with the body control techniques using a 0.5 μm SOI CMOS technology [27]. At the supply voltage of 1 V, the access time is 46 ns, which is much better than that of the SOI DRAM without body control.

Figure 5.31 shows the equalized body circuit architecture used in the DRAM [28]. As shown in this figure, the bodies of all NMOS devices of the surrounding logic gates of the equalized body circuit are connected to V_{NB} and the bodies of all PMOS devices are connected to V_{PB}. During the standby period, signal ACT is low, V_{NB} is low, and

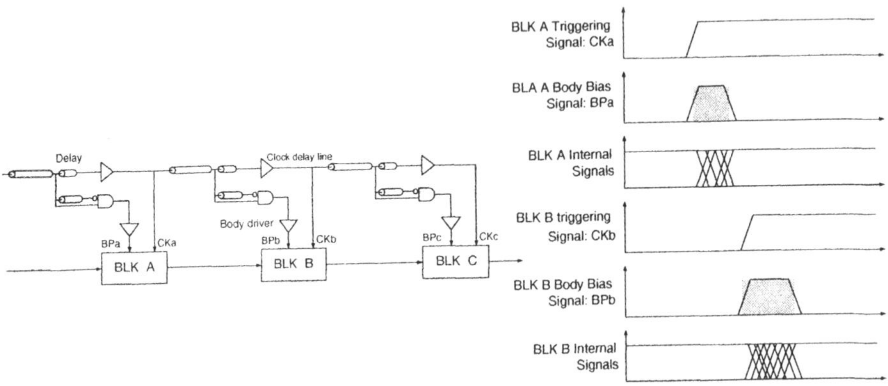

Fig. 5.29 BPTL circuit used in a 16-Mb SOI DRAM. (Adapted from Shimomura et al. [27].)

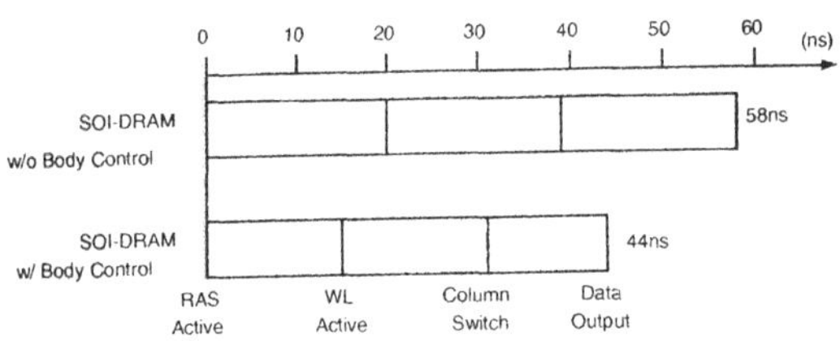

Technology	0.5 μm CMOS/SIMOX, Dual gate, Modified MESA isolation
Thickness (tox/SOI/BOX)	10nm / 100nm / 400nm
Cell size	1.35 x 2.8 = 3.78 mμ ^{2}
Chip size	17.5 x 8.8 = 154mm
RAS access time	46ns @ 1.0V
Active current	11.5mA @1.0V, Tc=100ns
Standby current	4.4 μA @1.0V

Fig. 5.30 Access time and performance of the 16-Mb SOI DRAM with the body control techniques using a 0.5 μm SOI CMOS technology. (Adapted from Shimomura et al. [27].)

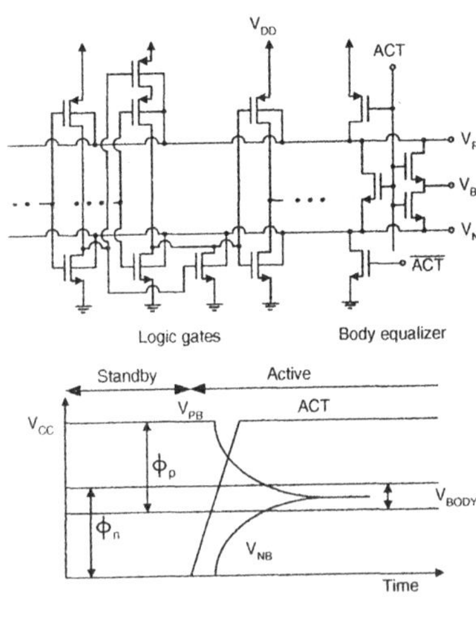

Fig. 5.31 Equalized body circuit architecture used in the DRAM. (Adapted from Yamauchi et al. [28].)

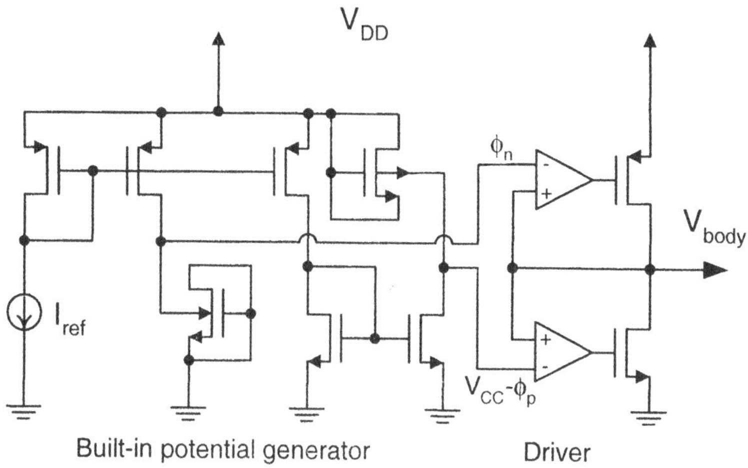

Fig. 5.32 Body voltage generator. (Adapted from Yamauchi et a. [28].)

V_{PB} is high. In the active period, ACT is high and both V_{NB} and V_{PB} are equalized to the same voltage level– V_{body}, which needs to be between ϕ_n and $V_{CC} - \phi_p$, to avoid the leakage current from the pn junction, where ϕ_n/ϕ_p is the turn-on voltage of the body-source diode in the NMOS/PMOS device. In order to effectively control the V_{body} potential, the body voltage generator, as shown in Fig. 5.32, is used. As shown in this figure, via current mirrors, the reference current I_{ref} is transformed to provide two potentials ϕ_n and $V_{CC} - \phi_p$. Via a comparator, the output of V_{body} is between ϕ_n and $V_{CC} - \phi_p$.

The driver circuit used in the output is also important for a DRAM circuit with low-voltage, low-power, and high-speed requirements. Fig. 5.33 shows the feedback body driver circuit with its transient waveforms [28], which is similar to the circuit described in the DTMOS section in Chapter 4 except for the addition of the extra devices PM1 and NM1. If we consider the NMOS portion, when the input switches from high to low, V_{NB} is pulled high by PM2 and the threshold voltage of the main NMOS device is lowered. Therefore, the output is pulled down quickly. If the input switches from low to high, V_{NB} is pulled to 0 V by NM1. Hence, the threshold voltage of the main NMOS device rises to reduce its leakage current, and hence its standby power consumption. In contrast, when the input switches from high to low, PM1 is used to make V_{PB} maintain high such that its leakage current can be reduced.

5.4 SRAM

Due to strong radiation hardness properties and small parasitic capacitances, SOI CMOS technology has been regarded as a good candidate for implementing SRAMs. Figure 5.34 shows the layout of an SOI CMOS SRAM cell [29]. As shown in this figure, in the SOI SRAM cell the drains of both NMOS and PMOS devices are

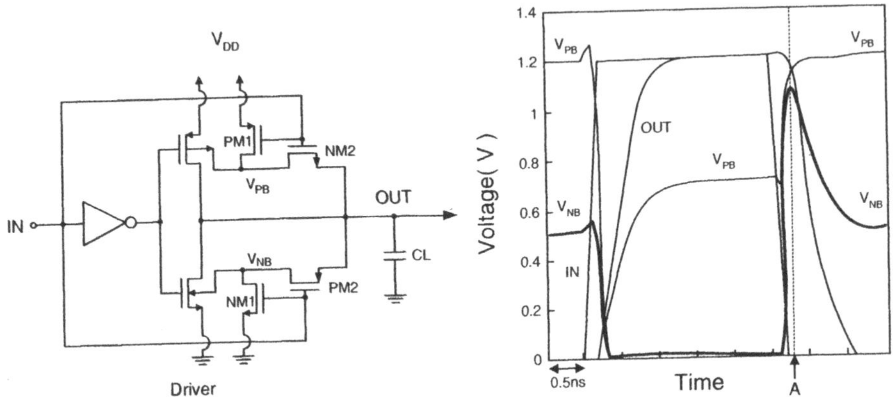

Fig. 5.33 Feedback body driver circuit with its transient waveforms. (Adapted from Yamauchi et al. [28].)

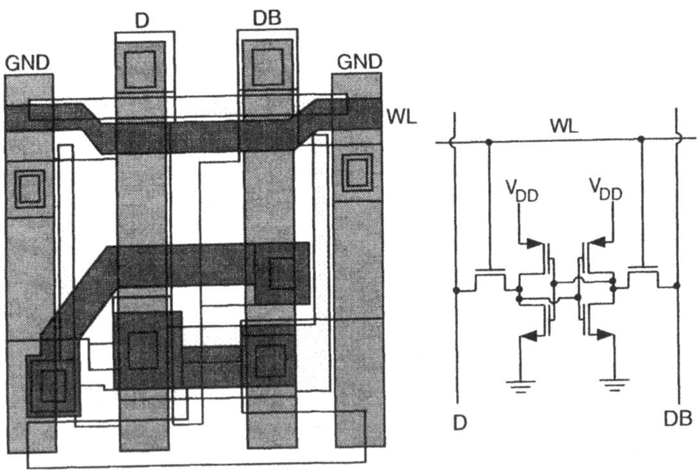

Fig. 5.34 Layout of an SOI CMOS SRAM cell. (Adapted from Kumagai et al. [29].)

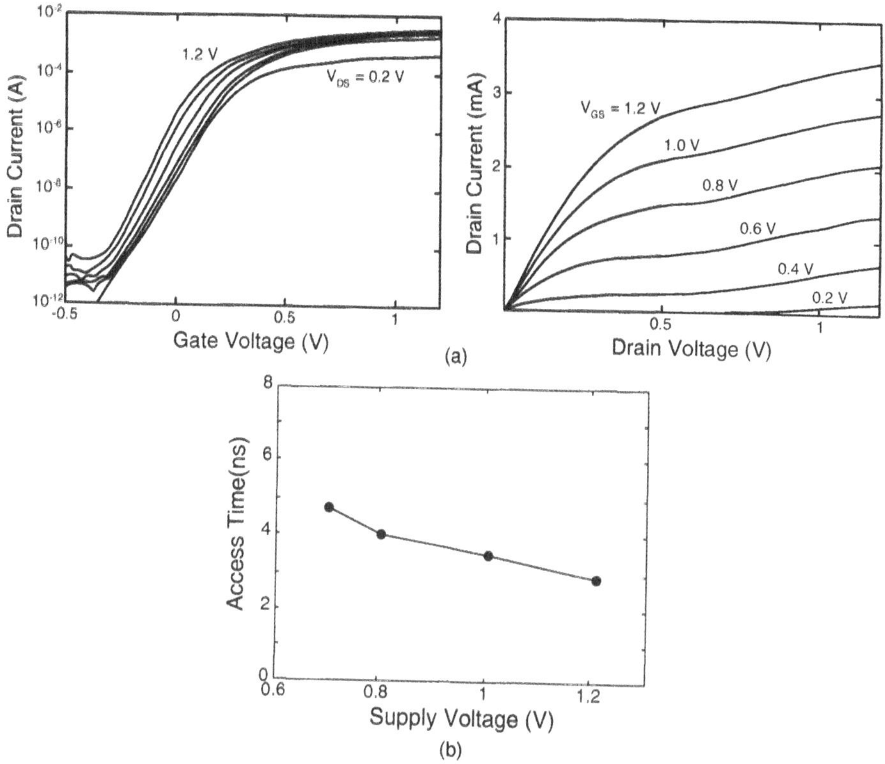

Fig. 5.35 (a) Characteristics of a 0.1 μm PD SOI NMOS device with a front gate oxide of 42 Å. (b) Access time versus supply voltage of the 512 K SRAM using a 0.2 μm SOI CMOS technology. (From Shahidi et al. [30]. ©1993 IEEE.)

connected directly without isolation problems, In addition, the access pass transistor shares the drain with the NMOS device in the inverter. By using the SOI technology, the layout area of the SRAM cell is much smaller as compared to the bulk case.

As shown in Fig. 5.35(a), the subthreshold slope of PD SOI MOS devices improves at a large drain voltage owing to the floating body effect, which can be taken advantage for realizing an SRAM cell under low-voltage, low-power requirements. However, during turn-on, the kink effects due to floating body effects degrade the output conductance, which can be improved by adding body contacts. As shown in Fig. 5.35(b), for a 512 K SRAM implemented by a 0.2 μm PD SOI CMOS technology, the access time at the supply voltage of 1 V is 3.5 ns. In addition, the power consumption is much smaller as compared the SRAM using bulk CMOS technology of the same level. From this figure, the potentials of SOI technology for realizing low-voltage, low-power SRAMs can be visualized.

Fig. 5.36 shows the memory cell and the sense amp in an early-stage SOI SRAM circuit [31]. As shown in this figure, in the memory cell, floating-body devices are used to have a high layout density. For the pass transistors, they are with their body

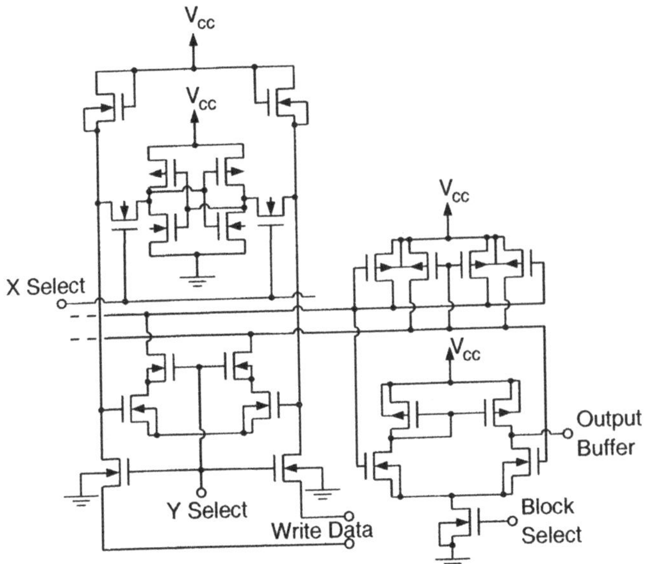

Fig. 5.36 Memory cell and sense amp in an early-stage SOI SRAM circuit. (Adapted from Isobe et al. [31].)

grounded. For sense amps and other biasing circuits, the body-tied-to-source scheme has been adopted such that the output conductance is not worsened to degrade the gain of the amplifier. In addition, the instability of the device due to floating body effects can also be avoided.

5.4.1 Floating Body Effects

The parasitic bipolar effect due to the floating body of the memory cell may impact the performance of an SRAM circuit. Figure 5.37 shows (a) the simplified block diagram of a 512-cell SRAM column with (b) the waveforms of the bit line (BL) and the cell node (A) during the transient of the write-0 to cell-0 operation of the SOI SRAM operating at $V_{DD} = 1.95$ V [32]. As shown in Fig. 5.37(b), 'max' refers to the case with data '1' stored in cells 1-511–node A is high. 'min' refers to the case with data '0' stored in cells 1-511– node A is low. When the write-0 to cell-0 operation is initiated, LWL is high, via the read/write switch, the input data DIN pulls down the BL. At this time, the pass transistor of cell 0 is turned on. Hence, storage node A is pulled low to accomplish the write-0 to cell-0 operation. The speed of the write-0 to cell-0 operation is affected depending on the storage node condition–'max' or 'min'. For the case with the 'max' condition– the storage node of cells 1-511 is high, the write-0 to cell-0 operation takes a longer time as compared to the 'min' case. For the case with 'max' condition, the drain voltage of pass transistor T1 is high. When its source voltage switches from high to low, a substantial amount of the leakage current exists in the parasitic bipolar device in pass transistor T1, which

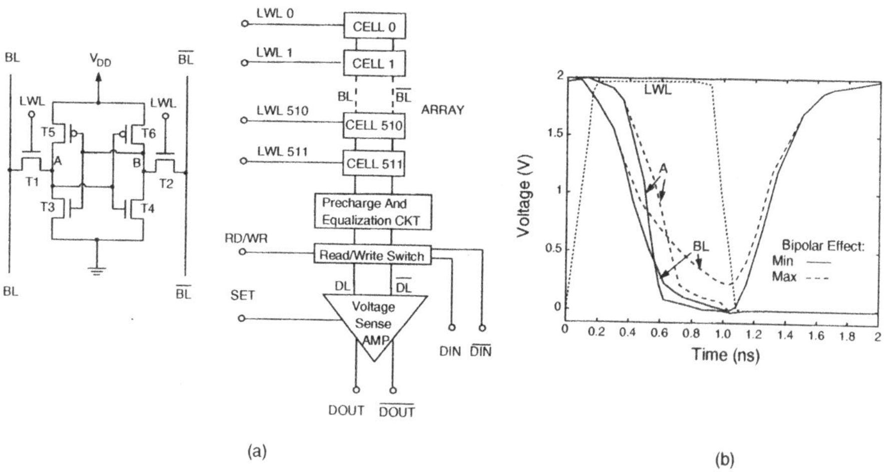

Fig. 5.37 (a) Simplified block diagram of a 512-cell SRAM column. (b) Waveforms of the bit line (BL) and the cell node (A) during the transient of the write-0 to cell-0 operation of the SOI SRAM operating at $V_{DD} = 1.95$ V. (Adapted from Kuang et al. [32].)

slows down the pull-low operation of the BL. Therefore, the write-0 operation to cell 0 is affected.

The radiation hardness of the SOI SRAM is much better as compared to the bulk SRAM. However, in addition to the α-particle induced soft errors, the parasitic bipolar devices of the PD SOI devices may also cause soft errors. In order to reduce the parasitic bipolar device induced soft errors, the body-fixed scheme should be used for the SRAM. Figure 5.38 shows the layouts of the SOI SRAM memory cells with the body-fixed and the body-floating schemes using a 0.35 μm SOI CMOS technology [33]. For the 128 kb SOI SRAM with the body-fixed scheme, its soft error rate is only one 300th of the SRAM with body-floating scheme. As shown in Fig. 5.38, for the body-fixed scheme, the body region is formed by the field-shielded gate to inhibit the leakage current at the transistor edge. Its body region is fixed at ground (for NMOS) and V_{DD} (for PMOS) to inhibit the function of the parasitic bipolar device. Due to the field-shield gate, LOCOS has been used to isolate the NMOS device from the PMOS device using the body-fixed scheme such that latchup can be avoided. In contrast, for the SOI SRAM with the body-floating scheme, N^+ and P^+ drain regions have been used to isolate the PMOS device from the NMOS device in order to inhibit latchup. Therefore, the layout area is smaller using the body-floating scheme. By using the body-fixed scheme in the SOI SRAM, its soft error rate is reduced owing to the inhibited parasitic bipolar device. However, the memory cell occupies a larger layout area.

The floating body effect of the PD devices may also affect the performance of a sense amp in a PD SOI SRAM. Figure 5.39 shows the SRAM sense amp using a clocked dual-slope technique [34]. As shown in this figure, during sensing, device M_{narrow} in the dual-slope SRAM sense amp is turned on first. Therefore, the current

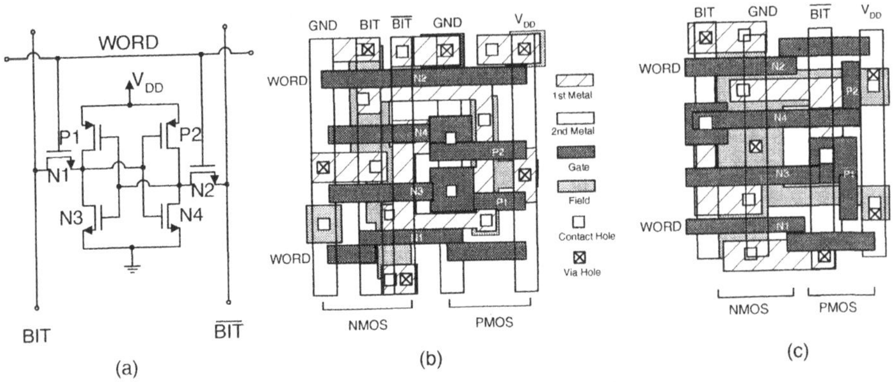

Fig. 5.38 Layouts of the SOI SRAM memory cell with body-fixed and body-floating schemes using a 0.35 μm SOI CMOS technology. (From Wada et al. [33]. ©1998 IEEE.)

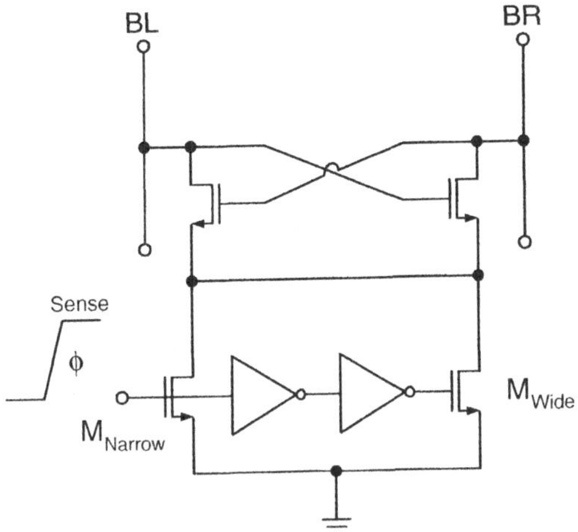

Fig. 5.39 Dual-slope SRAM sense amp. (Adapted from Chuang et al. [34].)

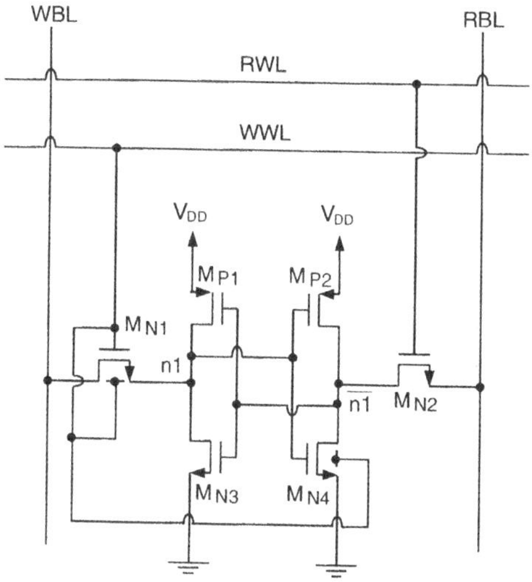

Fig. 5.40 A 0.7 V two-port 6T SRAM memory cell structure with the single-BL simultaneous read-and-write access (SBLSRWA) capability using PD SOI DTMOS techniques. (Adapted from Liu & Kuo [35].)

increases slowly such that the differential voltage across the cross-coupled pair grows. After the differential voltage reaches ~ 150 mV, device M_{wide} also turns on with a much larger current such that the differential voltage is amplified further. Initial development of the differential voltage across the bit lines is critical. Due to the floating body effect of the PD devices, the SOI dual-slope sense amp may encounter more difficulties. Because of the floating body effect, the body voltage of the PD SOI devices in the sense amp is not certain. Thus the related threshold voltage is uncertain. The reduced threshold voltage of the sense transistor due to the floating body effect causes a large amount of current in device M_{narrow}. Consequently, the differential voltage across the cross-coupled pair cannot be developed. The sense margin and the sense speed of the sense amp are worsened. A wrong logic state may even be generated at the output of the sense amp. These floating-body induced problems in the dual-slope sense amp can be avoided by adding body contacts to the PD SOI devices.

5.4.2 Two-Port SRAM

SOI floating body techniques can also be used in the design of low-voltage two-port SRAM circuits. Figure 5.40 shows a 0.7 V two-port 6T SRAM memory cell structure with the single-BL simultaneous read-and-write access (SBLSRWA) capability using PD SOI DTMOS techniques [35]. As shown in this figure, in this two-port 6T SRAM memory cell, independent write word line (WWL) and read word lines (RWL) have been used. In addition, the conventional BL pair has been divided into write bit line (WBL) and read bit line (RBL). In order to facilitate single-BL write access, dynamic

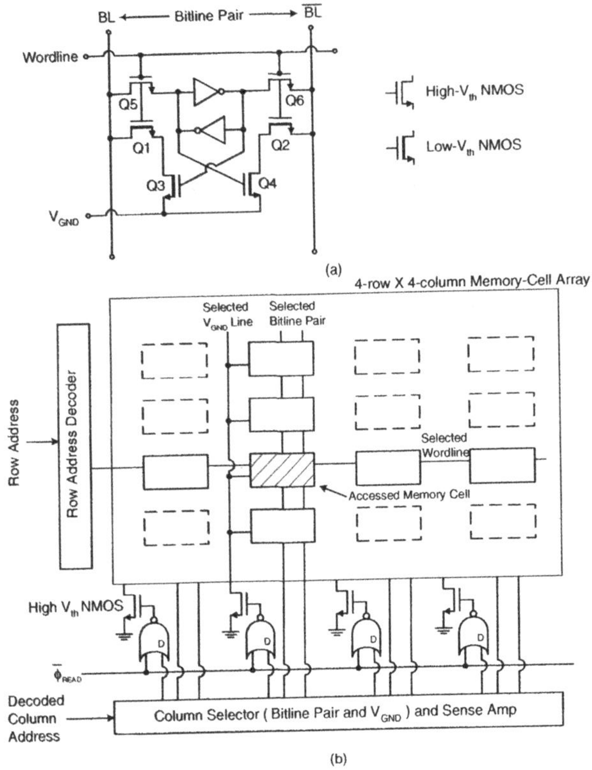

Fig. 5.41 (a) A 10T SRAM cell and (b) architecture of the 1 V 128 K embedded 100-MHz SRAM using SOI MTCMOS techniques. (Adapted from Shibata et al. [36].)

threshold techniques have been used for transistors MN4 and MN1. The body of latch transistor MN4 is controlled by the WWL. Both the body and the gate of pass transistor MN1 are connected to the WWL. The right side of the latch (MP2, MN4) in the memory cell is connected to the RBL and the left side (MP1, MN3) is connected to the WBL. Owing to the dynamic threshold characteristics of transistors MN1 and MN4, their reduced threshold voltage during the write access facilitates the single-BL write operation.

5.4.3 MTCMOS Techniques

SOI MTCMOS techniques have been used to design low-voltage, high-speed SRAMs. By using SOI MTCMOS techniques, the low-threshold devices have been used to enhance the speed during operation and the high-threshold devices have been used to reduce the subthreshold current during standby for the purpose of lowering power dissipation. Figure 5.41(a) shows a 10T SRAM cell used in the 1 V 128 K embedded 100-MHz SRAM using SOI MTCMOS techniques [36]. As shown in this figure, high-threshold NMOS devices Q5 and Q6 are used to reduce the subthreshold current

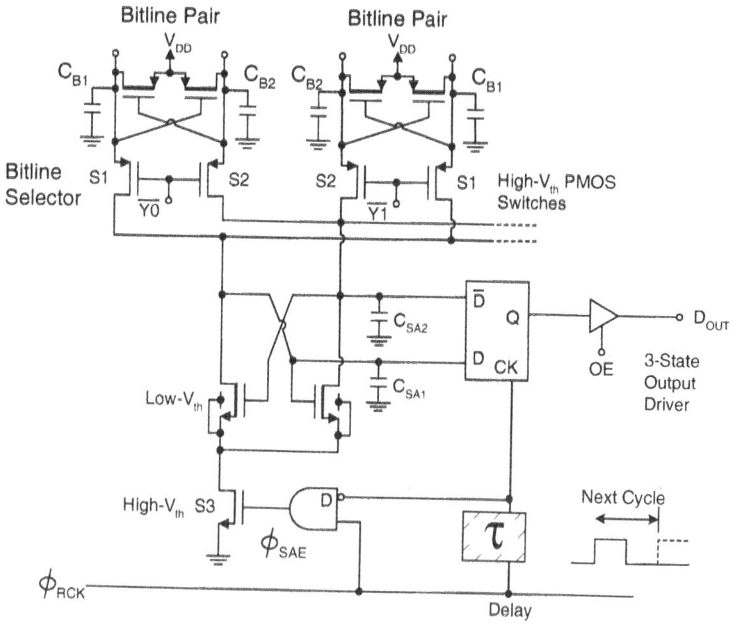

Fig. 5.42 Selector-merged MTCMOS sense amp in the readout circuit with the cross-coupled pullup pairs at the bitlines in the 1 V 128 K embedded 100-MHz SRAM. (From Shibata et al. [36]. ©2000 IEEE.)

to achieve low power dissipation during standby. Four low-threshold NMOS devices (Q1-Q4) are used to enhance the current driving capability such that the readout operation can be speeded up. Fig. 5.41(b) shows the architecture of the 1 V 128 K embedded 100-MHz SRAM using SOI MTCMOS techniques with the row decoder, the memory array, and the sense amps [36]. As shown in this figure, virtual ground line V_{GND} has been used. During the read cycle, the virtual ground line V_{GND} in the selected column has been switched to the ground line such that a BL discharge path can be provided before the switch-high of the WL. Hence, the BL discharge delay time can be minimized. During another cycle, the virtual ground line V_{GND} is floating to reduce power dissipation. For the purpose of reducing the subthreshold current to achieve lower power dissipation, high-threshold NMOS devices have been used in the flip-flop structures.

Figure 5.42 shows the selector-merged MTCMOS sense amp in the readout circuit with the cross-coupled pullup pairs at the bitlines in the 1 V 128 K embedded 100-MHz SRAM [36]. As shown in this figure, a latch-type cross-coupled pair is used as a sense amp. Pass transistors S1 and S2, which are implemented by high-threshold PMOS devices and controlled by the decoded column address signals $\overline{Y0}$ and $\overline{Y1}$, are used to determine which pair of bit lines are to be connected to the sense amp. Switch S3, which is made of a high-threshold NMOS device, is used as a switch for the sense amp. In addition, switch S3 also provides a high common-source impedance

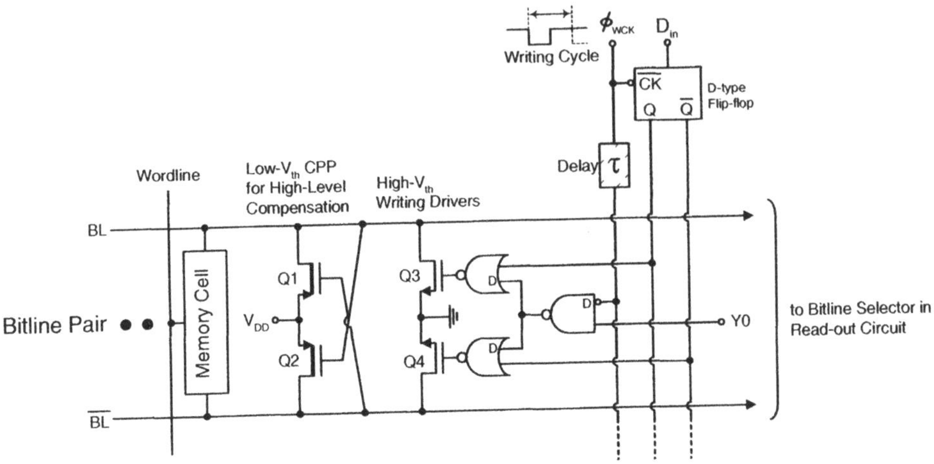

Fig. 5.43 Precharged BL scheme used in the write-in circuit in the 1 V 128 K embedded 100-MHz SRAM using SOI MTCMOS techniques. (Adapted from Shibata et al. [36].)

such that a high sensitivity can be obtained for the sense amp. In order to avoid the threshold mismatch of the sense amp due to the different biasing conditions for the PD SOI NMOS cross-coupled pairs from the floating body effect, the body of the cross-coupled devices is connected to the common source node. At the BL pairs, the cross-coupled pull up pair circuit (CPP) has been used such that the high BL can be maintained at the V_{DD} level to avoid accidental turn-off of the PMOS switches. Note that the memory cell can only pull the BL up to $V_{DD} - V_{TN}$, where V_{TN} is the threshold voltage of the NMOS device at the low supply voltage and PMOS switches S1 and S2 may not be turned on. With the adoption of CPP, the BL can be pulled up to V_{DD} and PMOS switches S1 and S2 can be turned on when necessary. Before the readout operation, during the precharge period, the turn-on of the decoded column address signals $\overline{Y0}$ and $\overline{Y1}$ selects the pair of the bit lines to be connected to the sense amp. Based on the clock signal ϕ_{RCK}, switch S3 may turn on for sensing or turn off for disabling the sense amp. When the sense amp is disabled, the readout data is store in the D-type flip-flop.

Figure 5.43 shows the precharged bitline scheme used in the write-in circuit in the 1 V 128 K embedded 100-MHz SRAM [36]. As shown in this figure, the adoption of the SOI MTCMOS techniques is to have a high-speed write-in operation at the low power dissipation. As shown in this figure, for each BL pair, there are writing drivers Q3 and Q4 made of high-threshold NMOS devices to cut off the conduction path of the subthreshold currents from the precharged BL to ground. In addition, the CPP made of low-threshold PMOS devices has been used to increase the current driving capability to shorten the writing time.

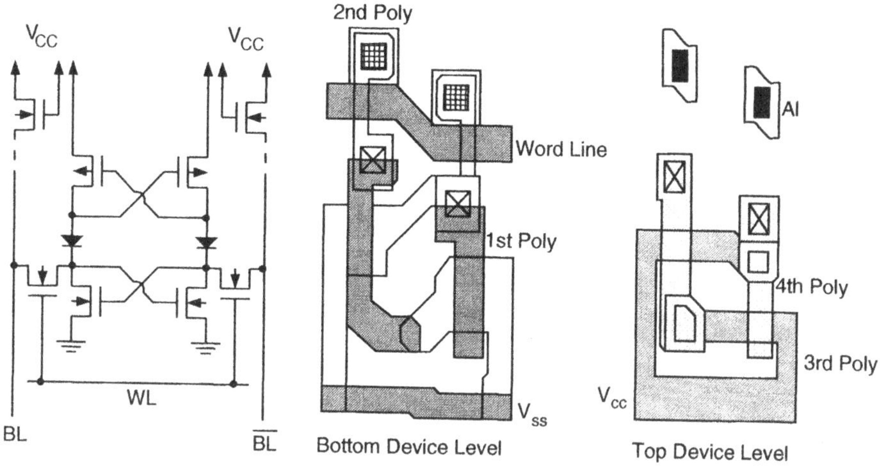

Fig. 5.44 SRAM cell using a bulk CMOS technology with laser-recrystallized SOI PMOS loads. (Adapted from Takao et al. [37].)

5.4.4 Advanced SOI SRAM

Figure 5.44 shows an SRAM cell using a bulk CMOS technology with laser recrystallized SOI PMOS loads [37]. As shown in this figure, the bottom level of the SRAM cell including the pass transistor and the NMOS devices in the cross-coupled latch is made of bulk NMOS devices. The top level–the PMOS devices of the cross-coupled latch is implemented by the SOI PMOS devices, whose thin film is formed by laser recrystallization of the CVD polysilicon layer. In this memory cell, the NMOS device is based on the N^+ polysilicon gate. The PMOS device, which is based on P^+ polysilicon gate, is connected to the drain of the NMOS device. Therefore, an equivalent diode structure exists.

Figure 5.45(a) shows the cross-section of the bulk CMOS technology with the SOI PMOS device used for realizing the SRAM described in Fig. 5.44. As shown in this figure, the bottom level is the standard bulk NMOS device. The top level is the SOI PMOS device. As shown in Fig. 5.45(b), compared to the other loads made of thin-film transistors (TFT) for the SRAM cells, the load made of the laser-recrystallized SOI PMOS device has an even smaller turn-off leakage and a larger turn-on current, which result in a much better performance of the SRAM circuit.

Three-dimensional IC technology has also been used to realize SRAM. As shown in Fig. 5.46, in the 3D SRAM structure, the bottom layer is the memory cell made of bulk MOS devices and the top layer is the SOI devices made by zone melting recrystallization (ZMR) techniques [39]. The top layer is used for address decoders, sense amps, I/O buffers, and chip controllers, etc.

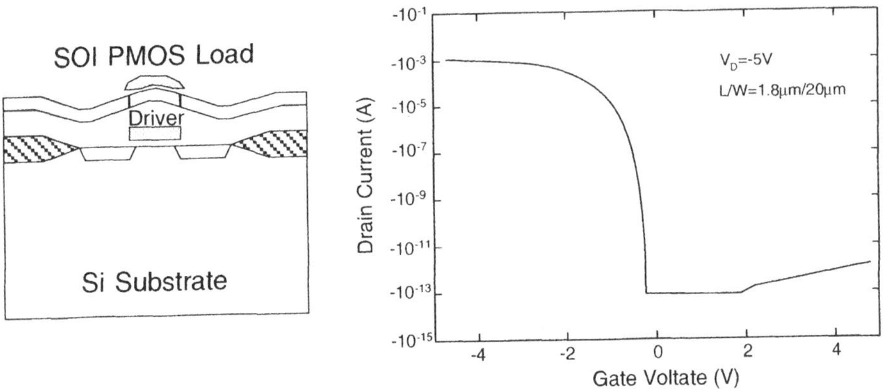

(a) (b)

Fig. 5.45 (a) Cross-section of the bulk CMOS technology with SOI PMOS load. (b)Drain current characteristics of the SOI PMOS load with a channel width of 20 μm and a channel length of 1.8 μm, biased at the drain voltage of -5 V. (Adapted from Takao et al. [38].)

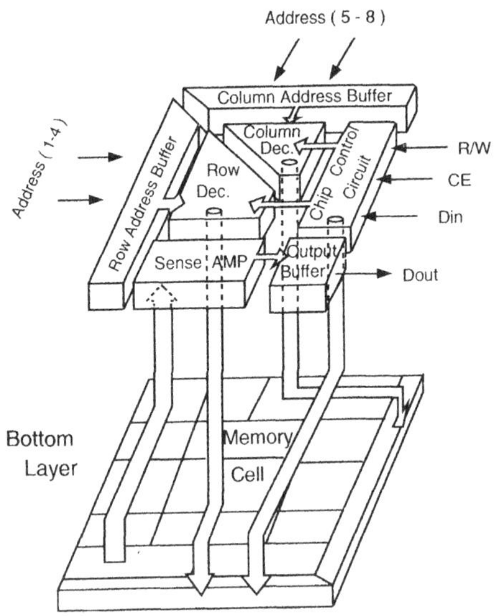

Fig. 5.46 3D SRAM structure. (Adapted from Nishimura et al. [39].)

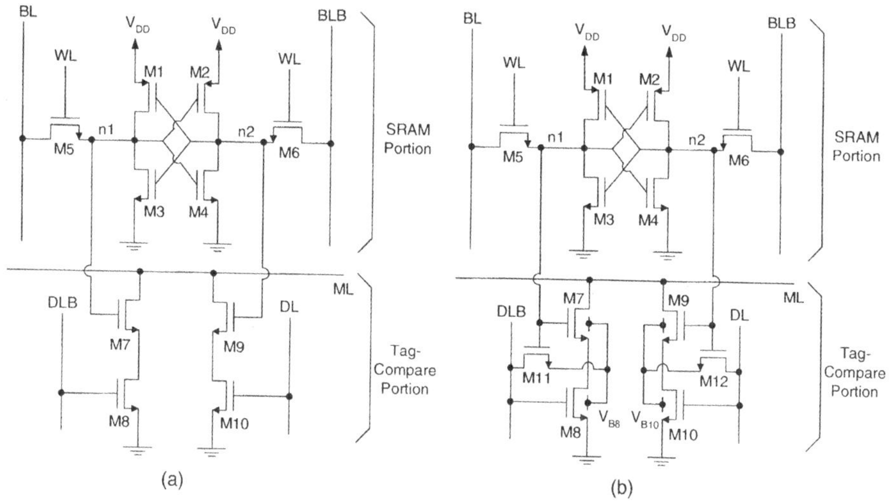

Fig. 5.47 (a) Conventional content addressable memory (CAM) cell [41]. (b) Low-voltage CAM cell using PD SOI DTMOS techniques. (Adapted from Liu et al. [42].)

5.5 CAM

Content addressable memory (CAM), which is derived from SRAM, is broadly used in VLSI systems to facilitate operation of fast comparison and validation of patterns [40]. In a conventional 10-T CAM cell [41] as shown in Fig. 5.47(a), CAM is composed of two portions: the SRAM portion (M1-M6) and the tag-compare portion (M7-M10) for performing the XOR operation of the data stored in the SRAM cell with the input data at the digit lines. If a logic-1 is stored at the internal node n1, which is different from the logic state of the data on the digit line (DL), then the match line (ML) is pulled down to ground, indicating a miss. Using the PD SOI DTMOS techniques, Fig. 5.47(b) shows a low-voltage CAM cell structure with a fast tag-compare capability [42]. As shown in this figure, in addition to the 10 transistors for the SRAM cell portion and the tag-compare portion as in the conventional CAM cell, two auxiliary transistors M11 and M12, which are controlled by the internal storage nodes of the SRAM cell for controlling the bodies of transistors M7-M10 have been added to the tag-compare portion. Digit lines DL and DLB are connected to the drain of M12 and M11. In addition, digit lines DL and DLB also control the turn-on of M10 and M8. By using this arrangement, either M11 or M12 turns on depending on the data stored in the SRAM cell. If the internal storage node n1 in the memory cell is high, both M7 and M11 turn on and both M9 and M12 are off. Prior to the tag-compare operation, both digit lines DL and DLB are low and the ML is precharged to high. Since M11 is on, the bodies of M7 and M8 are tied to low via M11. During the tag-compare operation, if logic-0 exists on the digit line DL (logic 1 exists on the complementary digit line DLB), M8 turns on. Due to the logic 1 status on the complementary digit line DLB, the threshold voltages of both M7 and

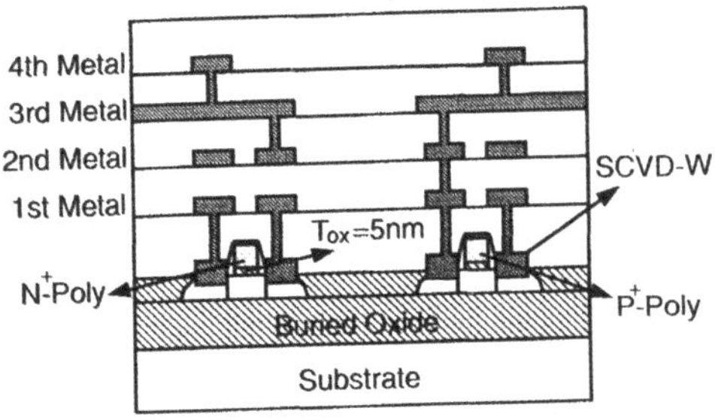

	NMOS	PMOS
L(μm)	0.25	0.25
W(μm)	8.05	12.25
t_{ox}(nm)	5.0	5.0
t_{si}(nm)	50	50
t_{BOX}(nm)	100	100
R_{SD}(Ω/sq.)	3	4
V_T(V)	0.20	-0.25

Fig. 5.48 Cross-section of a 0.25 μm FD SOI CMOS technology for implementing a 300 kG gate array. (Adapted from Sato et al. [43].)

M8 are lowered since the bodies of M7 and M8 are high. As a result, the conductance of M7 and M8 increases, and the match line ML is discharged (indicating a miss) faster than in the conventional case. When the tag-compare procedure is over, the complementary digit line DLB is restored back to low. Therefore, M7 remains on and M8 turns off with their threshold voltages restored to their zero-biased value since their bodies are tied to zero by M11. This implies the leakage current in M8 stays low as in the case with DTMOS techniques. Using the DTMOS techniques, the CAM cell using PD SOI DTMOS techniques can work at a supply voltage of $V_{DD} = 0.7$ V with a faster speed performance.

5.6 GATE ARRAY

SOI CMOS technology has been used to integrate gate arrays. In this section, gate arrays using FD, PD, DTMOS, and MTCMOS technologies are described.

5.6.1 FD Gate Array

Since body contacts are not required for the FD SOI devices, the layout density is high, which is especially useful for implementing gate arrays. Figure 5.48 shows the cross-section of a 0.25 μm FD SOI CMOS technology with a front gate oxide of 50 Å, a thin film of 500Å, a buried oxide of 1000 Å, and tungsten-deposited source/drain for implementing a 300 kG gate array. For NMOS devices, N^+ polysilicon has been used. For PMOS devices, P^+ polysilicon is adopted. The channel width of the NMOS/PMOS device is 8.05/12.25 μm [43]. Figure 5.49 shows the layout of the cell in a typical gate array using a 0.25μm FD SOI CMOS technology [44]. As shown in this figure, in the cell, there are 10 CMOS devices of normal and small sizes. Normal-size devices are for implementing logic gates. Small-size devices are used for the load devices and the transmission gate devices used in the SRAM. Each gate array cell can be used to form a 2-input NAND gate or a two-port 6T SRAM cell.

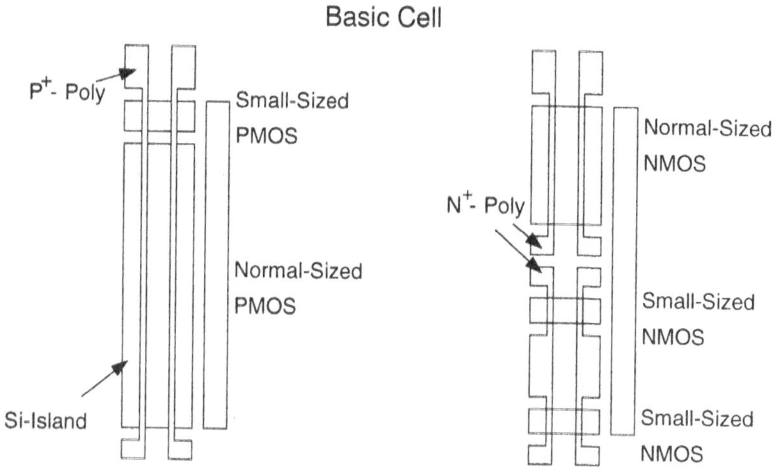

Fig. 5.49 Layout of the cell of a typical gate array using the 0.25 μm FD SOI CMOS technology. (Adapted from Kado et al. [44].)

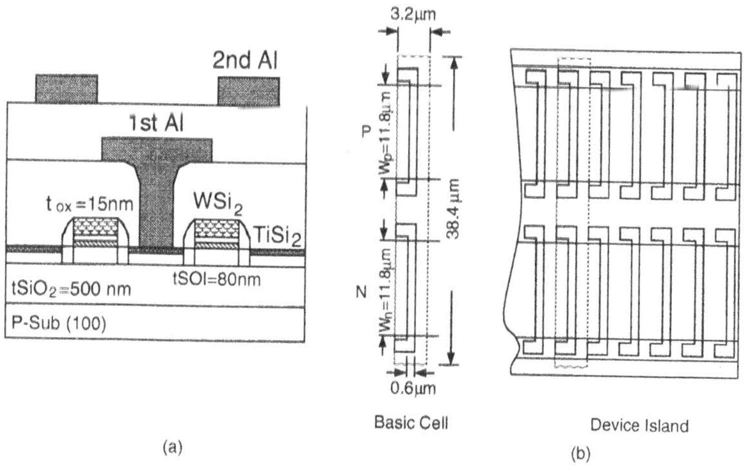

Fig. 5.50 (a) Cross-section of a 0.6 μm gate-array FD SOI CMOS technology with a front gate oxide of 150 Å, a thin film of 500 Å, and a buried oxide of 5000 Å. (b) Layout of the gate-array cell in the sea of gate (SOG) configuration. (Adapted from Yamaguchi et al. [45].)

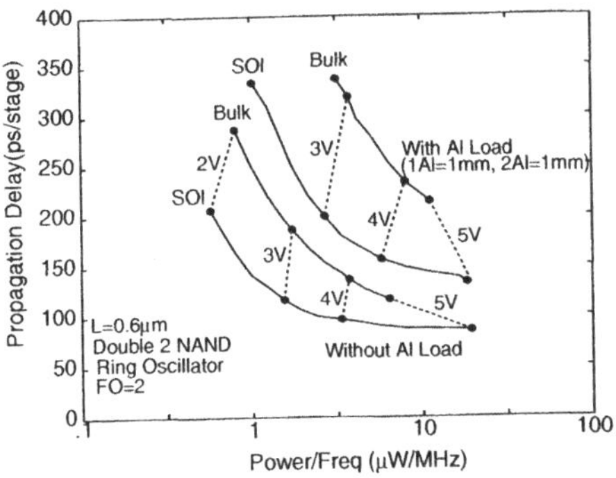

(a)

	C (SOI)	C (bulk)	C(SOI)/C(bulk)
Active Gate (F.O.=1) C_{ox}	36.6fF	37.6fF	0.97
N^+ Junction (1 drain) C_{Jn}	9.5fF	18.9fF	0.50
P^+ Junction (1 drain) C_{Jp}	7.6fF	21.6fF	0.35
Polysilicon (10μm²) C_{Poly}	0.43fF	0.98fF	0.44
1st Aluminum (1mm) C_{1Al}	72.6fF	123.2fF	0.59
2nd Aluminum (1mm) C_{2Al}	63.9fF	98.4fF	0.65

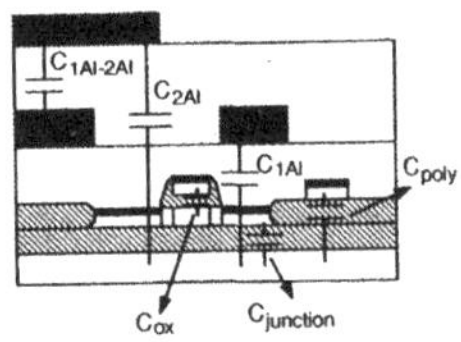

(b)

Fig. 5.51 (a) Propagation delay per stage versus normalized power consumption (by the oscillating frequency) of a 2-input NAND ring oscillator using 0.6 μm bulk and FD SOI CMOS technologies with and without 1mm aluminum loads. (b) Capacitances in 0.6 μm bulk and FD SOI CMOS technologies. (From Yamaguchi et al. [45]. ©1993 IEEE.)

Figure 5.50 shows (a) the cross section of a 0.6 μm gate-array FD SOI CMOS technology with a front gate oxide of 150 Å, a thin film of 800 Å, and a buried oxide of 5000 Å and (b) the layout of the gate-array cell in the sea of gate (SOG) configuration [45]. As shown in this figure, NMOS and PMOS devices with an identical aspect ratio are placed in individual silicon islands separated by LOCOS. Between active MOS devices in an island, inactive MOS devices have been used for isolation– for the NMOS device with the gate grounded or for the PMOS device with the gate connected to V_{DD}, which are the so-called gate isolation. Figure 5.51(a) shows the propagation delay per stage versus normalized power consumption (by the oscillating frequency) of a 2-input NAND ring oscillator using 0.6 μm bulk and FD SOI CMOS technologies with and without 1 mm aluminum loads and (b) the capacitances in 0.6 μm bulk and FD SOI CMOS technologies [45]. As shown in this figure, at the identical normalized power consumption, the speed of the SOI circuit is 1.4 times faster, which is due to the one-half of the parasitic capacitance of the SOI device as compared to the bulk one.

5.6.2 PD Gate Array

Although FD devices offer many advantages, PD SOI devices have advantages in the threshold voltage uniformity and source/drain resistances. PD SOI CMOS gate arrays have also been developed. Figure 5.52 shows (a) the layout and (b) the cross-section of 0.35 μm bulk and PD SOI CMOS technologies for implementing gate arrays [46]. As shown in this figure, the sea-of-gate (SOG) arrangement has been adopted. In addition, field-shield isolation techniques have been used with the neutral body in the PD device connected to ground or V_{DD} via the thin film under the field-shield gate. The arrangement of this PD SOI CMOS gate array is compatible to the bulk one.

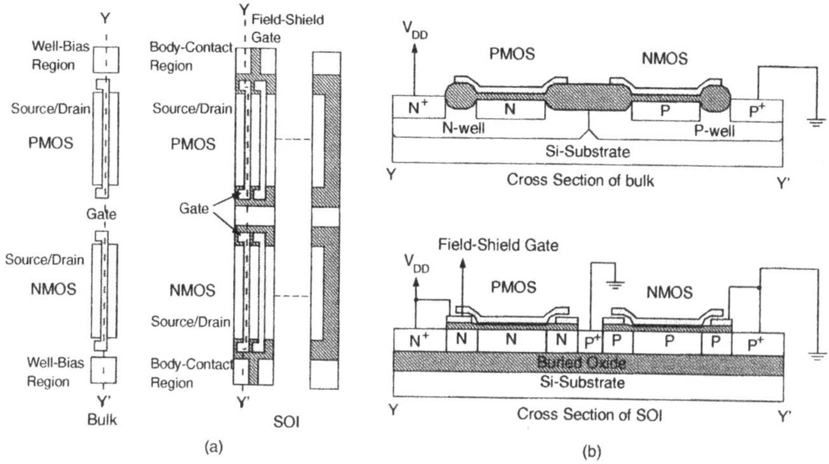

Fig. 5.52 (a) Layout and (b) cross-section of 0.35 μm bulk and PD SOI CMOS technologies for implementing gate arrays. (Adapted from Ueda et al. [46].)

Thus, the layout of the circuits for the bulk gate arrays can be used directly for the PD SOI CMOS gate array. Since impact ionization is more serious for the NMOS device, in order to avoid the floating body effects, both sides of the NMOS devices are with body contacts. In contrast, for the PMOS devices only one side is with body contacts. Figure 5.53 shows the macro-cell placement for the PD SOI CMOS gate array. As shown in this figure, metal 2 and metal 3 are placed vertically and horizontally, respectively. Also shown is the placement for an inverter. Both sides of the active MOS devices are surrounded by the inactive MOS devices for isolation–the gate of the inactive PMOS device is connected to V_{DD} and the gate of the inactive NMOS device is connected to ground.

For PD SOI gate arrays, body contacts are required. Adding body contacts may result in a sacrifice in the layout density. Figure 5.54 shows the propagation delay versus input frequency of a 2-input NAND gate using a 1-V 0.35 μm 560 kG PD SOI CMOS gate array with field-shield isolation [47]. As shown in this figure, with the body contacted, the instability in the speed performance of circuit due to floating body effects can be avoided.

5.6.3 DTMOS/MTCMOS Gate Arrays

By using the body control techniques of the SOI DTMOS schemes, the gate array can work with a high speed performance at a low power supply voltage. Figure 5.55 shows (a) the basic cell structure and (b) a 2-input NAND gate in a 0.5 V 0.32 μm SOI DTMOS gate array[48]. As shown in the figure, conventional LOCOS have been used to isolate the NMOS device from the PMOS device. In order to avoid the leakage current of the parasitic transistor at the edge of the LOCOS, P^+ isolation has been added at the edge of the NMOS device. In addition, gate isolation between adjacent

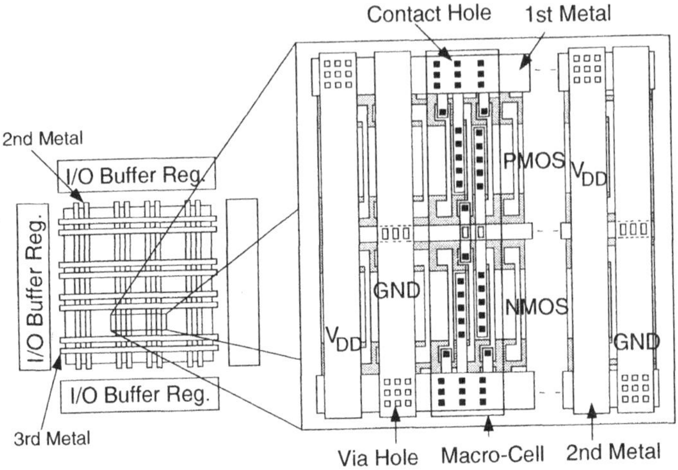

Fig. 5.53 Macro-cell placement for the PD SOI CMOS gate array. (From Ueda et al. [46]. ©1997 IEEE.)

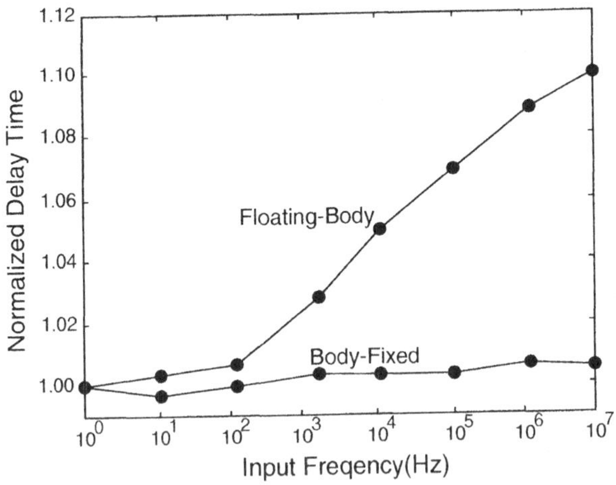

Fig. 5.54 Propagation delay versus the input frequency of a 2-input NAND gate using a 1-V 0.35 μm 560 kG PD SOI CMOS gate array with field-shield isolation. (Adapted from Mashiko et al. [47].)

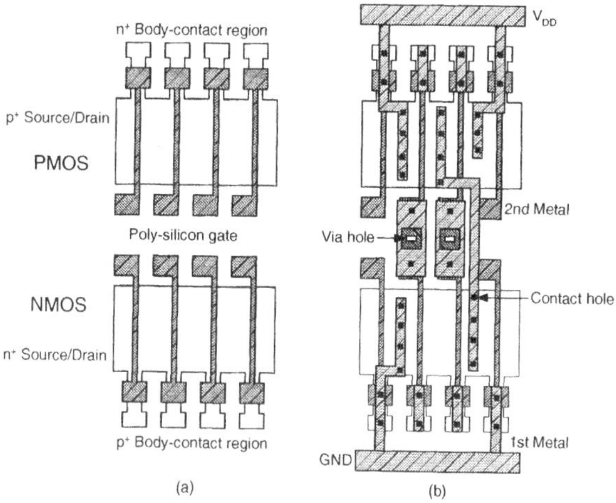

Fig. 5.55 (a) Basic cell structure and (b) layout of a 2-input NAND gate in a 0.5 V 0.32 μm SOI DTMOS gate array. (Adapted from Hirota et al. [48].)

PMOS and NMOS devices have been added to avoid interference. The cell layout area of the SOI DTMOS gate array is about the same as that of the conventional SOI gate array with the body fixed scheme. Fig. 5.55(b) shows the layout of a 2-input NAND gate in the 0.5 V 0.32 μm SOI DTMOS gate array. At the power supply voltage of 0.5V, its propagation delay is 1.8 ns, which is much faster than that implemented by the SOI gate array with the body fixed scheme (3.5 ns).

In addition to DTMOS, SOI MTCMOS techniques have also been used in the gate array design. Figure 5.56 shows (a) the SOI MTCMOS circuit scheme and (b) the layout pattern of the basic cell and its 2-input NAND circuit using 0.25 μm SOI MTCMOS gate array [49]. As shown in this figure, the layout of a fundamental SOI MTCMOS gate-array cell includes the power-switch MOS devices for the power supply lines V_{DD}/V_{DDV} and GND/GNDV and the logic gate portion for the circuit function. As shown, the layout of the SOI MTCMOS gate-array cell is composed of the multithreshold basic cells (MVBC) and the low-threshold basic cells (LVBC). The MVBC includes the high-threshold PD devices used for power switches and low-threshold FD devices. The LVBC includes the low-threshold FD devices. For every three cells, two are MVBC and one is LVBC. This SOI MTCMOS gate array is based on a four-layer metalization SOI technology. Layer-3 metal is used for V_{DD} and GND power lines. Layer-1 metal is used for V_{DDV} and GNDV power lines. In order to save layout area, the high-threshold power-switch PD devices are placed as a subblock under the layer-2 metal. As shown in Fig. 5.57, in order for the SOI MTCMOS gate array to use the low power supply voltage of 0.5 V, the body of the power-switch high-threshold PD devices offered by MVBC is connected to the sleep control signal via a gate-drain-connected reverse biased MOS diode to form a variable high-threshold PD SOI MOS device. During the standby mode at 0.5 V, the improved power switch has an identical small leakage current as the conventional power switch

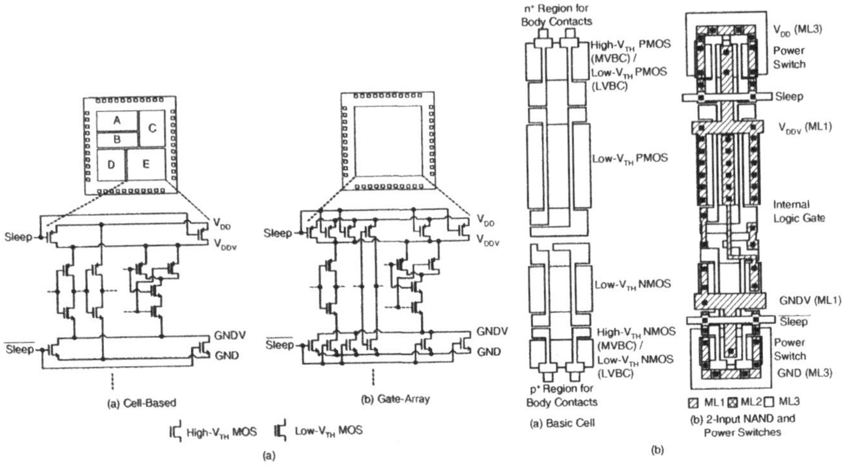

Fig. 5.56 (a) SOI MTCMOS circuit scheme. (b) Layout pattern of the basic cell and its 2-input NAND circuit using a 0.25 μm SOI MTCMOS gate array. (From Urano et al. [49]. ©1997 IEEE.)

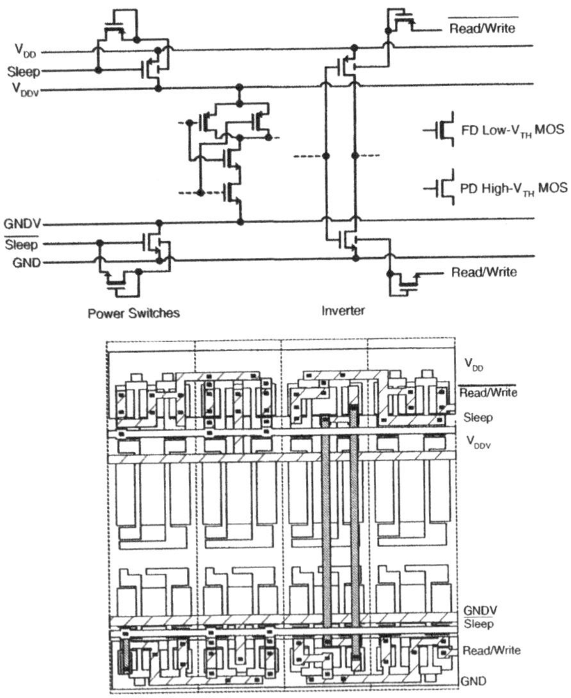

Fig. 5.57 Circuit diagram and layout pattern of variable high-threshold circuit schemes using the SOI MTCMOS gate array. (Adapted from Urano et al. [49].)

CMOS		NMOS	PMOS
SOI silicon	μm	0.1	0.1
Buried Oxide	μm	0.4	0.4
Gate Oxide	Å	105	105
Channel Length	μm	0.41	0.50
Threshold Voltage	mV	530	460
I_{DSAT} @ 1.2V	μA/μm	65	22
sub-Vt slope	mV/dec	70	73
BVDSS	V	3.2	6.3

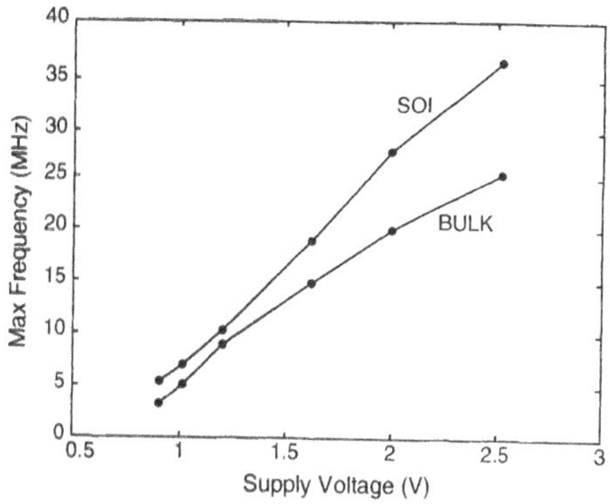

Fig. 5.58 Maximum operating frequency versus power supply voltage of an FD SOI microcontroller using the 0.5 μm FD SOI CMOS technology. (Adapted to Huang et al. [53].)

but it has a three times larger supply drain current. Its layout is also based on typical LVBC and MVBC except that the connection has been changed to form the variable high-threshold power switch PD device. As shown in Fig. 5.57, a data hold circuit using a body control concept is shown. Only when this circuit is active is its body voltage adjusted dynamically such that the threshold voltage of the device can be minimized to obtain the top speed. During standby, the body voltage is recovered to its original value such that the leakage current is decreased.

5.7 CPU

Owing to its intrinsic capabilities, SOI CMOS VLSI technology is especially useful for realizing low-voltage, low-power VLSI CPUs. Recently, SOI microprocessor chips having multimillions of transistors with a speed performance over 550 MHz have been reported [50]-[52]. Based on a 0.5 μm FD SOI CMOS technology, a microcontroller, which is composed of a CPU, an SRAM, and a ROM operating at sub-1V, was first reported in 1995 [53]. Fig. 5.58 shows the maximum operating frequency versus the power supply voltage of the FD SOI microcontroller[53]. As shown in this figure, at a power supply voltage of 0.9 V, the operating frequency of this FD SOI microcontroller is 5.7 MHz, which is 1.9 times faster as compared to the bulk counterpart.

Figure 5.59 shows the macrostructure overview of a 1.5 V 64-bit microprocessor using a 0.25 μm FD SOI CMOS technology [54] with 4-layer metal interconnects. The CMOS devices in this 0.25 μm FD SOI CMOS technology have a front gate oxide of 40 Å, a thin film of 460 Å, a buried oxide of 2000 Å, and an N^+ polysilicon gate for NMOS and a P^+ polysilicon gate for the PMOS, using shallow trench isolation. As shown in this figure, the macrostructure of the SOI CPU, which uses the four-way super-scalar with a 96 K set-associative SRAM cache, is divided into three streams– the instruction stream, the execution stream, and the memory stream.

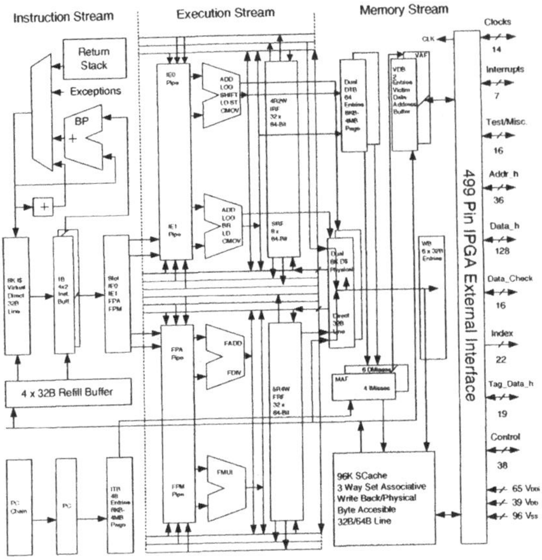

0.25μm 4-Metal FD-SOI CMOS	
SOI Thickness	46nm
BOX Thickness	200nm
Isolation	Shallow Trench Isolation
Gate Oxide	4nm GNOX
Transistor	Dual Gate Co-Silicide
V_{TN}/V_{TP}	0.36V/-0.36V
Subthreshold Swing	74mV/dec for nMOSFET

	Bulk μC	SOI μC
Technology	0.25μm Bulk	0.25μm FD-SOI
Speed @1.6V	476MHz	625MHz
Capacitance	12.3nF	9.2nF
Power	71mW/MHz	21mW/Hz

Fig. 5.59 Macrostructure overview of the 1.5 V 64-bit 600-MHz microprocessor using a 0.25 μm FD SOI CMOS technology with 4-layer metal interconnects. (From Park et al. [54].) ©1999 IEEE.

In order to achieve high-speed performance, intensive CMOS dynamic logic circuits such as dynamic-level latch, single-phase multistage domino logic, multistage cascode voltage switch logic (CVSL) circuits with large clock drivers, and precise clock skew control network have been used. Since FD devices do not have floating body effects as for the PD devices, no dynamic circuit output failure due to the dynamic leakage from the turn-on of the parasitic bipolar device occurs. This SOI CPU has 9.66 million transistors, occupying a die area of 209 mm^2. As shown in this figure, the speed performance of this FD SOI CPU is 625 MHz, which is 1.3 times faster as compared to the bulk counterpart, owing to smaller junction capacitances of the FD SOI devices. The overall chip capacitance of this FD SOI CPU is 9.2 nF, which is only 75% of that of the bulk counterpart. In addition, the power dissipation of the FD SOI CPU is 21 mW/MHz, which is 30% lower than the bulk counterpart's figure.

In addition to FD technology, PD SOI CMOS technology has also been used to integrate microprocessors. Although PD devices have kink effects due to their floating body, microprocessors integrated by PD SOI CMOS technology still offer a better performance as compared to the bulk counterpart. Figure 5.60 shows the dynamic power dissipation versus the operating frequency of the 1.65 V SOI CMOS microprocessor using a 0.3 μm PD SOI CMOS technology where CMOS devices are with a front gate oxide of 60 Å, a thin film of 2000 Å, and a buried oxide of 2000 Å, with shallow trench isolation and three layer metal interconnects. Also shows in this figure are the results for the bulk counterpart [55]. The speed performance of the microprocessor implemented by the PD SOI CMOS technology is 20% faster as compared to the bulk counterpart owing to the smaller junction capacitances. In

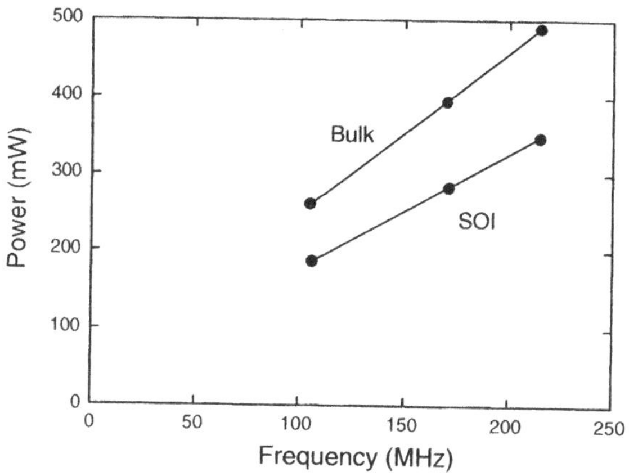

L_{gate}/L_{eff}	0.3μm/0.25μm
t_{ox}	6nm
$\lvert V_{th}\rvert$	425mV
t_{si}	200nm
t_{Box}	200nm
Isolation	STI (bulk : LOCOS)
metal	4 (μ C=3)

Fig. 5.60 Dynamic power dissipation versus operating frequency of the 1.65 V SOI CMOS microprocessor using a 0.3 μm PD SOI CMOS technology, where CMOS devices are with a front gate oxide of 60 Å, a thin film of 2000 Å, and a buried oxide of 2000 Å, with shallow trench isolation and three layer metal interconnects. (Adapted from Mistry et al. [55].)

addition, the dynamic power dissipation of the PD SOI CMOS microprocessor, which is proportional to CV^2f (where C is the load capacitance, V is the supply voltage, and f is the operating frequency) is 30% less than that of the bulk one. At a lower supply voltage, the low power advantages of the SOI is even larger.

The SOI DTMOS techniques with body connected to gate have been used for pass gate logic. Figure 5.61 shows the body bias-controlled (BC) SOI pass gate logic using PD SOI DTMOS techniques [56]. As shown in this figure, there are two types of BC SOI pass gate logic. The first type, which has an inverter type buffer, is mainly for driving a large load with a long wiring or a large fan-out. The second type, which has a body-controlled PMOS feedback buffer, is targeted for lower power dissipation. Figure 5.62 shows (a) the block diagram of the SOI 32-bit ALU and (b) the maximum operating frequency versus the supply voltage of the ALU implemented by BC SOI pass gate logic and the bulk counterpart based on a 0.3 μm PD SOI CMOS technology, where CMOS devices are with a front gate oxide of 80 Å and a thin film of 1000 Å [56]. As shown in this figure, the maximum operating frequency of the 32-bit ALU implemented by BC SOI pass gate logic at the supply voltage of 0.5 V is 200 MHz at a power consumption of 20 mW. The drawback of the circuit is that due to the DTMOS configuration, the circuit cannot work with a power supply voltage > 0.7 V.

Fig. 5.63 shows (a) the performance of the 1.8-V SOI 550-MHz 64-bit microprocessor using a 0.2 μm SOI CMOS technology with a front gate oxide of 35 Å and 6-layer copper interconnects and (b) the speed performance of key circuits used [15]. As shown in this figure, the transistor count of the 64-bit SOI microprocessor is 34 million, which is similar to the bulk counterpart. The power consumption of the SOI microprocessor is 24 W, which is 10% higher than the bulk counterpart. The die size of the SOI CPU is 139 mm^2, which is similar to the bulk counterpart. The maximum operating frequency of the SOI microprocessor is 550 MHz, which is 22% higher than

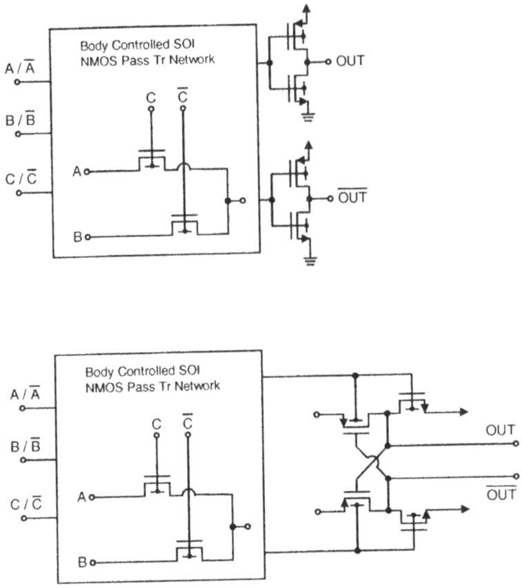

Fig. 5.61 Body bias-controlled (BC) SOI pass gate logic using PD SOI DTMOS techniques. (Adapted from Fuse et al. [56].)

the bulk counterpart. Fig. 5.63(b) shows the comparison of the performance of the key circuits used in this microprocessor using SOI and bulk technologies. As shown, when the key circuits are transformed from bulk into SOI, static circuits are most improved in speed performance, followed by array circuit. Dynamic circuits have the smallest improvement. In addition, the more complicated circuits such as XOR have the best improvement. Therefore, the advantages of the SOI CMOS technologies can be fully exploited for complicated circuits, which are important for next-generation VLSI systems.

5.8 EMBEDDED MEMORY

Recent trends on integrating digital VLSI systems have been toward embedded memory architecture–memory is integrated on a chip with a microprocessor or logic, which can be used to improve the mismatch of bandwidth performance among individual memory chips, logic, and microprocessor [28]. In addition, when memory and logic are integrated on the same chip, the I/O count is much smaller. Much smaller load capacitances also reduce the power dissipation of the overall circuits during operation. With the embedded memory architecture, since internal transistors or function logic gates can be shared, the chip count in the whole system is reduced and thus electromagnetic interference (EMI) among parts becomes smaller. Owing to the superior isolation properties of the SOI devices, SOI technology is especially suitable for integration of the microprocessor or logic systems with the embedded memory on the same chip– SOI is suitable for integrating systems on a chip (SOC) .

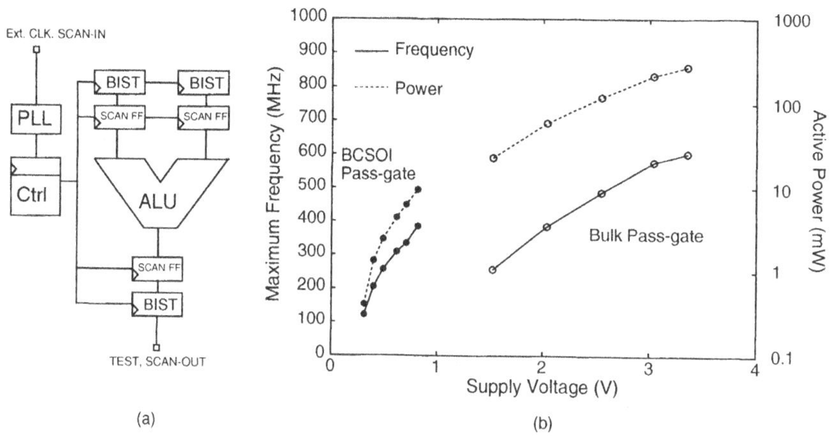

Process Technology	0.3μm SOI CMOS partially-depleted device 3-level metal L_g =0.3μm T_{ox} =8nm T_{Si} =100nm
Transistor Count	20k (ALU,4k)
Supply Voltage	0.5V
Operation Frequency	200MHz
Power Consumption	20mW at 200MHz

Fig. 5.62 (a) Block diagram of the SOI 32-bit ALU. (b) Maximum operating frequency versus the supply voltage of the ALU implemented by BC SOI pass gate logic and the bulk counterpart based on a 0.3 μm PD SOI CMOS technology, where CMOS devices are with a front gate of 80 Å and a thin-film of 1000 Å. (From Fuse et al. [56]. 1997 IEEE.)

Figure 5.64 shows the hierarchical configuration of the embedded SRAMs with logic using SOI MTCMOS techniques [36]. As shown in this figure, via the SOI MTCMOS circuit techniques, the hierarchical configuration of embedded SRAM structure can be realized for a power supply of 1 V. In the circuit, high- and low-threshold SRAMs have been included. Low-threshold SRAMs are used for high-speed operation in the active mode. High-threshold SRAMs are for data back-up during the sleep mode. With a high threshold, during the sleep mode the standby power is small.

Figure 5.65 shows the block diagram of the embedded SRAM using the size configurable architecture with abutting nine kinds of leaf cells[36]. As shown in the figure, in the embedded SRAM architecture, there are memory-cell array and eight adjacent peripheral circuit blocks, which include control circuits, row-address decoders, data I/O circuits, and five power-line elements to form power-line rings. The length of the layout of each circuit block is identical to the adjacent blocks for forming a compact circuit structure. Each block is called a leaf cell.

	CMOS7S	CMOS7S SOI
Core Clock Frequency	450MHz	550MHz
Supply Voltage	1.8V	1.8V
Power	22W	24W
Transistors	34M	34M
Die Size	139mm^2	139mm^2
Poly pitch/2	0.22μm	0.22μm
L_{eff} (NMOS)	0.12μm	0.12μm
T_{OX}	3.5nm	3.5nm
Metalization	6 layers Cu	6 layers Cu
Contacted M2-M4 pitch	0.81μm	0.81μm

(a)

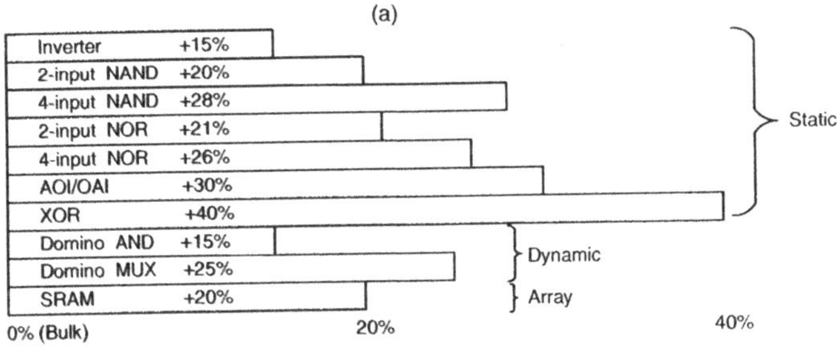

(b)

Fig. 5.63 (a) Performance of the 1.8-V SOI 550-MHz 64-bit microprocessor using a $0.2\mu m$ SOI CMOS technology with a front gate oxide of 35Å and 6-layer copper interconnects. (b) Speed performance of key circuits. (Adapted from Aipperspach et al. [15].)

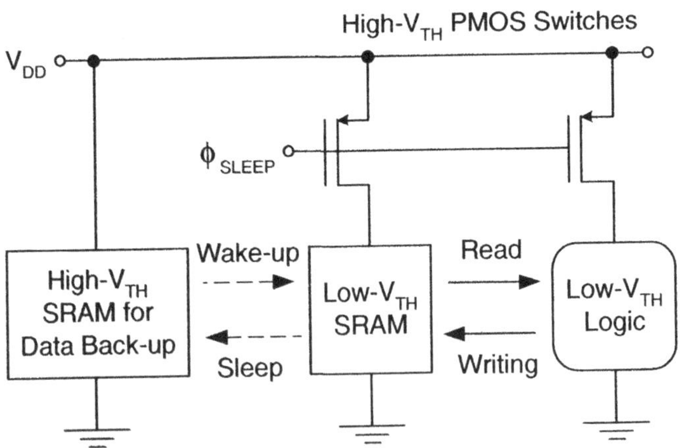

Fig. 5.64 Hierarchical configuration of the embedded SRAMs with logic using SOI MTCMOS techniques. (Adapted from Shibata et al. [36].)

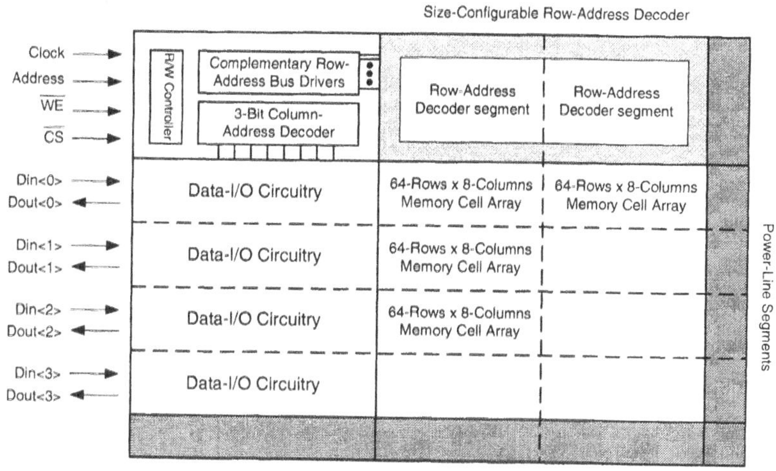

Fig. 5.65 Block diagram of the embedded SRAM using the size configurable architecture with abutting nine kinds of leaf cells. (Adapted from Shibata et al. [36].)

In the control circuit block, the row-address-bus driver and the column-address decoder are used to work with the row-address decoder block for the embedded SRAM. The operation of the row-address decoder and the data I/O circuit block is determined by the input data signals. A $0.35\,\mu m$ SOI MTCMOS technology has been used to integrate the embedded SRAM in terms of important components such as the address decoder, the memory cell, and the sense amp, etc. At a power supply voltage of 1.2 V, the power consumption during the standby mode is $0.2\mu W$ and 14 mW at the operating frequency of 100 MHz during the active mode.

5.9 MULTIPLIER AND DSP

Multipliers and digital signal processing (DSP) related circuits are important components for VLSI systems. In this section, multipliers and DSP related circuits using SOI CMOS devices are described.

5.9.1 Multiplier

Figure 5.66 shows the organization of a 0.5 V 54b × 54b multiplier using SOI MTCMOS techniques with triple-threshold devices [57]. As shown in this figure, FD low-threshold, FD medium-threshold, and PD high-threshold devices are used. The low-threshold FD devices are used along the critical path of the multiplier circuit. Medium-threshold FD devices are used in the noncritical path related areas. High-threshold PD devices are used as power switches. In addition to power switches, the multiplier is made of the Booth encoder, the partial product generator, Wallace tree, and the 108b adder. In the Booth encoder and the partial product generator, which

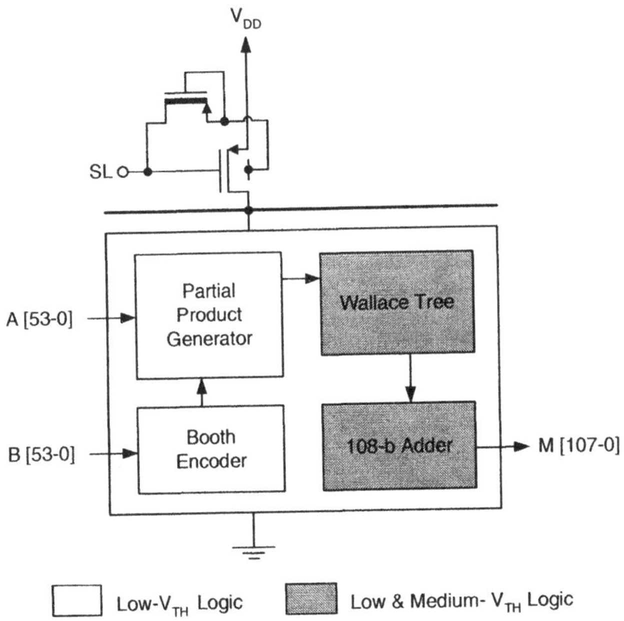

Fig. 5.66 Organization of a 0.5V 54b × 54b multiplier using SOI MTCMOS techniques with triple-threshold devices. (Adapted from Fujii et al. [57].)

are along the critical path, low-threshold FD devices are used. In the Wallace tree and the 108b adder areas, as shown in Fig. 5.66, depending on the internal structure, some devices located in the critical path, low-threshold devices are used. Some devices not locating in the critical path, medium-threshold devices are adopted to reduce leakage current and standby power consumption.

As shown in Fig. 5.67, in the Wallace tree structure, only the full-adder in the critical path is implemented by the low-threshold FD devices. As for the 108b adder circuit, transmission-gate multiplexers, as shown in Fig. 5.68(b), are used. Compared to the conventional source-controlled circuit, as shown in Fig. 5.68(a), the source-controlled transmission-gate multiplexer, as shown in Fig. 5.68(b), is more complicated. In the source-controlled transmission-gate multiplexer, most of the critical signals and CC are connected to the source/drain of the devices. In contrast, in the conventional gate-controlled transmission-gate multiplexer, they are connected to the gate of the devices. For SOI devices, the parasitic capacitances at the source/drain are much smaller than those at the gate. Thus, using the source-controlled circuit, the load capacitances for CC to drive are smaller as compared to the conventional gate-controlled circuit. Note that CC is located in the critical path of the 108b adder. By adopting the source-controlled transmission-gate multiplexer for the CC, the speed performance of the 108b adder is enhanced by 10%.

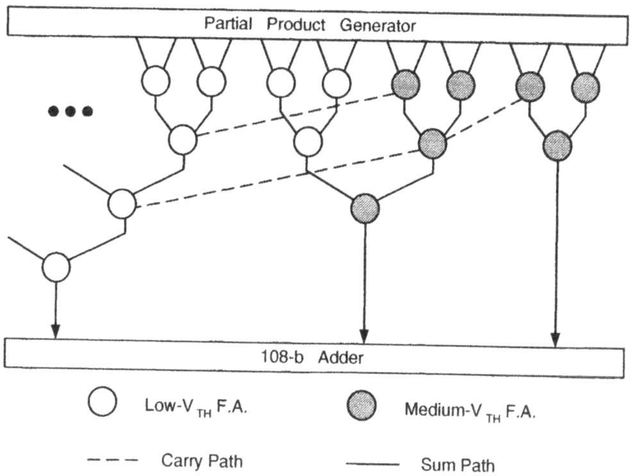

Fig. 5.67 Wallace tree structure with medium-threshold and low-threshold full adders. (Adapted from Fujii et al. [57].)

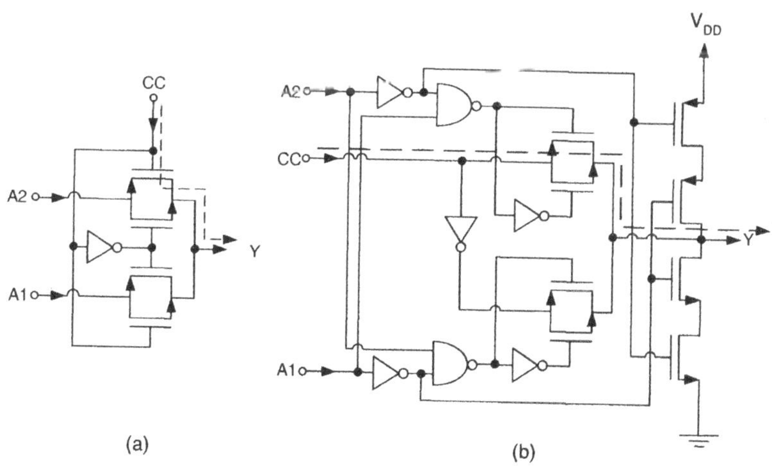

Fig. 5.68 (a) Gate-controlled and (b) source-controlled transmission-gate multiplexer. (Adapted from Fujii et al. [57].)

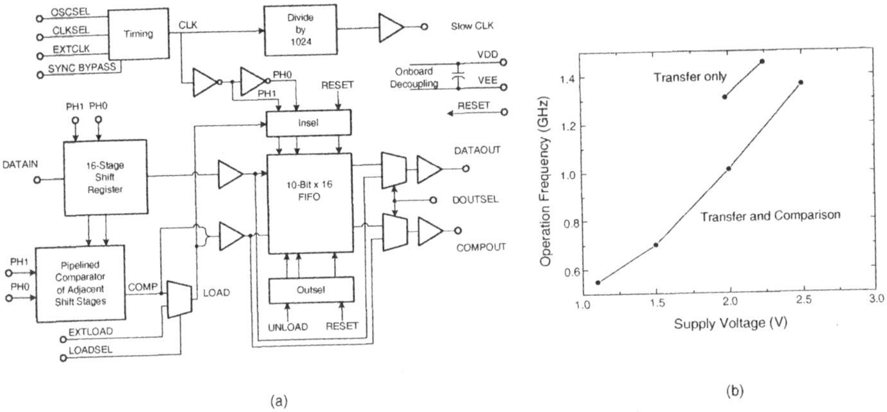

Fig. 5.69 (a) Block diagram of a DSP chip for low-power, high-speed pulse processing using a 0.25 μm SOI technology. (b) Operating frequency versus supply voltage performance. (Adapted from Berger et al. [58].)

5.9.2 DSP

DSP techniques have been used in the wideband compressor receiver for digitizing fast analog pulses, which converts the real-time analog signal into a series of narrow analog pulses in the time domain. Each narrow analog pulse represents a specific input frequency. Via filtering, detection, and data thinning, the analog pulse stream is transformed into digital signals in the digital domain. To accomplish the series of pulse processing, shift registers, binary comparators, FIFO storage, and data drivers are required.

Figure 5.69 shows the block diagram of a digital signal processing (DSP) chip for low-power high-speed pulse processing using a 0.25 μm SOI technology [58]. As shown in this figure, this SOI DSP circuit includes a shift register based on clocked master-slave D flip-flops, a 10b 16-word FIFO storage using an SRAM array, the word reading and writing system circuit (INSEL, OUTSEL) based on a 16-bit shift register, and a comparator made of pass-transistor logic. This SOI DSP circuit is integrated using 0.25 μm FD SOI CMOS technology. Compared to PD technology, the FD SOI technology has a smaller body effect, a smaller threshold voltage, and smaller parasitic capacitances at source/drain, which can be taken advantage of to implement the DSP chip. As shown in Fig. 5.69(b), at a supply voltage of 2.5 V, a 1.3 GHz operating frequency has been achieved. Using the DSP chip, a pulse processor with a 20-Gb/s data stream capability has been implemented.

5.10 FREQUENCY DIVIDER

Advanced personal communication systems require very low-power VLSI systems operating in the giga hertz frequency regime using a battery. A low-power multi-

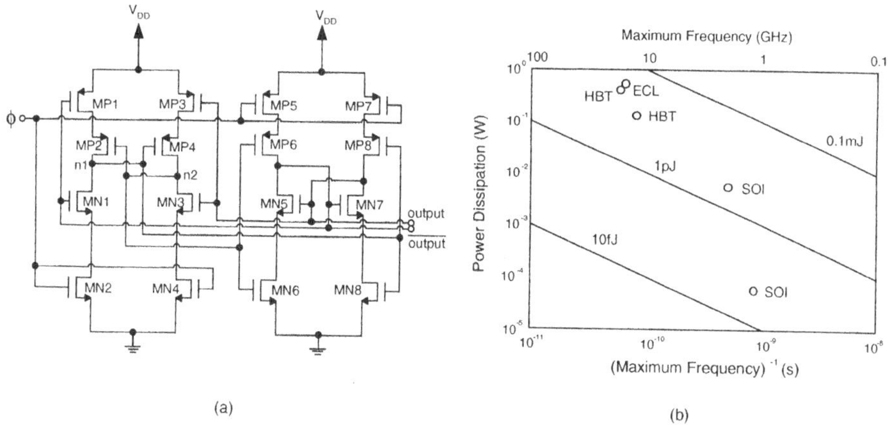

Fig. 5.70 (a)SOI dynamic master-slave 1/2 frequency divider. (b)Power dissipation versus inverse of maximum operating frequency. (Adapted from Fujishima et al. [59].)

giga-hertz frequency divider is important for realizing the phase-locked loop circuit used in the advanced personal communication system. In this section, a high-speed low-power frequency divider circuit using SOI CMOS technology is described.

Figure 5.70 shows an SOI dynamic master-salve 1/2 frequency divider circuit [59]. As shown in this figure, transistors MN5, MN6, MP5, and MP6 determine the logic level of the output node. Transistors MN7, MN8, MP7, and MP8 determine the logic level of the $\overline{\text{output}}$ node. In timing period (I), the input (ϕ) and the output are low. MP1, MP4, MP8, and MN2-MN6 are off. Node 1 is high and node 2 is low. In timing period (II), the input (ϕ) switches from low to high, which turns on MN2. Via MN1 and MN2, node 1 is pulled low. Therefore, MP8 is turned on and MN8 is off. Since MP7 is off because of the input, at this time, the pullup (MP7-8) and the pulldown (MN7-8) paths for the output disappear. Therefore, the output node maintains low. When node 1 changes from high to low, MP4 turns on, which pulls node 2 from low to high. Therefore, MN6 turns on and MP6 turns off. Since MN5 is off due to the low state of the output node, the $\overline{\text{output}}$ node maintains high due to the disappearance of the pullup path (MP5-6) and the pulldown path (MN5-6). This master-salve flip-flop circuit is designed with a symmetry to its center. When entering into timing period (III) from (II), nodes 1 and 2 maintain their logic state as the output node and the $\overline{\text{output}}$ node in timing period (II). In addition, the output node and the $\overline{\text{output}}$ node proceed with the change in their logic states as nodes 1 and 2 in timing period (II). Via the operation of the SOI dynamic master-slave circuit, divide-by-2 frequency divide operation is achieved. Figure 5.70(b) shows the power dissipation versus the inverse of maximum operating frequency of this SOI dynamic master-slave 1/2 frequency divider. Also shown in this figure are the results for the heterojunction bipolar transistor (HBT) and the emitter-coupled logic (ECL) frequency dividers. As shown, although the maximum operating frequency of the SOI circuit is lower than that of HBT or ECL, the SOI circuit has a superior capability of operation in low-voltage (< 1 V) and low-power consumption ($< 70\ \mu$W). Thus, the potential of the

SOI circuits for applications in the high-frequency low-power portable equipments is very high.

5.11 SUMMARY

In this chapter, fundamental SOI CMOS static and dynamic digital logic circuits have been described. In addition, DRAM and SRAM circuits using SOI CMOS technology have been analyzed. SOI cache memory and content addressable memory (CAM) have been depicted, followed by SOI gate arrays. SOI CPU and embedded memory have been introduced, and finally SOI multipliers/digital signal processing (DSP) circuits, and SOI frequency dividers have been described.

REFERENCES

1. C. Tretz, C. T. Chuang, L. Terman, M. Pelella, and C. Zukowski, "Performance Comparison of Differential Static CMOS Circuit Topologies in SOI Technology," *SOI Conf. Dig.*, 123–124 (1998).
2. T. Fuse, Y. Oowaki, M. Terauchi, S. Watanabe, M. Yoshimi, K. Ohuchi, and J. Matsunaga, "0.5V SOI CMOS Pass-Gate Logic," *ISSCC Dig.*, 88–89 (1996).
3. N. Lindert, T. Sugii, S. Tang, and C. Hu, "Dynamic Threshold Pass-Transistor Logic for Improved Delay at Lower Power Supply Voltages," *IEEE J. Sol. St. Ckts.*, **34**(1), 85–89 (1999).
4. B.-T. Wang and J. B. Kuo, "A Novel Low-Voltage Silicon-On-Insulator (SOI) CMOS Complementary Pass-Transistor Logic (CPL) Circuit using Asymmetrical Dynamic Threshold Pass-Transistor (ADTPT) Technique," *MWSCAS Dig.*, (2000).
5. L. G. Heller, W. R. Griffin, J. W. Davis, and N. G. Thomas, "Cascode Voltage Switch Logic : A Differential CMOS Logic Family," *ISSCC Dig.*, 16–17 (1984).
6. K. M. Chu and D. L. Pulfrey, "Design Procedures for Differential Cascode Voltage Switch Circuits," *IEEE J. Sol. St. Ckts.*, **21**(6), 1082–1087 (1986).
7. P.-F. Lu, C.-T. Chuang, J. Ji, L. F. Wagner, C.-M. Hsieh, J. B. Kuang, L. L.-C. Hsu, M. M. Pelella, Jr., S.-F. S. Chu, and C. J. Anderson, "Floating-Body Effects in Partially Depleted SOI CMOS Circuits," *IEEE J. Sol. St. Ckts.*, **32**(8), 1241–1252 (1997).
8. A. G. Dickinson and J. S. Denker, "Adiabatic Dynamic Logic," *IEEE J. Sol. St. Ckts.*, **30**(3), 311–315 (1995).
9. V. K. De and J. D. Meindl, "Complementary Adiabatic and Fully Adiabatic MOS Logic Families for Gigascale Integration," *ISSCC Dig.*, 298–299 (1996).

10. Y. Moon and D.-K. Jeong, "An Efficient Charge Recovery Logic Circuit," *IEEE J. Sol. St. Ckts.*, **31**(4), 514–522 (1996).

11. C. C. Yeh, J. H. Lou, and J. B. Kuo, "1.5V CMOS Full-Swing Energy Efficient Logic (EEL) Circuit Suitable for Low-Voltage and Low-Power VLSI Applications," *Elec. Let.*, **33**(16) 1375–1376 (1997).

12. D. Suvakovic and C. Salama, "Two Phase Non-Overlapping Clock Adiabatic Differential Cascode Voltage Switch Logic (ADCVSL)," *ISSCC Dig.*, 364–365 (2000).

13. H. Weste and K. Eshraghian, "Principles of CMOS VLSI Design: A System Perspective," *Addison-Wesley*, (1985).

14. J. B. Kuo, T. Y. Chiang, and B. T. Wang, "A 0.7V Manchester Carry Look-Ahead Circuit using PD (Partially-Depleted) SOI CMOS Dynamic Threshold (DTMOS) Techniques Suitable for Low-Voltage CMOS VLSI Systems," *IEEE Trans. Circuits and Systems*, (to be published).

15. A. G. Aipperspach, D. H. Allen, D. T. Cox, N. V. Phan, and S. N. Storino, "A 0.2μm, 1.8-V, SOI, 550-MHz, 64-b PowerPC Microprocessor with Copper Interconnects," *IEEE J. Sol. St. Ckts.*, **34**(11), 1430–1435 (1999).

16. S. Bobba and I. N. Hajj, "Design of Dynamic Circuits with Enhanced Noise Tolerance," *ASIC/SOC Conf. Dig.*, 54–58 (1999).

17. S. B. Park, Y. W. Kim, Y. G. Ko, K. I. Kim, I. K. Kim, H.-S. Kang, J. O. Yu, and K. P. Suh, "A 0.25μm, 600-MHz, 1.5-V, Fully Depleted SOI CMOS 64-Bit Microprocessor," *IEEE J. Sol. St. Ckts.*, **34**(11), 1436–1445 (1999).

18. Y. Yamaguchi and Y. Inoue, "SOI DRAM: Its Features and Possibility," *SOI Conf. Dig.*, 122–124 (1995).

19. H.-S. Kim, S.-B. Lee, D.-U. Choi, J.-H. Shim, K.-C. Lee, K.-P. Lee, K.-N. Kim, and J.-W. Park, "A High Performance 16M DRAM on a Thin Film SOI," *Symp. VLSI Tech. Dig.*, 143–144 (1995).

20. J. G. Fossum, M.-H. Chiang, and T. W. Houston, "Design Issues and Insights for Low-Voltage High-Density SOI DRAM," *IEEE Trans. Elec. Dev.*, **45**(5), 1055–1062 (1998).

21. J. A. Mandelman, J. E. Barth, J. K. DeBrosse, R. H. Dennard, H. L. Kalter, J. Gautier, and H. I. Hanafi, "Floating-Body Concerns for SOI Dynamic Random Access Memory (DRAM)," *SOI Conf. Dig.*, 136–137 (1996).

22. S. Tomishima, F. Morishita, M. Tsukude, T. Yamagata, and K. Arimoto, "A Long Data Retention SOI-DRAM with the Body Refresh Function," *Symp. VLSI Ckts. Dig.*, 198–199 (1996).

23. H.-S. Kim, D.-U. Choi, S.-H. Lee, S.-K. Lee, J.-K. Park, K.-N. Kim, and J.-W. Park,"Data Retention Times in SOI-DRAMs," *Symp. VLSI Tech. Dig.*, 126–127 (1996).

24. Y.-H. Koh, J.-H. Choi, J.-W. Yang, M.-H. Nam, W.-C. Lee, J.-W. Lee, and M.-R. Oh, "64Mbit SOI-DRAM Technologies Using Body-Contacted (BC) Structure," *SOI Conf. Dig.*, 170–171 (1997).

25. S. Kuge, F. Morishita, T. Tsuruda, S. Tomishima, M. Tsukude, T. Yamagata, and K. Arimoto, "SOI-DRAM Circuit Technologies for Low Power High Speed Multigiga Scale Memories," *IEEE J. Sol. St. Ckts.*, **31**(4), 586–591 (1996).

26. K. Suma, T. Tsuruda, H. Hidaka, T. Eimori, T. Oashi, Y. Yamaguchi, T. Iwamatsu, M. Hirose, F. Morishita, K. Arimoto, K. Fujishima, Y. Inoue, T. Nishimura, and T. Yoshihara, "An SOI-DRAM with Wide Operating Voltage Range by CMOS/SIMOX Technology," *IEEE J. Sol. St. Ckts.*, **29**(11), 1323–1329 (1994).

27. K. Shimomura, H. Shimano, N. Sakashita, F. Okuda, T. Oashi, Y. Yamaguchi, T. Eimori, M. Inuishi, K. Arimoto, S. Maegawa, Y. Inoue, S. Komori, and K. Kyuma, "A 1-V 46-ns 16-Mb SOI-DRAM with Body Control Technique," *IEEE J. Sol. St. Ckts.*, **32**(11), 1712–1720 (1997).

28. T. Yamauchi, F. Morisita, S. Maeda, K. Arimoto, K. Fujishima, H. Ozaki, and T. Yoshihara, "High-Performance Embedded SOI DRAM Architecture for the Low-Power Supply," *IEEE J. Sol. St. Ckts.*, **35**(8), 1169–1178 (2000).

29. K. Kumagai, T. Yamada, H. Iwaki, H. Nakamura, H. Onishi, Y. Matsubara, K. Imai, and S. Kurosawa, "A New SRAM Cell Design Using 0.35μm CMOS/SIMOX Technology," *SOI Conf. Dig.*, 174–175 (1997).

30. G. G. Shahidi, T. H. Ning, T. I. Chappell, J. H. Comfort, B. A. Chappell, R. Franch, C. J. Anderson, P. W. Cook, S. E. Schuster, M. G. Rosenfield, M. R. Polcari, R. H. Dennard, and B. Davari, "SOI for a 1-Volt CMOS Technology and Application to a 512Kb SRAM With 3.5ns Access Time," *IEDM Dig.*, 813–816 (1993).

31. M. Isobe, Y. Uchida, K. Maeguchi, T. Mochizuki, M. Kimura, H. Hatano, Y. Mizutani, and H. Tango, "An 18ns CMOS/SOS 4K Static RAM," *IEEE J. Sol. St. Ckts.*, **16**(5), 460–465 (1981).

32. J. B. Kuang, S. Ratanaphanyarat, M. J. Saccamango, L. L.-C. Hsu, R. C. Flaker, L. F. Wagner, S.-F. S. Chu, and G. G. Shahidi, "SRAM Bitline Circuits on PD SOI: Advantages and Concerns," *IEEE J. Sol. St. Ckts.*, **32**(6), 837–844 (1997).

33. Y. Wada, K. Nii, H. Kuriyama, S. Maeda, K. Ueda, and Y. Matsuda, "A 128Kb SRAM with Soft Error Immunity for 0.35μm SOI-CMOS Embedded Cell Arrays," *SOI Conf. Dig.*, 127–128 (1998).

34. C.-T. Chuang and P.-F. Lu, and C. J. Anderson, "SOI for Digital CMOS VLSI: Design Considerations and Advances," *Proc. of IEEE*, **86**(4), 689–720 (1998).

35. S. C. Liu and J. B. Kuo, "A Novel 0.7V Two-Port 6T SRAM Memory Cell Structure with Single-Bit-Line Simultaneous Read-and-Write Access (SBLSRWA) Capability using Partially-Depleted SOI CMOS Dynamic-Threshold Technique," *SOI Conf. Dig.*, 75–76 (1999).

36. N. Shibata, H. Morimura, and M. Harada, "1-V 100-MHz Embedded SRAM Techniques for Battery-Operated MTCMOS/SIMOX ASICs," *IEEE J. Sol. St. Ckts.*, **35**(10), 1396–1407 (2000).

37. Y. Takao, H. Shimada, N. Suzuki, Y. Matsukawa, Y. Kobayashi, and N. Sasaki, "A Low-Power SRAM Utilizing High ON/OFF Ratio Laser-Recrystallized SOI PMOSFET Load," *Symp. VLSI Ckts. Dig.*, 95–96 (1991).

38. Y. Takao, H. Shimada, N. Suzuki, Y. Matsukawa, and N. Sasaki, "Low-Power and High-Stability SRAM Technology Using a Laser-Recrystallized p-Channel SOI MOSFET," *IEEE Trans. Elec. Dev.*, **39**(9), 2147–2152 (1992).

39. T. Nishimura, Y. Inoue, K. Sugahara, M. Nakaya, Y. Horiba, and Y. Akasaka, "A Three Dimensional Static RAM," *Symp. VLSI Tech. Dig.*, 30–31 (1985).

40. J. B. Kuo and J. H. Lou, "Low-Voltage CMOS VLSI Ckts.," *John Wiley: New York*, ISBN 0471321052 (1999).

41. H. Kadota, J. Miyake, Y. Nishimichi, H. Kudoh, and K. Kagawa, "An 8-kbit Content Addressable and Reentrant Memory," *IEEE J. Sol. St. Ckts.*, **20**(5), 951–957 (1985).

42. S. C. Liu, F. A. Wu, and J. B. Kuo, "A Novel Low-Voltage Content-Addressable-Memory (CAM) Cell with a Fast Tag-Compare Capability using Partially Depleted (PD) SOI CMOS Dynamic-Threshold (DTMOS) Techniques, *IEEE J. Sol. St. Ckts.*, **36**(4), 2001.

43. Y. Sato, Y. Kado, T. Tsuchiya, T. Kosugi, H. Ishii, and K. Nishimura, "300 KG Gate-Array LSI Using $0.25\mu m$ Ultra-Thin-Film Fully-Depleted CMOS/SIMOX with Tungsten-Deposited Source/Drain," *SOI Conf. Dig.*, 168–169 (1997).

44. Y. Kado, H. Inokawa, Y. Okazaki, T. Tsuchiya, Y. Kawai, M. Sato, Y. Sakakibara, S. Nakayama, H. Yamada, M. Kitamura, S. Nakashima, K. Nishimura, S. Date, M. Ino, K. Takeya, and T. Sakai, "Substantial Advantages of Fully-Depleted CMOS/SIMOX Devices as Low-Power High-Performance VLSI Components Compared with its Bulk-CMOS Counterpart," *IEDM Dig.*, 635–638 (1995).

45. Y. Yamaguchi, A. Ishibashi, M. Shimizu, T. Nishimura, K. Tsukamoto, K. Horie, and Y. Akasaka, "A High-Speed $0.6\mu m$ 16K CMOS Gate Array on a Thin SIMOX Film," *IEEE Trans. Elec. Dev.*, **40**(1), 179–186 (1993).

46. K. Ueda, K. Nii, Y. Wada, I. Takimoto, S. Maeda, T. Iwamatsu, Y. Yamaguchi, S. Maegawa, K. Mashiko, and H. Hamano, "A CAD-Compatible SOI/CMOS Gate

Array Having Body-Fixed Partially-Depleted Transistors," *ISSCC Dig.*, 288–289 (1997).

47. K. Mashiko, K. Ueda, K. Nii, Y. Wada, T. Hirota, S. Maeda, T. Iwamatsu, Y. Yamaguchi, T. Ipposhi, S. Maegawa, and H. Hamano, "A 0.35μm 560KG SOI/CMOS Gate Array using Field-Shield Isolation Technique," *SOI Conf. Dig.*, 166–167 (1997).

48. T. Hirota, K. Ueda, Y. Wada, K. Mashiko, and H. Hamano, "0.5V 320MHz 8b Multiplexer/Demultiplexer Chips Based on a Gate Array with Regular-Structured DTMOS/SOI," *ISSCC Dig.*, 188–189 (1998).

49. M. Urano, T. Douseki, T. Hatano, H. Fukuda, M. Harada, and T. Tsuchiya, "An Ultra-Low-Voltage MTCMOS/SIMOX Gate Array," *ASIC Conf. Dig.*, 7–11 (1997).

50. D. H. Allen, A. G. Aipperspach, D. T. Cox, N. V. Phan, and S. N. Storino, "A 0.2μm 1.8V SOI 550MHz 64b PowerPC Microprocessor with Copper Interconnects," *ISSCC Dig.*, 438–439 (1999).

51. M. Canada, C. Akrout, D. Cawthron, J. Corr, S. Geissler, R. Houle, P. Kartschoke, D. Kramer, P. McCormick, N. Rohrer, G. Salem, and L. Warriner, "A 580MHz RISC Microprocessor in SOI," *ISSCC Dig.*, 430–431 (1999).

52. Y. W. Kim, S. B. Park, Y. G. Ko, K. I. Kim, I. K. Kim, K. J. Bae, K. W. Lee, J. O. Yu, U. Chung, and K. P. Suh, "A 0.25μm 600MHz 1.5V SOI 64b ALPHA Microprocessor," *ISSCC Dig.*, 432–433 (1999).

53. W. M. Huang, K. Papworth, M. Racanelli, J. P. John, J. Foerstner, H. C. Shin, H. Park, B. Y. Hwang, T. Wetteroth, S. Hong, H. Shin, S. Wilson, and S. Cheng, "TFSOI CMOS Technology for Sub-1V Microcontroller Circuits," *IEDM Dig.*, 59–62 (1995).

54. S. B. Park, Y. W. Kim, Y. G. Ko, K. I. Kim, I. K. Kim, H.-S. Kang, J. O. Yu, and K. P. Suh, "A 0.25μm, 600-MHz, 1.5-V, Fully Depleted SOI CMOS 64-Bit Microprocessor," *IEEE J. Sol. St. Ckts.*, **34**(11), 1436–1445 (1999).

55. K. Mistry, G. Grula, J. Sleight, L. Bair, R. Stephany, R. Flatley, and P. Skerry, "A 2.0V, 0.35μm Partially Depleted SOI-CMOS Technology," *IEDM Dig.*, 583–586 (1997).

56. T. Fuse, Y. Oowaki, T. Yamada, M. Kamoshida, M. Ohta, T. Shino, S. Kawanaka, M. Terauchi, T. Yoshida, G. Matsubara, S. Yoshioka, S. Watanabe, M. Yoshimi, K. Ohuchi, and S. Manabe," A 0.5V 200MHz 1-Stage 32b ALU Using a Body Bias Controlled SOI Pass-Gate Logic," *ISSCC Dig.*, 286–287 (1997).

57. K. Fujii and T. Douseki, "A 0.5-V, 3-mW, 54 x 54-b Multiplier with a Triple-Vth CMOS/SIMOX Circuit Scheme," *SOI Conf. Dig.*, 73–74 (1999).

58. R. Berger, W. G. Lyons, and A. Soares, "A 1.3-GHz SOI CMOS Test Chip for Low-Power High-Speed Pulse Processing," *IEEE J. Sol. St. Ckts.*, **33**(8), 1259–1261 (1998).

59. M. Fujishima, K. Asada, Y. Omura, and K. Izumi, "Low-Power 1/2 Frequency Dividers Using 0.1μm CMOS Circuits Built with Ultrathin SIMOX Substrates," *IEEE J. Sol. St. Ckts.*, **28**(4), 510–512 (1993).

Problems

1. For a true single-phase-clocking (TSPC) dynamic latch as shown in Fig. 4.43 of Ref. 1.3, how does the leakage current from the floating body affects the performance of the TSPC dynamic latch if PD SOI CMOS devices are used?

2. As shown in Figs. 5.40 and 5.47, DTMOS techniques have been used to improve the SRAM cell and tag cell in a cache memory using PD SOI CMOS devices at a low supply voltage. How can the DTMOS techniques be used to improve the performance of a true-single-phase-clocking (TSPC) dynamic latch as described in problem 1, using PD SOI CMOS devices at a supply voltage of 0.8 V?

3. Use the ADTPT CPL circuit described in Fig. 5.6 to design an exclusive OR logic gate. What are the drawbacks of the ADTPT CPL circuit?

4. Use the DTMOS techniques to improve the performance of a semidynamic DCVSPG-domino logic circuit as shown in Fig. 4.73 in Ref. 1.3, using PD SOI CMOS devices operating at 0.8 V.

5. Compare the tradeoffs in performance of SOI CMOS gate arrays using PD and FD devices.

6. In Fig. 5.11, is there any other way to avoid the leakage current due to the floating body of the PD SOI CMOS devices in this dynamic logic gate?

6

SOI CMOS Analog Circuits

SOI CMOS technology has been used to integrate analog circuits. In this chapter, operational amplifiers (op amps), filters, analog-to-digital converters (ADC) and digital-to-analog converters (DAC), sigma-delta ADC, RF circuits, low noise amplifiers (LNA), mixers, voltage-control oscillator (VCO), and high-temperature analog circuits using SOI CMOS technology are described.

6.1 SOI OP AMPS

In this section, SOI CMOS op amps are described. Starting from a single-stage op amp, a single-stage folded-cascode op amp is analyzed. Then, the performance comparison of op amps using bulk and SOI CMOS technologies is presented, and finally the advanced cascode op amp.

6.1.1 Single-Stage Op Amp

Figure 6.1 shows an SOI CMOS single-stage op amp with a symmetrical topology, which has a good capability to drive a large capacitive load [1]. As shown in this figure, this SOI CMOS single-stage op amp is composed of an input differential pair (M1, M2) and three current mirrors (M3-M8). Owing to a small threshold voltage, the current driving capability of SOI devices is large, which is suitable for high-speed operation of the op amp. In addition, the small parasitic capacitances at the source/drain may also help realization of the SOI CMOS op amps for high-speed operation.

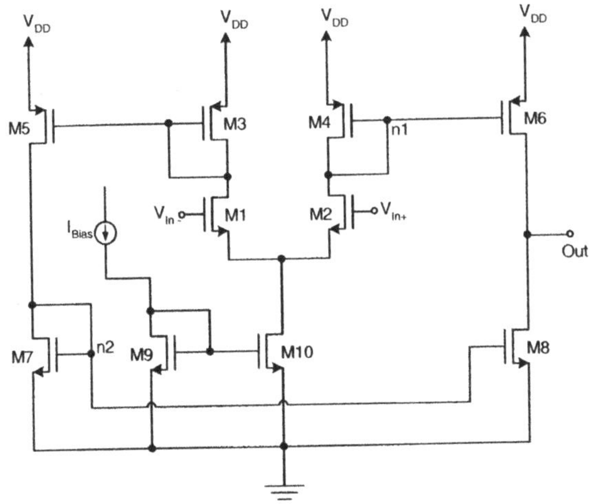

Fig. 6.1 SOI CMOS single-stage op amp with a symmetrical topology, which has a good capability to drive a large capacitive load. (Adapted from Eggermont et al. [1].)

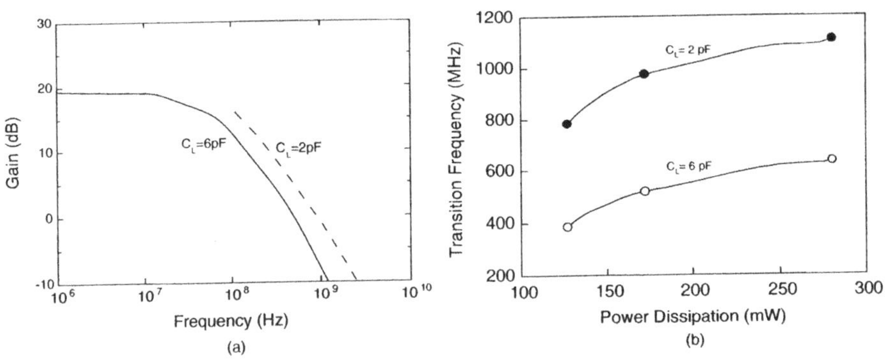

Fig. 6.2 (a) Frequency response of the small-signal gain and (b) transition frequency (f_T) versus power dissipation of the SOI CMOS single-stage op amp described in Fig. 6.1 using 0.7 μm SOI CMOS devices, operating at a supply voltage of $V_{DD} = 4$ V. (Adapted from Eggermont et al. [2].)

Figure 6.2 shows (a) the frequency response of the small-signal gain and (b) the transition frequency (f_T) versus the power dissipation of the SOI CMOS single-stage op amp described in Fig. 6.2 using 0.7 μm SOI CMOS devices, operating at a supply voltage of $V_{DD} = 4$ V [2]. As shown in this figure, with an output load of 2 pf, this SOI single-stage op amp has a transition frequency of 1.1 GHz at the power dissipation of 280 mW. The DC small-signal voltage gain of this op amp is not large, which is due to the short-channel devices (a channel length of 0.7 μm) used, biased in the strong inversion region. In addition, the poles due to the internal nodes do not affect the overall frequency response, which is determined by the dominant pole at the output node. This op amp is suitable for realizing switched capacitor circuits and sigma-delta converters, etc.

For the SOI CMOS single-stage op amp as shown in Fig. 6.1, the analysis of the frequency response is described below. The open-loop gain of the single-stage SOI op amp is

$$A_{vo} = V_{early} \frac{g_{m1}}{I_{D1}} \tag{6.1}$$

where g_{m1} is the transconductance of device M1 and I_{D1} is the drain biasing current of M1. V_{early} the equivalent Early voltage is expressed as $V_{early} = \frac{V_{e6}V_{e8}}{V_{e6}+V_{e8}}$, where V_{e6} and V_{e8} are the Early voltages of M6 and M8, respectively. The transition frequency (f_T) is:

$$f_T = \frac{(W/L)_6}{(W/L)_4} \frac{g_{m1}}{2\pi C_L} \tag{6.2}$$

under condition that the nondominant pole due to the PMOS current source (M4, M6), $\omega_{p1} = g_{m4}/C_1$ is several times larger than the transition frequency, where $(W/L)_6$ and $(W/L)_4$ are aspect ratios of devices M6 and M4, respectively. C_L is the load capacitance. C_1, the total capacitance at the internal node 1, is composed of the intrinsic gate-source capacitances of M4 and M6 (C_{gs4}, C_{gs6}), the gate-source overlap capacitances of M4 and M6 (C_{gso4}, C_{gso6}), the drain-substrate capacitances of M4 and M2 (C_{bd4}, C_{bd2}), and the gate-drain overlap capacitances of M2 and M6 (C_{gdo2}, C_{gdo6}): $C_1 = C_{gs4}+C_{gs6}+C_{gso4}+C_{gso6}+C_{gdo6}+C_{bd4}+C_{bd2}+C_{gdo2}$. Note that the gate-source capacitance can be approximated as $C_{gs} = \frac{2}{3}C_{ox}WL+C_{gso}W$, where C_{ox} is unit-area gate oxide capacitance, W/L is the channel width/channel length of the device. The gate-drain capacitance can be approximated as $C_{gd} = C_{gdo}W$. The body-drain capacitance can be approximated as C_bW, where C_b is the unit-width drain-substrate capacitance. Similarly, the nondominant pole is associated with the internal node 2, which is due to the NMOS current source, $\omega_{p2} = g_{m7}/C_2$, where g_{m7} is the transconductance of the NMOS device M7. C_2, the total capacitances at the internal node 2, is composed of the intrinsic gate-source capacitances of devices M7 and M8 (C_{gs7}, C_{gs8}), the gate-source overlap capacitances of M7 and M8 (C_{gso7}, C_{gso8}), the body-substrate capacitances of M7 and M5 (C_{bd7}, C_{bd5}), and the gate-drain overlap capacitances of M5 and M8 (C_{gdo5}, C_{gdo8}): $C_2 = C_{gs7} + C_{gs8} + C_{gso7} + C_{gso8} + C_{bd7} + C_{bd5} + C_{gdo5} + C_{gdo8}$. The nondominant pole ω_{p2} should also be several times larger than the transition frequency. Due to the buried oxide structure of the SOI devices, the drain-substrate capacitances in C_1 and C_2 are very small, which results in the nondominant poles at high frequencies.

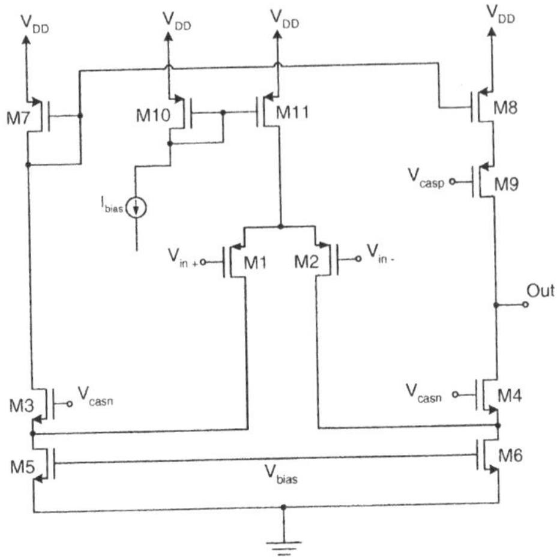

Fig. 6.3 SOI CMOS folded-cascode op amp with a symmetrical topology. (Adapted from Eggermont et al. [3].)

6.1.2 Single-Stage Folded-Cascode Op Amp

In addition to the single-stage SOI CMOS op amp described in the previous subsection, in this subsection a single-stage cascode op amp is described. Figure 6.3 shows an SOI CMOS folded-cascode op amp with a symmetrical topology [3]. As shown in this figure, the PMOS input differential device M1/M2 biased by a current source (M11) is connected to a common-gate device M3/M4 biased by M5/M6. Owing to the cascoded load (M3/M4), this folded-cascode op amp has a superior high-frequency performance. In addition, owing to its folded structure, this op amp is suitable for operation at a low power supply voltage. As for the single-stage op amp, the DC small-signal voltage gain of this single-stage folded-cascode op amp is $A_v = V_{early}(g_{m1}/I_{D1})$ and its transition frequency is $f_T = g_{m1}/2\pi C_L$. In this folded-cascode op amp, the second pole $\omega_p = g_{m4}/C_1$, which is critical for high-speed performance, is determined by the common-gate device M4 and the input device as described in Section 6.1.1. Due to a high electron mobility of the common-gate NMOS device M4, a high second pole (ω_p) leads to a good high-frequency performance. Compared to the single-stage op amp presented in Fig. 6.1, owing to the cascoded common-gate NMOS device, the folded-cascode op amp has a higher DC small-signal voltage gain. However, due to the more complicated poles and zeroes involved, the transition frequency of this folded-cascode op amp may not be higher than that of the single-stage op amp.

6.1.3 Op Amp: SOI versus Bulk

Figure 6.4 shows the g_m/I_D versus $I_D/(W/L)$ of SOI and bulk NMOS and PMOS devices [4]. As shown in the figure, at the same drain current per aspect ratio level,

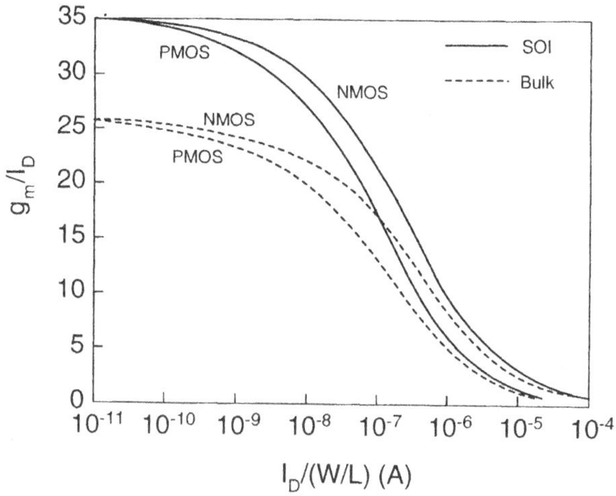

Fig. 6.4 g_m/I_D versus $I_D/(W/L)$ of SOI and bulk NMOS and PMOS devices. (Adapted from Silveria et al. [4].)

g_m/I_D of the SOI device is better than that of the bulk ones. When the device is biased in the subthreshold region, the small-signal gain is at the peak. When the biasing current increases, the voltage gain decreases. On the other hand, since the transition frequency of the device is a function of the transconductance, the biasing current should not be too small. Otherwise its transition frequency may not be acceptable.

Figure 6.5 shows (a) the single-stage op amp with the symmetric topography and (b) the comparison of the performance of the single-stage op amp with an identical transition frequency, using bulk and SOI CMOS devices for various designs with an identical channel length [4]. As shown in this figure, for the designs using the bulk and SOI CMOS devices with an identical die area, the voltage gain, the power consumption, and the phase margin of the SOI op amp (SOI2) are much better than those of the bulk device. For the SOI and bulk op amp designs with identical gain and power consumption, the phase margin of the SOI op amp (SOI3) is much better than that of the bulk device. In addition, the die area is much smaller. From the above analysis, compared to bulk, the SOI CMOS devices are suitable for op amp designs with a better performance.

Figure 6.6(a) shows the two-stage SOI CMOS op amp using body-tied-to-source SOI devices [5]. As shown in the figure, the two-stage SOI CMOS op amp has a quiescent current control and a class AB output stage. Devices M1-M5 form the input stage and devices M11 and M12 comprise the output stage, where the gate of device M12 is controlled by the output of the input stage. The total static supply current, which is determined by the aspect ratio of the transistors and the externally imposed biasing current, is temperature independent. As for the bulk CMOS two-stage op amps, the compensation capacitor C_C has been added between the op amp output and the first stage output to lower the dominant pole such that the gain decays to 0 dB before reaching the first nondominant pole for a good stability in frequency

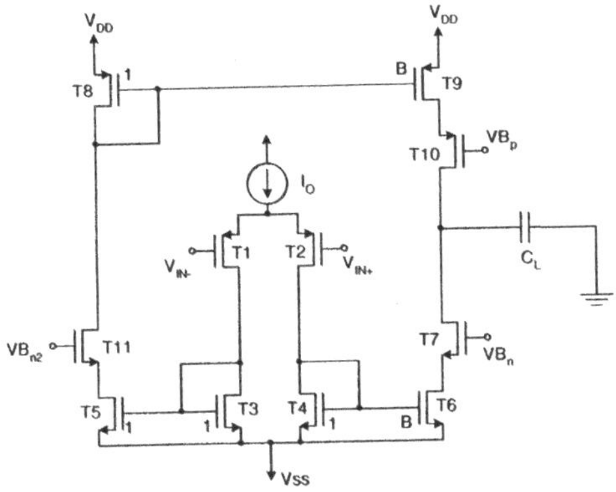

(a)

	Bulk	SOI1	SOI2	SOI3
$(g_m/I_D)_1$(1/V)	19	28	27	19
$(g_m/I_D)_2$(1/V)	17	30	27	17
A_0 (dB)	95.6	103.9	102.7	95.6
I_{tot} (μA)	2.9	2	2	2.9
PM (°)	72.9	72.0	79.0	84.0
ΣW.L (μm²)	2628	4900	2881	1808

(b)

Fig. 6.5 (a) Single-stage op amp with the symmetric topography. (b) Comparison of the performance of the single-stage op amp with an identical transition frequency, using bulk and SOI CMOS devices for various op amp designs with an identical channel length. (Adapted from Silveria et al. [4].)

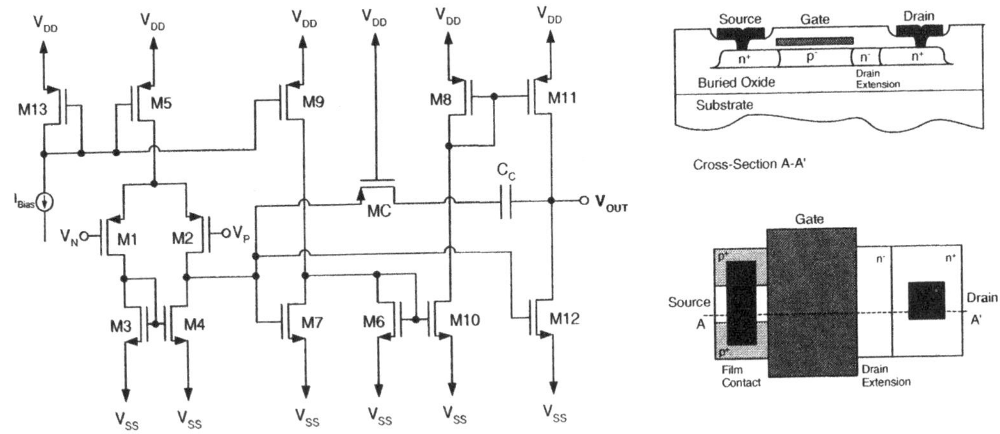

(a) (b)

Fig. 6.6 (a) Two-stage SOI CMOS op amp. (b) Body-tied-to-source SOI device. (Adapted from Verbeck et al. [5].)

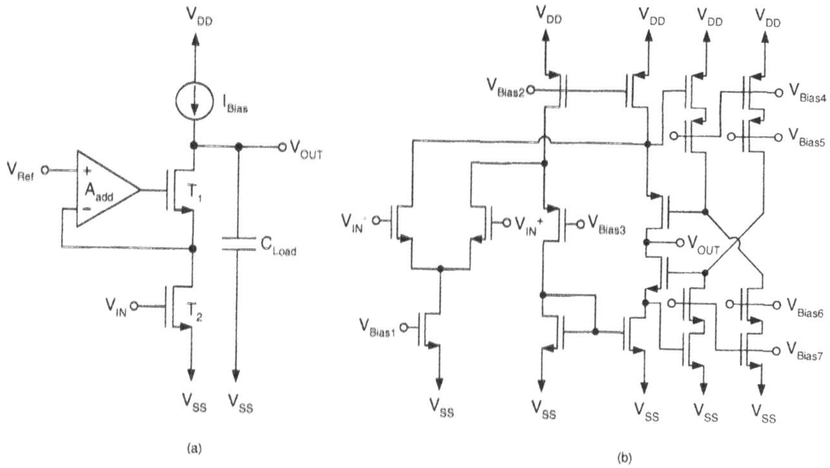

Fig. 6.7 (a) Principle of the gain boosting scheme. (b) SOI folded cascode op amp with the p and n cascoded gain-boosting stages. (From Gentinne et al. [6]. ©1993 IEEE.)

response. In addition, pass transistor MC has been added to push the right half plane (RHP) zero to a very high frequency to avoid high-frequency feedforward problems, which are generated due to the limited transconductance of M12.

While designing the two-stage op amp using PD SOI CMOS devices, the notorious kink effects due to the floating body should be avoided such that the output conductance is not degraded. As shown in Fig. 6.6(b), in this SOI two-stage CMOS op amp, the PD SOI CMOS devices are designed with their body tied to the source to eliminate the kink effects. In addition, in the drain region, a lightly doped drift region that served as a drain extension has been added in order to increase the drain-source breakdown voltage. While operating in the saturation region, its output conductance can be improved and a large DC gain for the op amp can be obtained.

6.1.4 Advanced Cascode Op Amp

Figure 6.7 shows (a) the principle of the gain boosting scheme and (b) the SOI folded cascode op amp with the p- and n- cascoded gain-boosting stages [6]. As shown in this figure, at the output of the cascode output stage, a high-gain feedback loop has been added to maintain the gate-to-source voltage of the cascoded transistor. Via the feedback loop the output impedance and the DC gain can be enlarged. This gain-boosting stage can be designed by a transistor or by another amplifier depending on the needs of the circuit. As shown in Fig. 6.7(b), via n-type and p-type cascoded transistors to form the gain-boosting stage, the folded cascode op amp has been designed, where most transistors are operating in the moderate inversion region. Thus the DC gain of this circuit is sensitive to its biasing condition since its output conductance easily fluctuates with the changes in the biasing current.

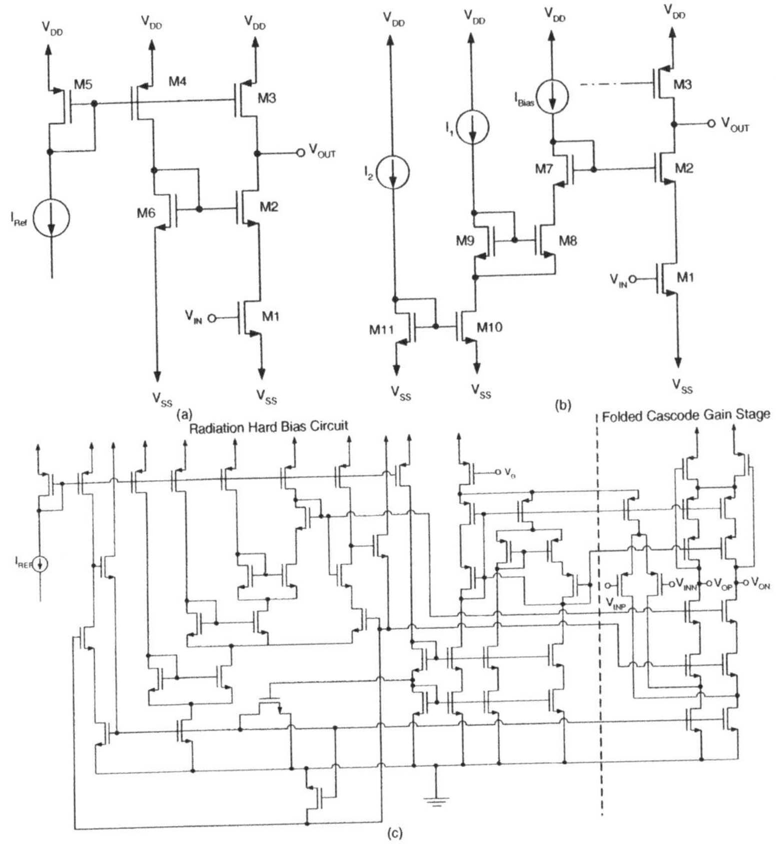

Fig. 6.8 (a) Conventional cascode biasing configuration. (b) Radiation-hardened cascode biasing configuration for use in an amp. (c) Fully-differential single-stage SOI op amp with additional cascoding to reduce the impact of the floating body effect, and the radiation-hardened biasing circuit. (From Edwards et al. [7]. ©1999 IEEE.)

For the SOI device susceptible to irradiation, its threshold voltage may fluctuate, which may change the performance of the related op amp substantially. Figure 6.8 shows (a) the conventional cascode biasing configuration and (b) the radiation-hardened cascode biasing configuration for use in an amp [7]. As shown in Fig. 6.8(a), in the conventional cascode biasing configuration subject to irradiation due to the difference in the gate-source biasing voltage of devices M2 and M6, the extra charge caused by irradiation leads to a decrease in the threshold voltage of device M6, which is more serious as compared to device M2. Thus, the gate voltage of device M2 and the drain voltage of device M1 are affected, which may make device M1 operate out of the saturation region, and hence the output resistance is degraded. In order to resolve this radiation-induced problems, a stacked pair of devices have been adopted in the design as shown in Fig. 6.8(b). By biasing devices M8-M10 in the triode region such that the total drain-source voltage drops are about equal to V_{DSAT} and the IV behavior

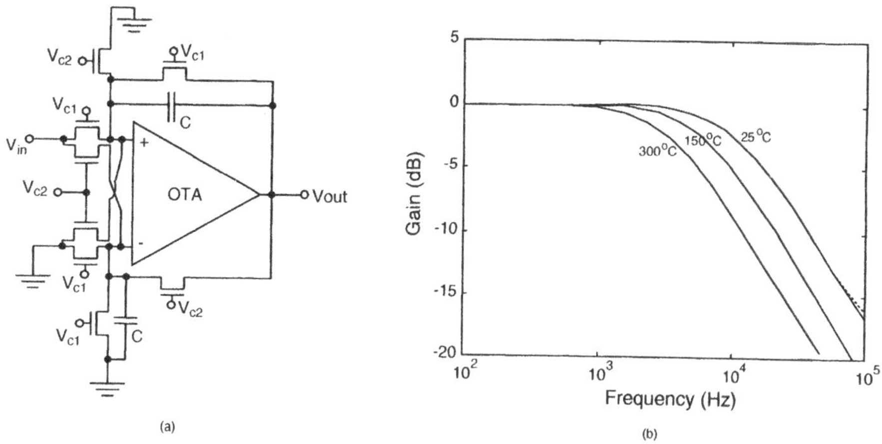

Fig. 6.9 (a) SOI continuous-time first-order filter using an SOI op amp and a highly linear 4-MOSFET structure, with a capacitance of $C = 2$ pF and MOS devices with a channel width of 4 μm and a channel length of 266 μm, operating at the power supply voltages of ± 2.5 V. (b) Frequency response of the filter at 150°C. (Adapted from Dessard et al. [8].)

of devices M2 and M7 is identical after radiation. Consequently, the loss in the drain voltage of device M1 due to irradiation can be compensated–the radiation-hardened protection is achieved. Fig. 6.8(c) shows the fully differential single-stage SOI op amp with an additional cascoding to reduce the impact of the floating body effect and the radiation-hardened biasing circuit [7]. As shown in this figure, a triode region common-mode feedback is used to reduce the kink effects of the PD devices. At the same time, in the output stage, a double NMOS cascode structure with a single PMOS cascode current source is used to stabilize the output resistance and the DC gain. In order to avoid the damage caused by the cascode current source, the radiation-hardened biasing circuit in the left portion is adopted to ensure that the mismatch of the drain-source voltage of the devices does not occur.

6.2 FILTERS

Along with SOI op amps, SOI CMOS technology has been used to integrate filters. In this section, the continuous-time SOI filter and the SOI switched-capacitor filter are described.

6.2.1 Continuous-Time SOI Filter

Compared to PD, FD SOI devices have smaller body effects. Using the FD SOI MOS devices operating in the strong inversion region by controlling the gate voltage, an appropriate equivalent resistance can be obtained, which is helpful for use in a continuous-time filter. Figure 6.9 shows the SOI continuous-time first-order filter

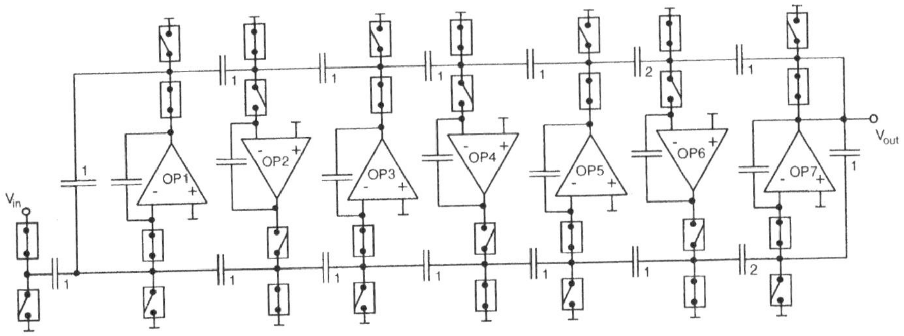

Fig. 6.10 SOI switched-capacitor seventh-order low-pass ladder filter. (From Verbeck et al. [5]. ©1996 IEEE.)

using an SOI op amp and a highly linear 4-MOSFET structure, with a capacitance of $C = 2$ pF and MOS devices with a channel width of 4 μm, operating at the power supply voltages of ± 2.5 V [8]. As shown in this figure, the op amp is in the Miller op amp configuration, which can drive a low impedance load to reduce distortion and power consumption. The 4-MOSFET structure is in two pairs of SOI MOS devices controlled by the gate voltage difference $\Delta V = V_{C1} - V_{C2}$ to provide an appropriate resistance value to form a low-pass filter response. By using the 4-MOSFET structure to realize an equivalent resistor, second and third harmonics can be lowered and a better linearity can be achieved. In addition, the influence of the threshold voltage in the performance of the MOS devices can be avoided. The transfer function of this continuous-time filter can be expressed as [9]:

$$\frac{V_{out}}{V_{in}} = \frac{\mu_n C_{ox} \frac{W}{L}(V_{C1} - V_{C2})}{\mu_n C_{ox} \frac{W}{L}(V_{C1} - V_{C2}) + sC} \tag{6.3}$$

From the above transfer function, the cut-off frequency of the filter can be adjusted by changing the biasing voltages V_{C1} and V_{C2}. In addition, the capacitor C, which is realized by the polysilicon-metal structure, also affects the performance of the filter. The bandwidth of the filter decreases at an elevated temperature due to the decrease in the mobility of the 4-MOSFET structure. At a fixed V_{C1}, by adjusting V_{C2} using an on-chip automatic tuning circuit, the temperature effect can be compensated.

6.2.2 SOI Switched-Capacitor Filter

Using the SOI op amps described in Section 6.1 [5] with body-tied-to-source SOI devices, Fig. 6.10 shows the SOI switched-capacitor seventh-order low-pass ladder filter. As shown in this figure, the switched-capacitor low-pass ladder filter technique used for the bulk CMOS technology has been adopted. Compared to the bulk one, this SOI switched-capacitor filter has much better stability with respect to temperature. At the temperature of 300°C, the performance of this SOI filter is still as good as expected. However, at 350°C, when the intrinsic carrier concentration is as large as

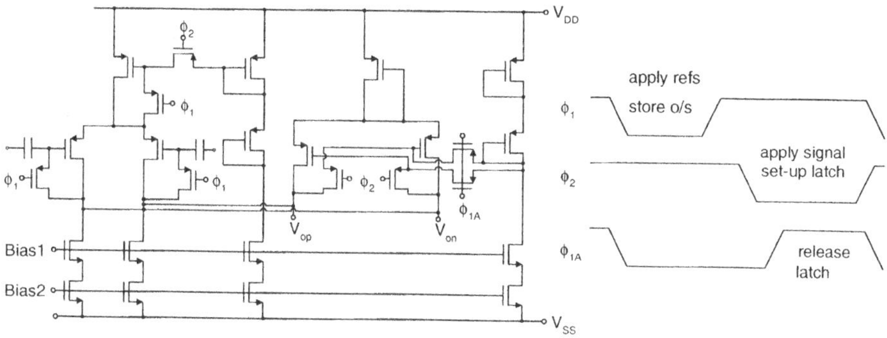

Fig. 6.11 Offset-compensated SOI comparator. (Adapted from Edwards et al. [7].)

the thin-film doping density, this SOI switched-capacitor filter cannot function any more.

6.3 ADC AND DAC

ADC and DAC are important components for building analog circuits. SOI CMOS technology has been used to integrate to ADC and DAC. In this section, SOI ADC and DAC are described.

6.3.1 ADC

Comparator is important in an ADC. Figure 6.11 shows the offset-compensated SOI comparator[7]. When the comparator circuit is susceptible to irradiation, the dose of the irradiation causes the imbalanced condition in the input differential pair, which results in an input offset voltage. In order to resolve the radiation induced imbalance problem, a radiation-hardened biasing circuit has been adopted in this offset-compensated SOI comparator. As shown in this figure, a folded p-channel regenerative structure to avoid the interference due to kink effects (PMOS SOI devices have much less kink effects) has been used. In the source portion of the n-channel devices, a radiation-hardened biasing circuit has been added. During the reset period (ϕ_1 = low, ϕ_{1A} = low), the two latches in the circuit form a current source to store the offset in the input capacitance. During the compare period (ϕ_2 = low, ϕ_{1A} = high), the differential pair becomes a current mirror. After the input signal is compensated by the offset voltage originally stored in the capacitance, it is imposed to the latch transistors. During every reset period, an updated offset voltage is stored. Thus, the comparator can be continuously compensated for the offset voltage due to irradiation.

Combining the offset-compensated SOI comparator, Fig. 6.12 shows a 4-bit SOI flash ADC, which has a radiation-hardened property. As shown in the figure, during operation the input signal is periodically reversed at a frequency higher than the change

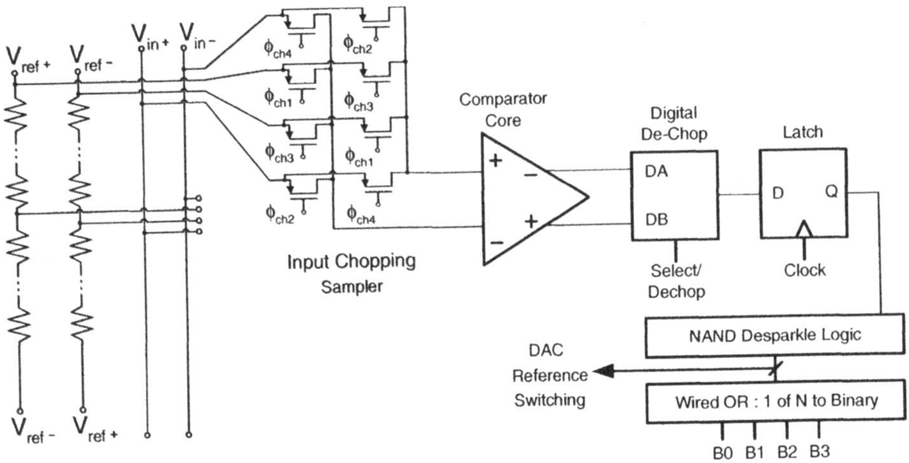

Fig. 6.12 4-bit SOI flash ADC. (Adapted from Edwards et al. [7].)

rate of the irradiation dose such that the radiation-induced offset can be compensated in the circuit. In addition, in the ADC a chopping sampling circuit has been used to provide a chopping operation of the input signal and the reference top signal to reduce the imbalance of the inputs to the next-state comparator such that the radiation-induced offset can be reduced.

Figure 6.13 shows another SOI flash ADC, which is designed to be against the radiation-induced effects [10]. In this ADC, before the comparator is connected to the decoder, which is different from the conventional approach, a stage of static master-slave flip-flops has been added. In addition, the PLA decoder circuit frequently used in a typical ADC has been replaced by a ROM. The modifications have been used to reduce the influence of the irradiation dose in the threshold voltage. As shown in Fig. 6.13(b), a dual auto-zero comparator has been used. Via the dual auto-zero approach, the irradiation-dose-induced threshold voltage shift can be lessened. In addition, when the output data of the comparator enters the flip-flops, the memory capability of the flip-flop, as shown in Fig. 6.13(c), enhances the stability of the output signal. Since the traditional PLA circuit in the decoder is sensitive to the fluctuation of the threshold voltage in this ADC, ROM has been used instead.

6.3.2 DAC

SOI CMOS technology has been used to integrate DAC. Figure 6.14 shows a 7-bit SOI current-steering DAC using a 0.7 μm SOI CMOS technology with the output waveforms [11][12]. As shown in the figure, the current switches are formed by two groups of transmission gates. When a specific current cell is selected, the corresponding transmission gate is turned on such that the current source is connected to the output node. If the cell is not selected, the other transmission gate bypasses the current to V_{DD}. No matter what the situation of the current switch is, a stable

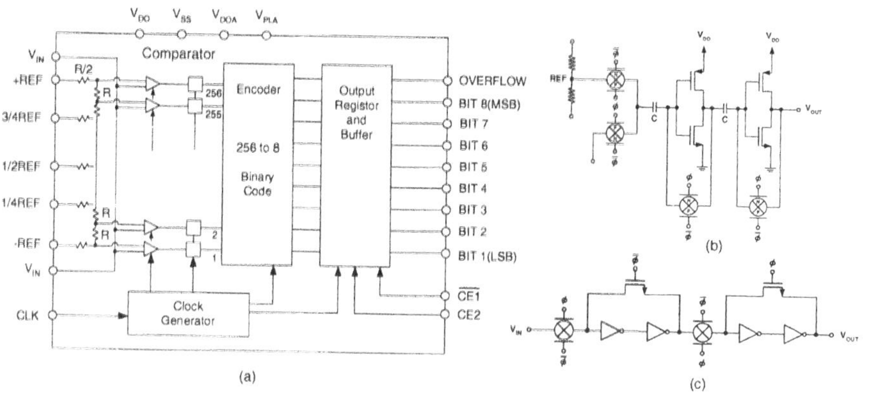

Fig. 6.13 (a) 8-bit 20 MHz SOI flash ADC. (b) Dual auto-zero comparator. (c) Comparator read flip-flop. (Adapted from Baille et al. [10].)

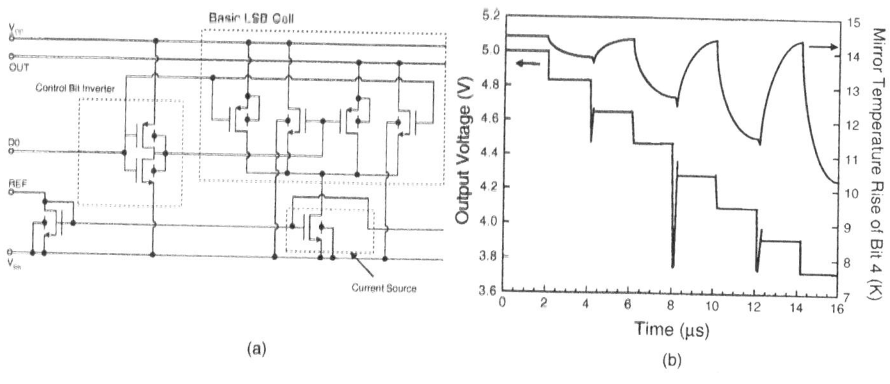

Fig. 6.14 (a) 7-bit SOI current-steering DAC using a 0.7 μm SOI CMOS technology. (b) Output waveforms of the DAC. (Adapted from Tenbroek et al. [12].)

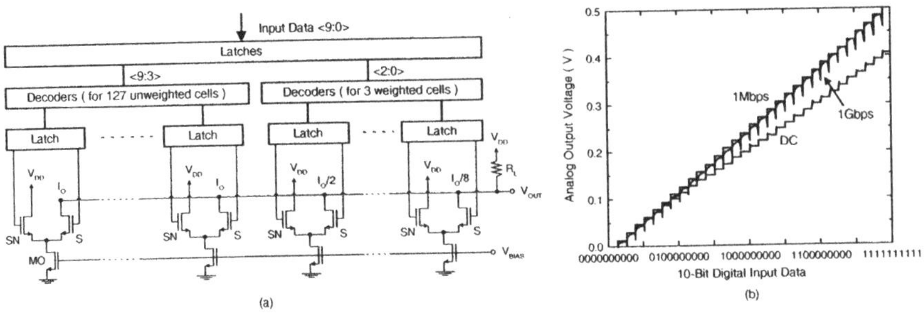

Fig. 6.15 (a) Kink-effect-free 10-bit SOI current-steering DAC using non-FD (NFD) SOI CMOS devices. (b) Its output-input transfer characteristics operating at DC, 1Mbps, and 1Gps. (Adapted from Chang et al. [13].)

current always flows through the current source. Note that the drain-source voltage of the NMOS device serving as the current source may switch between V_{out} and V_{DD}. Due to the unfixed value of drain-source voltage, power consumption of this NMOS device fluctuates, which leads to an instability in the lattice temperature of the device. Consequently, the output current of the device may be affected. As shown in Fig. 6.14(b), when various digital inputs are imposed, the drain-source voltage of the NMOS device may vary with the change in the output voltage. Thus, the lattice temperature of the device changes accordingly. As a result, the linearity of the DAC is worsened, which is more serious at a low switching frequency. At a high switching frequency, the lattice temperature of the device cannot respond in time. Thus, the drawback of this ADC is lessened for operation at a high switching frequency.

The kink effects due to the floating body of the non-FD (NFD) SOI CMOS devices also cause problems for the SOI DAC circuits. Figure 6.15 shows (a) a kink-effect-free 10-bit SOI current-steering DAC using non-FD (NFD) SOI CMOS devices [13] and (b) its output-input transfer characteristics operating at DC, 1Mbps, and 1Gbps. As shown in this figure, this DAC is composed of the latch, the decoder, the unweighted current cell and the weighted current cell. The kink-effect-free capability is achieved by the techniques used in the current cell circuit. The current cell in the lower portion of the circuit is based on the cascode configuration. Via controlling the gate voltage of the source transistor (MO) in the cascode current source, the source transistor can be kept away from the kink effect region. As a result, the output voltage of the current cell may have a kink-effect-free window, which is in the region of the drain-source voltage between the saturation voltage (V_{DSAT}) and the onset of the kink effect (V_{Dkink}). If the source transistor is biased in the kink-effect-free window, the output of the DAC may have a good linearity performance. Along with the increase of V_{BIAS}, the kink-effect-free window of the output voltage (V_{out}) widens. As long as the channel width of the source transistor is properly designed, the required output current (I_{out}) can be obtained. In this circuit, many transistors are connected at the output node. If bulk CMOS technology is adopted, the total parasitic capacitances at the output node may

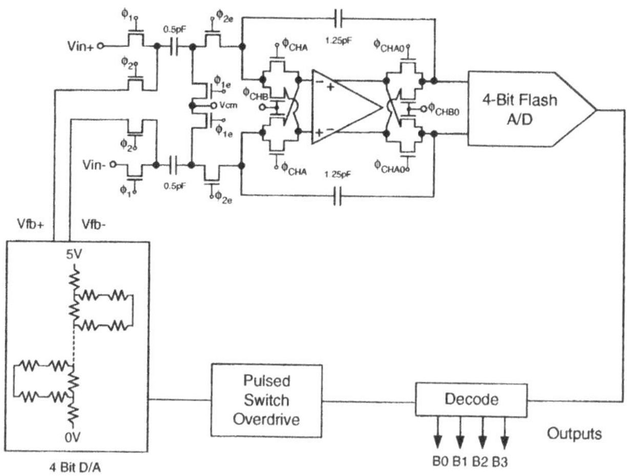

Fig. 6.16 First-order $\Sigma - \Delta$ modulator ADC using a 1.5 μm SOS CMOS technology. (Adapted from Edwards et al. [7].)

be large, and hence the speed performance of the circuit may be slow. In contrast, SOI technology with low parasitic capacitances is especially suitable to integrate this circuit with the output configuration. As shown in Fig. 6.15(b), when operating at DC, the output voltage cannot have a full swing, which is due to the degradation of the output conductance caused by the kink effects. At a higher frequency, since the holes cannot accumulate in time, kink effects are smaller. Therefore, the performance of the DAC becomes better.

6.4 SIGMA-DELTA ADC

ADC with a sigma-delta ($\Sigma - \Delta$) modulator, which is based on a switched-capacitor circuit, has been used to reduce the quantization noise in an ADC. Figure 6.16 shows a first-order $\Sigma - \Delta$ modulator ADC using a 1.5 μm SOS CMOS technology [7]. As shown in this figure, the sigma-delta ADC circuit contains a switched-capacitor integrator, a comparator, a flash ADC, and a ladder DAC. In order to avoid the increase in the layout area using the body contact scheme, a radiation-hardened cascoded biasing circuit with floating-body SOS devices described before has been used. In the $\Sigma - \Delta$ modulator, a fully differential approach has been adopted for the first-order switched-capacitor filter, which provides the function of modulator noise shaping with the chopper stabilization circuit to reduce the flicker noise in the SOS device. Owing to the identical gate voltage (in a time-averaged sense) of the input differential pair, the history effect-dependent radiation induced offset of the SOS devices can be reduced. Using the radiation-hardened cascoded biasing current, the resolution of the multi-bit $\Sigma - \Delta$ modulator is almost not affected.

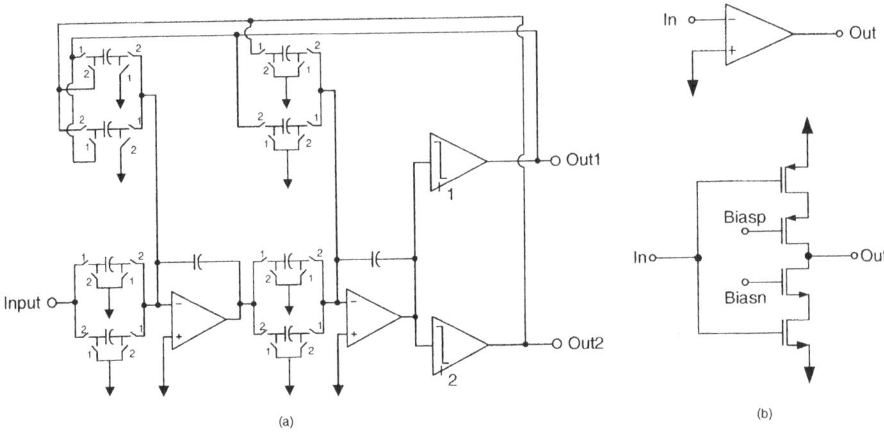

Fig. 6.17 (a) Second-order $\Sigma - \Delta$ modulator with a bilinear switched-capacitor filter using a 0.25 μm FD SOI CMOS technology. (b) Single-ended cascoded amplifier. (Adapted from Swaminathan et al. [14].)

Figure 6.17 shows (a) a second-order $\Sigma - \Delta$ modulator with a bilinear switched-capacitor filter using a 0.25 μm FD SOI CMOS technology with the single-ended cascoded amplifier [14]. As shown in this figure, the single-ended cascoded amplifier has been used in the second-order bilinear switched-capacitor integrator, which can operate during the phases of the clock to achieve doubling the sampling rate. The bilinear integrator has been used in the $\Sigma - \Delta$ modulator to reduce the error due to the transfer of the high-power noise at the output back to the input base via the feedback loop and the mismatch of the two sample paths. Note that the single-ended cascoded inverter amplifier cannot provide an appropriate analog ground voltage, which causes a DC output offset due to the mismatch between the analog ground and the mid-point potential of the circuit. In addition, this mismatch may also result in the gain degradation of the op amp, which limits the modulator noise shaping capability.

FD SOI CMOS technology has been used to integrate $\Sigma - \Delta$ modulators. Fig. 6.18(a) shows a first-order $\Sigma - \Delta$ modulator using an FD SOI CMOS technology [15]. The whole $\Sigma - \Delta$ modulator includes the switched-capacitor integrator, the comparator, and the latch circuit. Since the gain of the integrating amplifier is not the dominant factor in determining the performance of the $\Sigma - \Delta$ modulator. In order to reduce power consumption, a simple single-stage op amp has been used in the design. As shown in Fig. 6.18(b), operating at 300°C, this circuit still works. However, at high temperature, its SNR is worsened because of the harmonic distortions. In addition, at the high temperature, the leakage current of the SOI devices increases and the threshold voltage decreases. Therefore, the function of the switches degrades, which leads to the loss of the charge from the capacitors at an increased rate in the integrator. As a result, the minimum sampling frequency is worsened.

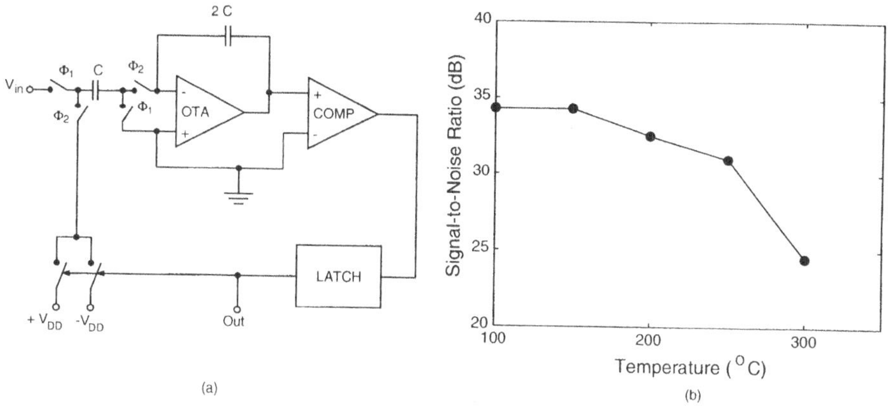

Fig. 6.18 (a) First-order $\Sigma - \Delta$ modulator using an FD SOI CMOS technology. (b) Signal-to-Noise ratio (SNR) versus temperature of this circuit. (Adapted from Viviani et al. [15].)

6.5 RF CIRCUITS

Wireless communication systems have been becoming a necessity in our daily life. For a wireless communication system, RF circuits are the key component. Owing to advantages, SOI CMOS technology has been used to integrate RF circuits. Figure 6.19 shows a typical radio frequency (RF) circuit in a wireless communication system [16]. As shown in this figure, the RF circuit is divided into two groups– the receive circuit and the transmit circuit. In the receive circuit, the received signal from the antenna is passed through the LNA. Then, via the down mixer with a clock generated by the VCO, after removing the high-frequency carrier it has become the IF signal. Now, after the demodulator, it has become the base-band signal for further processing. In the transmit circuit, the base-band signal is first modulated in the IF modulator. Then, via the mixer, it has become the RF signal to be transmitted. The key components in the RF circuits include LNA, mixer, VCO, and modulator. Owing to the low parasitics advantages and low-voltage capabilities, SOI CMOS technology has been used to integrate low-power RF circuits.

For implementing RF circuits in a wireless communication system such as a mobile phone, basically there are four technologies (1) deep-submicron CMOS, (2) SOI CMOS, (3) BiCMOS with SiGe-base HBT, and (4) GaAs as shown in Fig. 6.20 [17]. Among four technologies, GaAs technology does not have the capability to implement low-voltage digital VLSI circuits and mixed ADC. Using a SiGe-base structure, silicon HBTs can be used to design RF circuits. However, silicon bipolar devices have a poor linearity at low power and consume a substantial amount of power. BiCMOS technology with SiGe-base HBT may be too complicated. Deep-submicron silicon CMOS technology has been used to integrate RF circuits. However, the required low power supply voltage may be a major drawback to achieve high-speed requirements. Compared to other technologies, SOI technology can provide capabilities for realizing low-voltage digital VLSI logic circuits and analog circuits

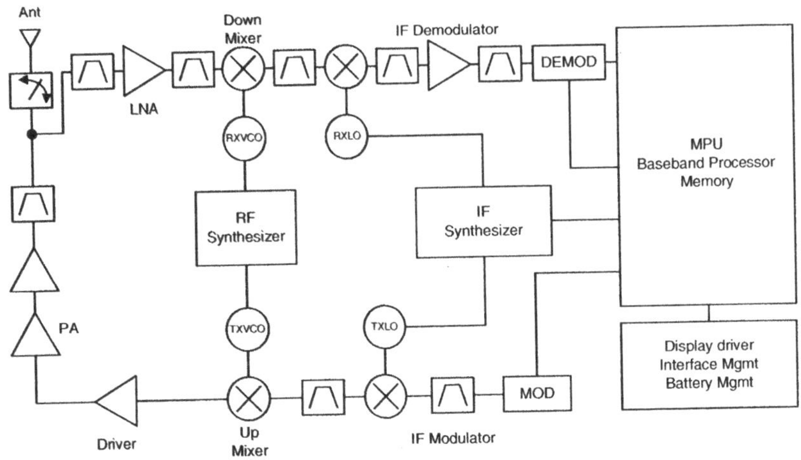

Fig. 6.19 Typical radio frequency (RF) circuit in a wireless communication system. (Adapted from Huang et al. [16].)

Function	Si CMOS, 0.18 μm	SOI CMOS, 0.35 μm	Si BiCMOS	GaAs
F_{MAX} (GHz)	30	60	50(SiGe)	60
Linearity	Good	Best	Poor(at low power)	Best
NF_{MIN}	<0.8	<0.8	<0.8	<0.5
RF Switches	Poor	Best	Poor	Best
Low Power Dig.	Yes	Yes	Yes	No
Passives Integration	Poor	Good	Poor	Good
A/D & D/A	Yes	Yes	Maybe	No
3V Swing	No	Yes	Yes	Yes
EEPROM/Flash	No	Yes	No	No
Isolation	Poor	Good-best	Poor	Best
Cost	Best	Good	Good	Poor

Fig. 6.20 Comparisons of four key technologies– (1) 0.18 μm CMOS, (2) 0.35 μm SOI CMOS, (3) BiCMOS with SiGe-base hetero-junction bipolar transistor(HBT), and (4) GaAs technologies, for implementing RF circuits. (Adapted from Reedy et al. [17].)

with a large voltage swing. In addition, the buried oxide layer of the SOI CMOS devices lowers the thin-film/substrate coupling such that the quality factor of the passive element such as inductor and the self-resonant frequency can be enhanced.

Owing to the low parasitic capacitances in the source/drain, high transconductance, excellent buried oxide isolation, and high resistivity substrate, SOI CMOS devices have been used in realizing microwave circuits. Figure 6.21(a) shows the small-signal equivalent circuit of an SOI CMOS device used in RF circuits [18]. As shown in this figure, the equivalent circuit includes the intrinsic device portion, the metal-insulator-metal (MIM) capacitances, and planar inductances. The SOI RF CMOS devices have a thin film of 1000 Å, a buried oxide of 4000 Å, a front gate oxide of 100 Å, a channel length of 0.2 μm/0.4 μm (NMOS/PMOS), and the LDD structure. In addition, self-aligned salicide, two-layer Al/Si metallization, Ti/N-barrier, and tungsten plugs have

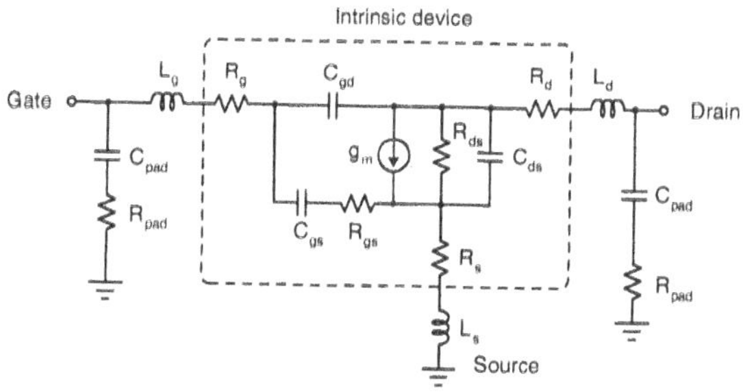

(a)

	NMOS			PMOS
t_{ox} (μm)	7nm	10nm	10nm	10nm
L_{eff} (μm)	0.2	0.2	0.2	0.4
W_{eff} (μm)	100	100	400	400
C_{gd} (pF)	0.04	0.04	0.16	0.35
C_{gs} (pF)	0.15	0.1	0.43	0.5
C_{ds} (pF)	0.02	0.03	0.09	0.13
R_{ds} (Ω)	440	420	110	80
R_{gs} (Ω)	1.2	1.6	1.8	2
R_G (Ω)	3.2	2.4	1.4	0.9
R_D (Ω)	4.3	7.5	1.4	6.2
R_S (Ω)	5	4.5	1.6	4.7
g_m (mS)	27	21	87	43
L_G (pH)	30	20	18	4.2
L_D (pH)	36	30	22	55
L_S (pH)	1.1	4.5	4.4	2.3
C_{pad} (fF)	50			50
R_{pad} (Ω)	50			450

(b)

Fig. 6.21 (a) Small-signal equivalent circuit of an SOI CMOS device used in RF circuits. (b) Extracted parameters of the SOI CMOS devices for RF circuits, biased at $V_{DS} = 2\ V/V_{GS} = 1\ V$(NMOS) and $V_{DS} = -2\ V/V_{GS} = -1.5\ V$ (PMOS). (Adapted from Eggert et al. [18].)

been used to reduce parasitic resistance of the interconnects and gate resistance. In order to avoid kink effects and parasitic edge transistor effects, the body-tied-to-source structure has been used. As shown in the figure, gate resistance, transconductance, parasitic capacitances, and channel length/width may affect the RF properties.

Among RF properties there are three key parameters—(1) unity-gain frequency (f_T), (2) cut-off frequency (f_{max}), and (3) minimum noise figure (F_{min}). The unity-gain frequency (f_T) is related to the transconductance (g_m) and the gate-source/gate-drain capacitances (C_{gs}/C_{gd}) as follows:

$$f_T = \frac{g_m}{2\pi(C_{gs} + C_{gd})} \tag{6.4}$$

The cut-off frequency (f_{max}) is referred to the drain-source resistance (R_{ds}), the gate resistance (R_g), the gate-source resistance (R_{gs}), the source resistance (R_s), the gate-drain capacitance (C_{gd}), and the unity-gain frequency (f_T) as follows:

$$f_{max} = \frac{1}{2} f_T \sqrt{\frac{R_{ds}}{R_g + R_{gs} + R_s + 2\pi f_T R_g C_{gd} R_{ds}}} \tag{6.5}$$

The minimum noise figure is related to the transconductance, the gate resistance, the source resistance, and the channel length (L_{eff}) as:

$$F_{min} = f L_{eff} \sqrt{g_m (R_g + R_s)} \tag{6.6}$$

where f is the input signal frequency. From the above equations, the unity gain frequency (f_T) is independent of the source resistance. The cut-off frequency (f_{max}) is strongly correlated to the source resistance. A lower source resistance leads to a higher cut-off frequency. For the SOI RF CMOS devices, the adoption of the Ti-salicide and the metal shunt has been used to lower the gate, the drain, and the source

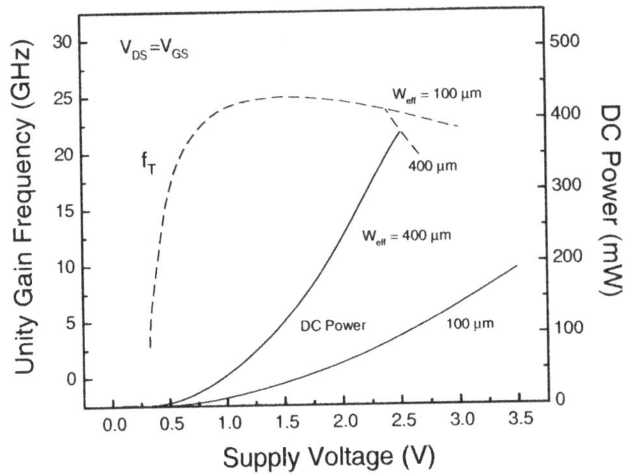

Fig. 6.22 Unity gain frequency (f_T) and DC power consumption versus supply voltage ($V_{DS} = V_{GS}$) of the SOI NMOS device with a front gate oxide of 100 Å and a channel length of 0.2 μm. (Adapted from Eggert et al. [18].)

resistances. Therefore, the cut-off frequency (f_{max}) has been improved several times. For SOI RF CMOS devices with a down-scaled channel length and gate oxide, during the evolution of the deep-submicron technology, its transconductance is improved. Thus, the maximum gain of the device can be improved.

When the channel width is scaled down along with the evolution of the SOI technology, a shrunk current leads to a reduced DC power consumption and reduced self-heating. Figure 6.22 shows the unity gain frequency (f_T) and the DC power consumption versus the supply voltage ($V_{DS} = V_{GS}$) of the SOI NMOS device with a front gate oxide of 100 Å and a channel length of 0.2 μm [18]. As shown in the figure, when the power supply voltage is scaled down from 3 to 1.2 V, its unity-gain frequency increases slightly. With a power supply voltage below 1.2 V, due to the decrease in V_{DD}, f_T falls sharply.

Figure 6.23 shows the comparisons of the FD and PD SOI CMOS devices for RF circuits in terms of (a) the unity-gain frequency (f_T) and the cut-off frequency (f_{max}) versus the gate voltage of 0.25 μm PD and FD SOI NMOS devices with a front gate oxide of 45 Å and a thin film of 400/1000 Å (FD/PD), biased at the drain voltage of 1V and (b) their extracted parasitic resistances, and (c) their capacitances, transconductances, and output conductances [19]. For the PD devices, Ti-salicide has been used to reduce the sheet resistance of the gate, the drain, and the source. For the FD devices, no Ti-salicide is used. Owing to Ti-salicide, the gate, the drain, and the source resistances of the PD devices have been reduced substantially as shown in Fig. 6.23(b). For low-voltage operation with a supply voltage of 1V, the cut-off frequency (f_{max}) of the PD device is much larger. Owing to the thinner thin film, the gate-source, the gate-drain, the drain-source capacitances (C_{gs}, C_{gd}, C_{ds}), and the transconductance (g_m) of the FD devices are better as shown in Fig. 6.23(c). Thus,

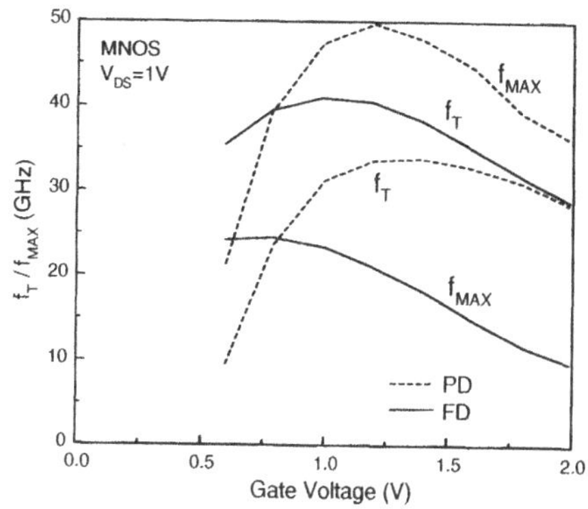

(a)

	With Salicide (PD)	Without Salicide (FD)
R_S (Ω)	1.9	2.8
R_D (Ω)	2.1	3.0
R_G (Ω)	17	83

(b)

	NMOS FD			NMOS PD		
V_D (V)	0.0	1.0	2.0	0.0	1.0	2.0
C_{gs} (fF)	38	81	84	42	86	92
C_{gd} (fF)	29	26	26	35	32	32
C_{ds} (fF)	19	23	28	21	23	24
g_m (mS)	-	29	30	-	24	27
g_{ds} (mS)	0.2	2.8	3.6	0.1	2.7	3.6

(c)

Fig. 6.23 (a) Unity-gain frequency (f_T) and cut-off frequency (f_{max}) versus gate voltage of 0.25 μm PD and FD SOI NMOS devices with a front gate oxide of 45 Å and a thin film of 400/1000 Å (FD/PD), biased at the drain voltage of 1 V. (b) Their extracted parasitic resistances. (c) Their capacitances, transconductances, and output conductances. (From Rozeau et al. [19]. ©1999 IEEE.)

the FD device has a better unity-gain frequency (f_T). If the FD device also adopts the Ti-salicide technology, at the drain-source/gate-source bias of 1 V, the unity-gain frequency (f_T) and the cut-off frequency (f_{max}) of the FD device are 45 and 60 GHz, respectively, which are much better as compared to the PD device.

In order to reduce floating body effects and terminal capacitances, the thin-film of the SOI devices has been gradually becoming thinner– PD devices have been changing into FD devices. More and more RF circuits have been designed using sub-0.25 μm FD SOI CMOS devices. Figure 6.24 shows (a) the cut-off frequency versus the drain voltage of sub-0.25 μm SOI CMOS devices with a front gate oxide of 72 Å, a thin film of 500 Å, and a buried oxide of 2000 Å, with (b) their key parameters [20]. As shown in this figure, with a smaller channel length, the transconductance increases slightly and the gate-source/gate-drain capacitance decreases substantially. As a result, the cut-off frequency increases for the device with a smaller channel length. Even at a supply voltage of 0.5 V, the cut-off frequency of the device does not decay to a large extent. the cut-off frequency of the NMOS device is almost not affected by the

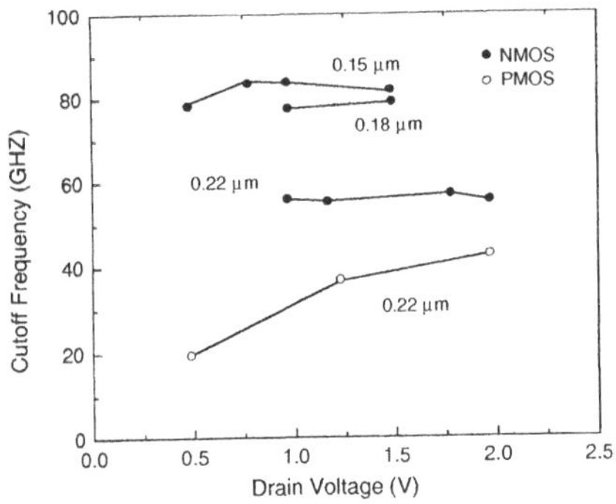

(a)

	Type ,Gate length (μm)			
	NMOS, 0.22	NMOS, 0.18	NMOS, 0.15	PMOS, 0.22
R_G (Ω)	350	550	610	205
C_{gs} (fF)	23	11	10	22
C_{gd} (fF)	7.5	5.0	4.0	11.0
C_{ds} (fF)	9.0	3.0	2.0	8.0
R_S (Ω)	11.2	12.1	13.5	25.0
R_D (Ω)	6.0	5.0	9.0	9.8
R_{gs} (Ω)	21.3	22.1	20.0	33.6
G_m (mS)	10.0	10.8	12.0	5.9
R_{ds} (Ω)	420	150	56	590
R_{OUT} (Ω)	583	424	250	622

(b)

Fig. 6.24 (a) Cut-off frequency versus drain voltage of sub-0.25μm SOI CMOS devices with a front gate oxide of 72 Å, a thin-film of 500 Å, and a buried oxide of 2000 Å. (b) Their key parameters. (Adapted from Chen et al. [20].)

change in the drain voltage. In contrast, the cut-off frequency of the PMOS device falls sharply when the drain voltage is reduced due to a decrease in the hole mobility.

6.6 LOW-NOISE AMPLIFIER (LNA)

LNA is an important component in the RF circuit. Owing to the low-voltage capability, SOI CMOS devices provide many advantages for designing LNA circuits. In this section, SOI LNA circuits are described.

6.6.1 Single-Stage LNA

Figure 6.25 shows (a) the SOI LNA using a 0.2 μm FD SOI CMOS device with (b) the frequency response in terms of gain and noise figure [21]. There is no input matching circuit in the LNA. Instead, an output capacitive load with two inductors has been used. The on-chip inductor (L_S), which is used to lower the noise figure under the 50 Ω input impedance, is based on a 3 μm-thick multilayer Al process. The quality factor of the on-chip inductor is 7 at 2 GHz. As shown in the figure, at 2 GHz, the dissipated current of the LNA is 10 mA for $V_{DD} = 1$ V and 4 mA for $V_{DD} = 0.5$ V. As shown, its noise figure is 3.5 dB (1 V) and 3.9 dB (0.5 V) at 2 GHz and its 1 dB compression point is -4 dBm (1 V) and -9 dBm (0.5 V). From the data, the SOI LNA is suitable to operate in the low-voltage environment.

In addition to FD devices, PD SOI devices may also be used for LNA. By controlling the floating body, the current driving capability of the PD SOI devices can be enhanced for use in LNA circuits. Figure 6.26(a) shows the field-shielded PD SOI CMOS devices, which use the field-shield isolation approach to produce PD devices with an active body control for use in the LNA circuit of the monolithic microwave integrated circuits (MMIC) operating at a supply voltage of 1 V [22]. Figure 6.26(b)

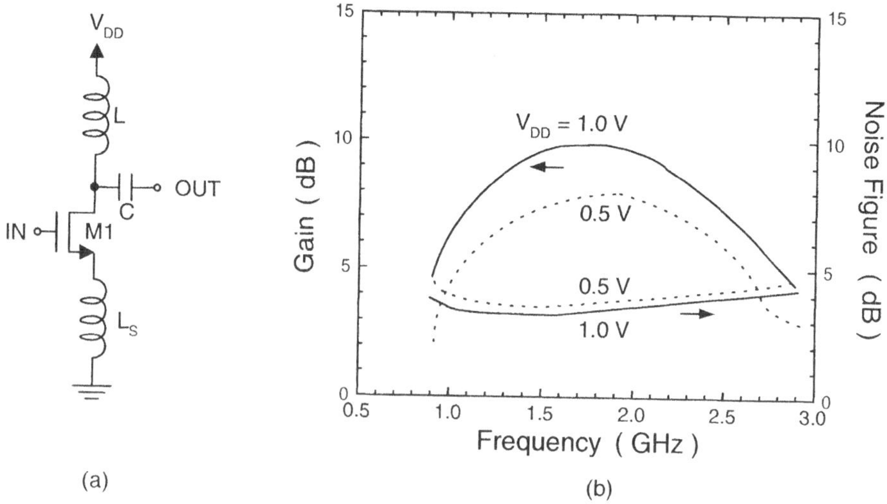

Fig. 6.25 (a) SOI low-noise amplifier (LNA) using a 0.2 μm FD SOI CMOS device. (b)Frequency response of the FD SOI LNA in terms of gain and noise figure. (Adapted from Harada et al. [21].)

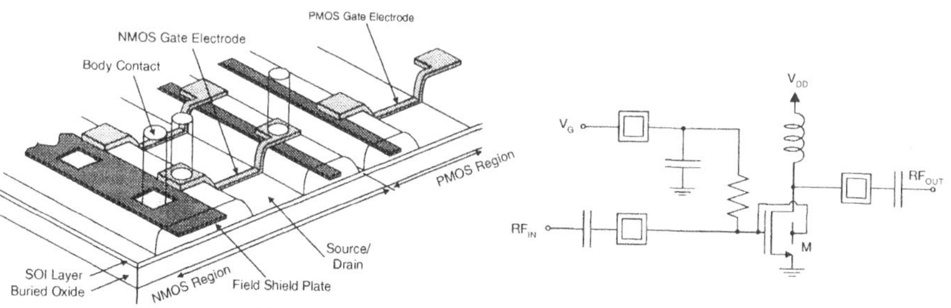

Fig. 6.26 (a) Field-shielded PD SOI CMOS devices. (b) SOI LNA circuit with the active-body PD SOI NMOS device operating at a supply voltage 1 V. (From Komurasaki et al. [22]. ©1998 IEEE.)

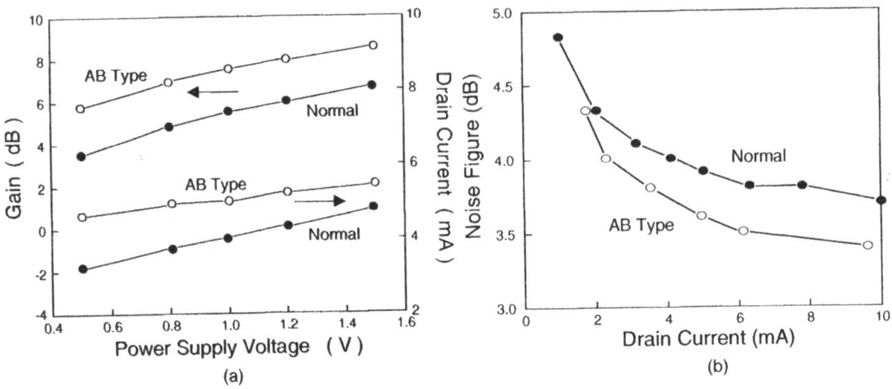

Fig. 6.27 (a) Gain and drain current versus drain voltage and (b) Noise figure versus drain current of the LNA using the PD SOI NMOS device with and without the active-body configuration. (Adapted from Komurasaki et al. [22].)

shows the LNA circuit using the field-shielded PD SOI CMOS devices, where the polysilicon field-shielded plate is used as the resistor and the MOS gate capacitance is served as the capacitor. In addition, 2.5 μm third-layer Al interconnects are served as the on-chip inductor to reduce loss. As shown in this figure, in the LNA circuit using PD devices with the active-body (AB) configuration, the field-shielded SOI devices use body contacts to connect the body to the gate–active body configuration, which is similar to the DTMOS techniques for the digital circuits. Owing to the active-body configuration (body tied to gate), the threshold voltage is reduced because of body effect and the drain current is increased. Even at a low supply voltage (0.5 V), the PD SOI device with the DTMOS configuration still maintains a high gain and a large drain current, which could drive a large capacitive load. As a result, the 1 dB compression point has been improved. In addition, with the body tied to the gate in the active body configuration, no floating body effects such as kink effects bother the LNA circuit.

Figure 6.27 shows (a) the gain and the drain current versus the drain voltage and (b) the noise figure versus the drain current of the LNA using the PD SOI NMOS device with and without the active-body configuration [22]. Note that "without active-body configuration" is referred to the body-tied-to-source scheme. As shown in this figure, with the active-body configuration (body-tied-to-gate), a larger drain current and thus a larger gain have been achieved. In addition, the noise figure at an identical drain current is smaller for the LNA with the active-body configuration.

LNA can be designed using SOI MMIC with on-chip matching. From the circuit design point of view, an LNA circuit is evaluated in terms of gain, noise figure, 1 dB compression point, and dielectric loss of the spiral inductor. In addition to gain, noise figure, and 1 dB compression point, the SOI substrate also provides a low-dielectric loss substrate to reduce the loss of the spiral inductors. Figure 6.28 shows (a) the plan and the cross sectional views of the spiral inductor with an outer size of 300 μm $\times$ 300 μm, a metal thickness of 3.1 μm, a stripeline width of 11 μm, and a line spacing of 10 μm and (b) the equivalent circuit and the insertion loss of

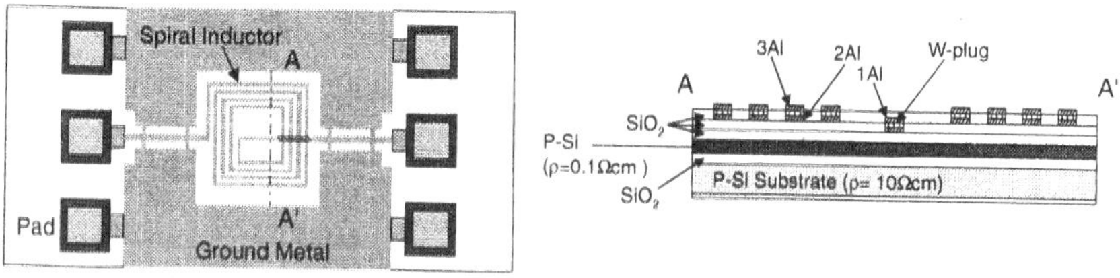

(a)

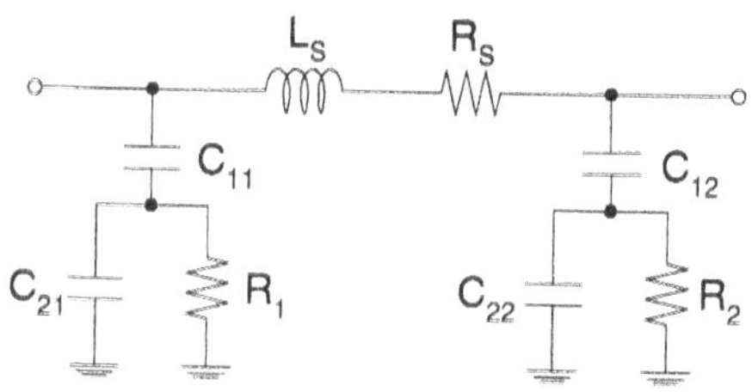

	SOI Substrate	Low Resistive Si Substrate
L_S (nH)	4.9	4.9
R_S (Ω)	4.9	4.4
C_{11} (pF)	0.23	0.51
C_{21} (pF)	0.12	0.12
R_1 (Ω)	170	172
C_{12} (pF)	0.29	0.52
C_{22} (pF)	0.12	0.13
R_2 (Ω)	203	201
Attenuation (dB)	2.1	2.6

(b)

Fig. 6.28 (a) Plan and cross-sectional views of the spiral inductor with an outer size of 300 $\mu \times 300\ \mu$m, a metal thickness of 3.1 μm, a stripeline width of 11 μm, and a line spacing of 10 μm, using a 0.35 μm SOI CMOS technology. (b) Equivalent circuit and insertion loss of the spiral inductor using SOI technology, operating at 2.1 GHz. (From Ono et al. [23]. ©1998 IEEE.)

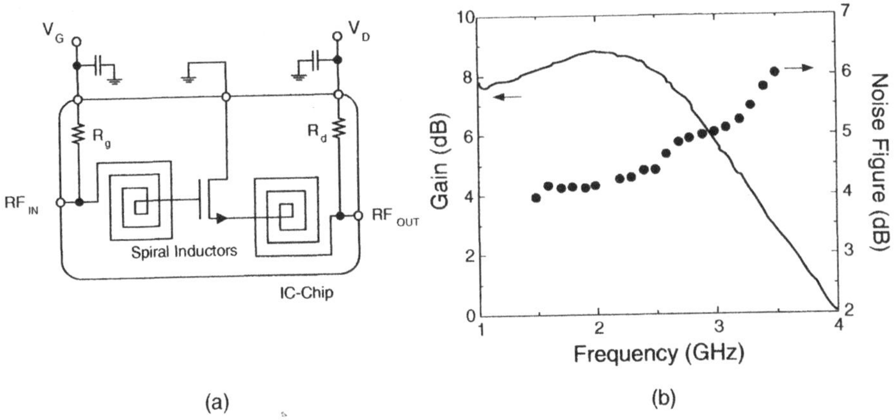

Fig. 6.29 (a) Schematic of the LNA. (b) Gain and loss versus frequency of the LNA using the 0.35 μm SOI CMOS devices biased at the drain voltage of 3 V and the drain current of 3 mA. (Adapted from Ono et al. [23].)

the spiral inductor using SOI technology, operating at 2.1 GHz [23]. As shown in Fig. 6.28(b), in the equivalent circuit of the spiral inductor, L_S is the series inductance and R_s is the series resistance. C_{11} and C_{12} are the parasitic capacitances and C_{21} and C_{22} are substrate capacitances. R_1 and R_2 are substrate resistances. By using the SOI substrate with a buried oxide of 4000 Å to design a spiral inductor, the parasitic capacitances (C_{11}, C_{12}) are about one-half of the values designed by the low-resistive silicon substrate and the loss of the spiral inductor made on the SOI substrate is ~ 80% of the low-resistive silicon substrate counterpart.

Based on the on-chip spiral inductor, the LNA circuit is as shown in Fig. 6.29(a), which is a single-stage amplifier with a 0.35 μm common-source enhancement-mode SOI NMOS device. In the input and the output portions of the LNA, the on-chip spiral inductors have been used as a matching part. R_g is the gate bias resistance. Owing to the reduced dielectric loss of the SOI spiral inductor, the performance of the SOI LNA has been improved.

Just as SOI, SOS CMOS technology is also suitable for integrating MMICs, and is different from SOI technology. Instead of the buried oxide layer/silicon substrate, the sapphire layer has been used for isolation. Just as SOI, thin-film SOS CMOS technology also has advantages in low parasitic capacitances, good radiation hardness, and no latchup. In addition, for SOS technology, the sapphire layer has a better thermal conductivity as compared to the buried oxide. Therefore, the self-heating of the SOS technology is much less as compared to the SOI devices. Owing to the compressive stress suffered by the silicon thin film, the hole mobility in the SOS CMOS devices has been increased, which is similar to the SiGe-channel PMOS devices [24] suitable for MMIC.

Figure 6.30 shows a single-stage LNA circuit using a 0.5 μm thin-film silicon-on-sapphire (TFSOS) CMOS technology with a silicon thin film of 1000 Å [25]. As shown in this figure, in this TFSOS LNA circuit, the designs of the spiral inductors and

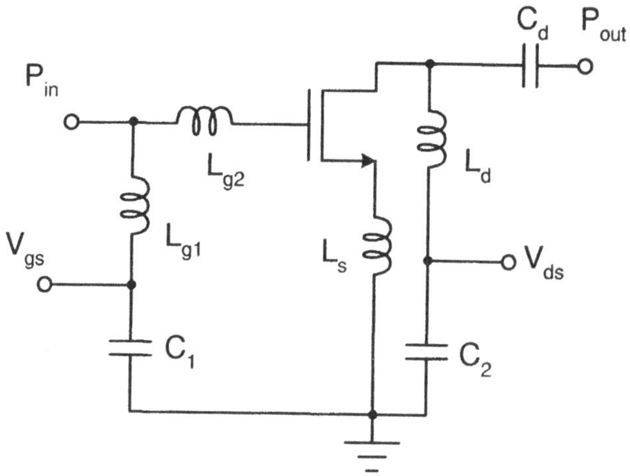

Fig. 6.30 Single-stage LNA circuit using a 0.5 μm thin-film silicon-on-sapphire (TFSOS) CMOS technology with a silicon thin film of 1000 Å. (Adapted from Johnson et al. [25].)

the MIM capacitors should match the MOS device to achieve a minimized noise figure. Usually the on-chip spiral inductor is realized using a multilayer metal technology with an increased thickness to reduce the noise figure. Since this TFSOS device is fully depleted, there is no body effect with good turn-off characteristics and a low output conductance, which result in a good linearity of the TFSOS LNA circuit.

6.6.2 Cascode LNA

As for bulk CMOS amplifiers, the cascoded configuration could help improve the gain. Figure 6.31 shows a 1.5 V 1.8 GHz single-stage cascoded LNA circuit using PD SOI CMOS technology [26] to enhance the performance. Specifically, using the cascoded configuration with a common-gate transistor, the gain and the linearity can be improved as compared to the common-source configuration. In this cascoded LNA circuit, transistors M1 and M2 and the inductor L_s are made on-chip. Off-chip inductors L_g and L_{load} are bond wired to achieve a high-quality factor. Inductor L_{load} is used to compensate for the output load such that a high gain can be obtained. Figure 6.32 shows (a) the gain versus the frequency and (b) the noise figure versus the power supply voltage of the cascoded SOI LNA circuit [26]. As shown in this figure, for the cascoded LNA circuit using the PD SOI devices with body floating, due to the reduced threshold voltage from the floating body effect, the gain is larger owing to a larger drain current. With the body-tied-to-source configuration, due to the body-contact resistance from the extra body contacts, the SOI cascoded LNA circuit shows a worse noise figure.

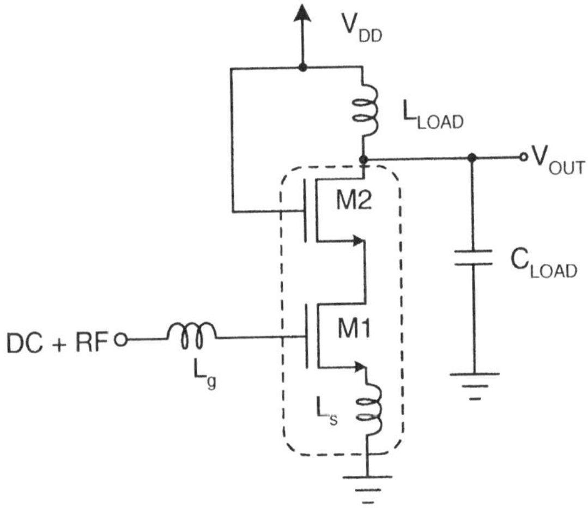

Fig. 6.31 1.5 V 1.8 GHz single-stage cascoded LNA circuit using a PD SOI CMOS technology. (Adapted from Jin et al. [26].)

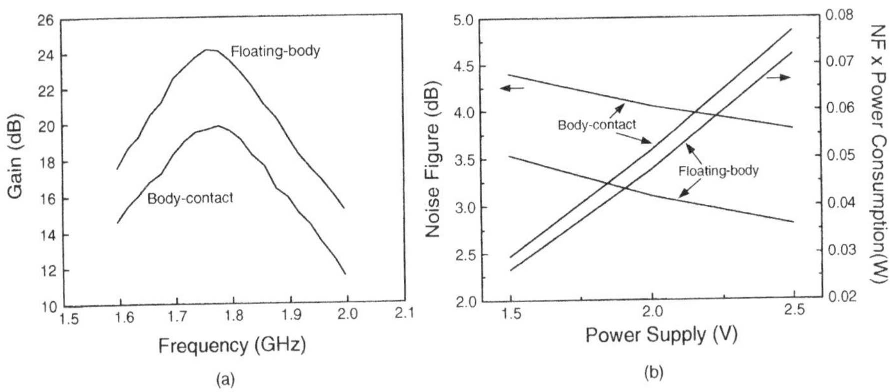

Fig. 6.32 (a) Gain versus frequency and (b) noise figure versus power supply voltage of this cascoded SOI LNA circuit. (Adapted from Jin et al. [26].)

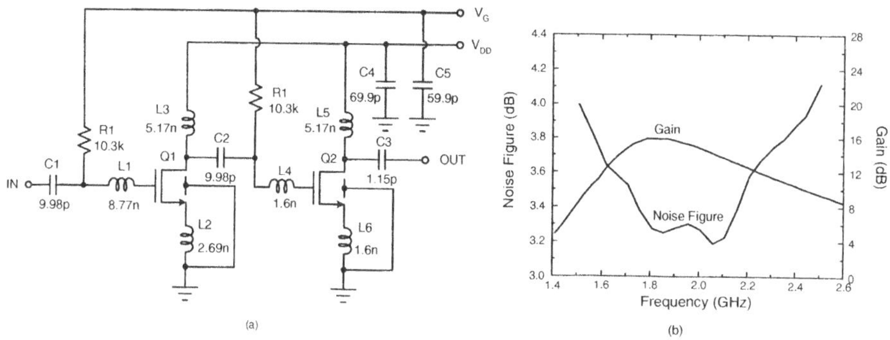

Fig. 6.33 (a) 1.8 GHz two-stage SOI LNA circuit. (b) Gain and noise figure versus frequency of the two-stage SOI LNA circuit. (Adapted from Vorwerk et al. [27].)

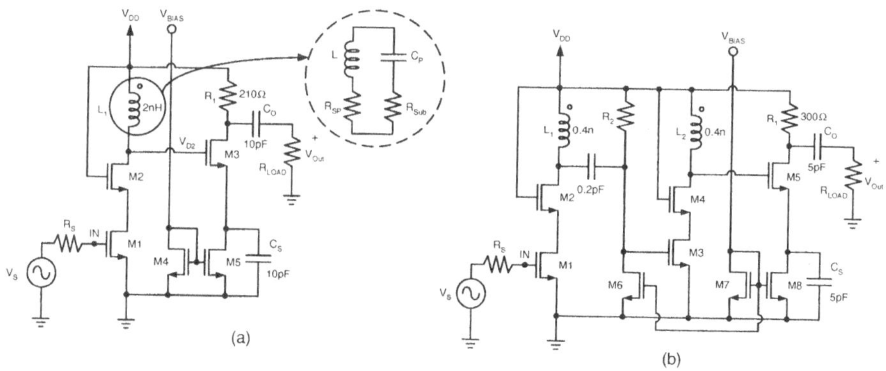

Fig. 6.34 (a) 4 GHz and (b) 13 GHz tuned amplifiers using SOI and SOS technologies. (Adapted from Ho et al. [28].)

6.6.3 Two-Stage LNA

Figure 6.33 shows (a) the 1.8 GHz two-stage SOI LNA circuit with (b) its gain and noise figure versus the frequency [27]. As shown in this figure, capacitor C1 and inductor L1 are designed to achieve a minimized noise figure. Capacitor C2 and inductor L4 are used to provide the matching between the output of the first stage and input of the second stage. The inductors are realized by the on-chip multilayer Al interconnects and the capacitors are implemented using the MIM structure. As shown in Fig. 6.33(b), at the power supply voltage of 2 V, this two-stage SOI LNA circuit has a center frequency of 1.8 GHz and a noise figure of 3.3 dB at a power dissipation of 24 mW.

Figure 6.34 shows (a) 4 GHz and (b) 13 GHz tuned amplifiers using SOI and SOS technologies[28]. As shown in this figure, the design of the tuned amplifiers is targeted to increase gain and linearity and to reduce noise figure. In Fig. 6.34(a), the 4 GHz tuned amplifier is made of the first-stage cascode amplifier and the second-stage common source amplifier. The cascode amplifier of the first stage determines

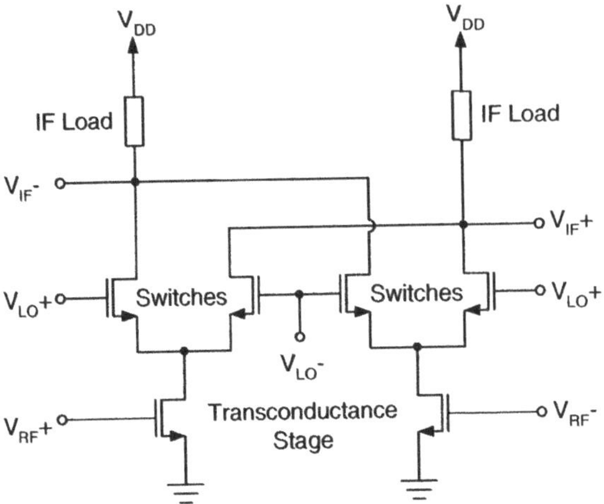

(a)

Power Supply (V)	IIP3 (dBm(RF))
2.0	25.4
2.5	24.9
3.0	16.5
3.5	17.6

(b)

Fig. 6.35 (a) Balanced active SOI CMOS chopping mixer. (b) Input RF power as a function of the supply voltage for the LO level equal to 10 dBm at 1.8 GHz and the IF frequency equal to 5 MHz. (Adapted from Demeus et al. [29].)

the resonant frequency of the tuned amplifier, which is specified by inductor L_1 and related capacitors C_{GS3}, C_{GD2}, C_{DSUB2}, and C_P, where C_{GS3} is the gate-source capacitance of device M3, C_{GD2} and C_{DSUB2} are the gate-drain and drain-substrate capacitances of device M2, and C_P is the equivalent parasitic capacitance of the spiral inductor. The gain of the second stage of the tuned amplifier is smaller than one, which is used to drive the 50 Ω output load. Owing to the small drain-substrate capacitance from the SOI and SOS devices, the tuned amplifiers implemented using SOI and SOS technologies provide a good performance. For the 13 GHz tuned amplifier, as shown in Fig. 6.34(b), an extra gain stage has been added after the input stage to provide a sufficient gain. In the 13 GHz tuned amplifier, its resonant frequency is determined by inductors L_1 and L_2 and their parasitic capacitances.

6.7 MIXER

In addition to LNA, mixer is also important in the RF circuits. In this section, mixers integrated by SOI CMOS technology are described. In an RF circuit, a mixer is used with a local oscillator (LO) to convert the input RF signal into the intermediate frequency (IF) signal for further processing.

Figure 6.35 shows (a) the balanced active SOI CMOS chopping mixer with (b) the performance in terms of the input RF power as a function of the supply voltage for LO level equal to 10 dBm at 1.8 GHz and the IF frequency equal to 5 MHz [29]. The performance of the RF mixer circuit can be evaluated in terms of the input-referred third-order intercept point (IIP3). As shown in Fig. 6.35(b), for the active mixer operating at 3 V and with the LO frequency of 1 GHz, its IIP3 is 10 dBm higher than

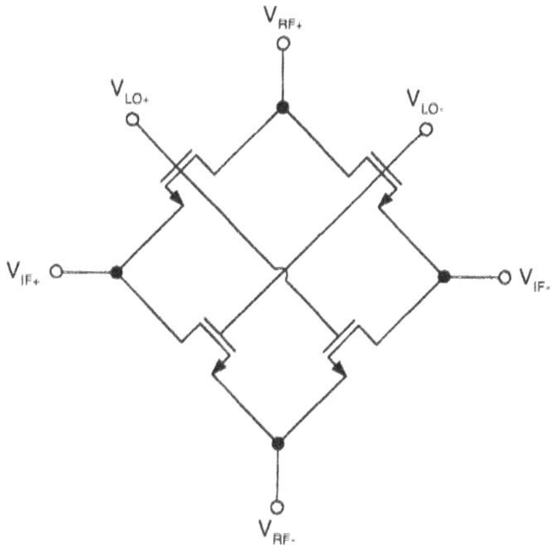

(a)

DC Gate Bias (V)	IIP3 (dBm(RF))
1.0	40.1
1.5	35.1
2.0	41.2

(b)

Fig. 6.36 (a) Resistive SOI CMOS mixer. (b) Input RF power as a function of the gate bias for the LO level equal to 10 dBm at 1.8 GHz and the IF frequency equal to 5 MHz. (Adapted from Demeus et al. [29].)

the value of the RF mixer implemented by bulk CMOS technology. Therefore, the FD SOI CMOS technology is suitable for implementing RF mixer circuits.

Figure 6.36 shows (a) the resistive SOI CMOS mixer with (b) the performance in terms of the input RF power as a function of the gate bias for the LO level equal to 10 dBm at 1.8 GHz and the IF frequency equal to 5 MHz [29]. Compared to the balanced active SOI CMOS chopping mixer, the resistive SOI CMOS mixer is suitable for applications at even higher frequencies. The resistive mixer implemented using a 1 μm SOI CMOS technology can work at the LO frequency up to 10 GHz. In addition, the frequency variation of the conversion gain, the power consumption, the intermodulation distortion, and the noise figure of the SOI resistive mixer are better than the active SOI counterparts. Owing to the small body effect of the FD SOI CMOS devices, the resistive mixer using the FD SOI CMOS technology has a good linearity and a low intermodulation distortion.

Figure 6.37(a) shows the block diagram of an RF receiver using a quadrature mixer [30]. As shown in this figure, after the LNA, via the quadrature mixer with the local oscillator, the RF signal is transformed into the IF signal. As shown, via the quadrature mixer the RF signal has become two differential signals. In the quadrature mixer two quadrature signals, which are generated by the polyphase filter with differential signals from the local oscillator, are used. Note that the polyphase filter is used to inhibit the image signals.

Figure 6.37(b) shows a 1 V SOI CMOS quadrature mixer circuit [30]. As shown in this figure, this quadrature mixer has a fully differential input-stage amplifier with a single-ended input and differential outputs. The current source of the input-stage amplifier, which is formed by a cascode stage of an LC tank and MOS device, provides an ideal current source at the resonant frequency of the LC tank. This mixer circuit is

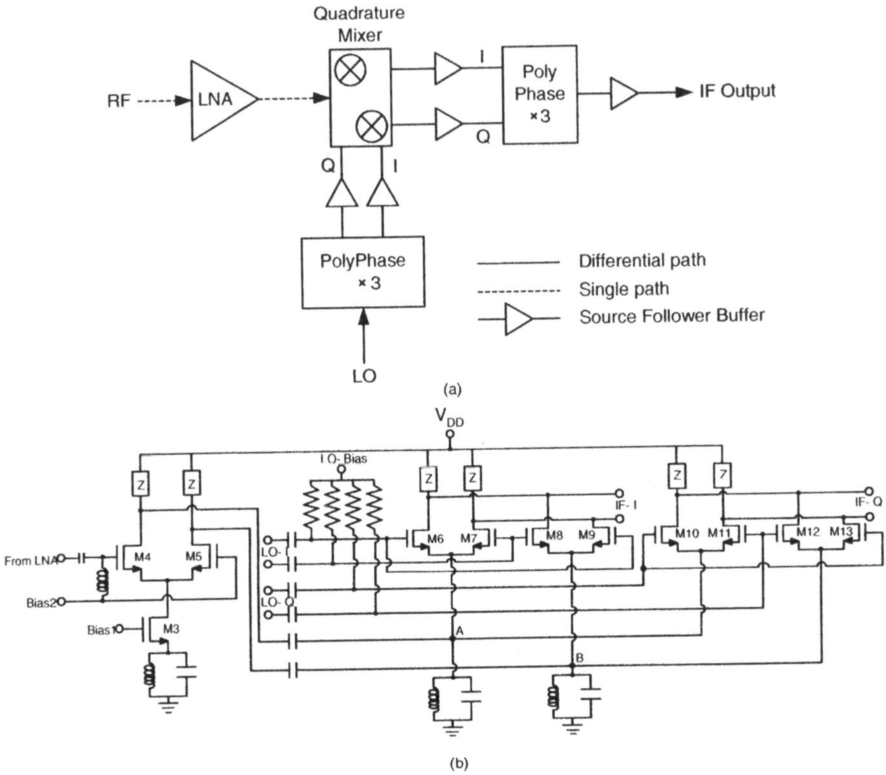

Fig. 6.37 (a) Block diagram of an RF receiver. (b) 1 V SOI CMOS quadrature mixer. (From Ugajin et al. [30]. ©2001 IEEE.)

based on an LC-tuned folded structure for low-voltage operation. Various differential pairs (M6/M7, M8/M9, M10/M11, and M12/M13) form the mixing stage, where the sources of M6/M7, M8/M9, M10/M11, and M12/M13 are connected together to form a common source configuration of M6/M7 and M10/M11 (node A) and M8/M9 and M12/M13 (node B). The local oscillator signal LO-I is provided to the differential pairs M6/M7 and M8/M9. The local oscillator signal LO-Q is for the differential pairs M10/M11 and M12/M13. After the amplifier, the RF signal from the input stage is imposed to the common source nodes A and B of the differential pairs. Via the differential pairs, the RF signal and the LO signal are processed in this mixing stage to transform into the IF output signals. In the mixing stage, the current source is formed by an LC tank circuit. Note that the phase error in the signals generated by the local oscillator is more serious than that in the RF signal. Prior to entering the mixer, the phase error needs to be removed from the LO signals in the polyphase filter.

Figure 6.38 shows an SOS CMOS RF mixer using Gilbert's cell circuit [25][31][32]. As shown in this figure, this mixer circuit has a doubly balanced configuration to increase isolation capability between the signals at the terminals. In a mixer circuit,

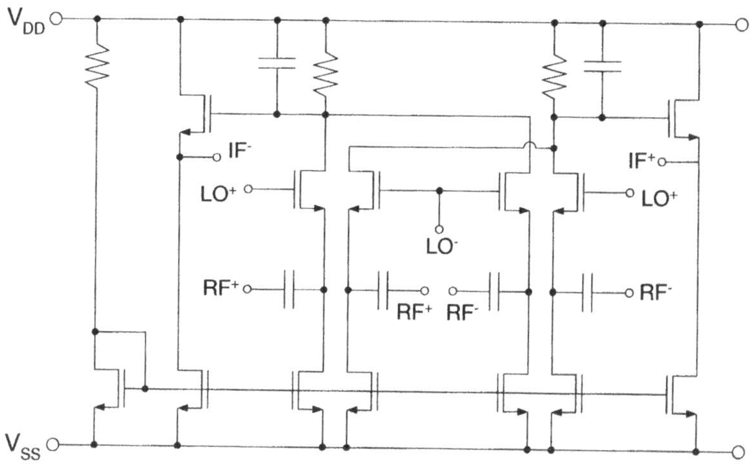

Fig. 6.38 SOS CMOS RF mixer using Gilbert's cell. (Adapted from Johnson et al. [25].)

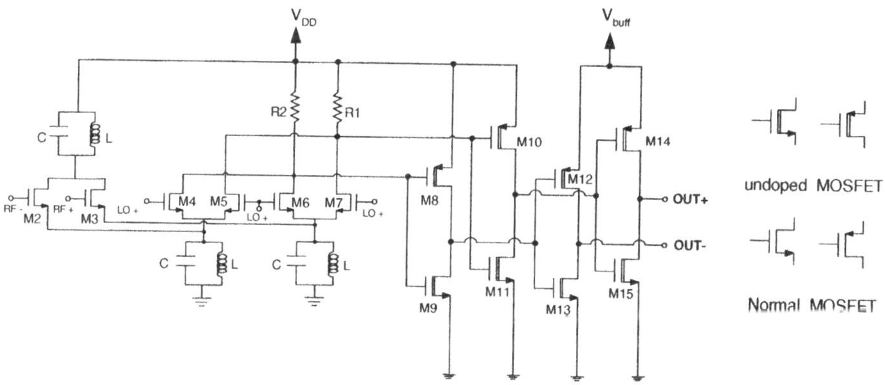

Fig. 6.39 0.5 V SOI RF mixer. (Adapted from Harada et al. [21].)

the RF signal is converted to the IF signal using the LO signal. In this mixer circuit, the LO signal is imposed at the gate of a device and the RF signal is at source. Thus, the isolation between the LO signal and the RF signal is important. The coupling between the thin film and the substrate of an SOI device may cause feedthrough of the LO and the RF signals to the output IF signal, which down-grades the linearity of the circuit. Owing to the advantages of the small coupling effect from the sapphire, the adoption of the SOS CMOS technology to realize this mixer circuit reduces the nonlinearity of the mixer circuit substantially.

Figure 6.39 shows a 0.5 V SOI RF mixer circuit [21]. As shown in this figure, the mixer circuit is made of the input mixing stage and the output stage. For the mixing stage, the traditional approach is to use Gilbert's cell based on three stacked transistors and the current source made of transistors, which is not suitable for sub-1 V operation. In this mixer circuit, an LC tank has been used to replace the traditional

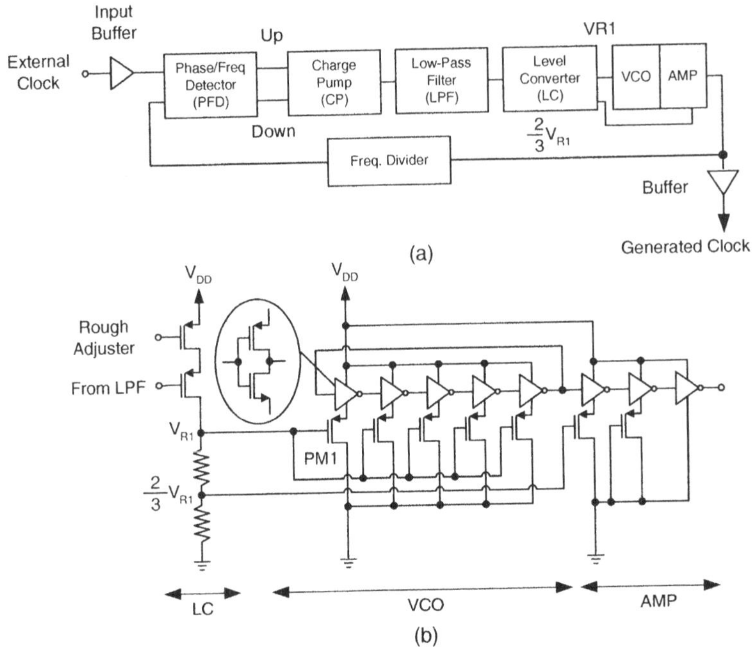

Fig. 6.40 (a)Block diagram of a PLL. (b) VCO used in a 2 V 3.5 Gbps optical transceiver using a 0.2 μm FD SOI CMOS technology. (Adapted from Ohtomo et al. [33].)

current source made of transistors. Due to high impedance of the LC tank at its resonant frequency, the LC tank can be regarded as a current source. In addition, the LC tank does not cause any DC voltage drop–suitable for low-voltage operation. RF and LO signals are imposed at the inputs to the input pairs M2/M3 and M4-M7, respectively. In the LO stage M4-M7 serve as switches, which do not consume extra current. In the output stage, a two-stage source follower has served as an output buffer. At a low power supply voltage, due to the asymmetric rise/fall signals, the traditional source follower may produce signal distortions to cause nonlinearity of the circuit. In this mixer, a complementary source follower using low-threshold MOS devices to overcome the asymmetric rise/fall signal problems at the low supply voltage has been used to achieve good linearity.

6.8 VOLTAGE-CONTROLLED OSCILLATOR–VCO

VCO is an important component in an RF circuit. VCO is frequently used in a phase-locked loop (PLL) for carrier and data recovery purposes. SOI CMOS technology has also been used to integrate VCO. Figure 6.40 shows (a) the block diagram of a phase-lock loop and (b) the voltage-controlled oscillator (VCO) used in a 2 V 3.5 Gbps optical transceiver circuit using a 0.2 μm FD SOI CMOS technology [33]. As

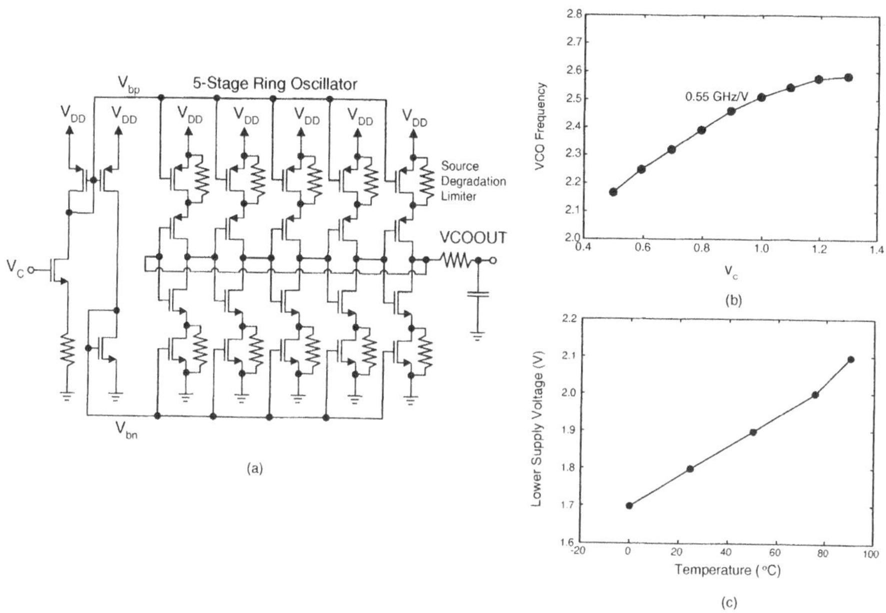

Fig. 6.41 (a) 2.5 GHz SOI VCO using a 0.18 μm SOI CMOS technology. (b) Operating frequency versus control voltage of the SOI VCO. (c) Lower-limit supply voltage versus temperature of the SOI VCO operating at 2.5 GHz. (From Yoshimura et al. [34]. ©1999 IEEE.)

shown in Fig. 6.40(a), the PLL is composed of the phase/frequency detector (PFD), the charge pump (CP), the low-pass filter (LPF), the level converter (LC), VCO, the amplifier (AMP), and the frequency divider. As shown in Fig. 6.40(b), the VCO is based on a ring oscillator made of an odd number of CMOS inverters. Via controlling the gate of the PMOS device PM1, the source voltage of the NMOS device in the CMOS inverters can be maintained at $V_{R1} + V_{tp}$, where V_{R1} is the gate voltage of the PMOS device PM1 and V_{tp} is the threshold voltage of the PMOS device PM1. Thus the oscillation frequency can be controlled via V_{R1}. The output voltage swing of the VCO is frequency dependent, which is equal to $V_{DD} - V_{R1} - V_{tp}$. At the output of the PLL, amplifier AMP made by cascading three inverter stages has been added.

Owing to small parasitic capacitances at the source/drain, SOI CMOS technology is suitable to integrate VCO and PLL operating at high frequency. Figure 6.41 shows (a) the 2.5 GHz SOI VCO using 0.18 μm SOI CMOS technology with (b) the operating frequency versus the control voltage of the SOI VCO and (c) the lower-limit supply voltage versus temperature of the SOI VCO at 2.5 GHz [34]. As shown in this figure, this SOI VCO is based on the ring oscillator made of CMOS inverters. The drawbacks of SOI technology such as floating body effects and self-heating do not cause stability or noise problems in this VCO design. Due to the relatively narrow frequency range of this VCO, floating body effects of the SOI device cause little effects on the circuit

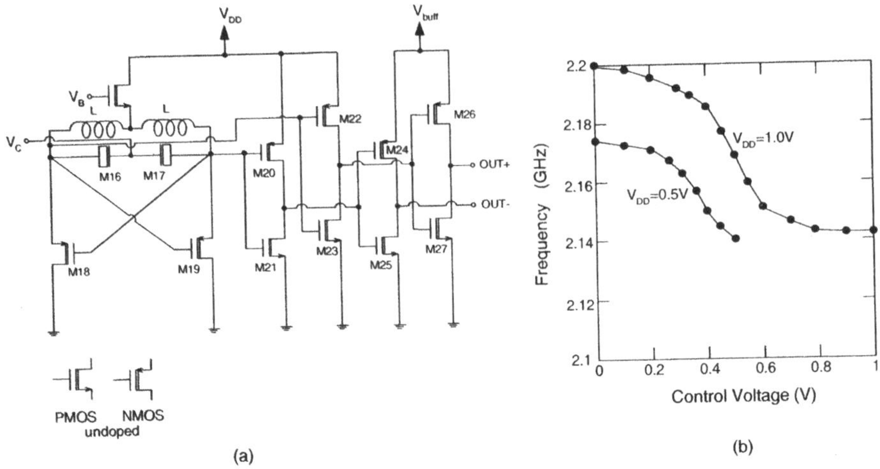

Fig. 6.42 (a)0.5 V 2 GHz SOI VCO based on integrated inductors using a 0.2 μm FD SOI CMOS technology. (b)Output frequency versus control voltage of the SOI VCO for supply voltages of 1 and 0.5 V. (Adapted from Harada et al. [21].)

operation. Since it only takes a few microseconds to reach thermal equilibrium of the ring oscillator, self-heating does not cause any significant influence in the lock-in process of this VCO. In addition, the buried oxide structure provides a good way to keep away the cross-talk noise via the substrate. In this VCO, the silicide polysilicon resistor has been used to reduce the output impedance of the current source device for high-speed oscillation. However, the frequency deviation caused by the silicide polysilicon resistor due to the elevation in the operation temperature may raise the lower-limit supply voltage required for operation as shown in Fig. 6.41(c). This problem is reduced by using a non-silicide polysilicon with a lower temperature coefficient. By using the 0.18 μm SOI CMOS technology, the output clock frequency of this circuit can be 16 times higher than the input frequency.

In this section, until now the VCO circuit was designed with ring oscillators. Figure 6.42(a) shows a 0.5 V 2 GHz SOI VCO based on integrated inductors using a 0.2 μm FD SOI CMOS technology. As shown in this figure, this SOI VCO is based on a cross-coupled SOI CMOS devices M18/M19, where undoped FD SOI devices with a smaller threshold voltage (< 0.5 V) have been used for low-voltage operation. In addition, source-follower buffers using undoped SOI devices with a smaller threshold have been used to reduce output impedance. For M16 and M17 normal FD devices are used. Fig. 6.42(b) shows the output frequency versus the control voltage of the SOI VCO for supply voltages of 1 and 0.5 V [21]. As shown in this figure, at the supply voltage of 0.5V, its operating frequency is over 2 GHz.

Until now the VCO circuits described in this section are for RF circuits. In fact, the PLL using the VCO circuit has also been used in a digital VLSI circuit to raise the internal clock frequency for high-speed operation. Figure 6.43 shows (a) the block diagram of the SOI PLL and (b) the 2 V SOI VCO using a 0.25 μm SOI CMOS

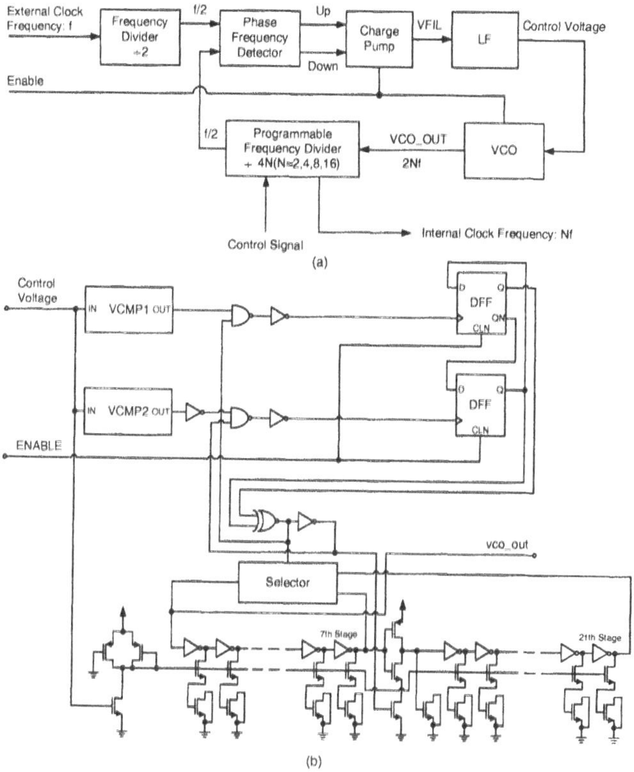

Fig. 6.43 (a) Block diagram of the SOI PLL. (b) 2 V SOI VCO using a 0.25 μm SOI CMOS technology. (Adapted from Sutoh et al. [35].)

technology [35]. As shown in this figure, this SOI VCO is composed of the ring oscillator and the control circuit. Depending on the required oscillation frequency, the stage number of ring oscillator can be selected. The control circuit is used to provide control signals for the ring oscillator to have an appropriate stage number. In the ring oscillator, between each inverter, there exist a pass transistor and a load capacitance. By controlling the gate voltage of the pass transistor, the oscillation frequency of the ring oscillator can be adjusted. In order to widen the range of the oscillation frequency, the load capacitance of the 8-21 stage ring oscillator is twice as compared to the 1-7 stage ring oscillator.

6.9 HIGH-TEMPERATURE CIRCUITS

Owing to the buried oxide isolation structure, at high temperatures the leakage current of SOI CMOS devices is about three to four orders of magnitude smaller than the bulk counterpart. In addition, when the operation temperature is changed, the variation of the threshold voltage is two to three times smaller compared to the bulk device. When

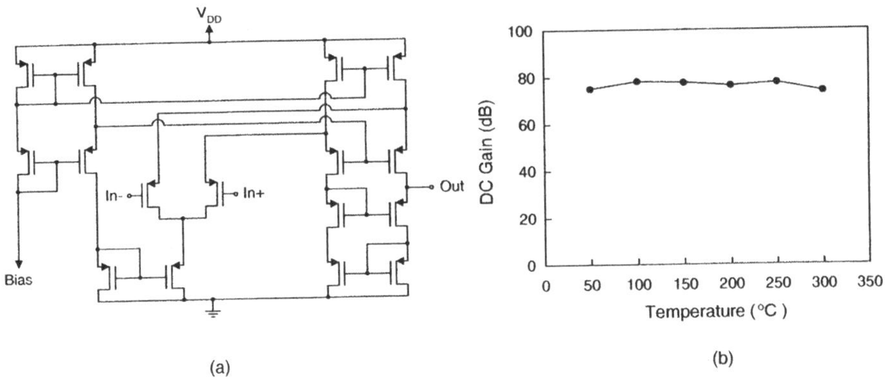

Fig. 6.44 (a) SOI folded cascoded op amp. (b) DC gain versus temperature of the op amp. (Adapted from Francis et al. [36].)

the temperature rises, the output conductance of the SOI devices improves. Compared to bulk, SOI CMOS devices are more suitable for designing analog circuits operating at a high temperature. In this section, high-temperature performance of an SOI op amp and reference voltage generators is described.

Figure 6.44 shows (a) the SOI folded cascoded op amp with (b) the DC gain versus temperature characteristics [36]. As shown in this figure, when operating at high temperatures, the transconductance of the SOI devices decreases. Owing to the increase of the output conductance at high temperatures, the DC gain sustains at high temperatures until 300°C as shown in Fig. 6.44(b). The offset voltage of the op amp is fixed at 0$\sim$2 mV regardless of the temperature change.

Figure 6.45(a) shows the reference voltage generator using SOI accumulation- and inversion-mode devices with a front gate oxide of 350 Å and a thin film of 700 Å doped with a p-/n-type doping density of 10^{17}cm^{-3} for the inversion-/accumulation-mode device [37]. As shown in this figure, M3 and M4 are used to provide an identical biasing current for M1 and M2. The reference voltage is generated by the difference of the threshold voltage between M1 and M2. M5-M8 are used as an output buffer. Due to an identical doping density of the inversion- and accumulation-mode devices biased at the same back gate bias of -5 V, the temperature dependences of their threshold voltages and drain currents are about identical. Thus, the reference voltage, which is generated by the difference in the threshold voltages of the inversion- and the accumulation-mode devices, is almost temperature independent ,as shown in Fig. 6.45(b).

Figure 6.46(a) shows another SOI reference voltage generator based on the threshold voltage difference of the two input devices of the differential pair, which receive different doses of the implant to their thin-film regions [38]. Owing to the difference in the threshold voltages of M1 and M2, a large offset voltage, which is almost temperature independent, has been generated. M12, R3, M13, M14, and M5 form a self-biasing circuit to produce a good power supply rejection to avoid the influence from the power supply. As shown in Fig. 6.46, the temperature coefficient of this SOI

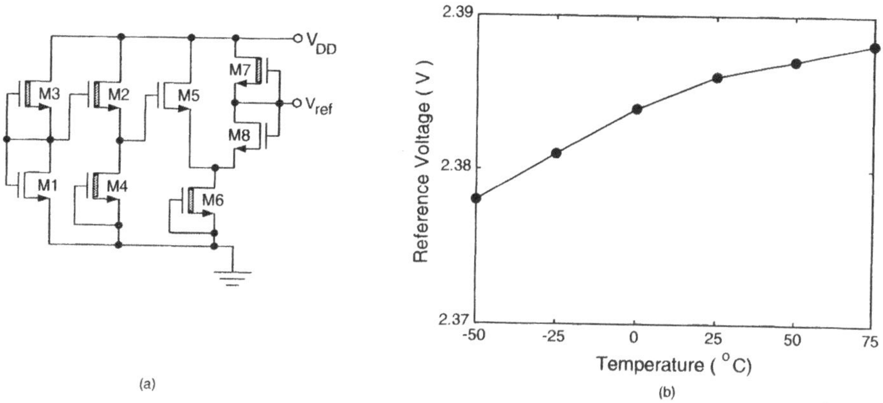

Fig. 6.45 (a) SOI reference voltage generator using SOI accumulation- and inversion-mode devices with a front gate oxide of 350 Å and a thin film of 700 Å doped with a p-/n-type doping density of 10^{17}cm^{-3} for the inversion-/accumulation-mode device. (b) Reference voltage versus temperature. (Adapted from Song & Kim [37].)

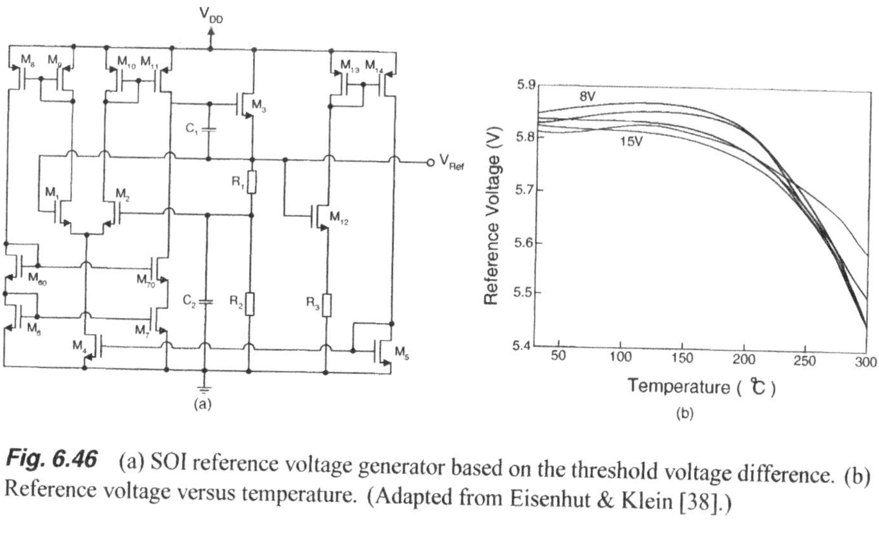

Fig. 6.46 (a) SOI reference voltage generator based on the threshold voltage difference. (b) Reference voltage versus temperature. (Adapted from Eisenhut & Klein [38].)

reference voltage generator is 50 ppm/C at the power supply voltage between 8 and 15 V.

6.10 SUMMARY

In this chapter, analog circuits using SOI CMOS technology were described. Op amps, filters, analog-to-digital converters (ADC), and digital-to-analog converters (DAC), sigma-delta ADC, RF circuits, low noise amplifiers (LNA), mixers, voltage-control oscillators (VCO), and high-temperature analog circuits using SOI CMOS technology were analyzed.

REFERENCES

1. J.-P. Eggermont, D. D. Ceuster, D. Flandre, B. Gentinne, P. G. A. Jespers, and J.-P. Colinge, "Design of SOI CMOS Operational Amplifiers for Applications up to 300°C," *IEEE J. Sol. St. Ckts.*, **31**(2), 179–186 (1996).
2. J. P. Eggermont, D. Flandre, R. Gillon, and J. P. Colinge, "A 1-GHz Operational Transconductance Amplifier in SOI Technology," *SOI Conf. Dig.*, 127–128 (1995).
3. J.-P. Eggermont, D. Flandre, J.-P. Raskin, and J.-P. Colinge, "Potential and Modeling of 1μm SOI CMOS Operational Transconductance Amplifiers for Applications up to 1 GHz," *IEEE J. Sol. St. Ckts.*, **33**(4), 640–643 (1998).
4. F. Silveria, D. Flandre, and P. G. A. Jespers, "A g_m/I_D Based Methodology for the Design of CMOS Analog Circuits and Its Application to the Synthesis of a Silicon-on-Insulator Micropower OTA," *IEEE J. Sol. St. Ckts.*, **31**(9), 1314–1319 (1996).
5. M. Verbeck, C. Zimmermann, and H.-L. Fiedler, "A MOS Switched-Capacitor Ladder Filter in SIMOX Technology for High Temperature Applications up to 300°C," *IEEE J. Sol. St. Ckts.*, **31**(7), 908–914 (1996).
6. B. Gentinne, J. P. Colinge, P. G. A. Jespers, and J. P. Eggermont, "Improvement of the Performances of SOI CMOS Operational Amplifiers by Means of a Gain-Boosting Stage," *SOI Conf. Dig.*, 184–185 (1993).
7. C. F. Edwards, W. Redman-White, M. Bracey, B. M. Tenbroek, M. S. L. Lee, and M. J. Uren, "A Multibit $\Sigma\Delta$ Modulator in Floating-Body SOS/SOI CMOS for Extreme Radiation Environments," *IEEE J. Sol. St. Ckts.*, **34**(7), 937–947 (1999).
8. V. Dessard, D. Baldwin, L. Demeus, B. Gentinne, and D. Flandre, "SOI Implementation of Low-Voltage and High-Temperature MOSFET-C Continuous-Time Filters," *SOI Conf. Dig.*, 24–25 (1996).

9. L. Demeus, A. Viviani, and D. Flandre, "High-Temperature Analog Instrumentation System in Thin-Film Fully-Depleted SOI CMOS Technology," *High Temperature Electronics Conf. (HITEC) Dig.*, 51–54 (1998).

10. F. Baille, G. Borel, B. Commere, F. Roy, C. Delmas, and C. Terrier, "A Multi MRad Hardened 8 Bit/ 20MHz Flash ADC," *IEEE Trans. Nuclear Science*, **39**(3), 401–404 (1992).

11. B. M. Tenbroek, M. S. L. Lee, W. Redman-White, C. F. Edwards, R. J. T. Bunyan, and M. J. Uren, "Measurement and Simulation of Self-Heating in SOI CMOS Analogue Circuits," *SOI Conf. Dig.*, 156–157 (1997).

12. B. M. Tenbroek, M. S. L. Lee, W. Redman-White, R. J. T. Bunyan, and M. J. Uren, "Impact of Self-Heating and Thermal Coupling on Analog Circuits in SOI CMOS," *IEEE J. Sol. St. Ckts.*, **33**(7), 1037–1046 (1998).

13. D. Chang, J. G. Fossum, S. K. Reynolds, and M. M. Pelella, "Kink-Free Analog Circuit Design with Floating-Body NFD/SOI CMOS: A Current-Steering D-A Converter," *SOI Conf. Dig.*, 158–159 (1997).

14. A. Swaminathan, N. Fong, P. Lauzon, H.-K. Yang, M. Maliepaard, C. Plett, and M. Snelgrove,"A Low Power $\Sigma\Delta$ Analog-to-Digital Modulator with 50MHz Sampling Rate in a 0.25μm SOI CMOS Technology," *SOI Conf. Dig.*, 14–15 (1999).

15. A. Viviani, D. Flandre, and P. Jespers, "A SOI-CMOS Micro-Power First-Order Sigma-Delta Modulator," *SOI Conf. Dig.*, 110–111 (1996).

16. W. M. Huang, D. J. Monk, D. C. Diaz, P. J. Welch, and J. M. Ford, "TFSOI–Can It Meet the Challenge of Single Chip Portable Wireless Systems?" *SOI Conf. Dig.*, 1–3 (1997).

17. R. Reedy, J. Cable, and D. Kelly, "Single Chip Wireless Systems Using SOI," *SOI Conf. Dig.*, 8–11 (1999).

18. D. Eggert, P. Huebler, A. Huerrich, H. Kueck, W. Budde, and M. Vorwerk, "A SOI-RF-CMOS Technology on High Resistivity SIMOX Substrates for Microwave Applications to 5GHz," *IEEE Trans. Elec. Dev.*, **44**(11), 1981–1989 (1997).

19. O. Rozeau, J. Jomaah, J. Boussey, C. Raynaud, J. L. Pelloie, and F. Balestra, "Comparison Between Fully- and Partially-Depleted SOI MOSFET's for Low-Power Radio-Frequency Applications," *SOI Conf. Dig.*, 22–23 (1999).

20. C. L. Chen, R. H. Mathews, J. A. Burns, P. W. Wyatt, D.-R. Yost, C. K. Chen, M. Fritze, J. M. Knecht, V. Suntharalingam, A. Soares, and C. L. Keast, "High-Frequency Characterization of Sub-0.25μm Fully Depleted Silicon-on-Insulator MOSFETs," *IEEE Elec. Dev. Let.*, **21**(10), 497–499 (2000).

21. M. Harada, T. Tsukahara, and J. Yamada, "0.5-1V 2GHz RF Front-End Circuits in CMOS/SIMOX," *ISSCC Dig.*, 378–379 (2000).

22. H. Komurasaki, H. Sato, N. Sasaki, K. Ueda, S. Maeda, Y. Yamaguchi, and T. Miki, "A Sub 1-V SOI CMOS Low Noise Amplifier for L-Band Applications," *Symp. Radio Frequency IC Conf. Dig.*, 153–156 (1998).

23. M. Ono, N. Suematsu, Y. Yamaguchi, K. Ueda, H. Komurasaki, S. Kubo, Y. Iyama, and O. Ishida, "L-Band On-Chip Matching Si-MMIC Low Noise Amplifier Fabricated in SOI CMOS Process," *Silicon Monolithic IC in RF Systems Dig.*, 90–93 (1998).

24. J. B. Kuo, M.-C. Tang, and J.-H. Sim, "An Analytical Back Gate Bias Dependent Threshold Voltage Model for SiGe-Channel Ultrathin SOI PMOS," *IEEE Trans. Elec. Dev.*, **40**(12), 2237–2244 (1993).

25. R. A. Johnson, P. R. de la Houssaye, C. E. Chang, P.-F. Chen, M. E. Wood, G. A. Garcia, I. Lagnado, and P. M. Asbeck, "Advanced Thin-Film Silicon-on-Sapphire Technology: Microwave Circuit Applications," *IEEE Trans. Elec. Dev.*, **45**(5), 1047–1053 (1998).

26. W. Jin, P. C. H. Chan, and C. Hai, "1.5-V 1.8-GHz SOI Low Noise Amplifiers for PCS Receivers," *SOI Conf. Dig.*, 16–17 (1999).

27. M. Vorwerk and D. Eggert. "A 1.8GHz Low-Noise Amplifier in CMOS/SIMOX Technology," *Electronics, Circuits and Systems Conf. Dig.*, 31–34 (1998).

28. Y.-C. Ho, K.-H. Kim, B. A. Floyd, C. Wann, Y. Taur, I. Lagnado, and K. K. O. "4- and 13-GHz Tuned Amplifiers Implemented in a 0.1μm CMOS Technology on SOI, SOS, and Bulk Substrates," *IEEE J. Sol. St. Ckts.*, **33**(12), 2066–2073 (1998).

29. L. Demeus, J. Chen, J.-P. Eggermont, R. Gillon, J.-P. Raskin, D. Vanhoenacker, and D. Flandre, "Advanced SOI CMOS Technology for RF Applications," *Signals, Systems and Electronics (ISSSE) Dig.*, 134–139 (1998).

30. M. Ugajin, J. Kodate, and T. Tsukahara, "A 1V 12mW 2GHz Receiver with 49dB Image Rejection in CMOS/SIMOX," *ISSCC Dig.*, 288–289 (2001).

31. B. Gilbert, "A Precise Four-Quadrant Multiplier with Subnanosecond Response," *IEEE J. Sol. St. Ckts.*, **3**(4), 365–373 (1968).

32. P. J. Sullivan, B. A. Xavier, and W. H. Ku, "Low Voltage Performance of a Microwave CMOS Gilbert Cell Mixer," *IEEE J. Sol. St. Ckts.*, **32**(7), 1151–1155 (1997).

33. Y. Ohtomo. T. Yoshida, M. Nishisaka, K. Nishimura, and M. Shimaya, "A Single-Chip 3.5Gb/s CMOS/SIMOX Transceiver with Automatic-Gain-Control and Automatic-Power-Control Circuits," *ISSCC Dig.*, 58–59 (2000).

34. T. Yoshimura, K. Ueda, T. Nakura, K. Kubo, K. Mashiko, S. Maeda, S. Maegawa, Y. Yamaguchi, and Y. Matsuda, "A 1.8V 2.5GHz PLL using 0.18μm SOI/CMOS Technology," *SOI Conf. Dig.*, 12–13 (1999).

35. H. Sutoh, K. Yamakoshi, and M. Ino, "A 0.25μm CMOS/SIMOX PLL Clock Generator Embedded in a Gate Array LSI with 5 to 400 MHz Lock Range," *CICC. Dig.*, 41-44 (1997).

36. P. Francis, A. Terao, B. Gentinne, D. Flandre, and J.-P. Colinge, "SOI Technology for High-Temperature Applications," *IEDM Dig.*, 353–356 (1992).

37. H.-J. Song and C.-K. Kim, "A Temperature-Stabilized SOI Voltage Reference Based on Threshold Voltage Difference Between Enhancement and Depletion NMOSFET's," *IEEE J. Sol. St. Ckts.*, **28**(6), 671-677 (1993).

38. C. Eisenhut and J. W. Klein, "SIMOX Voltage References for Applications up to 275°C Using the Threshold Voltage Difference Principle," *SOI Conf. Dig.*, 110–111 (1997).

Problems

1. For the single-stage SOI op amp, as shown in Fig.6.1, design the aspect ratios of all devices so that it can drive an output capacitive load of 1 pf with a unity gain bandwidth of 1.5 GHz using 0.25 μm SOI CMOS technology. The parasitic capacitance at the internal node 1 is assumed to be 50 fF. The power supply voltages are ± 1 V. What is the phase margin of your op amp? What is the output swing? What is the power dissipation?

2. For the folded-cascode SOI op amp as shown in Fig.6.3, repeat the design procedure for Problem 1.

3. Compare the tradeoffs between the performance of the basic switched capacitor, which is made of pass transistors and a capacitor using PD SOI and bulk CMOS technologies.

4. How does the leakage current from the floating body of the PD SOI NMOS device affect the accuracy of a switched-capacitor successive approximation ADC using PD SOI CMOS technology?

5. Calculate the performance in terms of f_{max} of a 0.13 μm FD SOI NMOS device with a front gate of 50 Å, and a thin film of 800 Å, targeted for RF applications as a function of the source/drain resistance. Various doping densities of the body and the source/drain region can be assumed.

7

PD SOI-Technology SPICE Models

In this chapter, PD SOI-Technology SPICE models using a concise BiCMOS approach are described. In following sections, the PD SOI SPICE models are described first, followed by the kink effects, the hysteresis behavior and the transient behavior using the PD SOI-Technology SPICE models. In the final portion of this chapter, in order to show the effectiveness of the PD SOI-Technology SPICE models, the behaviors of a static logic circuit, a dynamic logic circuit, and an SRAM critical path are analyzed using the PD SOI-Technology SPICE models.

7.1 PD SOI SPICE MODEL

Fig. 7.1 shows the cross-section of the partially depleted (PD) SOI NMOS device under study and the PD SOI-Technology SPICE models [1] using a concise BiCMOS approach. As shown in this figure, the PD SOI-Technology SPICE models are composed of the surface MOS device and the parasitic bipolar device connected in parallel, considering the neutral region in the thin film. The emitter of the parasitic BJT and the source of the surface MOS device are connected together (S). In addition, the collector of the parasitic BJT and the drain of the surface MOS device share the same terminal (D). The base of the parasitic BJT, which is connected to the substrate of the surface MOS device, becomes a floating body node (B). A capacitor (C_{buried}) has been adopted to account for the effect of the buried oxide. In the PD SOI-Technology SPICE models, conventional drain current and capacitance models for the surface MOS part [1] and Gummel-Poon model for the parasitic BJT part [2] have been used

[1] PD SOI-Technology SPICE models are available at http://ece.uwaterloo.ca/~jbkuo/Welcome.html.

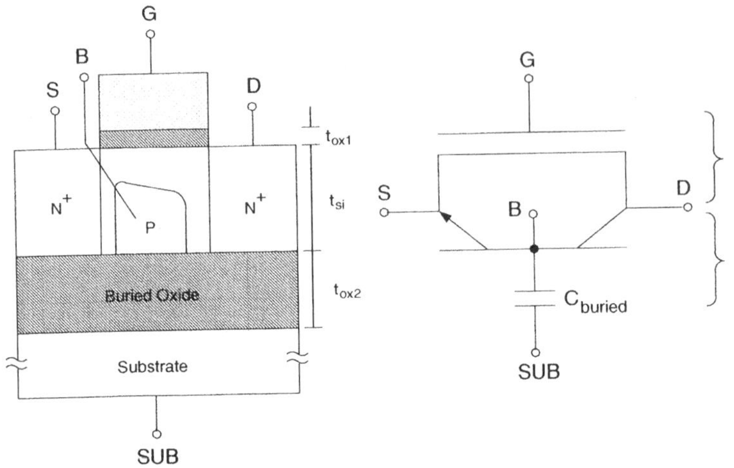

Fig. 7.1 Cross-section of the partially depleted (PD) SOI NMOS device under study and the PD SOI model used in the PD SOI-Technology SPICE models using a concise BiCMOS approach.

[3][4]. In the PD SOI-Technology SPICE models, one key component is the impact ionization current generated by the surface MOS device, which becomes the triggering base current of the parasitic BJT to reflect the floating body effects of the PD SOI MOS device. Using the concise PD SOI-Technology SPICE models, the potential of the floating body node (B) can be calculated to predict the floating-body behavior. In addition, in the PD SOI-Technology SPICE models, the lattice temperature model [5] has also been included. In the following sections, the surface MOS device model, the parasitic bipolar model, and the lattice temperature model used in the PD SOI-Technology SPICE program are described.

7.1.1 Surface MOS Device Model

Fig. 7.2 shows the surface MOS model used in the PD SOI-Technology SPICE models. As shown in this figure, the MOS portion is composed of the parasitic source/drain resistances (R_S/R_D), the surface channel current (I_{DS}), the related terminal capacitances (C_{ij}; $i, j = D, G, S, B$), the impact ionization current (I_{ii}), and the substrate to junction (source/drain) diode current (I_{BS}/I_{BD}). The surface channel current I_{DS} can be expressed as [1]:

$$I_{DS} = \frac{I_{DS0}}{1 + \frac{R_{DS} I_{DS0}}{V_{DSeff}}} \left(1 + \frac{V_{DS} - V_{DSeff}}{V_{A,MOS}}\right) \left(1 + \frac{V_{DS} - V_{DSeff}}{V_{ASCBE,MOS}}\right) \tag{7.1}$$

where I_{DS0} is the intrinsic unified saturation current from the subthreshold region to the strong-inversion region without considering the sheet resistance between the source and the drain junctions. R_{DS} is the parasitic resistance between the source and the drain. $V_{A,MOS}$ is the Early voltage due to channel length modulation and drain-

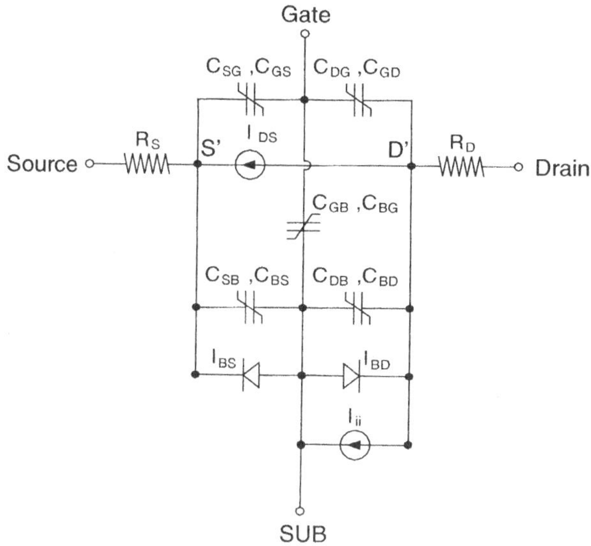

Fig. 7.2 The surface MOS device model used in the PD SOI-Technology SPICE models.

induced-barrier-lowering. $V_{ASCBE,MOS}$ is the Early voltage due to the substrate current induced body effect. V_{DSeff} is the unified saturation drain voltage. For a PD SOI CMOS device, impact ionization is important. The impact ionization current can be expressed as [1]:

$$\begin{aligned} I_{ii} &= \frac{\alpha_0}{L_{eff}}(V_{DS} - V_{DSeff})\exp(-\frac{\beta_0}{V_{DS} - V_{DSeff}}) \\ &\times \frac{I_{DS0}}{1 + \frac{R_{DS}I_{DS0}}{V_{DSeff}}}(1 + \frac{V_{DS} - V_{DSeff}}{V_{A,MOS}}) \end{aligned} \tag{7.2}$$

where L_{eff} is the effective channel length, α_0 and β_0 are the parameters of impact ionization current. Two substrate junction current (I_{BS}, I_{BD}) are modeled by diode-like equations [1]:

$$I_{B(D,S)} = \begin{cases} I_{sB(D,S)}[exp(\frac{V_{B(D,S)}}{n_j V_t}) - 1] & \text{for } V_{B(D,S)} < V_{tm}, \\ I_{sB(D,S)}[exp(\frac{V_{tm}}{n_j V_t}) - 1] + \frac{I_{sB(D,S)}}{n_j V_t} exp(\frac{V_{tm}}{N_j V_t}) & \\ \times (V_{B(D,S)} - V_{tm}) & \text{for } V_{B(D,S)} \geq V_{tm}, \end{cases} \tag{7.3}$$

where the n_j is the emission coefficient of the source (drain) junction, $V_t = \frac{kT}{q}$, and $I_{sB(D,S)}$ is the drain (source) / bulk saturation current including intrinsic drain (source) / bulk diode saturation current and sidewall saturation current of drain (source) / bulk diode. V_{tm} is the criterion voltage for checking the polarity of the diode. Considering the gate charge (Q_G), the source charge (Q_S), the drain charge (Q_D), and the bulk charge (Q_B) of MOS device biased in the triode region

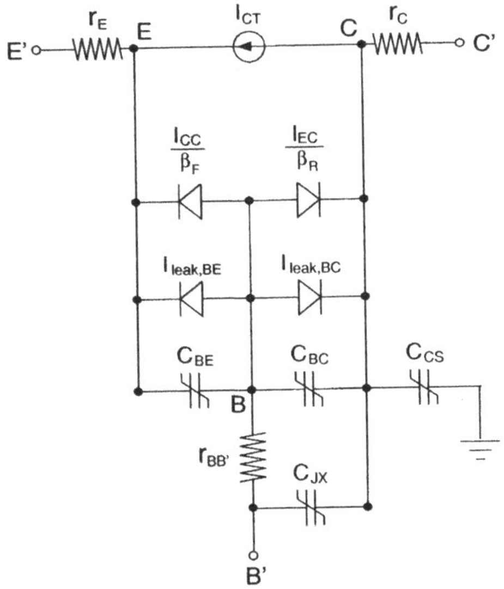

Fig. 7.3 Parasitic BJT model used in the PD SOI-Technology SPICE.

and the saturation regions, one obtains the capacitance model [1]:

$$\begin{aligned} C_{ij} &= \frac{\partial Q_i}{\partial V_j}, i, j = G, D, S, B \\ Q_G &= Q_{G0} + Q_{G,overlap} \\ Q_D &= Q_{D0} + Q_{D,overlap} \\ Q_S &= Q_{S0} + Q_{S,overlap} \\ Q_B &= Q_{B0}, \end{aligned} \tag{7.4}$$

where Q_{G0}, Q_{D0}, Q_{S0} and Q_{B0} are the intrinsic charges; $Q_{G,overlap}$, $Q_{D,overlap}$, and $Q_{S,overlap}$ are the overlap charges. According to charge neutrality, $Q_G + Q_D + Q_S + Q_B = 0$.

7.1.2 Parasitic BJT Model

As shown in Fig. 7.3, for the parasitic BJT model used in the PD SOI-Technology SPICE models, Gummel-Poon model has been adopted to include the current source I_{CT}, four diode currents ($\frac{I_{CC}}{\beta_F}$, $\frac{I_{EC}}{\beta_R}$, $I_{leak,BE}$, and $I_{leak,BC}$), the related terminal capacitances (C_{BE}, C_{BC}, C_{JX}, and C_{CS}) and parasitic resistance (r_E, r_C, and $r_{BB'}$). The DC current sources are expressed as [2]

$$\begin{aligned} I_{CT} &= \frac{I_{SS}}{q_b}\left[\exp\left(\frac{qV_{BE}}{n_F kT}\right) - \exp\left(\frac{qV_{BC}}{n_R kT}\right)\right] \\ I_{CC} &= \frac{I_{SS}}{q_b}\left[\exp\left(\frac{qV_{BE}}{n_F kT}\right) - 1\right] \end{aligned} \tag{7.5}$$

$$I_{EC} = \frac{I_{SS}}{q_b}\left[\exp\left(\frac{qV_{BC}}{n_R kT}\right) - 1\right]$$

where I_{SS} is the saturation current of bipolar transistor, V_{BE} is the voltage between base and emitter, V_{BC} is the voltage between base and collector. n_F and n_R are the forward and reverse current emission coefficient, respectively. β_F and β_R are the large-signal forward current gain and reverse current gain, respectively. q_b, the majority charge in the base, is expressed as [2]

$$\begin{aligned} q_b &= \frac{q_1}{2} + \sqrt{\left(\frac{q_1}{2}\right)^2 + q_2} \qquad (7.6) \\ q_1 &= 1 + \frac{V_{BE}}{V_{B,BJT}} + \frac{V_{BC}}{V_{A,BJT}} \\ q_2 &= \frac{I_{SS}}{I_{KF}}\left[\exp\left(\frac{qV_{BE}}{kT}\right) - 1\right] + \frac{I_{SS}}{I_{KR}}\left[\exp\left(\frac{qV_{BC}}{kT}\right) - 1\right] \end{aligned}$$

where I_{KF} is the forward knee current, and I_{KR} is the reverse knee current. $V_{A,BJT}$ and $V_{B,BJT}$ are the forward and reverse Early voltage of the bipolar transistor, respectively. q_1 models the effects of base-width modulation. q_2 models the effects of high-level injection. $I_{leak,BE}$ and $I_{leak,BC}$ are the junction leakage currents due to the generation and recombination mechanism in the base-emitter and base-collector junctions. The leakage currents are expressed as [2]

$$\begin{aligned} I_{leak,BE} &= I_{SE}\left[\exp\left(\frac{qV_{BE}}{n_{EL} kT}\right) - 1\right] \qquad (7.7) \\ I_{leak,BC} &= I_{SC}\left[\exp\left(\frac{qV_{BC}}{n_{CL} kT}\right) - 1\right] \end{aligned}$$

where I_{SE} and I_{SC} are the nonideal base-emitter and base-collector saturation currents, respectively. n_{EL} and n_{CL} are the nonideal low-current base-emitter and base-collector emission coefficients, respectively. Depending on Eqs.(7.5)~(7.7) and the polarity of V_{BE} and V_{BC}, one can obtain the bipolar transistor DC terminal current (I_B, I_C, and I_E) in a any operation mode (active, inverse, saturated, or off).

The terminal capacitances (C_{BE}, C_{BC}, C_{JX}, and C_{CS}) of the bipolar transistor are described as follow. Charge storage in the BJT is modeled by two types of capacitances: nonlinear junction and diffusion capacitances. The charge associated with the mobile carriers in a BJT is modeled by the diffusion capacitances. The fixed charges stored in the BJT depletion regions can be modeled by the junction capacitances. The base-emitter capacitance (C_{BE}) is expressed as

$$C_{BE} = \begin{cases} \tau_F \frac{dI_{CC}}{dV_{BE}} + C_{JE0}\left(1 - \frac{V_{BE}}{\phi_E}\right)^{-m_E} & \text{for } V_{BE} < FC \cdot \phi_E, \\ \tau_F \frac{dI_{CC}}{dV_{BE}} + \frac{C_{JE0}}{(1-FC)^{1+m_E}}\left[1 - FC(1+m_E) + \frac{m_E V_{BE}}{\phi_E}\right] & \text{for } V_{BE} \geq FC \cdot \phi_E, \end{cases} \qquad (7.8)$$

where the first term of C_{BE} is the diffusion capacitance and the second term is the junction capacitance. C_{JE0} is the junction capacitance of the base-emitter at zero-bias. ϕ_E is the base-emitter barrier potential. m_E is the base-emitter capacitance gradient factor. τ_F is the forward transit time. FC is a factor between 0 and 1. In order to find a better approximation of the distributed base-collector capacitance, the base-collector capacitance is divided into internal capacitance (C_{BC}) including junction and diffusion capacitances and external capacitance (C_{JX}) including only junction capacitance as shown in Fig. 7.3. The complete expression of the base-collector capacitance is expressed as

$$C_{BC} = \begin{cases} \tau_R \frac{dI_{EC}}{dV_{BC}} + X_{CJC}C_{JC0}\left(1 - \frac{V_{BC}}{\phi_C}\right)^{-m_C} & \text{for } V_{BC} < FC \cdot \phi_C \\ \tau_R \frac{dI_{EC}}{dV_{BC}} + \frac{X_{CJC}C_{JC0}}{(1-FC)^{1+m_C}}\left[1 - FC(1+m_C) + \frac{m_C V_{BC}}{\phi_C}\right] & \text{for } V_{BC} \geq FC \cdot \phi_C \end{cases} \tag{7.9}$$

$$C_{JX} = \begin{cases} C_{JC0}\left(1 - X_{CJC}\right)\left(1 - \frac{V_{BX}}{\phi_C}\right)^{-m_C} & \text{for } V_{BX} < FC \cdot \phi_C \\ \frac{C_{JC0}(1-X_{CJC})}{(1-FC)^{1+m_C}}\left[1 - FC(1+m_C) + \frac{m_C V_{BX}}{\phi_C}\right] & \text{for } V_{BX} \geq FC \cdot \phi_C \end{cases} \tag{7.10}$$

where the first term of C_{BC} is the diffusion capacitance and the second term is the junction capacitance. C_{JC0} is the capacitance of the base-collector junction at zero-bias. ϕ_C is the base-collector barrier potential. m_C is the base-collector capacitance gradient factor. τ_R is the reverse transit time. X_{CJC}, which is between 0 to 1, specifies the ratio of the partition of junction capacitances between the internal part and the external part. V_{BX} is the voltage between the internal collector and the external base. C_{CS} is the collector-substrate capacitance, which is expressed as

$$C_{CS} = \begin{cases} C_{JS0}\left(1 - \frac{V_{CS}}{\phi_S}\right)^{-m_S} & \text{for } V_{CS} < 0 \\ C_{JS0}\left(1 + \frac{m_S V_{CS}}{\phi_S}\right) & \text{for } V_{CS} > 0 \end{cases} \tag{7.11}$$

where C_{JS0} is the substrate capacitance at zero bias, ϕ_S is the collector-substrate barrier potential, and m_S is the collector-substrate capacitance gradient factor. In the PD SOI-Technology SPICE models, C_{JX} is usually important only at very high frequency and C_{CS} is less important as compared to C_{BE} and C_{BC}.

As shown in Fig. 7.3, r_E and r_C are the emitter and the collector resistances, respectively. Due to current crowding, the base resistance including the extrinsic and the intrinsic base resistances is current dependent. The base resistance ($r_{BB'}$) is expressed as

$$\begin{aligned} r_{BB'} &= r_{BM} + 3(r_B - r_{BM})\left(\frac{\tan(z) - z}{z \cdot \tan^2(z)}\right) \\ z &= \frac{-1 + \sqrt{1 + 144 I_B/\pi^2 I_{rB}}}{(24/\pi^2)\sqrt{I_B/I_{rB}}} \end{aligned} \tag{7.12}$$

where r_{BM} is the intrinsic base resistance, r_B is the extrinsic base resistance, I_B is the base current and the I_{rB} is the current where the base resistance falls halfway to its minimum value.

7.1.3 Lattice Temperature Model

Figure 7.4 shows the lattice temperature model used in the PD SOI-Technology SPICE. As shown in this figure, between the thin film and the substrate in the PD SOI

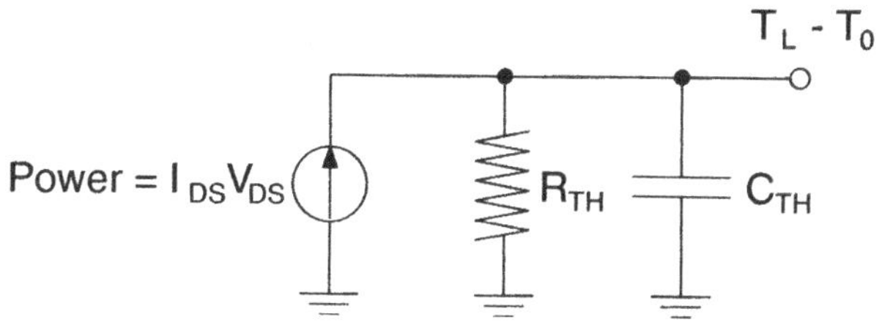

Fig. 7.4 Lattice temperature model used in the PD SOI-Technology SPICE.

CMOS devices, there is a buried oxide layer. A thick buried oxide may prevent the SOI device from dissipating the heat generated in the active region of the thin film. As a result, the lattice temperature may increase. Under this situation, thermal effect is important in determining the performance of the device. Lattice temperature can be computed by considering the equivalent circuit, which is composed of a current source, R_{TH} and C_{TH} as shown in Fig. 7.4 [5]. The current source is made of the power of the device—$I_{DS}V_{DS}$. R_{TH} and C_{TH} are the thermal resistance and the thermal capacitance between the device and the ambient, respectively. From Fig. 7.4, the lattice temperature can be obtained from the following equation:

$$-\text{Power} + \frac{1}{R_{TH}}(T_L - T_0) + C_{TH}\frac{d}{dt}(T_L - T_0) = 0 \tag{7.13}$$

where T_L is the lattice temperature and T_0 is the ambient temperature.

7.2 KINK EFFECTS

In this section, kink effects of the PD SOI CMOS devices are analyzed based on the compact PD SOI-Technology SPICE models. Figure 7.5 show (a) the drain current and (b) the body-source voltage characteristics of the PD SOI NMOS device with a channel length of 0.2 μm, a channel width of 50 μm, a front oxide of 70 Å, and a buried oxide of 3600 Å, biased in the strong inversion region, with its body node floating, based on the PD SOI-Technology SPICE model results, and the experimental data[6]. As shown in the figure, when the drain voltage reaches a certain value, there is an unsmooth transition in the drain current—the DC kink effect, which is due to the impact ionization current leading to the accumulation of holes in the neutral region. Therefore, the body-source voltage increases, which lowers the threshold voltage of the surface MOS portion. As a result, the parasitic BJT turns on to trigger the sudden increase in the drain current. As confirmed by the experimental data, with the inclusion of the body node, PD SOI-Technology SPICE can accurately predict the unsmooth transition in the drain current curves and the sudden increase in the body-source voltage curve at the onset of the kink effects– at a larger V_{GS}, the kink effect occurs at larger V_{DS}.

Figure 7.6 show (a) drain current (b) body-source voltage characteristics of a PD SOI NMOS device with a front gate oxide of 45 Å, a thin film of 1500 Å, a channel

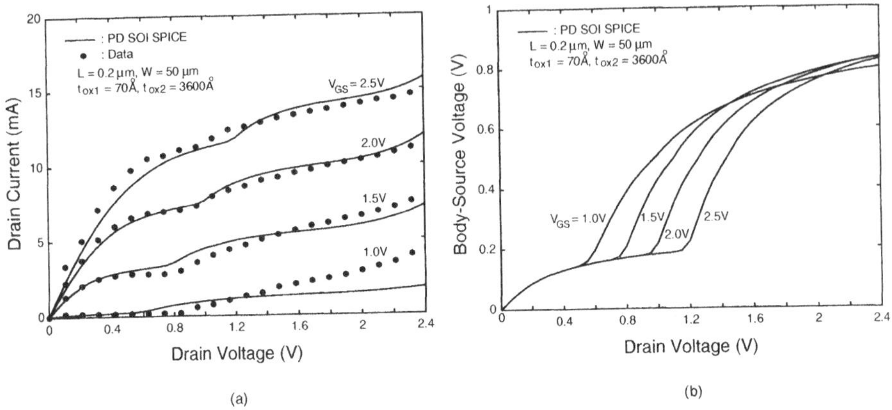

Fig. 7.5 (a) Drain current (b) body-source voltage characteristics of the PD SOI NMOS device with a front gate oxide of 70 Å, a buried oxide of 3600 Å, a channel length of 0.2 μm, and a channel width of 50 μm, biased in the strong inversion region, with its body node floating, based on the PD SOI-Technology SPICE model results, and the experimental data [6].

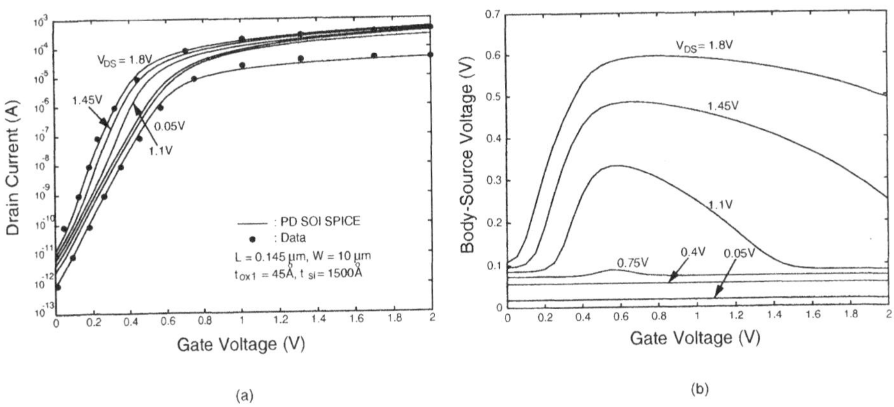

Fig. 7.6 (a) Drain current (b) body-source voltage characteristics of a PD SOI NMOS device with a front gate oxide of 45 Å, a thin film of 1500 Å, a channel length of 0.145 μm, and a channel width of 10 μm, biased in the subthreshold region, with its body node floating, based on the PD SOI-Technology SPICE model results, and the experimental data [7].

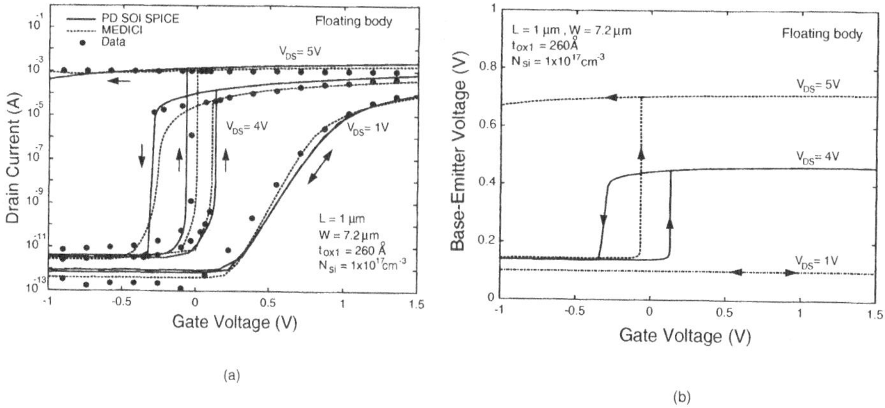

Fig. 7.7 (a) Subthreshold drain current (I_D) and (b) base-emitter voltage (V_{BE}) versus gate-source voltage (V_{GS}) of the parasitic bipolar device in the PD SOI NMOS device with a front gate oxide of 260 Å, a thin film doped with a p-type density of $10^{17}cm^{-3}$, a channel length of 1 μm, and a channel width of 7.2 μm, biased at $V_{DS} = 1, 4, 5$ V, with its body node floating, based on the PD SOI-Technology SPICE model results, MEDICI [9] simulation results, and the experimental data [8]. (a) I_D vs V_{GS}.

length of 0.145 μm, and a channel width of 10 μm, biased in the subthreshold region, with its body node floating, based on the PD SOI-Technology SPICE model results, and experimental data [7]. As shown in the figure, as predicted by the PD SOI-Technology SPICE model results and confirmed by the experimental data, when the drain voltage exceeds 1.1 V, a larger drain voltage leads to a steeper subthreshold slope due to the subthreshold kink effect. In the subthreshold kink effect region, when the drain voltage is increased, its impact ionization current also brings in the accumulation of holes in the neutral region. Therefore, the body-source voltage is raised and the threshold voltage of the surface MOS portion is lowered and its drain current increases suddenly– its subthreshold slope is steeper as compared to the case without kink effects. As shown in Fig. 7.5 and Fig. 7.6, and verified by the experimental data, PD SOI-Technology SPICE program predicts the effects well.

7.3 HYSTERESIS BEHAVIOR

Due to the floating body effect, the PD SOI NMOS device biased in the subthreshold region may have hysteresis behavior [8]. In this section, the hysteresis behavior of the PD SOI NMOS devices has been analyzed using the concise PD SOI-Technology SPICE models.

Figure 7.7 shows (a) the subthreshold drain current (I_D) and (b) the base-emitter voltage (V_{BE}) versus the gate-source voltage (V_{GS}) of the parasitic bipolar device in the the PD SOI NMOS device with a channel length of 1 μm, a channel width of 7.2 μm, a front gate oxide of 260 Å, and a p-type thin-film doped with a density of 10^{17}cm^{-3}, biased at $V_{DS} = 1, 4, 5$ V, with its body floating, based on the PD SOI-Technology SPICE model results, 2D MEDICI [9] simulation results, and the experimental data [8]. As shown in this figure, at $V_{DS} = 1$ V, consistent I_D vs V_{GS} characteristics can be seen. At $V_{DS} = 4$ V, when V_{GS} scans from -1 V toward the positive direction (positive scan), near $V_{GS} = 0.13$ V the drain current rises much more quickly as compared to the $V_{DS} = 1$ V case. On the other hand, when V_{GS} scans from 1.5 V toward the negative direction (negative scan), a quick fall at $V_{GS} = -0.3$ V can be seen. Steep rise and fall at different V_{GS}'s indicate that hysteresis occurs. At $V_{DS} = 5$ V, the steep rise occurs at $V_{GS} = -0.03$ V but there is no steep fall while scanning toward negative V_{GS}—latched. As shown in this figure, and verified by the experimental data (dots) [8] and MEDICI results (dashed lines), PD SOI-Technology SPICE models (solid lines) accurately predict the occurrence of the hysteresis and the latched conditions via monitoring V_{BE} of the parasitic BJT. As shown in Fig. 7.7(b), at $V_{DS} = 1$ V (without hysteresis), V_{BE} stays low regardless of the scanning direction of V_{GS}—the parasitic BJT never turns on. At $V_{DS} = 4$ V, the rise and the fall of the parasitic BJTs V_{BE} do not occur at the same locations, which is correlated to the drain current characteristics— the turn-on and the turn-off of the parasitic BJT occur at different locations. As triggered by the impact ionization current of the surface MOS device, the parasitic BJT is turned on mildly to $V_{BE} = 0.45$ V at $V_{GS} = 0.13$ V, for the $V_{DS} = 4$ V case. On the other hand, the parasitic BJT is turned off at a negative voltage $V_{GS} = -0.3$ V owing to the recombination of the minority carriers near the source in the thin film. The existence of hysteresis indicates that the turn-off of the parasitic BJT is more difficult as compared to turn-on. At $V_{DS} = 5$ V, V_{BE} rises to 0.7 V at $V_{GS} = -0.03$ V—the parasitic BJT strongly turns on during the positive scan of V_{GS}. However, during the negative scan of V_{GS}, no sudden fall in V_{BE} can be seen—parasitic BJT never turns off. With the extra body node (B) in the PD SOI-Technology SPICE models, the hysteresis and the latched conditions can be monitored via V_{BE}, which is difficult to be comprehended via other conventional approaches.

Figure 7.8 shows (a) the gate-source voltage (V_{GS}) at the onset of hysteresis during the positive scan (V_{GSH}) and the width of hysteresis (ΔV_H) and (b) V_{BE} versus V_{DS} of the PD SOI NMOS device with a front gate oxide of 260 Å, a thin film doped with a p-type density of 10^{17}cm^{-3}, a channel width of 7.2 μm, and channel lengths of 1 and 0.5 μm, biased with its body floating, based on the PD SOI-Technology SPICE model results and MEDICI simulation results. As shown in this figure, with a larger V_{DS}, the onset of hysteresis– V_{GSH} (solid lines for the PD SOI-Technology SPICE model results and solid dots/squares for MEDICI simulation results) occurs at a smaller V_{GS} during the positive scan, which indicates an earlier turn-on of the parasitic BJT as confirmed by the V_{BE} curves based on the PD SOI-Technology SPICE model results. In addition, with a larger V_{DS}, the width of hysteresis ΔV_H (dashed lines for the PD SOI-Technology SPICE model results and empty dots/squares for MEDICI simulation results) is larger, which implies that the parasitic BJT is more difficult to turn off than to turn on. From the figure, with a smaller channel length, the above trend is even more noticeable, which indicates a stronger action of the parasitic BJT.

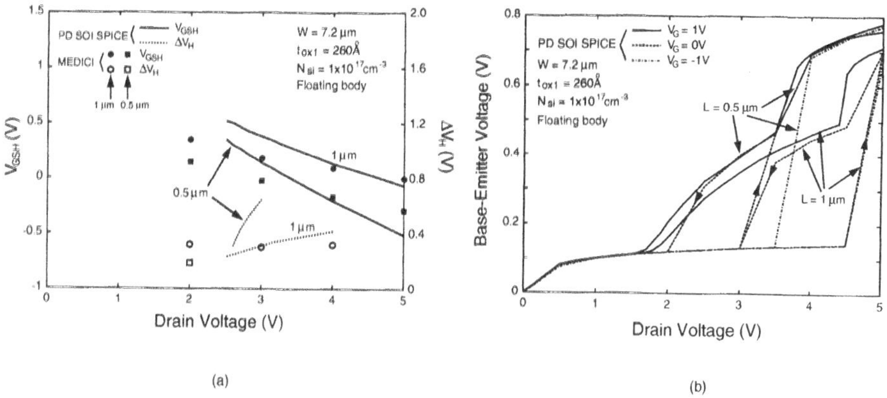

Fig. 7.8 (a) Gate-source voltage (V_{GS}) at the onset of hysteresis during the positive scan (V_{GSH})and the width of hysteresis (ΔV_H) and (b) V_{BE} versus V_{DS} of the PD SOI NMOS device with a front gate oxide of 260 Å, and a thin film doped with a p-type density of $10^{17}cm^{-3}$, a channel width of 7.2 μm, and channel lengths of 1 and 0.5 μm, biased with its body floating, based on the PD SOI-Technology SPICE model results and the MEDICI simulation results.

As shown in Fig. 7.8(b) for the $V_G = 0V$ case, different paths exist for turn-on (solid lines) and turn-off (dashed lines) of the parasitic BJT–hysteresis.

Figure 7.9 shows V_{GS} at the onset of hysteresis during the positive scan (V_{GSH}) and the width of hysteresis (ΔV_H) versus V_{DS} of the PD SOI NMOS device with a thin film doped with a p-type density of $10^{17}cm^{-3}$, a channel length of 1 μm, a channel width of 7.2 μm, and front gate oxide thicknesses of 260 and 200 Å, biased with its body floating. As shown in this figure, with a thinner front gate oxide, the onset of hysteresis occurs at a smaller V_{GSH}(solid lines and solid points)—the parasitic BJT is easier to turn on. In addition, with a thinner front gate oxide, the width of hysteresis ΔV_H (dashed lines and empty points) becomes narrower, which can be reasoned as follows. With a thinner front gate oxide, a larger impact ionization current in the surface MOS portion leads to an easier turn-on of the parasitic BJT. On the other hand, with a thinner front gate oxide, the stronger MOS action also makes the function of the parasitic BJT less important. As a result, the width of hysteresis is smaller.

Until now, the DC portion of the hysteresis behavior has been analyzed. Now, the transient portion is described. Figure 7.10 shows the drain current (I_D) and the base-emitter voltage (V_{BE}) of the parasitic BJT during the transients of the PD SOI NMOS device with a front gate oxide of 260 Å, a thin film doped with a p-type density of $10^{17}cm^{-3}$, a channel length of 1 μm, and a channel width of 7.2 μm, biased at $V_D = 5$ and 1 V and with its body floating, with an input step from +3 to −2 V imposed at its gate, based on the PD SOI-Technology SPICE model results and MEDICI simulation results. As shown in this figure, for the $V_{DS} = 5$ V case, after the input ramps down from +3 to −2 V, the drain current decreases accordingly. Since the parasitic BJT remains on, the drain current does not decrease to zero. Instead, it stays at ~ 2.5 mA at a certain time after the input step-down period. In contrast, for

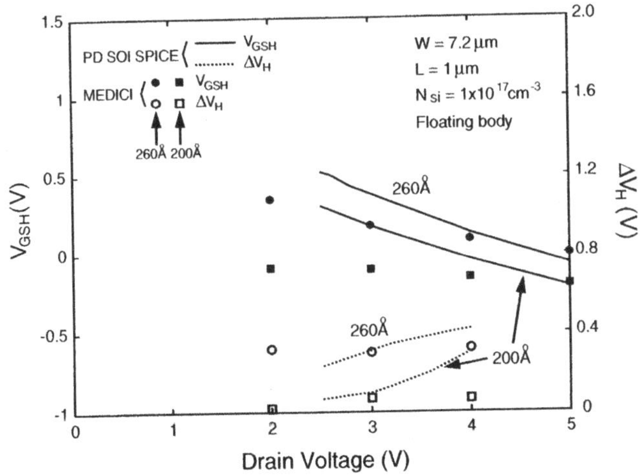

Fig. 7.9 V_{GS} at the onset of hysteresis during the positive scan (V_{GSH}) and the width of hysteresis (ΔV_H) vs V_{DS} of the PD SOI NMOS device with a thin film doped with a p-type density of $10^{17}cm^{-3}$, a channel length of 1 μm, a channel width of 7.2 μm, and front gate oxide thicknesses of 260 and 200 Å, biased with its body floating.

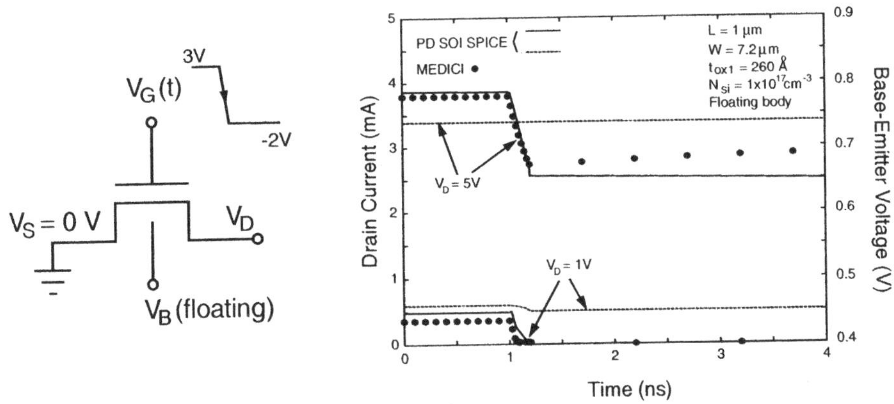

Fig. 7.10 Drain current (I_D) and base-emitter voltage (V_{BE}) of the parasitic BJT during the transients of the PD SOI NMOS device with a front gate oxide of 260 Å, a thin film doped with a p-type density of $10^{17}cm^{-3}$, a channel length of 1 μm, and a channel width of 7.2 μm, biased at V_S=0 V and V_D = 5 and 1 V, with an step from 3 to -2 V imposed at its gate and with its body floating, based on the PD SOI-Technology SPICE model results and MEDICI simulation results.

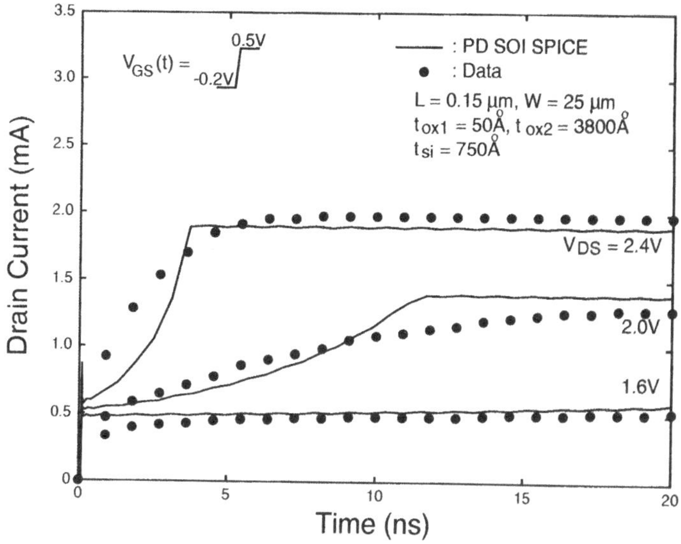

Fig. 7.11 Drain current characteristics during the turn-on transient of the PD SOI NMOS device with a front gate oxide of 50 Å, a thin film of 750 Å, a buried oxide of 3800 Å, a channel length of 0.15 μm, and a channel width of 25 μm, with its body node floating for various drain voltages, by imposing an input step from -0.2 to 0.5 V at its gate, based on SOI-technology SPICE model results and the experimental data [10].

the $V_{DS} = 1$ V case, the drain current does decrease to a much smaller value after the input ramp-down period, which indicates that the function of the parasitic BJT is reduced more as compared to the $V_{DS} = 5$ V case, as confirmed by the variation in the V_{BE} curves. The difference in the leakage currents of the PD SOI NMOS device after the input ramp-down period due to the non-turn-off of the parasitic BJT between the two cases ($V_{DS} = 5$ and 1 V) is strongly correlated to the trend on the DC situation. As shown in Fig. 7.10, the PD SOI-Technology SPICE models predict the transients of the PD SOI NMOS device well considering the latched and the hysteresis conditions due to the action of the parasitic BJT.

7.4 TRANSIENT BEHAVIOR

In this section, transient analysis of the PD SOI CMOS device using the PD SOI-Technology SPICE models is described. Figure 7.11 shows drain current characteristics during the turn-on transient of a PD SOI NMOS device with a channel length of 0.15 μm, a channel width of 25 μm, a front oxide of 50 Å, a buried oxide of 3800 Å, and a thin film of 750 Å, with its body node floating for various drain voltages, by imposing an input step from -0.2 to 0.5 V at its gate, based on PD SOI-Technology SPICE model results and experimental data [10]. As shown in Fig. 7.11, PD SOI NMOS device is biased at V_{GS}=0.5 V and V_{DS}=1.6, 2.0 and 2.4 V, the transient

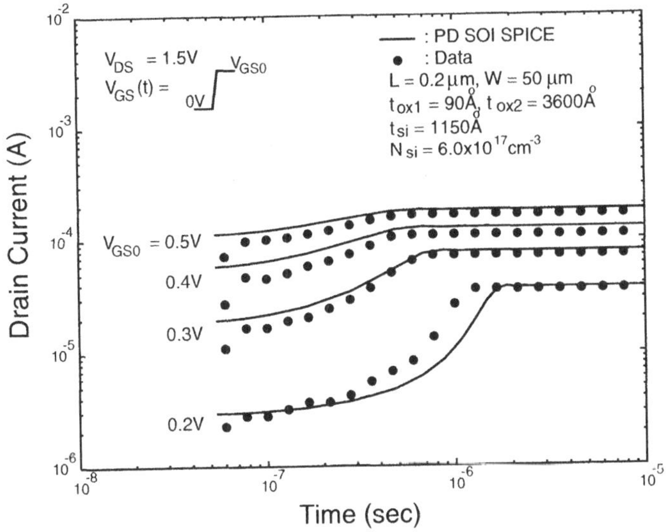

Fig. 7.12 Drain current waveforms during the turn-on transient of the PD SOI NMOS device with a front gate oxide of 90 Å, a buried oxide of 3600 Å, a thin film of 1150 Å doped with a p-type density of $6 \times 10^{17}\text{cm}^{-3}$, a channel length of 0.2 μm, and a channel width of 50 μm, biased at various gate voltage and with its body floating, based on the PD SOI-Technology SPICE model results and the experimental data [12].

behaviors are dominated by impact ionization current of device [11]. After the up step from -0.2 to 0.5 V at the input gate, due to a relatively large drain voltage, holes generated by strong impact ionization are accumulated at the floating body node. As a result, the body potential rises and the threshold voltage drops. Therefore, its drain current increases. This situation persists until the parasitic BJT turns on when the equilibrium is reached. As shown in this figure, when V_{DS} becomes larger, a larger impact ionization current leads to a quicker rise in the drain current to its equilibrium value.

Figure 7.12 shows drain current waveforms during the turn-on transient of a PD SOI NMOS device with a front gate oxide of 90 Å, a buried oxide of 3600 Å, a thin film of 1150 Å doped with a p-type density of $6 \times 10^{17}\text{cm}^{-3}$, a channel length of 0.2 μm, and a channel width of 50 μm, biased at various gate voltage and with its body floating, based on the PD SOI-Technology SPICE model results and the experimental data [12]. As shown in this figure, the influence of the gate voltage in the transient waveform of the drain current during turn-on can be observed [12]. At a smaller gate voltage, the rise in the drain current is larger and the time to reach steady state is longer due to a smaller current. Figure 7.13 shows switching characteristics of a PD SOI NMOS device with a front gate oxide of 700 Å, a buried oxide of 1 μm, a thin film of 5000 Å doped with a p-type density of $5 \times 10^{16}\text{cm}^{-3}$, a channel length of 2 μm, a channel width of 20 μm, biased at the drain voltage of 1 V, and with its body floating, with various input time intervals based on the PD SOI-Technology SPICE

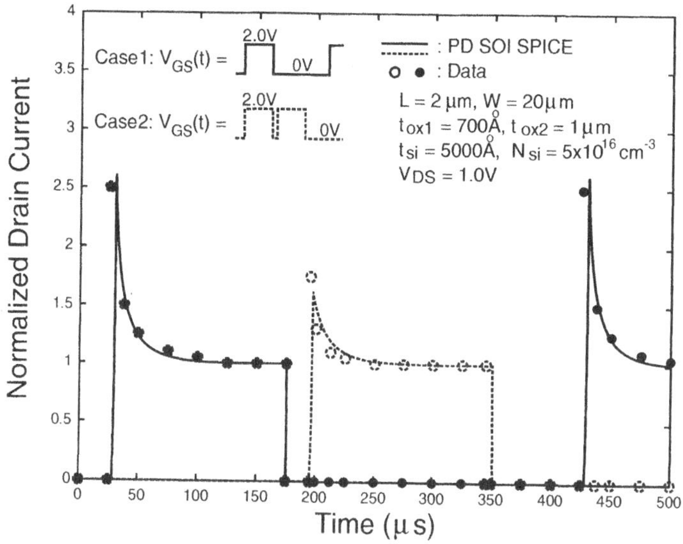

Fig. 7.13 Switching characteristics of a PD SOI NMOS device with a front gate oxide of 700 Å, a buried oxide of 1 μm, a thin film of 5000 Å doped with a p-type density of $5 \times 10^{16}\text{cm}^{-3}$, a channel length of 2 μm, and a channel width of 20 μm, biased at the drain voltage of 1 V and with its body floating, with various input time intervals based on the PD SOI-Technology SPICE model results and the experimental data [13].

model results and the experimental data [13]. From this figure, as verified by the experimental data, the concise BiCMOS approach can explain the sudden rise in the drain current when the gate voltage rises suddenly and the gradual decay due to the recombination of the body charge. In addition, the drain current overshoot is also related to the frequency of the pulse imposed at the gate. If two consecutive pulses are too close (case 2: input pulse frequency is 50 kHz), after the first pulse the body charge cannot be recovered in time before the arrival of the next pulse. Then, the drain current overshoot during the next pulse is less. On the other hand, if the input pulse frequency is small (case 1: input pulse frequency is 4 kHz), after the first pulse the body charge can reach the equilibrium before the next pulse comes. Then, the drain current overshoot during the next pulse remains the same shape. As shown in Fig. 7.11~Fig. 7.13, as verified by the experimental data, the BiCMOS equivalent circuit model used in the PD SOI-Technology SPICE predicts the turn-on transients well.

Figure 7.14 shows the leakage current of the PD SOI NMOS pass-transistor with a front gate oxide of 35 Å, a channel length of 0.11 μm, a channel width of 10 μm, biased with its body floating and with its source pulsed from 2.5 to 0 V, based on the PD SOI-Technology SPICE model results and the experimental data [14]. Before the sudden drop in the source voltage, the source and drain terminals stay at high (V_D=V_S=2.5 V), the floating body in the thin film is charged to high. The sudden drop in the source voltage turns on the body-source diode—the parasitic BJT is

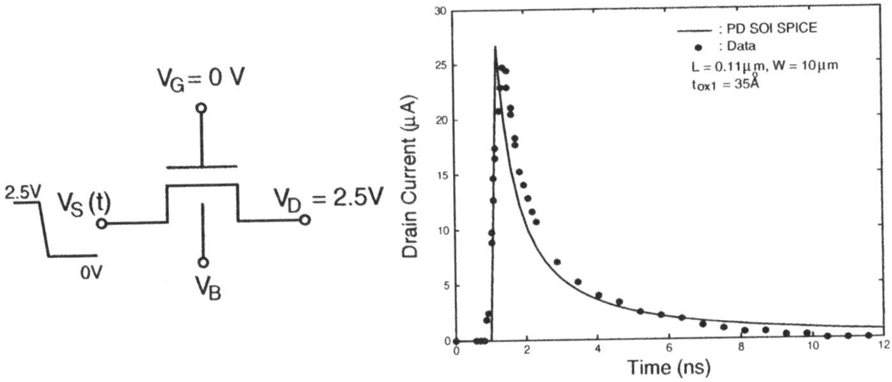

Fig. 7.14 Leakage current of the PD SOI NMOS pass-transistor with a front gate oxide of 35 Å, a channel length of 0.11 μm, and a channel width of 10 μm, biased with its body floating and with its source pulsed from 2.5 to 0 V, based on the PD SOI-Technology SPICE model results and the experimental data [14].

triggered. As a result, a sudden rise in the drain current due to the leakage current is generated and it decays when the body charge is recombined gradually.

7.5 STATIC LOGIC CIRCUIT

In this section, SOI CMOS static logic circuits are analyzed using the PD SOI-Technology SPICE models. Figure 7.15 shows the output transient waveforms of a body-grounded PD SOI NMOS device and an n-channel DTMOS device with a front gate oxide of 64 Å, a thin film of 1500 Å doped with a p-type density of $3 \times 10^{17}\text{cm}^{-3}$, a buried oxide of 4000 Å, and channel lengths of 0.2 and 0.5 μm, based on PD SOI-Technology SPICE model results and MEDICI simulation results. By connecting the body terminal to the gate terminal, the threshold voltage of the DTMOS device is a function of the gate voltage. DTMOS devices can provide a larger drain current owing to a reduced threshold voltage at a raised gate voltage. As shown in this figure, the DTMOS case shows a faster pull-down transient. Owing to incorporation of short-channel effect and dynamic-threshold behavior in the equivalent circuit model, the PD SOI-Technology SPICE can effectively predict transients of the PD SOI NMOS device with its body grounded and the DTMOS device with channel lengths of 0.2 and 0.5 μm.

Figure 7.16 shows the output transient waveforms of an inverter circuit using body-floating PD, body-tied PD and DTMOS PD SOI CMOS devices with a front gate oxide of 64 Å, a thin film of 1500 Å, a buried oxide of 4000 Å, a channel length of 0.3 μm, and a channel width of 10 μm, biased at $V_{DD} = 0.7$ V, based on the PD SOI-Technology SPICE model results and the MEDICI simulation results. As shown in this figure, the DTMOS one has the faster switching speed, followed by the body-floating one, the body-tied to V_{DD}/GND (for PMOS/NMOS) one has the slowest

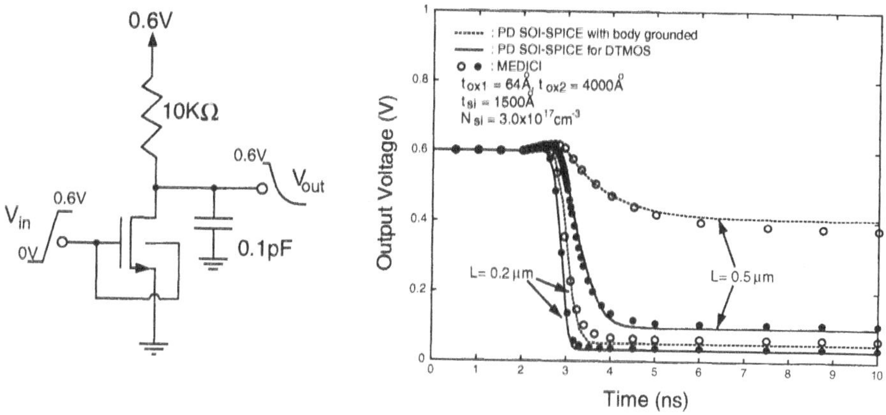

Fig. 7.15 Output transient waveforms of a body-grounded PD SOI NMOS device and an n-channel DTMOS device with a front gate oxide of 64 Å, a thin film of 1500 Å doped with a p-type density of 3×10^{17}cm³, a buried oxide of 4000 Å, and channel lengths of 0.2 and 0.5 μm, based on the PD SOI-Technology SPICE model results and the MEDICI simulation results.

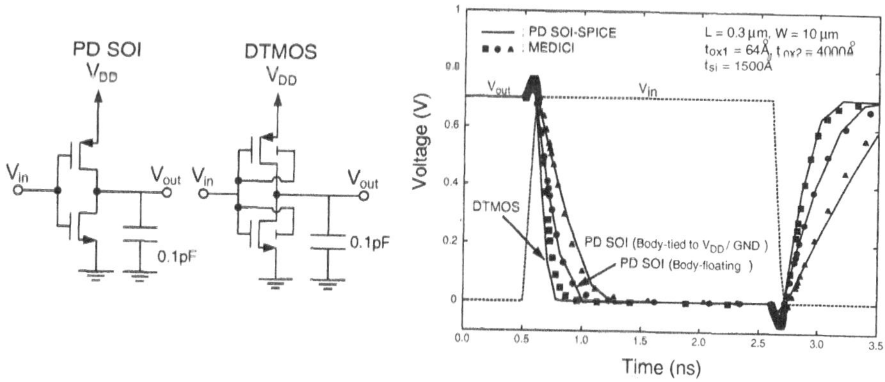

Fig. 7.16 Output transient waveforms of an inverter circuit using body-floating PD, body-tied PD and DTMOS PD SOI CMOS devices with a front gate oxide of 64 Å, a thin film of 1500 Å, a buried oxide of 4000 Å, a channel length of 0.3 μm, and a channel width of 10 μm, biased at $V_{DD} = 0.7$ V, based on the PD SOI-Technology SPICE model results and the MEDICI simulation results.

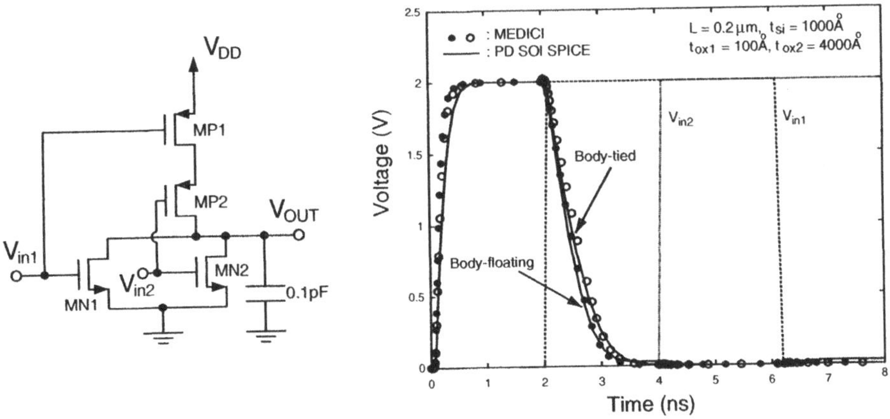

Fig. 7.17 Transient waveforms of a NOR circuit using body-floating and body-tied PD SOI CMOS devices with a front gate oxide of 100 Å, a thin film of 1000 Å, a buried oxide of 4000 Å, and a channel length of 0.2 μm, based on the PD SOI-Technology SPICE model results and the MEDICI simulation results.

speed due to body effect. It can be explained as follow. Due to the dynamic-threshold technique, the DTMOS case has a larger driving capability and a higher switching speed. Owing to the body-gate capacitance (C_{BG}) coupling effect, the threshold voltage of the body-floating PD SOI device can be reduced. Therefore, the driving capability of the body-floating PD SOI can be improved. But the driving capability of the body-floating PD SOI case is still less than the DTMOS one. For body-tied case, the body effect disappears and the threshold voltage remains unchanged. As a result, the driving capability is the worst. The fall time and the rise time of the DTMOS case are the shortest, followed by the body-floating one. The body-tied case has the longest fall and rise times.

Figure 7.17 show transient waveforms of a NOR circuit using the body-floating and the body-tied PD SOI CMOS devices with a front gate oxide of 100 Å, a thin film of 1000 Å, a buried oxide of 4000 Å, and a channel length of 0.2 μm, based on PD SOI-Technology SPICE model results and MEDICI simulation results. As shown in the Fig. 7.17, the body-floating case has a faster speed owing to a smaller threshold voltage. When the capacitor load is charged by V_{DD} via PD SOI PMOS device, the impact ionization effect of PMOS is not obvious. Therefore, the accumulation of charge in the neutral region of the thin film in the body-floating PD SOI PMOS is not obvious. The variation of the threshold voltage of the body-floating PD SOI PMOS is almost the same as the body-tied one. Therefore, the driving capability of these two cases are almost the same. As a result, the body-floating case and body-tied one have the same pull-up transient behavior. When the capacitor load is discharged via NMOS, due to the body-gate capacitance (C_{BG}) coupling effect and the impact ionization effect of body-floating PD SOI NMOS device, the parasitic BJT of body-floating PD SOI NMOS turns on. Owing to the parasitic BJT and variation of threshold voltage due to the accumulation of charge in the neutral region, the

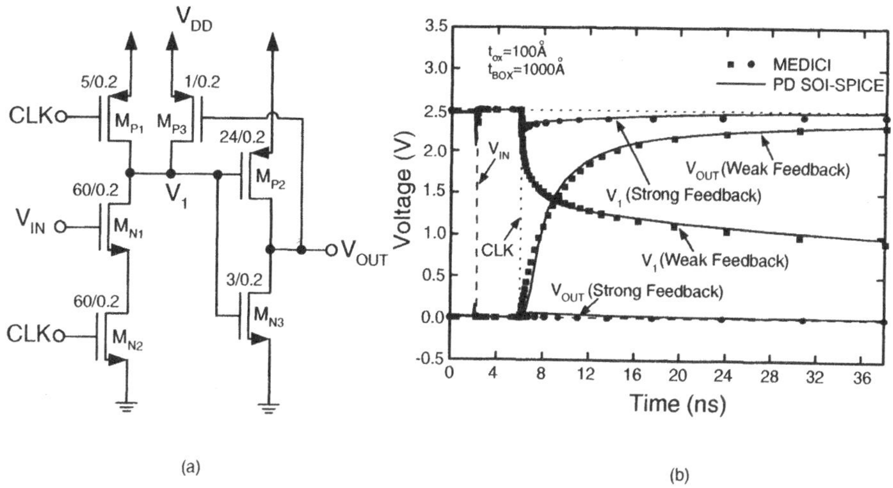

Fig. 7.18 (a) PD SOI dynamic inverter using PD devices with a front gate oxide of 100 Å and a thin film of 1000 Å and with a pullup PMOS feedback (MP3) as described in Chapter 4. (b)Transient waveforms of the PD SOI dynamic inverter with weak ($W/L = 1\ \mu m/0.2\ \mu m$) and strong ($W/L = 4\ \mu m/0.2\ \mu m$) pullup PMOS devices (MP3).

driving capability of body-floating PD SOI NMOS device increases. As a result, the pull-down transient of the body-floating case is much faster.

7.6 DYNAMIC LOGIC CIRCUIT

The PD SOI-Technology SPICE models have been used to study the leakage problems in SOI CMOS dynamic logic circuits. Figure 7.18(a) shows a PD SOI dynamic inverter using PD devices with a front gate oxide of 100 Å and a thin film of 1000 Å and with a pullup PMOS feedback as described in Chapter 4. The pullup PMOS feedback (MP3) is used to stabilize the internal node voltage (V1) when the input NMOS device (MN1) is off during the logic evaluation period. As described in Chapter 5, due to the leakage current from the parasitic bipolar device in the input NMOS device MN1, which has a floating body, the internal node voltage V1 may be unstable during the logic evaluation period. By adding a pullup PMOS feedback (MP3), the internal node voltage V1 can be secured. However, depending on the strength of the pullup PMOS feedback device (MP3), the leakage current of the parasitic bipolar device in the input NMOS device MN1 may still cause detrimental effects on the internal node voltage (V1). Using the PD SOI-Technology SPICE models, Fig. 7.18(b) shows the transient waveforms of the PD SOI dynamic inverter with weak ($W/L = 1\ \mu m/0.2\ \mu m$) and strong ($W/L = 4\ \mu m/0.2\ \mu m$) pullup PMOS device (MP3). As shown in the figure, after the precharge period, the internal node (V1) is precharged to high. During the logic evaluation period, the input voltage is low and the NMOS device MN1 is off. With a strong pullup PMOS device MP3 ($W/L = 4\ \mu m/0.2\ \mu m$), the internal node

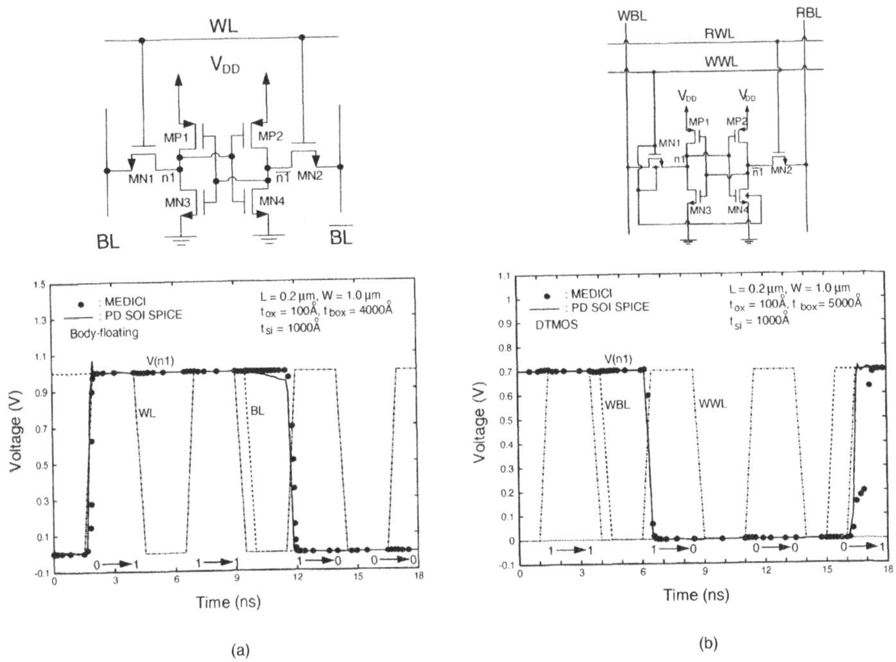

Fig. 7.19 Write transient waveforms for (a) a 6T SRAM memory cell using 0.2 μm body-floating PD SOI CMOS devices with a front gate oxide of 100 Å and a thin film of 1000 Å and (b) a two-port 6T SRAM memory cell with single-bit-line simultaneous read-and-write access (SBLSRWA) capability using DTMOS PD SOI CMOS devices with a front gate oxide of 100 Å and a thin film of 1000 Å [16], based on the PD SOI-Technology SPICE model results and the MEDICI simulation results.

voltage V1 is disturbed a little during the transition period and quickly back to the high value. With a weak pullup PMOS device MP3 ($W/L = 1\ \mu m/0.2\ \mu m$), the internal node voltage V1 is accidentally pulled low to an error state due to the leakage current of the parasitic bipolar device in the input NMOS device MN1. The leakage current effect due to the floating body of the PD SOI dynamic logic gate based on the PD SOI-Technology SPICE models confirms the results as described before [15]

7.7 SRAM CRITICAL PATH

In this section, transient analysis of the critical path in an SRAM during write access using PD SOI-Technology SPICE models is described. Figure 7.19 shows write transient waveforms for (a) a 6T SRAM memory cell using 0.2 μm body-floating PD SOI CMOS devices with a front gate oxide of 100 Å and a thin film of 1000 Å, and (b) a two-port 6T SRAM memory cell with single-bit-line simultaneous read-and-write access (SBLSRWA) capability using the DTMOS PD SOI CMOS devices with a front gate oxide of 100 Å and a thin film of 1000 Å [16], based on PD SOI-Technology

SPICE model results and MEDICI simulation results. As shown in Fig. 7.19(a), during the write access, the logic-0 is to be written to node n1, which stores logic-1 originally. Due to the node n1 voltage drop from 1 to 0 V, the accumulation of charge in the neutral region of body-floating PD SOI devices (MN1 and MN3) triggers the leakage current of these transistors. Therefore, during the write access, the node n1 is charged by the leakage current of transistor (MN1 and MN3). In contrast, during the write access, the logic-1 is to be written to node n1, which stores logic-1 originally. The node n1 voltage remains at 1 V. The leakage current of transistors (MN1 and MN3) is not obvious. As a result, the node n1 is not charged by the leakage current. Due to the leakage current of the body-floating PD SOI devices (MN1 and MN3), the write access time of logic-1 to logic-0 is much longer than the write access time of logic-1 to logic-1. From the MEDICI simulation result, the write access time of logic-1 to logic-0 is 1ns longer than the write access time of logic-1 to logic-1.

As shown in Fig. 7.19(b), the body terminal of the NMOS device MN4 in the latch is connected to the write work line (WWL) instead of floating as in the conventional SOI CMOS SRAM cell. By using the dynamic threshold technique, a 6T memory cell with the single-bit-line simultaneous read-and-write access for low-voltage (0.7 V) two-port SOI CMOS VLSI SRAM is realized. As verified by the PD SOI-Technology SPICE results, the four cases of the write access—(1) logic-0 → logic-0, (2) logic-0 → logic-1, (3) logic-1 → logic-1 and (4) logic-1 → logic-0, write operations can be done. Because the transistor MN3 is still the body-floating PD SOI NMOS device, the SBLSRWA memory cell write access time of logic-1 to logic-0 is still longer than the write access time of logic-1 to logic-1. Due to single-bit-line write-and-read, the driving capability of SBLSRWA memory cell is less than the conventional 6T SRAM memory cell one. From the MEDICI simulation result, the write access time of logic-1 to logic-0 is 1.4 ns longer than the write access time of logic-1 to logic-1.

Figure 7.20 shows write transient waveforms of an SRAM critical path with 42 body-floating and body-tied PD SOI CMOS devices based on the PD SOI-Technology SPICE model results. As shown in the figure, due to the smaller magnitude in the threshold voltage of the precharge body-floating SOI NMOS device connected to the bit lines, after precharge, the bit lines (BL) is set at a higher voltage for the body-floating case. Therefore, a slower sensing speed can be seen for the body-floating case. The current mirror sense amp is used in the SRAM architecture. Comparing the amplification effect of body-floating PD SOI current mirror sense-amplifier with body-tied, due to the accumulation of charge in the neutral region of body-floating PD SOI NMOS, the amplification effect of body-floating PD SOI current mirror sense amp is worse. Therefore, the amplification of signal lines (DB and $\overline{DB}$) in body-floating PD SOI SRAM is worse.

7.8 SUMMARY

In this chapter, PD SOI-Technology SPICE models using a concise BiCMOS approach have been described. The PD SOI SPICE model has been described first, followed by the kink effects, the hysteresis behavior, and the transient behavior using the PD

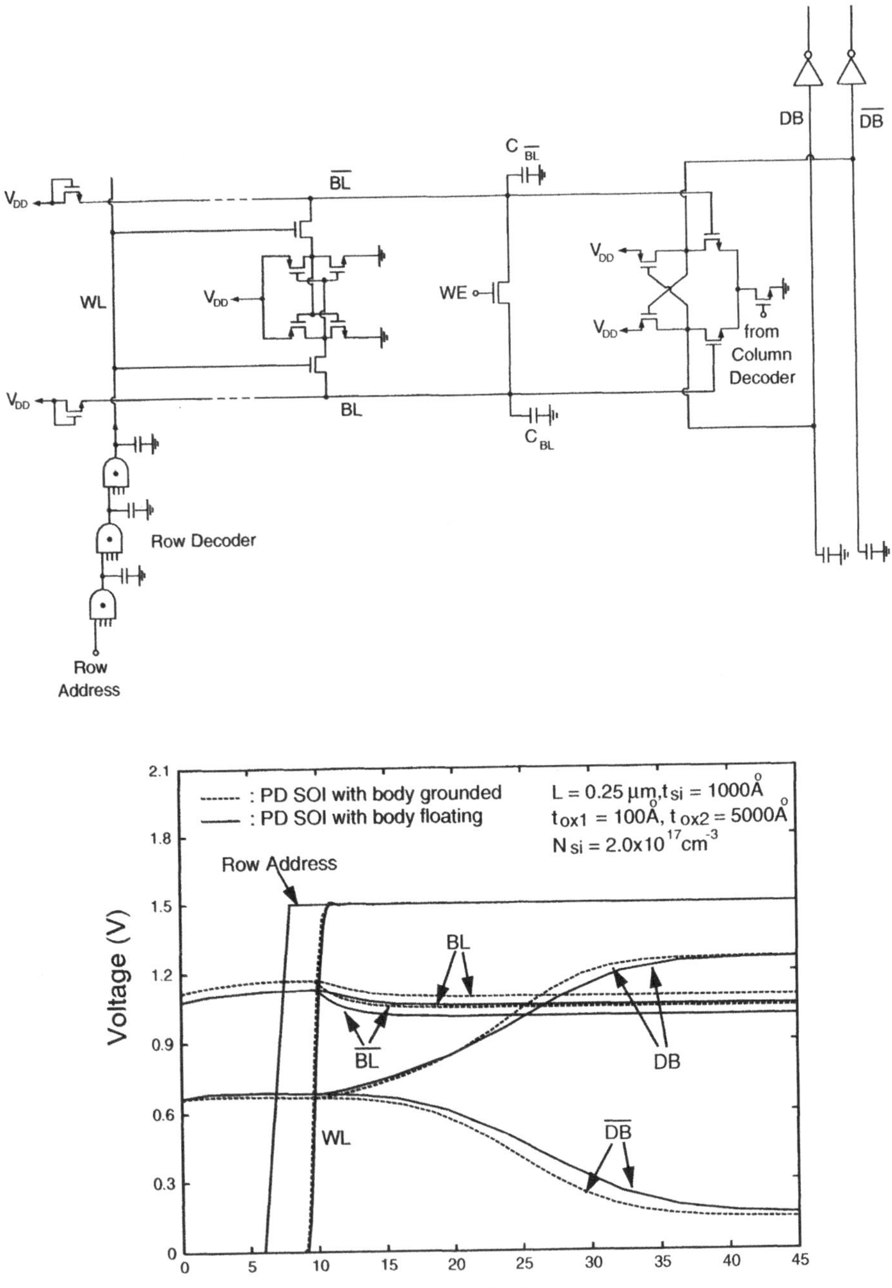

Fig. 7.20 Write transient waveforms of an SRAM critical path with 42 body-floating and body-tied PD SOI CMOS devices with a front gate oxide of 100 Å, a thin film of 1000 Å, a buried oxide of 5000 Å, with a channel length of 0.25 μm, based on the PD SOI-Technology SPICE model results.

SOI-Technology SPICE models. In the final portion of this chapter, a static logic circuit, a dynamic logic circuit, and an SRAM critical path have been analyzed using the PD SOI-Technology SPICE models.

REFERENCES

1. BSIM3v3 Manual, *UC Berkeley* (1996).
2. G. Massobrio and P. Antognetti, "Semiconductor Device Modeling with SPICE," *McGraw-Hill:New York* (1993).
3. J. B. Kuo, K. W. Su, and S. C. Lin, "Compact MOS/Bipolar Charge-Control Models of Partially-Depleted SOI CMOS Devices for VLSI Circuit Simulation—SOI-Technology (ST)-SPICE," *ESSDERC Dig.*, 480–483 (1999).
4. James B. Kuo, "SPICE Compact Modeling of PD-SOI CMOS Devices," *HKEDM Dig.*, (2000).
5. J. Bielefeld, G. Pelz, H. B. Abel, and G. Zimmer, "Dynamic SPICE-Simulation of the Electrothermal Behavior of SOI MOSFET's," *IEEE Trans. Elec. Dev.*, **42**(11), 1968–1974 (1995).
6. D. A. Antoniadis, "SOI CMOS Front-End Technology: Options and Tradeoffs," *SOI Conf. Dig.*, 1–3 (1995).
7. R. Chau, R. Arghavani, M. Alavi, D. Douglas, J. Greason, R. Green, S. Tyagi, J. Xu, P. Packan, S. Yu, and C. Liang, "Scalability of Partially Depleted SOI Technology for Sub-0.25μm Logic Applications," " *IEDM Dig.*, 591–594 (1997).
8. C.-E. D. Chen, M. Matloubian, R. Sundaresan, B.-Y. Mao, C. C. Wei, and G. P. Pollack, "Single-Transistor Latch in SOI MOSFET's," *IEEE Elec. Dev. Let.*, **9**(12), 636–638 (1988).
9. MEDICI: Two-Dimensional Semiconductor Device Simulation, *Technology Modeling Associates*, Palo Alto, CA (1996).
10. J. Gautier, K. A. Jenkins, and J. Y.-C. Sun, "Body Charge Related Transient Effects in Floating Body SOI NMOSFET's," *IEDM Dig.*, 623–626 (1995).
11. H. C. Shin, I.-S. Lim, M. Racanelli, W.-L. M. Huang, J. Foerstner, and B.-Y. Hwang, "Analysis of Floating Body Induced Transient Behaviors in Partially Depleted Thin Film SOI Devices," *IEEE Trans. Elec. Dev.*, **43**(2), 318–325 (1996).
12. A. Wei, M. J. Sherony, and D. A. Antoniadis, "Transient Behavior of the Kink Effect in Partially-Depleted SOI MOSFET's," *IEEE Elec. Dev. Let.*, **16**(11), 494–496 (1995).

13. K. Kato and K. Taniguchi, "Numerical Analysis of Switching Characteristics in SOI MOSFET's," *IEEE Trans. Elec. Dev.*, **33**(1), 133–139 (1986).

14. F. Assaderaghi, W. Ransch, A. Ajmera, E. Leobandung, D. Schepis, L. Wagner, H.-J. Wann, R. Bolan, D. Yee, B. Davari, and G. Shahidi, "A 7.9/5.5 psec Room/Low Temperature SOI CMOS," " *IEDM Dig.*, 415–418 (1997).

15. P.-F. Lu, C.-T. Chuang, J. Ji, L. F. Wagner, C.-M. Hsieh, J. B. Kuang, L. L.-C. Hsu, M. M. Pelella, Jr, S.-F. S. Chu, and C. J. Anderson, "Floating-Body Effects in Partially Depleted SOI CMOS Circuits," *IEEE J. Sol. St. Ckts.*, **32**(8), 1241–1252 (1997).

16. S. C. Liu and J. B. Kuo, "A Novel 0.7V Two-Port 6T SRAM Memory Cell Structure with Single-Bit-Line Simultaneous Read-and-Write Access (SBLSRWA) Capability using Partially-Depleted SOI CMOS Dynamic-Threshold Technique," *SOI Conf. Dig.*, 75–76 (1999).

Problems

1. For the TSPC dynamic latch made of PD SOI CMOS devices described in Problem 5.1, use SPICE simulation to study the effect of the leakage current from the floating body the device.

2. For the SOI CMOS dynamic logic circuit, as shown in Fig. 5.11, based on a 0.2 μm PD SOI CMOS technology, use SPICE simulation to find out the design criterion for MP1, MN1, and MN2, in terms of the minimum and maximum operating frequencies.

Index

Printed and bound by CPI Group (UK) Ltd, Croydon, CR0 4YY
16/09/2026
14963442-0005